Lecture Notes in Mathematics

For information about Vols. 1–521, please contact your bookseller or Springer-Verlag.

Vol. 522: C. O. Bloom and N. D. Kazarinoff, Short Wave Radiation Problems in Inhomogeneous Media: Asymptotic Solutions. V. 104 pages. 1976.

Vol. 523: S. A. Albeverio and R. J. Høegh-Krohn, Mathematical Theory of Feynman Path Integrals. IV, 139 pages. 1976.

Vol. 524: Séminaire Pierre Lelong (Analyse) Année 1974/75. Edité par P. Lelong. V, 222 pages. 1976.

Vol. 525: Structural Stability, the Theory of Catastrophes, and Applications in the Sciences. Proceedings 1975. Edited by P. Hilton. VI, 408 pages. 1976.

Vol. 526: Probability in Banach Spaces. Proceedings 1975. Edited by A. Beck. VI, 290 pages. 1976.

Vol. 527: M. Denker, Ch. Grillenberger, and K. Sigmund, Ergodic Theory on Compact Spaces. IV, 360 pages. 1976.

Vol. 528: J. E. Humphreys, Ordinary and Modular Representations of Chevalley Groups. III, 127 pages. 1976.

Vol. 529: J. Grandell, Doubly Stochastic Poisson Processes. X, 234 pages. 1976.

Vol. 530: S. S. Gelbart, Weil's Representation and the Spectrum of the Metaplectic Group. VII, 140 pages. 1976.

Vol. 531: Y.-C. Wong, The Topology of Uniform Convergence on Order-Bounded Sets. VI, 163 pages. 1976.

Vol. 532: Théorie Ergodique. Proceedings 1973/1974. Edité par J.-P. Conze and M. S. Keane. VIII, 227 pages. 1976.

Vol. 533: F. R. Cohen, T. J. Lada, and J. P. May, The Homology of Iterated Loop Spaces. IX, 490 pages. 1976.

Vol. 534: C. Preston, Random Fields. V, 200 pages. 1976.

Vol. 535: Singularités d'Applications Differentiables. Plans-sur-Bex. 1975. Edité par O. Burlet et F. Ronga. V, 253 pages. 1976.

Vol. 536: W. M. Schmidt, Equations over Finite Fields. An Elementary Approach. IX, 267 pages. 1976.

Vol. 537: Set Theory and Hierarchy Theory. Bierutowice, Poland 1975. A Memorial Tribute to Andrzej Mostowski. Edited by W. Marek, M. Srebrny and A. Zarach. XIII, 345 pages. 1976.

Vol. 538: G. Fischer, Complex Analytic Geometry. VII, 201 pages. 1976.

Vol. 539: A. Badrikian, J. F. C. Kingman et J. Kuelbs, Ecole d'Eté de Probabilités de Saint Flour V-1975. Edité par P.-L. Hennequin. IX, 314 pages. 1976.

Vol. 540: Categorical Topology, Proceedings 1975. Edited by E. Binz and H. Herrlich. XV, 719 pages. 1976.

Vol. 541: Measure Theory, Oberwolfach 1975. Proceedings. Edited by A. Bellow and D. Kölzow. XIV, 430 pages. 1976.

Vol. 542: D. A. Edwards and H. M. Hastings, Čech and Steenrod Homotopy Theories with Applications to Geometric Topology. VII, 296 pages. 1976.

Vol. 543: Nonlinear Operators and the Calculus of Variations, Bruxelles 1975. Edited by J. P. Gossez, E. J. Lami Dozo, J. Mawhin, and L. Waelbroeck, VII, 237 pages. 1976.

Vol. 544: Robert P. Langlands, On the Functional Equations Satisfied by Eisenstein Series. VII, 337 pages. 1976.

Vol. 545: Noncommutative Ring Theory. Kent State 1975. Edited by J. H. Cozzens and F. L. Sandomierski. V, 212 pages. 1976.

Vol. 546: K. Mahler, Lectures on Transcendental Numbers. Edited and Completed by B. Diviš and W. J. Le Veque. XXI, 254 pages. 1976.

Vol. 547: A. Mukherjea and N. A. Tserpes, Measures on Topological Semigroups: Convolution Products and Random Walks. V, 197 pages. 1976.

Vol. 548: D. A. Hejhal, The Selberg Trace Formula for PSL (2, $\mathbb{R}$). Volume I. VI, 516 pages. 1976.

Vol. 549: Brauer Groups, Evanston 1975. Proceedings. Edited by D. Zelinsky. V, 187 pages. 1976.

Vol. 550: Proceedings of the Third Japan – USSR Symposium on Probability Theory. Edited by G. Maruyama and J. V. Prokhorov. VI, 722 pages. 1976.

Vol. 551: Algebraic K-Theory, Evanston 1976. Proceedings. Edited by M. R. Stein. XI, 409 pages. 1976.

Vol. 552: C. G. Gibson, K. Wirthmüller, A. A. du Plessis and E. J. N. Looijenga. Topological Stability of Smooth Mappings. V, 155 pages. 1976.

Vol. 553: M. Petrich, Categories of Algebraic Systems. Vector and Projective Spaces, Semigroups, Rings and Lattices. VIII, 217 pages. 1976.

Vol. 554: J. D. H. Smith, Mal'cev Varieties. VIII, 158 pages. 1976.

Vol. 555: M. Ishida, The Genus Fields of Algebraic Number Fields. VII, 116 pages. 1976.

Vol. 556: Approximation Theory. Bonn 1976. Proceedings. Edited by R. Schaback and K. Scherer. VII, 466 pages. 1976.

Vol. 557: W. Iberkleid and T. Petrie, Smooth S^1 Manifolds. III, 163 pages. 1976.

Vol. 558: B. Weisfeiler, On Construction and Identification of Graphs. XIV, 237 pages. 1976.

Vol. 559: J.-P. Caubet, Le Mouvement Brownien Relativiste. IX, 212 pages. 1976.

Vol. 560: Combinatorial Mathematics, IV, Proceedings 1975. Edited by L. R. A. Casse and W. D. Wallis. VII, 249 pages. 1976.

Vol. 561: Function Theoretic Methods for Partial Differential Equations. Darmstadt 1976. Proceedings. Edited by V. E. Meister, N. Weck and W. L. Wendland. XVIII, 520 pages. 1976.

Vol. 562: R. W. Goodman, Nilpotent Lie Groups: Structure and Applications to Analysis. X, 210 pages. 1976.

Vol. 563: Séminaire de Théorie du Potentiel. Paris, No. 2. Proceedings 1975–1976. Edited by F. Hirsch and G. Mokobodzki. VI, 292 pages. 1976.

Vol. 564: Ordinary and Partial Differential Equations, Dundee 1976. Proceedings. Edited by W. N. Everitt and B. D. Sleeman. XVIII, 551 pages. 1976.

Vol. 565: Turbulence and Navier Stokes Equations. Proceedings 1975. Edited by R. Temam. IX, 194 pages. 1976.

Vol. 566: Empirical Distributions and Processes. Oberwolfach 1976. Proceedings. Edited by P. Gaenssler and P. Révész. VII, 146 pages. 1976.

Vol. 567: Séminaire Bourbaki vol. 1975/76. Exposés 471–488. IV, 303 pages. 1977.

Vol. 568: R. E. Gaines and J. L. Mawhin, Coincidence Degree, and Nonlinear Differential Equations. V, 262 pages. 1977.

Vol. 569: Cohomologie Etale SGA 4½. Séminaire de Géométrie Algébrique du Bois-Marie. Edité par P. Deligne. V, 312 pages. 1977.

Vol. 570: Differential Geometrical Methods in Mathematical Physics, Bonn 1975. Proceedings. Edited by K. Bleuler and A. Reetz. VIII, 576 pages. 1977.

Vol. 571: Constructive Theory of Functions of Several Variables, Oberwolfach 1976. Proceedings. Edited by W. Schempp and K. Zeller. VI. 290 pages. 1977

Vol. 572: Sparse Matrix Techniques, Copenhagen 1976. Edited by V. A. Barker. V, 184 pages. 1977.

Vol. 573: Group Theory, Canberra 1975. Proceedings. Edited by R. A. Bryce, J. Cossey and M. F. Newman. VII, 146 pages. 1977.

Vol. 574: J. Moldestad, Computations in Higher Types. IV, 203 pages. 1977.

Vol. 575: K-Theory and Operator Algebras, Athens, Georgia 1975. Edited by B. B. Morrel and I. M. Singer. VI, 191 pages. 1977.

Vol. 576: V. S. Varadarajan, Harmonic Analysis on Real Reductive Groups. VI, 521 pages. 1977.

Vol. 577: J. P. May, E_∞ Ring Spaces and E_∞ Ring Spectra. IV, 268 pages. 1977.

Vol. 578: Séminaire Pierre Lelong (Analyse) Année 1975/76. Edité par P. Lelong. VI, 327 pages. 1977.

Vol. 579: Combinatoire et Représentation du Groupe Symétrique, Strasbourg 1976. Proceedings 1976. Edité par D. Foata. IV, 339 pages. 1977.

continuation on page 333

Lecture Notes in Mathematics

Edited by A. Dold and B. Eckmann

756

H. O. Cordes

Elliptic Pseudo-Differential Operators – An Abstract Theory

Springer-Verlag
Berlin Heidelberg New York 1979

Author

H. O. Cordes
Department of Mathematics
University of California
Berkeley, CA 94720/USA

AMS Subject Classifications (1970): 35-02, 35A25, 35J30, 35J40, 35J45, 35J70, 35S99, 45E05, 45K05, 46F10, 46L05, 47B30, 47C10, 47F05

ISBN 3-540-09704-X Springer-Verlag Berlin Heidelberg New York
ISBN 0-387-09704-X Springer-Verlag New York Heidelberg Berlin

Printing and binding: Beltz Offsetdruck, Hemsbach/Bergstr.
2141/3140-543210

To the memory

of F.Rellich

Preface

The present volume essentially is an edited version of the notes after the first two quarters of a lecture on partial differential equations held in 1976/77 at Berkeley. It represents an approach to theory of linear elliptic partial differential equations with emphasis on the abstract link between normal solvability of a linear operator and the structure of certain commutative C^*-algebras, which are isomorphic to function algebras, by the Gel'fand-Naĭmark theorem.

We will require a rather large amount of functional analysis. In particular we use distribution theory (Chapters I and II), theory of Fredholm operators (Appendix AI) and theory of commutative C^*-algebras (Appendix AII). To be self-contained we also perhaps should have included a discussion of the spectral theorem for unbounded linear operators in Hilbert space, which is needed occasionally.

In chapter III we study an algebra of linear differential operators with C^∞-coefficients over $\mathbb{R}^n$, acting either on L^2-Sobolev spaces $\mathfrak{H}_s$, $s \in \mathbb{R}$, or on the locally convex spaces $\mathfrak{H}_\infty$, $\mathfrak{H}_{-\infty}$. This investigation is rather formal. In particular one seeks to adjoin convolution operators, like $(1-\Delta)^{-r}$, $r > 0$, with the n-dimensional Laplace operator Δ, and will have to derive a generalized Leibnitz formula for products $a(M)b(D)$ of a multiplication operator $a(M)$ and a convolution operator $b(D)$. This is accomplished in form of an asymptotic expansion which is a finite sum whenever $b(\xi)$ is a polynomial. We introduce an extension algebra of <u>finitely generated pseudo differential operators</u> for which a calculus is derived. A variety of compactness results are derived, preparatory for chapter IV.

In chapter IV we study the <u>Laplace comparison</u> algebras $\mathfrak{A}_s$, $-\infty<s\leq\infty$, of $L^2(\mathbb{R}^n)$, defined as C^*-operator-algebras with unit over $\mathfrak{H}_s$ generated by multiplications by bounded continuous functions with oscillation vanishing at infinity, and by the special convolutions $D_j(1-\Delta)^{-1/2}$, $j=1,\dots,n$. For an operator $A \in \mathfrak{A}_s$, $s \in \mathbb{R}$, an abstract symbol is defined as the continuous function $\sigma_A(x,\xi)$ associated to the coset $A^\vee$ of A mod $K(\mathfrak{H}_s)$ by the Gelfand-Naĭmark isometry $\mathfrak{A}_s/K(\mathfrak{H}_s) \cong C(\mathbb{M}_s)$ with the maximal ideal space $\mathbb{M}_s$. It proves that $\mathbb{M} = \mathbb{M}_s$ is independent of s, and the symbol σ of course is closely related to the 'conventional' symbol of a differential- (or pseudo-differential-) operator. An operator $A \in \mathfrak{A}_s$ is Fredholm if and only if its symbol does never vanish. This results in a variety of very sharp criteria regarding normal solvability of differential operators over $\mathbb{R}^n$, as well as on the spectrum, regularity etc. of such operators.

These results are extended to the case $s=\infty$ as well. In particular some effort is spent to answer questions about the image of $\mathfrak{A}_\infty$ in $C(\mathbb{M})$ under σ. This proves important for derivation of a sharp Garding inequality, for example.

While chapter IV in effect is devoted to singular elliptic problems over $\mathbb{R}^n$, with differential operators 'comparable' to the Laplace operator, we then turn to genuine elliptic boundary problems in chapter V. The approach there is two-fold. First one may apply Garding's inequality for results about the Dirichlet problem, using a conventional approach via apriori estimates. Second one again may generate a Laplace comparison algebra, for a subdomain of $\mathbb{R}^n$, and use it in a manner similar as in chapter IV. This is accomplished for a half space - and for s=0 only.

The general elliptic boundary problem is discussed, for an even order elliptic equation, with Lopatinskij-Shapiro type boundary conditions. It is no obstacle that a half space is noncompact. In fact the method seems to apply more easily for the singular problems.

The Laplace comparison algebra of the half space does not have compact commutators, but gives raise to an ideal chain $\mathfrak{A} \subset \mathfrak{L} \subset K(\mathfrak{H})$, with $\mathfrak{A}/\mathfrak{L} \cong C(\mathbb{M}_1)$ and $\mathfrak{L}/K(\mathfrak{H}) \cong C(\mathbb{M}_2, K(\mathfrak{f}))$, $\mathfrak{f} = L^2((0,\infty))$, $\mathfrak{H} = L^2(\mathbb{R}^{n+1}_+)$.

For reason of simplicity we are restricting the discussion to L^2-theory only, and only to the singular elliptic problem over $\mathbb{R}^n$, and to the half space for the elliptic boundary problem. Let us notice that far more general results have been established by the author and his associates and students. In particular*) R. Illner [1] discussed L^p-theory for $\mathbb{R}^n$. R. McOwen [2] discussed Laplace comparison algebras on complete Riemannian manifolds. M. Taylor [3] discussed a hypo-elliptic comparison algebra over $\mathbb{R}^n$ and E. Tomer [4] used a second order elliptic comparison operator different from Δ (the harmonic oscillator in quantum theory).

The following represents the author's personal approach to the subject, which exists for some time while the attempts to publish a comprehensive version have remained incomplete. We are indebted to E. Herman for joint efforts in the late 1960-s. To Lars Gårding we owe the chance of presenting the material in a lecture at Lund in 1970-71. We are indebted to J. Frehse and to Alexander v. Humboldt foundation for a free year to work on the subject. - Finally we are grateful to Mrs. I. Kreuder and the staff of SFB 72, Bonn for organizing the typing of the manuscript.

*) [1] R. Illner, On algebras of pseudo-differential operators in $L^p(\mathbb{R}^n)$; Commun. on partial diff. equ. 2 (1977) 359-393; [2] H.O. Cordes and R. McOwen, Remarks on singular elliptic theory for complete Riemannien manifolds; Pacific. Journ. Math. 70 (1977) 133-141; [3] M.E. Taylor, Gel'fand theory of pseudo-differential operators and hypo-elliptic operators; Trans. Amer. Math. Soc. 153 (1971) 495-510; [4] E. Tomer, On the C^*-algebra of the Hermite Operator; Thesis at Berkeley, to appear.

Contents

Chapter I: A Survey of Distribution Theory.

0. Introduction.

In the present chapter we will learn about the basics of the most important kind of generalized functions, called distributions. Perhaps it will be an aid for understanding the concept if we start with the remark that the emphasis is not so much on generalizing the topological function concept, but rather on a new function-integration theory, called distribution and distribution integral. The rigorous theory of Riemann and Lebesgue integral, while being a most satisfying logical complex has left some incompatibilities with practical needs. Some of these will be removed.

Physicists have known and used distributions (of a special kind) for a long time. For example, in Physics, mass or electrical charge, in the macroscopic world, will never be concentrated at single points. One will have a 'density function', a 'distribution function' $\mu(x) = \mu(x_1,x_2,x_3)$ defined and (non-negative) real-valued in (x_1,x_2,x_3)-space $\mathbb{R}^3$. The mass of a (measurable) subset $\Omega\subset\mathbb{R}^3$ is given by the integral $m_\Omega = \int_\Omega \mu(x)dx$. However, for a simple mathematical description of dynamics it is very useful to consider the idealization of point mass, which does not occur in nature but can be approximated arbitrarily by assigning large values to the distribution function μ near a given point x^0 and having it zero elsewhere.

A 'mass point' is a physical reality. A rigid body can be replaced in many respects, by a mass point at its center of gravity, etc.. However there is no distribution function for a mass point, if the integral in above formula for m_Ω is interpreted as Riemann or Lebesgue integral. Hence one introduces a generalized distribution function called Dirac delta function δ. It takes the values $\delta(0) = \infty$, $\delta(x) = 0$, $x\neq 0$. But the most important fact is that we have $\int_\Omega \delta(x)dx = 0$, as $0\notin\Omega$, $= 1$, as $0\in\Omega$, for open sets Ω. More generally,

$$\int \delta(x)\phi(x)dx = \phi(0) \tag{0.1}$$

for every continuous function ϕ over $\mathbb{R}^3$.

Here the function concept does not have to be generalized but a new integral concept must be introduced. Things get worse with the charge distribution of an electrical dipole. Physically, a dipole consists of a pair of point charges of equal magnitude but opposite sign, at distance zero. It is approximated by a pair of magnitude q at distance d. In the limit $d \to 0$, keeping the direction and the dipole moment $M=qd$ constant, an electrical field is reached which is not zero. The charge distribution μ_D of a 1-dimensional dipole, for simplicity, of dipole moment 1, pointing into the positive x-direction should have the property that

(0.2) $$\int \mu_D(x)\phi(x)dx = \phi'(0) = d\phi/dx(0) \quad ,$$

for every continuously differentiable ϕ over $\mathbb{R}$. Again this is incompatible with Riemann or Lebesgue integration. No assignment of values of the 'function' $\mu_D(x)$ at $x=0$ is reasonnable.

Another such incompatibility between integration theory and practical needs arises in theory of hyperbolic second order partial differential equations, as first explored by J. Hadamard [4]. There it was necessary to introduce a certain very general singular integral concept, called finite part. A finite part integral, from our later view point of chapter II, will be a special kind of distribution integral, just as well as the so-called Cauchy principal value which also occurs in other parts of analysis. The underlying generalized functions perhaps are not as simple as the above delta-function and the distribution μ_D. Thus we shall make no attempt of a description until later on (chapter II).

The above examples may serve as guide into the new concept: The name 'distribution' may be derived from the fact that the two generalized functions δ and μ_D above are modelled after mass or charge distributions of idealized physical objects. In abstract mathematical definition (in section 2) a distribution T no longer will be given by its values at every point $x \in \mathbb{R}^n$, but by the values of the integral

(0.3) $$\int T\phi dx \quad ,$$

with testing functions ϕ, infinitely differentiable and zero for large $|x|$, where the 'distribution integral' (0.3) simply is only an abstract linear functional over the space $\mathcal{D}$ of testing functions, continuous in a manner to be specified. Not all measurable functions are also distributions but only these which are locally integrable, and these are uniquely determined, modulo a nul set by their Lebesgue integrals (0.3).

The main reason we have here of presenting the material again, in view of a large number of existing books on the subject, is that of self-containedness and preselection of material most useful for us. We have tried to focus on the aspects of classical analysis, and refer to Laurent Schwartz [7] for all more sophisticated matters of topology. There perhaps the most natural introduction into distribution spaces, as a class of locally convex topological vector spaces can be found. Also there are detailed results of almost every kind, which had to be omitted here.

As another book emphasizing classical analysis we mention Gelfand-Silov [3]. Other works more oriented towards locally convex spaces are Treves [12] and Edwards [2]. It would be impossible to mention or even read all publications on the subject.

There are other kinds of generalized functions, like the hyper functions of Sato and the ultradistributions. None of these can be discussed.

Most important technical tools for partial differential equations are the Fourier transform of temperate distributions (section 4) and the convolution product (section 8). Section 2 defines the distribution concept and the distribution derivative. Section 1 and 4 are collections of well known facts on well known function spaces, on the convolution product of L^1-functions and the classical Fourier integral and Fourier inversion formula, partly with proofs. Section 5 considers distributions of compact support and their Fourier transform. Section 6 is a short introduction into the topology of distribution spaces. Section 7 discusses the tensor product which is needed for the convolution product. Finally section 9 establishes existence of the so-called distribution kernel for certain continuous linear maps between function and distribution spaces.

A somewhat more sophisticated discussion, with important applications will follow in chapter II where we investigate distributions with only one singularity, mainly regarding their Fourier transforms. This will have applications to theory of constant coefficient elliptic and hyperbolic partial differential equations.

1. Common facts about function spaces and functions.

In this section we are going to survey some facts about functions and function spaces to be used later on. Mostly we skip proofs or refer to basic texts, like [5], [8], [11]. Most readers will proceed to the following sections, and be referred back at the proper time.

L^p-spaces. If $\{X,\mu\}$ denotes some σ-finite measure space [5], then the spaces $L^p = L^p(X) = L^p(X,d\mu)$ are defined as the classes of μ-measurable complex-valued functions $u: X \to \mathbb{C}$ such that

$$(1.1) \qquad \|u\|_{L^p} = \{\int_X |u(x)|^p dx\}^{1/p} < \infty \quad .$$

In particular, for an open subset $\mathcal{D}$ of the Eucidean space $\mathbb{R}^n$ by $L^p(\mathcal{D})$ we usually mean the above collection with respect to the Lebesgue measure on $\mathcal{D}$, and will write $L^p(\mathcal{D},d\mu)$ if we refer to a measure $d\mu$ different from the Lebesgue measure.

In the above p will always denote a positive real number, satisfying $1 \leq p < \infty$. Since it is well known that

$$(1.2) \qquad \lim_{p\to\infty} \|u\|_{L^p} = \text{ess sup } \{|u(x)|: x \in X\} \quad ,$$

with

$$(1.3) \qquad \begin{aligned} &\text{ess sup } \{|u(x)|: x \in X\} = \sup \{\tau: \tau \in \mathbb{R}, \mu(E_\tau) > 0\} \quad , \\ &E_\tau = \{x: x \in X, |u(x)| \geq \tau\} \quad , \end{aligned}$$

whenever the limit of (1.2) is well defined (and finite), it is customary to introduce the space $L^\infty = L^\infty(X) = L^\infty(X,d\mu)$ as the class of all μ-measurable functions $u: X \to C$ such that

$$(1.4) \qquad \|u\|_{L^\infty} = \text{ess sup } \{|u(x)|: x \in X\} < \infty .$$

The spaces $L^p(X,d\mu)$, $1 \leq p \leq \infty$ are Banach spaces, and perhaps represent the most important class of function spaces, which are Banach spaces. It is evident, that the two expressions $\|u\|_{L^1}$ and $\|u\|_{L^\infty}$, defined by (1.1), for p=1, and by (1.4) are norms on the spaces L^1 and L^∞, respectively. More precisely we note that $\|u\|_{L^p} = 0$, $1 \leq p \leq \infty$, whenever u vanishes almost everywhere (i.e. with the exception of a set of measure zero). Therefore the spaces $L^p(X)$ are regarded as spaces of equivalence classes modulo the class of functions which are zero almost everywhere. Or, in other words, two functions which differ only on a set of measure zero will be regarded as equal. The completeness of L^1 is known as the Fischer-Riesz theorem, and the completeness of L^∞ also is discussed in standard textbooks [5], p. 107.

L^2-spaces. The investigation of $L^2 = L^2(X,d\mu)$ is facilitated by the observation that the expression $\|u\|_{L^2}$ may be written as

$$(1.5) \qquad \|u\|_{L^2} = \{(u,u)\}^{1/2} \quad , \quad (u,v) = \int_X \bar{u}(x)v(x)d\mu(x) \quad ,$$

where (u,v) is a positive definite sesqui-linear form on L^2, satisfying Schwarz' inequality: For $u,v \in L^2$ we have $\bar{u}v$ summable, so that (u,v) of (1.5) is defined, and

$$(1.6) \qquad |(u,v)| \le \|u\|_{L^2} \|v\|_{L^2} \quad ,$$

as follows from the definiteness of the form $(\sigma u+\tau v, \sigma u+\tau v)=F(\sigma,\tau)$, in the two complex numbers σ,τ. (1.6) simply is the condition that the discriminant of F is nonnegative. Particularly $F(\sigma,\tau)$ is finite for all $\sigma,\tau\in\mathbb{C}$, so that L^2 is a linear space. The triangle inequality (1.10) for $p=2$ follows simply from (1.6), and $\|u\|_{L^2}$ is a norm on L^2. Moreover, this is a Hilbert norm, associated to the inner product (u,v) of (1.5) on L^2, cf. [11]. The space L^2 is complete again, under the norm (1.5), by the Fischer Riesz theorem ([9], p. 38).
Therefore the space $L^2 = L^2(X,d\mu)$ is a Hilbert space. In particular, the space $L^2(\mathcal{D}) = L^2(\mathcal{D},dx)$ for an open subset D of $\mathbb{R}^n$ is a separable Hilbert space ([11], p. 65). Also the space

$$(1.7) \qquad l^2 = \{x=(x_1,x_2,\ldots) \ , \ x_j\in\mathbb{C} \ , \ \sum_{j=1}^{\infty} |x_j|^2 < \infty\}$$

is a special case of our present L^2-spaces, arising for the measure space $X = \mathbb{Z}_+ = \{m | m\in\mathbb{Z}, m>0\}$ of positive integers, with discrete measure.

For general p, $1<p<\infty$, we first notice a generalization of Schwarz' inequality (1.6), called Hoelder's inequality: For $1<p<\infty$ one defines the conjugate exponent q by the relation $1/p+1/q=1$, or $q=p/(p-1)$. Clearly then p and q are mutually conjugate exponents. Hoelder's inequality states that the function $u(x)v(x)$ is (absolutely) summable over X whenever $u\in L^p(X)$, $v\in L^q(X)$, and that

$$(1.8) \qquad \left|\int_X u(x)v(x)d\mu\right| \le \|u\|_{L^p} \|v\|_{L^q} \quad , \ u\in L^p(X), \ v\in L^q(X) \ .$$

It may be noted that $p=2$ implies $q=2$, so that 2 is its own conjugate exponent. For $p=q=2$ we obtain Schwarz' inequality from Hoelder's inequality. For the proof of Hoelder's inequality cf. [5], p. 175. If we define $q=\infty$ or $q=1$, for $p=1$ or $p=\infty$, respectively, then Hoelder's inequality also holds for these p and q, as immediately evident.

From Hoelder's inequality we may conclude that $L^p(X)$ is a linear space, and that $\|u\|_{L^p}$ defines a norm on it. First, for a pair of complex numbers f and g

and Hoelder's inequality (for a measure space of 2 discrete points) we conclude that

$$|f+g|^p \le |f|\cdot|f+g|^{p-1}+|g|\cdot|f+g|^{p-1}$$

$$\le (|f|^p+|g|^p)^{1/p}(2|f+g|^{q(p-1)})^{1/q} = 2^{1/q}|f+g|^{p-1}(|f|^p+|g|^p)^{1/p}$$

so that

$$(1.9)\qquad |f+g|^p \le 2^{p/q}(|f|^p + |g|^p) \ .$$

If we substitute $f = f(x)$, $g = g(x)$, for two functions $f,g\varepsilon L^p$, and then integrate over X, it follows that $f+g\varepsilon L^p$, whenever $f,g\varepsilon L^p$. Since $cf\varepsilon L^p$ for $f\varepsilon L^p$, $c\varepsilon\mathbf{C}$, holds trivially, we indeed conclude that $L^p=L^p(X)$ is a linear space.

Then note that

$$(\|f+g\|_{L^p})^p = \int_X |f+g|^p d\mu$$

$$\le \int|f|\cdot|f+g|^{p-1}d\mu+ \int|g|\cdot|f+g|^{p-1}d\mu$$

$$\le \|f\|_{L^p}\cdot(\int|f+g|^p d\mu)^{1/q} + \|g\|_{L^p}\cdot(\int|f+g|^p d\mu)^{1/q}$$

$$= (\|f\|_{L^p} + \|g\|_{L^p})(\|f+g\|_{L^p})^{p/q} \qquad ,$$

which implies the triangle inequality

$$(1.10)\qquad \|f+g\|_{L^p} \le \|f\|_{L^p}+\|g\|_{L^p} \ , \ f,g\varepsilon L^p(X) \ , \ 1<p<\infty \ .$$

Relation (1.10) often is referred to as Minkowski's inequality. As a consequence of inequality (1.10) the spaces $L^p(X)$, $1<p<\infty$ also are normed linear spaces. By the Riesz Fischer theorem they again are Banach spaces.

Note that there is an imbedding of the space $L^q=L^q(X)$, for the conjugate exponent $q=p/(p-1)$ into the adjoint space L^{p*} of continuous linear functionals over $L^p=L^p(X)$, for $1\le p\le\infty$. For $h\varepsilon L^q(X)$ define the functional $\langle h,u\rangle$, $u\varepsilon L^p$, by setting

$$(1.11)\qquad \langle h,u\rangle = \int_X hu\, dx \qquad , \qquad u\varepsilon L^p \ .$$

By Hoelder's inequality this defines a bounded linear functional over $L^p(X)$, and we get the estimate

$$(1.12)\qquad \sup\{|\langle h,u\rangle| : u\varepsilon L^p, \|u\|_{L^p} \le 1\} \le \|h\|_{L^p} \qquad ,$$

which shows that the embedding reduces the norm. In fact, the imbedding is an isometry, for $1<p\le\infty$, or $1\le q<\infty$, because a maximum is assumed, at left of (1.11) if we substitute $u=u_0/\|u_0\|_{L^p}$ with

$$(1.13)\qquad u_0 = \bar{h}|h|^{q-2},\ h(x)\neq 0 \ , \ = 0 \ , \ h(x) = 0 \ ,$$

which defines a function in L^p such that equality holds in (1.8) for f=h and $g=u_0$.

It is an important result of measure theory ([9], theorem 15.C for example) that the above isometry represents an isometric isomorphism between the two spaces L^{p^*}, and L^q, whenever $1\le p<\infty$. Accordingly, it follows that the spaces L^p, $1<p<\infty$, are reflexive Banach spaces. These spaces are separable as well, for $1\le p<\infty$ and for the case of $X=\mathcal{D}$ being an open subset of $\mathbb{R}^n$, and the Lebesgue measure dx.

It is known also, that neither the space L^1 nor the space L^∞ is reflexive. Moreover, the above imbedding $L^1 \to L^\infty$ maps onto a proper subspace of $L^{\infty*}$.

The spaces $C(\mathbf{X})$, $CB(\mathbf{X})$, and $CO(\mathbf{X})$. For a compact Hausdorff space $\mathbf{X}$ we consider the space $C(\mathbf{X})$ of all continuous complex-valued functions $f: \mathbf{X} \to \mathbb{C}$, under the norm

$$(1.14)\qquad \|f\| = \|f\|_C = \sup\{|f(x)|: x\in\mathbf{X}\} \quad .$$

Convergence of a sequence f_m of functions in $C(\mathbf{X})$, under this norm $\|\cdot\|_C$ amounts to uniform convergence of the sequence f_m, so that the pointwise limit $f(x) = \lim_{m\to\infty} f_m(x)$ is continuous. This in particular implies completeness of the normed space $C(\mathbf{X})$ so that it is a Banach space. Similarly, the collection of all bounded continuous complex-valued functions over a locally compact Hausdorff space $\mathbf{X}$ again is a Banach space, under the norm (1.14), denoted by $CB(\mathbf{X})$, and also its closed subspace $CO(X)$ of all functions $f\in CB(\mathbf{X})$ satisfying $\lim_{x\to\infty} f(x) = 0$ (i.e. $|f(x)|<\varepsilon$, outside a sufficiently large compact set of $\mathbf{X}$, for any $\varepsilon>0$.).

On the other hand, for a locally compact noncompact Hausdorff space $\mathbf{X}$ the space $C(\mathbf{X})$ of all continuous complex-valued functions over $\mathbf{X}$ will again be denoted by $C(\mathbf{X})$. This space will be locally convex, with a collection of generating semi-norms given by

$$(1.15)\qquad \{\|f\|_A : A \text{ compact}, A\subset X\}\ ,\quad \|f\|_A = \sup\{|f(x)|: x\in A\} \quad .$$

The norm (1.14) will be considered as the natural norm for $CB(\mathbf{X})$ and $CO(\mathbf{X})$, and the system (1.15) will always be used for the topology of $C(\mathbf{X})$, unless stated otherwise.

The spaces $C(\mathbf{X})$, $CB(\mathbf{X})$ and $CO(\mathbf{X})$ also are algebras, with product given by pointwise multiplication : $(uv)(x) = u(x)\cdot v(x)$. For more detailed discussion see appendix AII.

The spaces $C^\tau(\mathbf{X})$. For a compact metric space $\mathbf{X}$, with metric $d(x,y)$, $x,y\in\mathbf{X}$, we call a continuous function $f: \mathbf{X} \to \mathbb{C}$ Hoelder continuous, with exponent τ, $0<\tau\le 1$, whenever there exists a constant $c\ge 0$ such that

$$(1.16)\qquad |f(x)-f(y)| \le c\, d(x,y)^\tau \ ,\quad x,y\in X \ .$$

The class of all Hoelder continuous functions over $\mathbf{X}$ with a given exponent τ is denoted by $C^\tau(\mathbf{X})$. This space $C^\tau(\mathbf{X})$ is a Banach space, under the norm

$$(1.17) \qquad \|f\|_{C^\tau} = \|f\|_C + h_\tau(f) \quad ,$$

with

$$(1.18) \qquad h_\tau(f) = \sup\{|f(x)-f(y)|\,(d(x,y))^{-\tau} : x,y\in X,\ x\neq y\} \quad .$$

In particular, for a nonempty subset $\mathcal{D}$ of the Euclidean space $\mathbb{R}^n$ it is customary to consider the spaces $C^\tau(\mathcal{D})$ with respect to the Euclidean distance $d(x,y)=|x-y|$.

If X is only locally compact then $C^\tau(X)$ denotes the class of all functions in $C(X)$ whose restriction to A is in $C^\tau(A)$ for every compact $A\subset X$, and will regard $C^\tau(X)$ as a locally convex space with generating system of semi-norms given by

$$(1.19) \qquad \{\|f\|_{C^\tau(A)} : A \ X,\ A \text{ compact}\} \ ,$$

where $\|f\|_{C^\tau(A)}$ denotes the norm (1.17) for f replaced by $f|A$.

For $\tau=1$ a function satisfying (1.16) will also be called Lipschitz continuous, and the space of all Lipschitz continuous functions will be denoted by $LC(X)$.

The spaces $C^k(X), C^\infty(X)$. For $X=\mathcal{D}$, an open set of $\mathbb{R}^n$, we introduce the space $C^k(\mathcal{D})$, consisting of all functions $f: \mathcal{D} \to \mathbb{C}$, which are continuous in $\mathcal{D}$ together with all their partial derivatives up to the order k including. Here k may denote any integer $0,1,2,\ldots$. By $C^\infty(\mathcal{D})$ we denote the intersection

$$(1.20) \qquad C^\infty(\mathcal{D}) = \ C^k(\mathcal{D}) \ ,$$

i.e. the class of continuous functions with partial derivatives of arbitrary order existing (and continuous). For a nonvoid open set $\mathcal{D}\subset\mathbb{R}^n$ the spaces C^k, $k=0,1,2,\ldots\infty$, are not Banach spaces, but are locally convex spaces indeed.

For a region $\mathcal{R}$ of $\mathbb{R}^n$ with smooth boundary one introduces the class $C^k(\mathcal{R})$ of all functions, defined and complex-valued in $\mathcal{R}$, which have an extension to an open set $\mathcal{D} \supset \mathcal{R}$ which is in $C^k(\mathcal{D})$. For a compact region $\mathcal{R}$ one then will have $C^0(\mathcal{R})= C(\mathcal{R})$, and then all spaces $C^k(\mathcal{R})$ are Banach spaces with norm defined to be the maximum over $\mathcal{R}$ and the absolutes of the values of all derivatives up to order k, including.

The support of a complex valued function $f: \mathfrak{X} \to \mathbb{C}$ is defined to be the smallest closed set $S\subset\mathfrak{X}$ such that $f=0$ outside of S where $\mathfrak{X}$ is a topological space. The support of a function f is denoted by supp f, and the class of all functions of compact support in $C(\mathfrak{X}), C^\tau(\mathfrak{X}), C^k(\mathcal{D}), C^k(\mathcal{R})$ is denoted by $C_0(\mathfrak{X}), C_0^\tau(\mathfrak{X}), C_0^k(\mathcal{D})$, etc..

It will be useful in the spaces $C^k(\Omega), \Omega\subset\mathbb{R}^n$, to employ multi-index notation, as follows. We shall write

$$x^\alpha = \prod_1^n x_j^{\alpha_j}, \quad D^\alpha = \prod_1^n D_j^{\alpha_j}, \quad u^{(\alpha)} = i^{|\alpha|} D^\alpha u,$$

$$D_j = -i\partial/\partial x_j, \quad |\alpha| = \sum_1^n \alpha_j, \quad \alpha! = \prod_1^n \alpha_j!,$$

$$\binom{\alpha}{\beta} = \prod_1^n \binom{\alpha_j}{\beta_j},$$

with α denoting a multi-index: $\alpha = (\alpha_1,\ldots\alpha_n)$ of dimension n, where α_j, $j = 1,\ldots,n$, are nonnegative integers. Similarly $\beta = (\beta_1,\ldots,\beta_n)$; the class of all multi-indices $\alpha = (\alpha_1,\ldots,\alpha_n)$ will be denoted by $\mathbb{Z}_+^n$.

In particular we shall use the notations $\alpha + \beta$ and $\alpha - \beta$ for the multi-indices $(\alpha_j + \beta_j)$, and $(\alpha_j - \beta_j)$, respectively, with subtraction only meaningful if $\beta \le \alpha$, that is $\beta_j \le \alpha_j$, $j = 1,\ldots,n$. We shall write $\alpha < \beta$, if $\alpha \le \beta$ and $\alpha \neq \beta$. Also $\binom{p}{q}$, in (1.21), will denote the corresponding binomial coefficient, and p! will denote the factoriel, for nonnegative integers p, q:

(1.22) $$p! = \prod_{l=1}^{p} l, \quad \binom{p}{q} = p!/(q!(p-q)!), \quad 0! = 1, \quad 0 \le q \le p, \quad p, q = 0,1,\ldots$$

With the multi-index notation, Leibnitz' formula for differentiation of a product, will take the forms

(1.23)
$$(uv)^{(\alpha)} = \sum_{\beta\le\alpha} \binom{\alpha}{\beta} u^{(\beta)} v^{(\alpha-\beta)},$$
$$u\, v^{(\alpha)} = \sum_{\beta\le\alpha} (-1)^{|\beta|} \binom{\alpha}{\beta} (u^{(\beta)} v)^{(\alpha-\beta)}$$

The last formula (1.16) also shall be denoted as the adjoint Leibnitz formula.

Finally we are going to consider integral operators as bounded operators of L^p.

Lemma 1.1. Given two σ-finite measure spaces $\{X,\mu\}$ and $\{Y,\nu\}$, and let the complex-valued function $k(x,y)$ be defined and measurable over $\{X \times Y, \mu \times \nu\}$. Let the two functions k_1, k_2, defined by

(1.24)
$$k_1(s) = \int_X |k(x,s)|\, d\mu_x, \quad s \in Y,$$
$$k_2(t) = \int_Y |k(t,y)|\, d\nu_y, \quad t \in X,$$

be in $L^\infty(Y,d\nu)$ and in $L^\infty(X,d\mu)$, respectivley.
Then, if the Banach spaces $\mathfrak{X}^p \mathfrak{Y}^p$, $1 \le p \le \infty$, are defined by

(1.25) $$\mathfrak{X}^p = L^p(X,d\mu), \quad \mathfrak{Y}^p = L^p(Y,d\nu),$$

it follows that the integral expression

(1.26) $$Ku = v, \quad v(x) = \int_Y k(x,y)u(y)\,d\nu_y, \quad u \in \mathfrak{Y}^p,$$

defines a (continuous) linear map in $L(\mathfrak{Y}^p,\mathfrak{X}^p)$, for $1 \le p \le \infty$, and the operator norm

(1.27) $$||K||_{\mathfrak{Y}^p,\mathfrak{X}^p} = \sup \{||Ku||_{\mathfrak{X}^p} \mid ||u||_{\mathfrak{Y}^p} < 1\},$$

satisfies the inequality

(1.28) $$||K||_{\mathfrak{Y}^p,\mathfrak{X}^p} \le \{||k_1||_{\mathfrak{Y}^\infty}\}^{1/p}\{||k_2||_{\mathfrak{X}^\infty}\}^{1/q} ,$$

with the conjugate exponent q of p.

As a sceleton of a proof we mention the chain of estimates, below which has only absolutely convergent integrals, readily interchangeable. The case $p = \infty$ is evident.

$$(||Ku||_{L^p})^p = \int d\mu_x |\int k(x,y)u(y)d\nu_y|^p$$

$$\int d\mu_x \; |\int k(x,y)|^{1/q}\{|k(x,y)|^{1/p}|u(y)|\}d\nu_y|^p$$

$$\int d\mu_x\{|\int k(x,y)|d\nu_y\}^{p/q}\{\int|k(x,y)|u(y)|^p d\nu_y\}$$

$$(||k_2||_{\mathfrak{X}^\infty})^{p/q}\int d\nu_y |u(y)|^p\{\int d\mu_x|k(x,y)|\}$$

$$||k_1||_{\mathfrak{Y}^\infty}(||k_2||_{\mathfrak{X}^\infty})^{p/q}(||u||_{\mathfrak{Y}^p})^p.$$

Estimate (1.28) results from (1.29) by taking the power 1/p.

2. Definition of distributions and elementary properties.

At the basis of the distribution concept there is the lemma, below.

Lemma 2.1. Let a function be defined (complex valued) and measurable in Ω, an open subset of $\mathbb{R}^n$. Let also f be locally integrable.

Then, if

$$\int_\Omega f(x)\phi(x)dx = 0, \quad \phi \in C_0^\infty(\mathbb{R}^n), \tag{2.1}$$

it follows that $f(x) = 0$ almost everywhere in Ω.

Remarks. 1) In all the following, we mean "complex - valued function" when we say "function", unless explicitely stated otherwise.

2) ' f locally integrable in Ω' or '$f \in L^1_{loc}(\Omega)$' means that every point $x^0 \in \Omega$ has a neighborhood N such that $\int_N |f|dx < \infty$. In other words,

$$L^1_{loc}(\Omega) = \{f: f|A \in L^1(A), A \subset \Omega, A \text{ compact}\}. \tag{2.2}$$

Sketch of proof. If $Q \subset \Omega$ denotes a rectangular box

$$Q = \{x: \alpha_j < x_j < \beta_j, j = 1,\dots,n \} \tag{2.3}$$

with $\alpha_j < \beta_j$, α_j, β_j finite, and if χ_Q denotes the characteristic function of $Q[\chi_Q = 1, x \in Q, = 0$ elsewhere$]$ then (2.1) and a suitable approximation ϕ_j of χ_Q by a sequence of C_0^∞-functions may be used in conjunction with Lebesgue integral theorem to show that

$$\int_Q f dx = \int_\Omega f\chi_Q dx = \lim_{j\to\infty} \int f\phi_j dx = 0. \tag{2.4}$$

Next, for any open set $O \subset \Omega$ we have $O = \bigcup_{k=1}^\infty Q_k$, where the rectangular boxes Q_j, Q_k, for $j \neq k$, have only boundary points in common. It follows that

$$\int_O f dx = \sum_k \int_{Q_k} f dx = 0. \tag{2.5}$$

using additivity of the Lebesgue integral. Next a measurable set E Ω may be represented as $E = \bigcup_j O_j + N$ with a null set N and an increasing sequence of open sets O_j. It follows that

$$\int_E f dx = \lim \int_{O_k} f dx = 0, \tag{2.6}$$

at least if E is bounded and measurable.

Suppose now f does not vanish almost everywhere then there must exist a set E_0 of positive measure in which $|f| \geq p > 0$. It may be assumed that E_0 is bounded and that $f \geq 0$ on E_0 without loss of generality. Accordingly we get

$$0 < pm(E_0) \leq \int_{E_0} f dx = 0, \tag{2.7}$$

a contradiction, and the lemma is proven.

Lemma 2.1 may be expressed by saying that a locally integrable function f is characterized (modulo a null function) by the values of the linear functional

$$(2.8)\qquad \langle f,\phi\rangle = \int f\phi dx \quad , \phi \in C_0^\infty(\Omega).$$

Two functions $f, g \in L^1_{loc}$ for which the linear functionals (2.8) coincide over the space $C_0^\infty(R^n)$ must coincide almost everywhere in Ω since we get

$$(2.9)\qquad \int (f - g)\phi dx = 0, \ \phi \in C_0^\infty(\Omega).$$

A distribution T over Ω is defined to be a continuous linear functional

$$(2.10)\qquad T: C_0^\infty(\Omega) \to \mathbb{C}$$

The value of this functional at $\phi \in C_0^\infty(\Omega)$ (is a complex number and) will be denoted by $\langle T, \phi\rangle$. Also it is the convention to denote the space $C_0^\infty(\Omega)$ by $\mathcal{D} = \mathcal{D}(\Omega)$. Then members of $\mathcal{D}$ are called testing functions.

Moreover it has become generally accepted to also use the notation

$$(2.11)\qquad \langle T,\phi\rangle = \int_\Omega T(x)\,\phi(x)dx,$$

where the right hand side is referred to as a distribution integral. Let us emphazise, however, that the notation (2.11) is purely formal and merely intends to preserve intuition gained by dealing with functions and integrals. Please remember that neither does the value T(x) at $x \in \Omega$ need to have any meaning, nor can we in general interprete a distribution integral as a Lebesgue-Stieltjes or Daniell integral related to a measure. The notation will save many words and will be very suggestive, nevertheless.

Regarding continuity of the linear functional T the space $\mathcal{D}$ will be equipped with its "natural topology" defined as the inductive limit topology corresponding to

$$(2.12)\qquad \mathcal{D}(\Omega) = \bigcup_j \mathcal{D}(\Omega, K_j)$$

with a sequence $\{K_j\}$ of compact sets $K_j \subset \Omega$ each K_j being the closure of its interior Ω_j and such that

$$(2.13)\qquad \Omega = \bigcup_j \Omega_j,$$

and

$$(2.14)\qquad \mathcal{D}(\Omega,K) = \{u: u \in \mathcal{D}, \text{ supp } u \subset K\} .$$

For readers interested in theory of topological vector spaces we spell out some more details of this, although we do not intend to use this, in the following (c.f. our alternate approach, below, as well as the criteria for continuity in section 6).

Th spaces $\mathcal{D}(\Omega, K)$, for any compact subset K of Ω are naturally Frechet spaces with systems of generating semi-norms given by

(2.15) $$||u||_{\Omega,k} = \sup\{|u^{(\alpha)}(x)| : x \in \Omega, |\alpha| \le k\}, \quad k = 0,1,2,\ldots \quad .$$

Using the natural injection maps from $\mathcal{D}(\Omega,K_j)$ to $\mathcal{D}(\Omega)$ we may regard the union $\mathcal{D}(\Omega)$ of (2.12) as an inductive limit space. Recall that the inductive limit topology is defined as the strongest topology in which all above injection maps still are continuous (c.f. [1],also [2] for more details regarding the inductive limit topology of $\mathcal{D}(\Omega)$).

It also is to be observed that the topology obtained is independent of the sequence K_j chosen subject to above requirements.

We shall use the following alternate definition of continuity of a linear functional T over $\mathcal{D}$ [It proves to be equivalent with continuity under the above inductive limit topology but we shall not prove this here and shall never use it either].

Definition 2.2. A linear functional $T: \mathcal{D}(\Omega) \to \mathbb{C}$ will be called continuous if we have

(2.16) $$\langle T,\phi_k\rangle \to 0, \quad k \to \infty$$

for every sequence $\{\phi_k\}$ of testing functions satisfying the two conditions, below.

(i) $\operatorname{supp} \phi_k \subset K \subset \Omega$, with a given fixed compact set K independent of k.
(ii) For every $\alpha \in \mathbb{Z}^n_+$ we have $\phi_k^{(\alpha)}(x) \to 0$, as $k \to \infty$, uniformly in Ω.

A sequence $\{\phi_k\}$ of testing functions $\phi_k \in \mathcal{D}(\Omega)$ satisfying the above conditions (i) and (ii) will be said to converge to zero in $\mathcal{D}(\Omega)$ [Notation: '$\phi_k \to 0$ in $\mathcal{D}$']. Thus we may say that a linear functional $T: \mathcal{D} \to \mathbb{C}$ is continuous if and only if $(T,\phi_k) \to 0$ whenever $\phi_k \to 0$ in $\mathcal{D}$.

If f is a locally integrable function over Ω then a distribution is defined by (the right hand side of)(2.8), as an immediate consequence of (i) and (ii), above, and Lebesgue integral theorem, concerning continuity of this functional. By lemma 2.1 there is a 1-1-correspondence between the functions in $L^1_{loc}(\Omega)$ and the corresponding functionals (2.8) which will be used to identify f and the distributions (2.8). Therefore the locally integrable functions form a class of distributions.

Only the L^1_{loc}-functions have become distributions, in this way. For a function which is not L^1_{loc} there may be distributions assigned similarly, by so-called singular integrals, in special cases. However we shall see that such more general correspondence usually is not bi-unique. Accordingly an identification between non-locally integrable functions and distributions will not be attempted. For functions in $C^\infty(\mathbb{R}^n - \{0\})$, not in L^1_{loc} c.f. chapter II.

We have at least one distribution so far, which is not a function: the linear functional of (0.1) trivially is a distribution called the Dirac measure. [Again continuity is trivial, using Definition 1.2.]

Lemma 2.3. Let T be a distribution over Ω. If $a \in C^{\infty}(\Omega)$ is an infinitely differentiable function, and if $\alpha \in Z_+^n$ is a multi-index, then the two linear functionals T_a and $T^{(\alpha)}$ defined by setting

$$\langle T_a,\phi\rangle = \langle T,a\phi\rangle , \quad \phi \in \mathcal{D} , \tag{2.17}$$

and

$$\langle T^{(\alpha)},\phi\rangle = (-)^{|\alpha|}\langle T,\phi^{(\alpha)}\rangle , \quad \phi \in \mathcal{D} \tag{2.18}$$

define distributions.

Proof. The only point to be discussed is continuity of the two functionals, again. It is a consequence of the fact that $\phi_k \to 0$ in $\mathcal{D}$ implies $\phi_k^{(\alpha)} \to 0$ in $\mathcal{D}$ as well as $a\phi_k \to 0$ in $\mathcal{D}$. The first is evident, and the second is a simple consequence of Leibnitz formula (1.23). Q.e.d.

The distribution T_a will be called the (pointwise) product between the distribution T and the C^{∞}-function a, and will be denoted by aT or by Ta. It is evident that this notation of product agrees with the conventional pointwise product

$$(af)(x) = a(x)f(x) \tag{2.19}$$

whenever $T = f \in L^1_{loc}$.

Similarly the distribution $T^{(\alpha)}$ will be referred to as the $(\alpha)^{th}$ derivative of the distribution T. Again if $T = f \in L^1_{loc}$ admits contiuous derivatives $f^{(\alpha)}$ in the usual sense [henceforth referred to as "classical derivatives" or "derivatives in the classical sense"] for all $|\alpha| \le k$, then we get $T^{(\alpha)} = f^{(\alpha)}$, $|\alpha| \le k$, simply by a partial integration in the functional (2.18):

$$\langle T^{(\alpha)},\phi\rangle = (-)^{|\alpha|}\int f\phi^{(\alpha)}dx = \int f^{(\alpha)}\phi dx = \langle f^{(\alpha)},\phi\rangle , \quad \phi \in \mathcal{D}, \tag{2.20}$$

[there will be no boundary terms due to the compact support of ϕ]. The same will be true for generalized derivatives, for example in the sense of L^p, as will be left to the reader to confirm (c.f. III, 1-7).

Accordingly the two concepts of derivative [for distributions and differentiable functions] agree whenever they are both defined.

It is seen therefore that a distribution has derivatives of all orders, which are distributions again. We also use the notation $D^{\alpha}T = (-i)^{|\alpha|}T^{(\alpha)}$, if convenient and notice that Leibnitz' formula remains correct for distributions and their derivatives. By $(1.23)_2$ we get

$$\text{(2.21)}\quad \begin{aligned}\langle (aT)^{(\alpha)},\phi\rangle &= (-)^{|\alpha|}\langle T, a\phi^{(\alpha)}\rangle = \sum_{\beta\le\alpha}\binom{\alpha}{\beta}(-)^{|\alpha-\beta|}\langle T,(a^{(\beta)}\phi)^{(\alpha-\beta)}\rangle \\ &= \langle \sum_{\beta\le\alpha}\binom{\alpha}{\beta}a^{(\beta)}T^{(\alpha-\beta)},\phi\rangle \quad , \ \phi \in \mathcal{D},\end{aligned}$$

confirming formula $(1.23)_1$ for the more general case of $u = a \in C^\infty(\Omega)$ and $v = T$, a distribution over Ω. Similarly formula $(1.23)_2$ may be verified, again for a C^∞-function and a distribution.

In case where a distinction is to be made between the two kinds of derivative we shall refer to the above $T^{(\alpha)}$ as the distribution derivative of T.

Suppose we know that the distribution derivatives of T are functions. Is it true then that T itself must be a differentiable function with classical derivatives equal to its distribution derivatives? Note lemma 2.4, below, regarding this question, which will be important, at times.

First let us introduce some further simple conveniences. In the following we shall denote distributions with the same symbols as functions - that is u, v, f, g, etc. rather than T, T_1, etc. Also the class of all distributions over Ω will be denoted by $\mathcal{D}'(\Omega)$ [the conjugate space of $\mathcal{D}(\Omega)$].

Let Ω_1 and Ω_2 be two open subsets of $\mathbb{R}^n$, and let $\Omega \subset \Omega_j$, $j = 1,2$, be contained in each of them (and open). We shall say that $u_1 \in \mathcal{D}'(\Omega_1)$ and $u_2 \in \mathcal{D}'(\Omega_2)$ coincide over Ω (or are equal over Ω) (notation: $u_1 = u_2$ on Ω) if

$$\text{(2.22)}\qquad \langle u_1,\phi\rangle = \langle u_2,\phi\rangle \ , \ \phi \in C_0^\infty(\Omega) = \mathcal{D}(\Omega) \subset \mathcal{D}(\Omega_1)\cap\mathcal{D}(\Omega_2).$$

In fact for $\Omega \subset \Omega_1$ we will have a trivial imbedding $\mathcal{D}(\Omega) \to \mathcal{D}(\Omega_1)$ by extending $\phi \in \mathcal{D}(\Omega)$ zero in $\Omega_1 - \Omega$, and therfore may speak of the restriction $u_1|\Omega$ of the distribution $u_1 \in \mathcal{D}'(\Omega_1)$ to $\Omega \subset \Omega_1$, (defined as the restriction of the functional to $\mathcal{D}(\Omega)$). Then we may say that the two distributions $u_j \in \mathcal{D}'(\Omega_j)$ coincide over Ω if and only if their restrictions to Ω are equal:

$$\text{(2.23)}\qquad u_1|\Omega = u_2|\Omega \ .$$

It may happen, that the restriction $u_1|\Omega$ is a function, while $u_1 \in \mathcal{D}'(\Omega_1)$ is not. In that case we may say that $u \in \mathcal{D}'(\Omega_1)$ is $L^1_{loc}(\Omega)$ [meaning that $u_1|\Omega \in L^1_{loc}(\Omega)$]. Similarly, if $X \subset L^1_{loc}(\Omega)$ is a subset of $L^1_{loc}(\Omega)$ we shall say that $u_1 \in \mathcal{D}'(\Omega_1)$ is in X if $u_1|\Omega \in X$. Particularly it makes sense then to speak of distributions $u_1 \in \mathcal{D}'(\Omega_1)$ which are k times continuously differentiable in $\Omega \subset \Omega_1$, $k = 0,1,2,\ldots$. We shall say so if $u_1 \in C^k(\Omega)$, in the above sense.

An important point to be made with the above conventions is this: While we already observed that in general the value of a distribution u at a point $x \in \Omega$ is not defined (or, in other words, the restriction of u is to the set $\{x\}$ containing one single point is not meaningful), the restriction to open sets is quite meaningful indeed. Correspondingly we may speak of properties of a distribution near a point, because the restriction $u|N$ is defined for any open neighborhood N of x.

For example it makes no sense to say that $u = 0$ at a point x but we can say that $u = 0$ near x, (if $u = 0$ for some neighborhood N of x) or that u is C^∞ near x (if $u \in C^\infty(N)$ for some open neighborhood N of x). Correspondingly we define the support and the singular support of $u \in \mathcal{D}'(\Omega)$ as the complement of the set of all points near which u vanishes (near which u is C^∞), respectively. These are both closed subsets of Ω, they will be denoted by supp u and sing supp u respectively. If u is a function the two notions of support (as defined here and in section 1) clearly coincide.

Lemma 2.4. Suppose a distribution $u \in \mathcal{D}'(\Omega)$ has all ist distribution derivatives $u^{(\alpha)}$, $|\alpha| = k$ of a given order k in $C(\Omega)$. Then we have $u \in C^k(\Omega)$.

Proof. It suffices to prove the lemma for $k = 1$, since then a induction argument will prove the lemma for arbitrary k. [For $k = 0$ the assertion is trivial. Let it be satisfied for all $|\beta| = k - 1$, $k \geq 1$. Then consider $v = u^{(\beta)} \in \mathcal{D}'(\Omega)$ for such $|\beta| = k - 1$ and observe that by assumption we have $v^{(\gamma)} \in C(\Omega)$ for all $|\gamma| = 1$ so that $v = u^{(\beta)} \in C^1(\Omega)$. Hence all distribution derivatives $u^{(\beta)}$, for $|\beta| = k - 1$ are in $C(\Omega)$ so that $u \in C^{k-1}$, by induction hypothesis. But since we also proved that every $u^{(\beta)}$, $|\beta| = k - 1$ is in C^1, it follows that $u \in C^k(\Omega)$, which gives the lemma for general k, once we get it for $k = 1$].

First we shall assume that

$$\Omega = Q_n = \{ x \in \mathbb{R}^n : x = (x_1,\dots,x_n),\ |x_j| < 1,\ j = 1,\dots,n\}$$

without loss of generality. We denote $u_j = iD_j u$, $\nabla u = (u_1,\dots,u_n)$ and note the integrability condition (with $\phi_{x_j} = \partial\phi/\partial x_j$)

$$(2.24)\qquad \langle u_j, \phi_{x_l}\rangle = \langle u_l, \phi_{x_j}\rangle\ ,\ \phi \in C_0^\infty(Q_n)\ j, l = 1,\dots,n.$$

By assumption the u_j are continuous functions, thus we may write (2.24) in the form

$$(2.25)\qquad \int u_j \phi_{x_l}\, dx = \int u_l \phi_{x_j}\, dx\ ,\ j, l = 1,\dots,n,\qquad \phi \in D(Q_n),$$

with Riemann integrals, not distribution integrals.

It is possible to construct a sequence ϕ^k of functions in $\mathcal{D}(Q_n)$ such that for every $u \in C(Q_n)$ we get

$$(2.26)\qquad \lim_{k\to\infty} \int u\phi^k dx = \int_R dtdt' u(\tilde{z},t,t')\ ,$$

where $\tilde{z} = (z_1,\dots,z_{n-2})$, and where

(2.27) $R = [x'_{n-1}, x''_{n-1}] \times [x'_n, x''_n] \subset \Omega_2$

denotes a rectangle in the (x_{n-1}, x_n)-plane.

Here we assume $n>2$, for a moment. For explicite construction of ϕ_k let $\delta^k \varepsilon C_0^\infty(\mathbb{R}^{n-2})$ have support in the ball $|\tilde{x}-\tilde{z}|<1/k$, $\tilde{x}=(x_1,\dots,x_{n-2})$ and let $\int \delta^k d\tilde{x}=1$. Also let $\eta^k \varepsilon C_0^\infty(\mathbb{R}^2)$ be constant in the set M_k of all points $\hat{x}=(x_{n-1},x_n)$ having distance from the boundary Γ of the rectangle R larger than 1/k, with the constant equal 1 or 0 inside and outside R, respectively. Then define $\phi^k(x)=\delta^k(\tilde{x})\eta^k(\hat{x})$, and confirm (2.26), using the theorem of the means for integrals.

For a function u which is even continuously differentiable in Ω_n we get (for j=n-1 or n)

(2.28) $$\int_{\Omega_n} u\phi^k_{x_j} dx = -\int u_{x_j} \phi^k dx \to \int_R u_{x_j}(\tilde{z},\tilde{t})d\tilde{t} = (\int_{\Gamma''_j} - \int_{\Gamma'_j})u(\tilde{z},\tilde{t})ds$$

where the integration is with respect to arc length and where Γ'_j, Γ''_j are the parts of Γ with $x_j=x'_j$ and $x_j=x''_j$, respectively.

By a continuity argument we conclude that equality between the first and last term of (2.28) remains true also in case of u being only a continuous function.

Now we use (2.25) and the above and find that

(2.29) $$\int_{\hat{\Gamma}} \nabla u \cdot dx=0,\ \nabla u\cdot dx=u_1dx_1+\dots+u_ndx_n\ ,\ \hat{\Gamma}=\{\tilde{z}\}\times\Gamma\ .$$

Similarly it follows that (2.29) holds for any rectangular countour x_k = given fixed, $k\neq j,l$ $(x_j,x_l)\ \varepsilon\ R$, with some rectangle R in the (x_j,x_l)-plane, having its sides parallel to the coordinate axes.

Thus we conclude that the line integral

(2.30) $$w(z) = \int_0^z \nabla u\cdot dx$$

is independent of the path Γ connecting 0 and $z\ \varepsilon\ \Omega_n$, as long as $\Gamma\subset\Omega_n$ is composed of finitely many straight line segments, parallel to one of the axes. Especially

(2.31) $$w(x) = \int_0^{x_1} u_1(t,x_2,\dots,x_n)dt + \int_0^{x_2} u_2(0,t,x_3,\dots,x_n)dt + \dots + \int_0^{x_n} u_n(0,0,\dots,0,t)dt\ ,$$

which shows that $\partial w/\partial x_1(x) = u_1(x)\ \varepsilon\ C^0(\Omega_n)$. Since the variable x_1 is not distinguished from the others we similarly get $\partial w/\partial x_j = u_j$. It follows that $w\varepsilon C^1(\Omega_n)$, and $\nabla u = \nabla w$, in Ω_n.

To complete the proof we will show that the distribution v=u-w is constant in Ω_n. Notice that we have

(2.32) $$\langle v,\phi_{x_j}\rangle = 0\ ,\quad \phi\ \varepsilon\ \mathcal{D}(\Omega_n)\ ,\quad j=1,\dots,n.$$

We require the lemma, below.

Lemma 2.5. Given a fixed $\omega \varepsilon C_0^\infty((-1,1))$, $\int_{-1}^{1} \omega dt = 1$, and let $\omega_n(x) = \prod_1^n \omega(x_j)$. Then for each $\phi \varepsilon C_0^\infty(Q_n)$ we have a decomposition

$$(2.33) \qquad \phi = c\,\omega_n + \sum_{j=1}^{n} \partial\psi_j/\partial x_j \quad , \quad c = \int_{Q_n} \phi dx \ , \ \psi_j \ \varepsilon \ C_0^\infty(Q_n) \ .$$

Proof. The second condition (2.33) is a consequence of the first and third condition, by partial integration. Let $\tilde{x} = (x_1,\ldots,x_{n-1})$, and define $\mu(\tilde{x}) = \int_{-\infty}^{\infty} \phi(\tilde{x},t)dt$. Then $\mu \varepsilon C_0^\infty(Q_{n-1})$, and $\eta(x) = \phi(x) - \mu(\tilde{x})\omega(x_n) \ \varepsilon \ C_0^\infty(Q_n)$ satisfies $\int_{-\infty}^{\infty} \eta(\tilde{x},t)dt = 0$, $\tilde{x} \ \varepsilon \ Q_{n-1}$, hence is of the form $\eta = \partial\psi_n/\partial x_n$, with $\psi_n \ \varepsilon \ C_0^\infty(Q_n)$.

By induction assume (2.33) for n-1, then

$$\mu(\tilde{x}) = c\,\omega_{n-1}(\tilde{x}) + \sum_{j=1}^{n-1} \partial\tilde{\psi}_j/\partial x_j \quad , \quad \tilde{\psi}_j \ \varepsilon \ C_0^\infty(Q_{n-1}) \ .$$

Set $\psi_j(x) = \tilde{\psi}_j(\tilde{x})\omega(x_n)$, to arrive at (2.33), q.e.d.

For the final argument in the proof of lemma 2.4, use (2.32) and (2.33) to conclude $\langle v,\phi\rangle = \langle v,\omega_n\rangle\langle 1,\phi\rangle$, $\phi \varepsilon C_0^\infty(Q_n)$, so that v equals the constant function $c_0 = \langle v,\omega_n\rangle$ in Q_n , q.e.d.

It is necessary next, in this section, to come to the point of distributions as generalized solutions of linear partial differential equations again. Let

$$(2.34) \qquad P(x,D) = \sum_{|\alpha|\leq N} a_\alpha(x) D^\alpha$$

denote a partial differential operator with coefficients $a \varepsilon C^\infty(\Omega)$, for an open set $\Omega \subset \mathbb{R}^n$. If u is any distribution in $\mathcal{D}'(\Omega)$ then the expression P(x,D)u is meaningful - it is a distribution in $\mathcal{D}'(\Omega)$, since differentiation of u and multiplication of $u^{(\alpha)}$ by $a \varepsilon C^\infty(\Omega)$ all are legitimate and again define distributions over Ω.

Accordingly, if $f \varepsilon \mathcal{D}'(\Omega)$ is given (and particularly also if f is a function in $L^1_{loc}(\Omega) \subset \mathcal{D}'(\Omega)$) we may look for distributions $u \varepsilon \mathcal{D}'(\Omega)$ satisfying the equation

$$(2.35) \qquad P(x,D)u = f \quad .$$

In that case we shall call u a <u>(distribution) solution of the differential equation</u> (2.35). Solutions of (2.35) in the conventional sense will be referred to as <u>classical solutions</u>. It is clear that classical solutions are special distribution solutions.

Also in the case where f is a (differentiable) function, so that the differential equation (2.35) does not involve any distribution coefficients and could be discussed without the distribution concept, distribution solutions may exist, which are not classical solutions.

In the theory of the subsecuent chapters the existence problem for solutions of partial differential equations (2.35) generally will be approached by showing existence of distribution solutions. We shall see that this amounts to a strong simplification of the theory. Only after proving existence of distribution solutions we shall look for criteria insuring that the distribution solutions are differentiable functions. It then follows by the above remarks that such differentiable distribution solution must be a classical solution.

Problems. 1) Discuss in all details why the functional $\delta^{(\alpha)}$, defined as $\langle\delta^{(\alpha)},\phi\rangle = (-1)^{|\alpha|}\phi^{(\alpha)}(0)$, $\phi\in\mathcal{D}$, defines a distribution. 2) For an invertible linear map $\ell(x) = gx + z$, with non-singular real $n\times n$-matrix $g=((g_{jk}))$, and $z\in\mathbb{R}^n$, and a distribution $T\in\mathcal{D}'(\mathbb{R}^n)$ the "composition" $T\circ\ell=T_\ell$ defined as the functional $\langle T_\ell,\phi\rangle = |\det g|^{-1}\langle T,\phi_{\ell^{-1}}\rangle, \phi\in\mathcal{D}$, with $\phi_{\ell^{-1}}(x) = \phi(\ell^{-1}(x))$, $\phi_{\ell^{-1}} = \phi\circ\ell^{-1}$. Show that T_ℓ is a distribution which agrees with the ordinary composition, defined for functions, as $T = f\in L^1_{loc}$. Calculate explicit formulas for $\delta^{(\alpha)}\circ\ell$. 3) Using that $\delta^{(\alpha)}$ has its support at the point 0, show that $\delta^{(\alpha)}\circ\kappa$ can be defined as in 2) for any diffeomorphism $\kappa:\mathcal{O}\to N(0)$ between a neighbourhood $N(0)$ and some open set $\mathcal{O}\subset\mathbb{R}^n$. Calculate the distribution $\delta\circ\kappa$ explicitly in terms of δ and the Jacobian determinante of κ. 4) Calculate explicitly the distribution derivatives of the function $|x|\in L^1_{loc}\subset\mathcal{D}'(\mathbb{R})$. 5) Calculate explicitly the distribution derivative $d/dx \log|x|$ (which must be a distribution in $\mathcal{D}'(\mathbb{R})$). 6) Show that a distribution $T_+\in\mathcal{D}'(\mathbb{R})$ is defined by $\langle T_+,\phi\rangle = \lim_{\varepsilon\to 0,\varepsilon>0}\int_{\mathbb{R}}\phi(x)dx/(x+i\varepsilon)$. 7) Show that a distribution is defined by the map $\phi\to\int_{|x|=1}\phi\, dS$, with the surface integral dS of the unit ball $\{|x|=1\}\subset\mathbb{R}^n$. Obtain the support of this distribution.

3. The convolution algebra $L^1(R^n)$; the Fourier integral and the space S

Formally, for two complex-valued functions u,v over $\mathbb{R}^n$, we define the convolution product $u*v = w$ by

$$(3.1)\qquad w(x) = (2\pi)^{-n/2} \int_{\mathbb{R}^n} dy\, u(x-y)v(y).$$

At present, the integral in (3.1) will be a Lebesgue integral, but we shall admit distribution integrals later on.

If u and v are in $L^1(R^n)$, then it follows that

$$\int_{\mathbb{R}^n} |w(x)|dx = (2\pi)^{-n/2} \int_{\mathbb{R}^n} dx \left|\int_{\mathbb{R}^n} u(x-y)v(y)dy\right|$$

$$(3.2)\qquad \leq (2\pi)^{-n/2} \int_{\mathbb{R}^n\times\mathbb{R}^n} dx\, dy\, |u(x-y)v(y)|$$

$$= (2\pi)^{-n/2} \int_{\mathbb{R}^n} |u(x)|dx \cdot \int_{\mathbb{R}^n} |v(y)|dy \quad ,$$

where a substitution of integration variable was used. The integrals are absolutely convergent, hence their interchange is justified.

If we introduce the norm $\|u|_{L^1}$ by

$$(3.3)\qquad \|u|_{L^1} = (2\pi)^{-n/2} \|u\|_{L^1} ,$$

then (3.2) amounts to

$$(3.4)\qquad \|u*v|_{L^1} \leq \|u|_{L^1} \cdot \|v|_{L^1} , \quad u,v \in L^1(\mathbb{R}^n) .$$

It is readily confirmed that the operation $[u,v] \to u*v$ introduces an associative and commutative algebra product on the Banach space $L^1 = L^1(\mathbb{R}^n)$ with norm (3.5), and that L^1, with this norm, is a commutative Banach algebra (without unit) (Appendix AII).

Especially we verify that

$$u*v = v*u,\ u*(v*w) = (u*v)*w ,$$
$$(3.5)\qquad (cu)*v = c(u*v),\ (u+v)*w = u*w + v*w ,$$
$$u,v,w \in L^1,\ c \in \mathbf{C} ,$$

which is a matter of interchange of absolutely convergent integrals, and of simple changes in integration variables, at the most.

For $a \in L^1(\mathbb{R}^n)$ we formally define the integral operator K_a by

$$(3.6)\qquad K_a u = a*u \quad ,$$

where u denotes a complex-valued function over R^n, at present. The operator K_a, often referred to as a convolution operator (with characteristic $a \in L^1$) certainly is well defined for $u \in L^1$. In fact, K_a is a bounded operator in $L(L^1)$ with operator norm satisfying

$$(3.7)\qquad \|K_a\| = \|K_a\|_{L^1} \le \|a|_{L^1} ,$$

as a consequence of (3.4).

As a consequence of lemma 1.1, the convolution operator K_a also is a bounded operator of every $L^p(\mathbb{R}^n)$, $1 \le p \le \infty$, and we have the estimate

$$(3.8)\qquad \|K_a\|_{L^p} = \sup \{ \|K_a u\|_{L^p} : u \in L^p(\mathbb{R}^n),\ \|u\|_{L^p} \le 1 \} \le \|a|_{L^1} .$$

Indeed K_a is an integral operator, and the functions k_1, k_2 of (1.24) both are constants, equal to $\|a|_{L^1}$, for all s,t.

For a function $u \in L^1(\mathbb{R}^n)$ we define the <u>Fourier integral</u> u , and the conjugate (or inverse) Fourier integral u by the formulas

$$u^\wedge(x) = (2\pi)^{-n/2} \int_{\mathbb{R}^n} e^{-ixy} u(y)dy ,$$

$$(3.9)\qquad u^\vee(x) = (2\pi)^{-n/2} \int_{\mathbb{R}^n} e^{ixy} u(y)dy ,$$

$$xy = \sum_{j=1}^{n} x_j y_j ,\ x,y \in \mathbb{R}^n .$$

Both integral (3.9) exist absolutely, for every $x \in \mathbb{R}^n$. The functions $u^\wedge$, $u^\vee$, $x \in \mathbb{R}^n$, are bounded and continuous over $\mathbb{R}^n$, and we get

$$(3.10)\qquad \begin{aligned} \|u^\wedge\|_C &\le \|u|_{L^1} , \\ \|u^\vee\|_C &\le \|u|_{L^1} , \qquad u \in L^1(\mathbb{R}^n) . \end{aligned}$$

In studying the Fourier integral it is convenient to introduce the space

$$(3.11)\qquad S = \{u : u \in C_0^\infty(R^n),\ \|x^\alpha u^{(\beta)}\|_C < \infty ,\ \alpha,\beta \in Z_+^n\} .$$

It is clear that $C_0^\infty(\mathbb{R}^n) \subset S$. The space S is a Frechet space, with topology defined by the (countably many) semi-norms

$$(3.12)\qquad \|x^\alpha u^{(\beta)}\|_C ,\quad \alpha,\beta \in Z_+^n ,$$

as immediately evident .

<u>Lemma 3.1.</u> We have $S \subset L^1(\mathbb{R}^n)$. Moreover, $u \in S$ implies $x^\alpha u^{(\beta)} \in L^1$, and

$$(3.13)\qquad \|x^\alpha u^{(\beta)}\|_{L^1} \le c \sum_{|\gamma|\le n+1} \|x^{\alpha+\gamma} u^{(\beta)}\|_C \ , \ u\varepsilon S,\ \alpha,\beta\varepsilon \mathbb{Z}_+^n \ ,$$

with $c=(n+1)!\int(1+|x|)^{-n-1}dx$.

Moreover, the (conjugate) Fourier transform $u^\wedge$ ($u^\vee$) of $u\varepsilon S$ is in $C^\infty(\mathbb{R}^n)$, and we have

$$(3.14)\qquad (x^\alpha D^\beta u)^\wedge = (-1)^{|\alpha|} D^\alpha x^\beta u^\wedge,\ (x^\alpha D^\beta u)^\vee = (-1)^{|\beta|} D^\alpha x^\beta u^\vee \ .$$

The proof of lemma 3.1 is a matter of the estimate

$$(3.15)\qquad \begin{aligned} |x^\alpha u^{(\beta)}| &= |(1+|x|)^{-n-1}|\cdot|(1+|x|)^{n+1} x^\alpha u^{(\beta)}| \\ &\le (n+1)!(1+|x|)^{-n-1} \sum_{|\gamma|\le n+1} |x^{\alpha+\gamma} u^{(\beta)}| \ . \end{aligned}$$

Relations (3.14), and the differentiability of $u^\wedge$ and $u^\vee$ follow from partial integration under the Fourier integral, and from interchanging of differentiation with the Fourier integral, using the identity

$$(3.16)\qquad D_z^\beta z^\alpha e^{-ixz} = (-1)^{|\alpha|+|\beta|} D_x^\alpha x^\beta e^{-ixz} \ ,$$

with D_x and D_z denoting the differentiations with respect to x and z, respectively. One also will have to use the fact that $u\varepsilon S$, and all its derivatives, vanish stronger than any power of $(1+|x|)$, as follows from the fact that $(1+|x|^{2m})u^{(\beta)}$ is a sum of expressions $cx^{2\alpha}u^{(\beta)}$ with $|\alpha|\le m$, and with multinomial coefficients c.

Lemma 3.2. We have $u^\wedge\varepsilon S$, $u^\vee\varepsilon S$, for all $u\varepsilon S$. Moreover, for all $\alpha,\beta\varepsilon Z_+^n$ we have

$$(3.17)\qquad \|x^\alpha (u^\wedge)^{(\beta)}\|_C \le c_{\alpha\beta} \sum_{\alpha'\le\alpha,\beta'\le\beta,|\gamma|\le n+1} \|x^{\beta'+\gamma} u^{(\alpha')}\|_C \ ,$$

with constants $c_{\alpha\beta}$ independent of $u\varepsilon S$, and the corresponding formula, with $u^\wedge$ replaced by $u^\vee$.

Proof. From (3.10), (3.13) and (3.14) we conclude that

$$(3.18)\qquad \|D^\beta x^\alpha u^\wedge\|_C \le c \sum_{|\gamma|\le n+1} \|x^{\beta+\gamma} u^{(\alpha)}\|_C \ .$$

By Leibnitz' formula (1.23), in its second form, we get

$$(3.19)\qquad \begin{aligned} \|x^\alpha D^\beta u^\wedge\|_C &\le \sum_{\beta'\le\beta} \binom{\beta}{\beta'} \|D^{\beta-\beta'}(x^\alpha)^{(\beta')} u^\wedge\|_C \\ &\le \alpha!\ \beta! \sum_{\alpha'\le\alpha,\beta'\le\beta} \|D^{\beta'} x^{\alpha'} u^\wedge\|_C \ . \end{aligned}$$

Combining these formulas we get (3.17), q.e.d.

Corollary 3.3. The Fourier integral and its conjugate define two continuous linear maps $S \to S$ of the Frechet space S.

The proof is a consequence of estimate (3.17).

Lemma 3.4. (Fourier inversion formula) We have

$$u = u^{\wedge\vee} = u^{\vee\wedge} \quad , \quad u\in S \quad . \tag{3.20}$$

Proof. (For this proof c.f. Hörmander [6], theorem 1.7.1.) Note that it is sufficient to prove $u = u^{\wedge\vee}$ only, because the other identity will follow by taking complex conjugates.

Let $\mu\in S$, then (with absolutely convergent integrals),

$$\begin{aligned} \int u^{\wedge}(z)\mu(z)e^{ixz}dz &= (2\pi)^{-n/2}\iint dz\, dy\mu(z)u(y)e^{-i(y-x)z} \\ &= \int dy\ u(y)\mu^{\wedge}(y-x) = \int u^{\wedge}(y)u(x+y)dy \quad , \quad u\in S \ . \end{aligned} \tag{3.21}$$

In this formula we replace μ by μ_ε, with $\mu_\varepsilon(x)=\mu(\varepsilon x)$, $\varepsilon>0$, and with a fixed $\mu\in S$. Note that $\mu_\varepsilon \in S$, for all $\varepsilon>0$, and that $\mu_\varepsilon^{\wedge}(z)=\varepsilon^{-n}\mu^{\wedge}(z/\varepsilon)$, by a change of variable in the Fourier integral. It follows that

$$\int_{\mathbb{R}^n} u^{\wedge}(z)\mu(\varepsilon z)e^{ixz}dz = \int_{R^n}\mu^{\wedge}(y)u(x+\varepsilon y)dy \tag{3.22}$$

by another change of integration variable, at right. If $\varepsilon\to 0$, $\varepsilon>0$, then we obtain

$$(2\pi)^{n/2}\mu(0)u^{\wedge\vee}(x) = u(x)\int_{\mathbb{R}^n}\mu^{\wedge}(y)dy \ , \tag{3.23}$$

where the limit under the integral is justified, since $u^{\wedge},\mu^{\wedge}\in S \subset L^1$, while $u,\mu \in S \subset CB(\mathbb{R}^n)$, using lemma 3.2.

To obtain the constants $\mu(0)$ and $\mu\, dy$ explicitly we may substitute $\mu(x) = \exp(-|x|^2/2)$, which defines a function in S. Its Fourier integral may be calculated explicitly:

$$\begin{aligned} \int e^{-|z|^2/2-ixz}dz &= e^{-|x|^2/2}\int \exp\Big(-\sum_{j=1}^{n}(ix_j+z_j)^2/2\Big)dz \\ &= e^{-|x|^2/2}\Big(\int_{-\infty}^{+\infty}e^{-t^2/2}dt\Big)^n = (2\pi)^{n/2}e^{-|x|^2/2} \quad , \end{aligned}$$

using Cauchy's integral formula. Hence

$$\mu(0)=1, \int\mu^{\wedge}dx = \int e^{-|x|^2/2}dx = \Big(\int e^{-t^2/2}dt\Big)^n = (2\pi)^{n/2} \ ,$$

which implies the statement of lemma 3.4, if substituted into (3.23), q.e.d.

Lemma 3.5. The space S, and its subspace $C_0^\infty(\mathbb{R}^n) \subset S$ are dense in every space $L^p(\mathbb{R}^n)$, $1\le p<\infty$.

Proof. It is sufficient to show that C_0^∞ is dense in L^p. Any $u \in L^p$ may be approximated by step functions (taking on only finitely many distinct values $\neq 0$, each on a bounded measurable set). A step function may be approximated by a finite sum $\sum c_j \chi_{G_j}$, with bounded open sets G_j, and their characteristic functions χ_{G_j}. A function χ_{G_j} may be approximated by a finite sum $\sum \chi_{Q_l}$, with closed cubes Q_l and each χ_{Q_l} may be approximated by a continuous function. All above approximations are understood in the L^p-norm, and up to any degree of accuracy. It follows the existence of a continuous function u_ε with compact support such that $\|u-u_\varepsilon\|_{L^p} \le \varepsilon/2$. Let $\phi \in C_0^\infty(\mathbb{R}^n)$, and let $0 \le \phi \le 1$, $\int_{\mathbb{R}^n} \phi(x)dx = 1$, and define

$$(3.24) \qquad u_{\varepsilon,\tau} = \tau^n \phi_\tau * u_\varepsilon , \qquad \phi_\tau(x) = \phi(\tau x) , \quad \tau > 0 \quad .$$

Show that (i) $u_{\varepsilon,\tau} \in C_0^\infty(\mathbb{R}^n)$, and (ii) $\lim_{\tau\to\infty} \|u_\varepsilon - u_{\varepsilon,\tau}\|_{L^p} = 0$, so that

$$\|u-u_{\varepsilon,\tau}\|_{L^p} \le \|u-u_\varepsilon\|_{L^p} + \|u_\varepsilon - u_{\varepsilon,\tau}\|_{L^p} \le \varepsilon, \ \tau \ge N(\varepsilon), \text{ q.e.d.}$$

In section 1 we already mentioned the space $CO(\mathbb{X})$. Here we use $CO(\mathbb{R}^n)$, which is a Banach space and a Banach algebra (without unit), under pointwise multiplication $(uv)(x) = u(x)v(x)$. The space $CO(\mathbb{R}^n)$ consists of all functions in $CB(\mathbb{R}^n)$ having the limit zero, as $|x| \to \infty$. Evidently, $C_0^\infty(\mathbb{R}^n) \subset \mathcal{S} \subset CO(\mathbb{R}^n)$, and the spaces C_0^∞ and $\mathcal{S}$ are dense in CO, under its norm $\|\cdot\|_C$ of (1.14). Indeed, a continuous function f over $\mathbb{R}^n$, with limit zero, as $|x| \to \infty$, may be approximated by continuous functions with compact support, which may be approximated by C_0^∞-functions, by a construction as in the proof of lemma 3.5, above.

Theorem 3.6. We have the commutative diagram

$$(3.25) \qquad \begin{array}{ccc} \mathcal{S} & \longrightarrow & L^1 \\ {\scriptstyle u\to u^\vee}\uparrow\downarrow{\scriptstyle u\to u^\wedge} & & \downarrow{\scriptstyle u\to u^\wedge} \\ \mathcal{S} & \longrightarrow & CO \end{array}$$

of continuous algebra isomorphisms, where the horizontal arrows denote the natural injections, and where the algebra product is convolution "$*$" in the upper row, but pointwise multiplication "$\cdot$" in the lower row. Specifically:

(i) The space $\mathcal{S}$ with "$*$" is a subalgebra of L^1 ;

(ii) the space $\mathcal{S}$ with "$\cdot$" is a subalgebra of CO ;

(iii) we have

(3.26) $\quad (u*v)^\wedge = u^\wedge . v^\wedge \;, \quad (u*v)^\vee = u^\vee \cdot v^\vee \;, \quad u,v\in S \;;$

(iv) we have

(3.27) $\quad \|u^\wedge\|_C \le \|u\|_{L^1} \quad , \quad u\in L^1 \;,$

i.e., the isomorphism $u \to u^\wedge$ is a contraction. The topology of S is stronger than the relative topology of L^1 and CO each;

(vi) the mapping $L^1 \to$ CO defined by the Fourier transform is injective;

There is a corresponding diagram (3.25) and (i)-(vi) are also valid with "$\wedge$" and "$\vee$" interchanged.

Proof. Clearly S is an algebra under "$\cdot$", by Leibnitz' formula, and $S \subset$ CO is evident, so that we get (ii). Use (3.10) and that $u^\wedge \in S \subset$ CO for $u\in S$, and that S is dense in L^1, by lemma 3.5, to conclude that $u^\wedge \in$ CO for all $u\in L^1$. Next, formula (3.26) follows by direct evaluation of the Fourier integral:

(3.28)
$$\begin{aligned}(2\pi)^n (u*v)^\wedge(x) &= \int dz\, e^{-ixz} \int u(z-y)v(y)\,dy \\ &= \int dy v(y) \int dz\, e^{-ixz} u(z-y) = \int dy\, v(y) \int dt\, e^{-ix(y+t)} u(t) \\ &= (2\pi)^n u^\wedge(x)\, v^\wedge(x) \quad .\end{aligned}$$

All integrals converge absolutely, hence may be interchanged.
Thus (iii) follows, and (i) is a consequence of (3.26), and the Fourier inversion formula. Then (iv) is a consequence of (3.10).
The natural imbeddings $S \to L^1$ and $S \to$ CO are clearly continuous, hence (v) results.
Finally, (vi) follows from (4.11) and the fact that only $u=0$, among all functions in L^1 satiesfies the relation $\int u\phi dx = 0$ for all $\phi\in S$ (lemma 2.1). This completes the proof of theorem 3.6.

It is not true that the map $L^1 \to$ CO, defined by the Fourier integral, also is surjective. (Use problem 4), below, for a discussion.)

Most of the results discussed in the present section possess generalizations to other spaces, such as the circle, the (poly)-torus, and the (poly)-cylinder.

In fact, the convolution algebra L^1 has been studied for general locally compact abelian groups, and their Haar measure. This will be beyond our present interest, however.

Problems: 1) Verify the following formulas, giving explicit one-dimensional Fourier integrals. a) $((1+x^2)^{-1})^\wedge = c_0 e^{-|\xi|}$ (determine the constant c_0 explicitly); b) $\sqrt{2\pi}(1/\sinh x)^\wedge = -i\pi\tanh(\xi\pi/2)$; c) $\sqrt{2\pi}(1/\cosh x)^\wedge = \pi/\cosh(\pi\xi/2)$. For all of the integrals use complex curve integral techniques. Note the function b) is not L^1 at $x=0$; therefore interpret the Fourier integral as the Cauchy principal value $\lim_{\varepsilon\to 0,\varepsilon>0} \int_{|x|\geq\varepsilon}$. 2) The Hankel transform H_ν is defined by the one-dimensional integral (with the Bessel function J_ν, c.f. [9])

(3.29) $$H_\nu \omega(\rho) = \int_0^\infty \omega(r) J_\nu(r\rho)(r\rho)^{1/2} dr,\quad \omega\varepsilon L^1((0,\infty),\ r^{\nu+1/2}dr),\quad 0\leq\nu<\infty.$$

In particular, H_ν is its own inverse: $H_\nu^2\omega = \omega$, as $\omega\varepsilon C_0^\infty((0,\infty))$. Show that a Fourier transform Fu in essence is a Hankel transform, if u depends only on $r=|x|$, in the sense that $f(x) = u(|x|)$ has its Fourier transform given by

(3.30) $$f^\wedge(\xi) = v(|\xi|),\quad \rho^{(n-1)/2} v(\rho) = H_{n/2-1}(r^{(n-1)/2}u(r))(\rho).$$

(Hint: Express $\int_{|x|=1} e^{-it\cos(x,\xi)} ds_x$ by Bessel functions c.f. [9]).

3) Using the formulas given in [9] calculate the Fourier transform of the function $\lambda_s(x) = (1+x^2)^{-s}$, $x\varepsilon\mathbb{R}^n$, $s>n/2$. Show that

(3.31) $$\lambda_s^\wedge(\xi) = \Theta_s(|\xi|)\ ,\quad \Theta_s(t) = 2^{1-s}/\Gamma(s)\, t^{s-n/2} K_{s-n/2}(t)\quad ,$$

with $\Gamma(s)$ and $K_\sigma(t)$ denoting the Gamma function and the modified Hankel function, respectively. 4) Find a function $f\varepsilon CO(\mathbb{R})\cap L^1(\mathbb{R})$ such that the inverse Fourier transform $f^\vee$ is not in $L^1(\mathbb{R})$.

4. Temperate distributions and temperate functions; Fourier transform of distributions and Fourier multipliers.

The continuous linear functionals over the Frechet space S will be called temperate distributions. The space of temperate distributions is denoted by S', following a general convention. The value of $u\varepsilon S'$ at $\phi\varepsilon S$ is denoted by $\langle u,\phi\rangle$, again.

A linear functional u over S is continuous if we have $\langle u,\phi_k\rangle \to 0$ as $k\to\infty$, whenever the sequence $\{\phi_k\}$ of functions in S satisfies the conditions

(4.1) $\quad x^{\alpha}\phi_k^{(\beta)} \to 0 \quad \text{as} \quad k \to \infty \quad ,$

uniformly over $\mathbb{R}^n$, for each pair of multi-indices α,β.

If a sequence satisfies (4.1), we shall say that it converges to zero in S [it is evident that this indeed amounts to convergence to zero in the Frechet topology of S, with the system of semi-norms given by (3.11)]. Thus we again may express continuity of a functional by requiring that $\langle u,\phi_k\rangle \to 0$ whenever $\phi_k \to 0$ in S and it is clear that this indeed amounts to continuity of u in the topology of the semi-norms (3.11).

Please notice that we have $\mathcal{D}(\mathbb{R}^n) = C_0^{\infty}(\mathbb{R}^n) \subset S$, and that $\phi_k \to 0$ in $\mathcal{D}$ (c.f. section 2) implies $\phi_k \to 0$ in S. It follows that every temperate distribution $u\varepsilon S'$ induces a distribution $v = u|\mathcal{D}\varepsilon\mathcal{D}'$, which is uniquely determined by u, in view of the lemma, below.

Lemma 4.1. The space $\mathcal{D} = \mathcal{D}(\mathbb{R}^n)$ is dense in S. That is, for every $\phi\varepsilon S$ a sequence $\phi_k\varepsilon\mathcal{D}$ may be constructed, such that $\phi-\phi_k \to 0$ in S.

Proof. The desired sequence may be obtained in the form $\phi_k = \phi\chi_k$, with $\chi_k(x) = \chi(x/k)$, $\chi\varepsilon C_0^{\infty}(\mathbb{R}^n)$ being equal to one near zero. We shall have $\phi-\phi_k = \phi\omega_k$ where $\omega_k(x) = \omega(x/k)$ with $\omega=1-\chi$ vanishing near zero and equal to one near ∞. It follows that ω_k vanishes in $|x|\le ck$ with some $c>0$ independent of k. This and Leibnitz' formula (1.23), and the fact that $\phi^{(\alpha)}(x)$ becomes very small as $|x|$ gets large, for all x, will indeed supply (4.1) for $\phi-\phi_k$, q.e.d..

The above amounts to the fact that there is a natural injection $S' \to \mathcal{D}'$ induced by $u \to v = u|\mathcal{D}$. This injection will be used to identify the temperate distribution u with the distribution v. With this convention temperate distributions are special distributions, and the space S' is a subspace of $\mathcal{D}'$.

Let L^1_{pol} denote the linear space of all measurable complex-valued functions u over $\mathbb{R}^n$ such that $(1+|x|)^{-k}u\varepsilon L^1$, for a suitable integer k. For $u\varepsilon L^1_{pol}$ a continuous linear functional over $\mathcal{S}$ may be defined by

$$(4.2) \qquad \langle u,\phi\rangle = \int_{\mathbb{R}^n} u(x)\phi(x)\,dx \;, \quad \phi\varepsilon\mathcal{S} \;.$$

We get

$$(4.3) \qquad |\langle u,\phi\rangle| \le (n+k+1)!\,\{\int |u(x)|\,(1+|x|)^{-k}dx\} \sum_{|\alpha|\le n+k+1} \|x^{\alpha}\phi\|_{C} \;,$$

so that continuity of (4.1) over $\mathcal{S}$ is implied.

Clearly we have $L^1_{pol} \subset L^1_{loc}(\mathbb{R}^n)$, so that (4.2) defines a distribution, which also is a temperate distribution. The functions of L^1_{pol} shall be regarded as special temperate distributions by identifying $u\varepsilon L^1_{pol}$ with its functional (4.2). We shall refer to $u\varepsilon L^1_{pol}$ as a function of <u>polynomial growth</u> (in the sense of L^1).

In particular, all the spaces $L^p(\mathbb{R}^n)$, $1\le p\le\infty$, are linear subspaces of the space L^1_{pol}, due to

$$(4.4) \qquad \|u(1+|x|)^{-n-1}\|_{L^1} \le \|u\|_{L^p}\,\|(1+|x|)^{-n-1}\|_{L^q} \le \infty \;, \quad u\varepsilon L^p \;,$$

as follows from Hoelder's inequality (1.8). Therefore all functions in L^p are special temperate distributions.

As another important subspace of L^1_{pol} we define $\mathcal{T}$ to be the linear space of all $u\varepsilon C^{\infty}(\mathbb{R}^n)$ such that all derivatives $u^{(\alpha)}$, $\alpha\varepsilon\,\mathbb{Z}^n_+$ are in L^1_{pol}, that is,

$$(4.5) \qquad \|(1+|x|)^{m_\alpha}u^{(\alpha)}\|_{L^1} < \infty \;, \quad \alpha\varepsilon\mathbb{Z}^n_+ \;.$$

The space L^1_{pol} clearly is the union $\bigcup_k X_k$ of the Banach spaces

$$(4.6) \qquad X_k = \{a:\ a\varepsilon L^1_{loc} \;,\ (1+|x|)^{-k}a\varepsilon L^1(\mathbb{R}^n)\} \;.$$

Accordingly, we will endow it with the corresponding inductive limit topology [strongest locally convex topology in which all the injections $X_k \to L^1_{pol}$ still are continuous]. Then, if $\{\rho_\iota\}_{\iota\varepsilon I}$ denotes a generating system of semi-norms for L^1_{pol}, we shall obtain a corresponding generating system for the topology of $\mathcal{T}$ in the form

$$(4.7) \qquad \{\rho_\iota(u^{(\alpha)}) : \iota\varepsilon I \;,\ \alpha\varepsilon\mathbb{Z}^n\} \;.$$

Note that we have the relations

$$(4.8) \qquad \|(1+|x|)^{-j-n-1}a^{(\alpha)}\|_{L^1} \le c\,\|(1+|x|)^{-j}a^{(\alpha)}\|_{L^{\infty}}$$

as well as

$$(4.9) \qquad \|(1+|x|)^{-j}a^{(\alpha)}\|_{L^\infty} \le c \sum_{|\beta|\le n}' \|((1+|x|)^{-j}a^{(\alpha)})^{(\beta)}\|_{L^1}$$

with constants c independent of a. More precisely, if $a\in C^\infty(\mathbb{R}^n)$ has the expression at right of (4.8) (or (4.9)) finite, then also the left hand side is finite and we have the corresponding estimate. Estimate (4.8) is a consequence of the fact that $\|(1+|x|)^{-n-1}\|_{L^1} < \infty$. For derivation of (4.9) c.f. lemma 4.7.

The above estimates make evident that T may be simply described as the space of all $C^\infty(\mathbb{R}^n)$-functions of polynomial growth.

$$(4.10) \qquad T = \{a: a\in C^\infty(\mathbb{R}^n), a^{(\alpha)} = O((1+|x|)^{k_\alpha})\} .$$

Leibnitz' formula (1.23) implies that T is an algebra under "$\cdot$", and that T is invariant under differentiation. Moreover, since S precisely consists of those $u\in T$ satisfying (4.5) for every $\alpha\in Z_+^n$ and any arbitrary positive or negative m, it follows that S is an ideal of T. Note also that the linear map a: $S \to S$, defined by $\phi \to a\cdot\phi$, for $a\in T$, is continuous.

The functions of T will be called temperate functions. Since temperate distributions are special distributions all the notions of section 2 like distribution derivative, support, singular support etc. are meaningful again. However, temperate distributions also have a Fourier transform, as seen below. This will let appear them as objects of special interest, in the following.

Lemma 4.2 a) We have

$$(4.11) \qquad \int_{\mathbb{R}^n} uv^\wedge dx = \int_{\mathbb{R}^n} u^\wedge v\, dx \ , \qquad \int_{\mathbb{R}^n} uv^\vee dx = \int_{\mathbb{R}^n} u^\vee v\, dx \ , \quad u,v\in L^1 ,$$

and

$$(4.12) \qquad \int_{\mathbb{R}^n} u^{(\alpha)}\phi\, dx = (-1)^{|\alpha|}\int_{\mathbb{R}^n} u\phi^{(\alpha)}dx \ , \quad \phi\in S , \qquad u^{(\beta)} \in CB \cap L^1_{pol} \ , \quad \beta \le \alpha .$$

b) the mappings $S \to S$, defined by $u \to u^\wedge$, $u \to u^\vee$, $u \to u^{(\alpha)}$, $u \to au$, respectively, for a multi-index α and a function $a\in T$, are all continuous.

Proof. Relations (4.11) and (4.12) result from interchanging absolutely convergent integrals and from partial integration, using that $u^\wedge$, $v^\wedge$, $u^\vee, v^\vee \in CO$, by theorem 3.6. Assertion b) has been discussed above for "$\wedge$" and "$\vee$" (corollary 3.3), and for "$a\cdot$", and is trivial for D^α, q.e.d.

Definition 4.3. If $u\varepsilon S'$ is a temperate distribution, then its (distribution) derivative $u^{(\alpha)}$ of (arbitrary) order (α) and its (conjugate) Fourier transform $u^{\wedge}(u^{\vee})$ are defined as the continuous linear functionals over S given by

$$<u^{(\alpha)},\phi> = (-1)^{|\alpha|}<u,\phi^{(\alpha)}> \quad ,$$

$$(4.13) \qquad <u^{\wedge},\phi> \quad = <u,\phi^{\wedge}> \quad ,$$

$$<u^{\vee},\phi> \quad = <u,\phi^{\vee}> \quad , \qquad \phi\varepsilon S \quad .$$

For a temperate function $a\varepsilon T$ and a temperate distribution $u\varepsilon S'$ the (distribution) product $a\cdot u$ is defined to be the continuous linear functional over S given by

$$(4.14) \qquad <a\cdot u,\phi> = <u,a\phi> \quad , \quad \phi\varepsilon S \ .$$

It is evident from lemma 4.2 that $u^{(\alpha)}$, $u^{\wedge}$, $u^{\vee}$ and $a\cdot u$ are elements of S', and that all operations defined (in definition 4.3) are compatible with the corresponding operations-product, derivative, (conjugate) Fourier transform - previously defined for functions and (or) general distributions.

We shall frequently use the notation $\hat{u}=Fu$, $\check{u}=\bar{F}u=F^{-1}u$ for the (conjugate) Fourier transform, and $D^{\alpha}u=(-i)^{|\alpha|}u^{(\alpha)}$ for the derivative. All four operations are (bi-)linear and satisfy all the conventional rules. The last three lines of (4.15) are also valid for $a,b\varepsilon C^{\infty}(\Omega)$, $u\varepsilon \mathcal{D}'(\Omega)$.

$$D^{\alpha}D^{\beta}u = D^{\alpha+\beta}u \ , \ \bar{F}Fu = F\bar{F}u = u \ ,$$

$$F(x^{\alpha}D^{\beta}u) = (-1)^{|\alpha|}D^{\alpha}x^{\beta}Fu \ , \ \bar{F}(x^{\alpha}D^{\beta}u) = (-1)^{|\beta|}D^{\alpha}x^{\beta}\bar{F}u \quad ,$$

$$(4.15) \qquad a(bu) = (ab)u \ , \ (au)^{(\alpha)} = \sum_{\beta\le\alpha}\binom{\alpha}{\beta}a^{(\beta)}u^{(\alpha-\beta)} \quad ,$$

$$a\,u^{(\alpha)} = \sum_{\beta\le\alpha}(-1)^{|\beta|}\binom{\alpha}{\beta}\,(a^{(\beta)}u)^{(\alpha-\beta)} \quad ,$$

$$a,b\varepsilon T \ , \ u\varepsilon S' \ .$$

The theorem below summarizes a few properties of the distribution and function spaces discussed. As far as topologies in S' are concerned, the weak topology is determined by the convergence

$$(4.16) \qquad \lim u = u^{0} \Longleftrightarrow \lim <u,\phi> = <u^{0},\phi> \quad , \quad \phi\varepsilon S.$$

For more details regarding distribution topologies c.f. section 6.

Theorem 4.4. (i) The space T with the above topology and the product "$\cdot$" is a commutative topological algebra with unit ;

(ii) S is an ideal of T, and the product $a\cdot\phi$ also is continuous as a map $T \times S \to S$, under the topology of S ;

(iii) the natural injections of the commutative diagram below are all continuous:

(4.17)
$$\begin{array}{ccccc} L^p & \to & L^1_{pol} & & \\ & \nearrow & & \searrow & \\ S & \to & T & \to & S' \end{array} \quad ;$$

(iv) the Fourier transform and its inverse are weakly and strongly continuous maps from S' to S' ;

(v) the differential operator D^α is continuous, as an operator of each space S, T, S' to itself, respectively, and under weak topology of S'.

The proof of theorem 4.4 will not be discussed in details.

Lemma 4.5. (Parsevals relation). The restrictions F/L^2 and $\bar{F}/L^2$, with the Hilbert space $L^2=L^2(\mathbb{R}^n)\subset L^1_{pol}\subset S'$, constitute a pair of (bounded) unitary operators of L^2 to itself inverting each other.

Proof. In the first and second formula (4.11) substitute $u=\bar{\phi}^\vee=\overline{\phi^\wedge}$ $u^\wedge=\bar{\phi}$, $v=\phi\varepsilon S$, and $u=\bar{\phi}^\wedge$, $u^\vee=\bar{\phi}$, $v=\phi$, respectively, to obtain

$$(4.18)\qquad (\|\phi^\wedge\|_{L^2})^2 = (\|\phi^\vee\|_{L^2})^2 = (\|\phi\|_{L^2})^2 ,\ \phi\varepsilon S.$$

Since S is dense in L^2, by lemma 3.5, (4.18) implies that the restrictions $F|S$ and $\bar{F}|S$ may be extended to be unitary operators of L^2 each, inverting each other, by continuous extension in L^2. Since the injection of L^2 into S' is continuous, by (4.17), and since the Fourier transform is continuous in S', it follows that the above extensions coincide with $F|L^2$ and $\bar{F}|L^2$, respectively, and the lemma is proven.

For a complex-valued function a over $\mathbb{R}^n$ we define the formal Fourier multiplier $a(D)$ by setting

$$(4.19)\qquad a(D)u = (a\,u^\wedge)^\vee = \bar{F}(a\cdot Fu) \quad .$$

Formally $a(D)$ acts as a linear operator on functions or distriutions. The notation (4.19) is in agreement with that used before for polynomials; for any polynomial $P(x) = \sum_{|\alpha|\le N} c_\alpha x^\alpha$ in the n variables $x=(x_1,\dots,x_n)$, with complex coefficients c_α we get

$$(4.20)\qquad P(D)u = \sum_{|\alpha|\le N} c_\alpha D^\alpha u = (P\,u^\wedge)^\vee \quad ,$$

with the function $P=P(x)\varepsilon T$, using (4.15).

Formal Fourier multipliers will be used just like differential expressions for the definition of linear operators with domain still to be fixed, as a suitable function or distribution space.

Lemma 4.6. If $a\varepsilon L^\infty(\mathbb{R}^n)$, then $a(D)$ defines a bounded linear operator of $L^2(\mathbb{R})$ (to itself), and its operator norm is given by

$$(4.21)\qquad \|a(D)\|_{L^2} = \sup\{\|a(D)u\|_{L^2} : \|u\|_{L^2}\le 1\} = \|a\|_{L^\infty}\ .$$

If $a\in T$, then $a(D)$ defines a continuous operator from S to S, and also from S' to S', with weak and strong topology each.

The proof is a consequence of Parsevals relation (lemma 4.5), and of theorem 4.4.

Let $T^\wedge$ be defined as the space of all (conjugate) Fourier transforms of functions in T. Clearly $T^\wedge$ is a space of temperate distributions. For $u\in T^\wedge$ and $v\in S'$, or also for $u,v\in L^2$, we may define a <u>convolution product</u> $u*v$ by

$$(4.22)\qquad u*v = (u^\wedge \cdot v^\wedge)^\vee\ .$$

As a consequence of (3.26) and the properties of the Fourier transform the above convolution product agrees with the previously introduced convolution product, defined for L^1-functions only (c.f.(3.1)), whenever both are defined. Similarly for $u\in L^1$, $v\in L^p$, ore vice versa, for $1\le p\le\infty$. It is evident, that the above space $T^\wedge$, with the product "$*$", is a commutative topological algebra with unit, which contains S as a proper ideal.

Note that we get

$$(4.23)\qquad a^\wedge(D)u = K_a u\ ,\qquad u\in S\ ,$$

with the convolution operator K_a of (3.6) whenever $a\in L^1$. Accordingly formal Fourier multipliers $a(D)$ may be regarded as generalized convolutions, and we choose to use both notations $a(D)$ or $K_{a^\vee}$ synonymously, if ever convenient, for $a\in S'$, with the understanding that we mean the operator $b(D)$, whenever the convolution operator K_b is not defined as an integral operator, in the sense of section 3. In particular, we have seen that the operator K_b is defined on L^2, if only $b\in L^\infty$, and on S or S', if $b\in T^\wedge$. For another concept of convolutions for distributions more closely related to the convolution integral (3.1) c.f. section 8. We will postpone further discussion of Fourier multipliers and convolutions until then.

In this section we finally discuss some estimates (already used) concerning various norms, related to the space T.

<u>Lemma 4.7.</u> We have

$$(4.24)\qquad |a(x)| \le \sum_{\alpha_j=0,1} \|a^{(\alpha)}\|_{L^1}\ ,$$

for any $a\in C^\infty(\mathbb{R}^n)$ with the corresponding derivatives in $L^1(\mathbb{R}^n)$.

<u>Proof.</u> Let at first be n=1. We get

$$a(x) = a(y) + \int_y^x a'(t)dt$$

and integrate this from zero to 1 with respect to the variable y.

$$a(x) = \int_0^1 a(y)dy + \int_0^1 dy \int_y^x dt\, a'(t) \quad .$$

In the last integral a partial integrations yields

$$a(x) = \int_0^1 a(y)dy + \int_1^x a'(y)dy + \int_0^1 y a'(y)dy$$

Now if we estimate we get

$$(4.25) \qquad |a(x)| \leq (\|a\|_{L^1} + \|a'\|_{L^1}) \ , \ x \geq 1 \ .$$

The right hand side of (4.25) is translation invariant, therefore the condition $x \geq 1$ may be omitted and the estimate is true in general.

Note that (4.25) coincides with (4.26) for n=1. The case of general n>1 follows easily by induction. Q.E.D.

Problems: 1) Calculate the Fourier transform of the distribution T_+ defined in 2, problem 6 (In particular show that T_+ is temperate). 2) Show that the function e^x, $x \in \mathbb{R}$, is not a temperate distribution (That is, no extension of the functional over $\mathcal{D}(\mathbb{R})$ associated to the L^1_{loc}-function e^x to $\mathcal{S}(\mathbb{R})$ exists which is continuous.) (Hint: construct a sequence $u_j \in \mathcal{D}(\mathbb{R})$ with $u_j \to 0$ in $\mathcal{S}$, such that $\langle e^x, u_j \rangle$ does not tend to zero.) 3) Show that a function $u \in C^\infty(\mathbb{R})$, which is periodic with period 2π is a temperate distribution, and calculate its Fourier transform. (Hint: The function u has a Fourier series $u(x) = \sum_{-\infty}^{\infty} c_j e^{ijx}$, with well known coefficients c_j. Show that the Fourier transform may be taken "term by term" in that series.) 4) If ℓ is a linear transformation as in 2, problem 2, calculate $(T \circ \ell)^\wedge$, for $T \in \mathcal{S}'(\mathbb{R}^n)$. (Show that $T \circ \ell$ is temperate, as T is temperate.) 5) Show that a temperate distribution has all of its derivatives temperate distributions. Also show that $u \in \mathcal{S}'$, $a \in \mathcal{T}$ implies $au \in \mathcal{S}'$.

5. Distributions with compact support.

We already mentioned a third space of testing functions for distributions in the introduction, the space $C^{\infty}(\Omega)$, for an open set $\Omega \subset \mathbb{R}^n$ again - henceforth denoted by $\mathcal{E}=\mathcal{E}(\Omega)$. Ignoring the topological aspect we define continuity of $u: \mathcal{E} \to \mathbb{C}$ by asking that $\langle u,\phi_k\rangle \to 0$ whenever $\phi_k \to 0$ in $\mathcal{E}$. Here $\phi_k \to 0$ in $\mathcal{E}$ means that $\phi_k^{(\alpha)} \to 0$ uniformly for each compact set $K \subset \Omega$ and every multi-index α. The collection of all continuous linear functionals over $\mathcal{E}$ is denoted by $\mathcal{E}'$. Again it follows that $\mathcal{D} \subset \mathcal{E}$ and that $\mathcal{D}$ is dense in $\mathcal{E}$. If $\Omega=\mathbb{R}^n$ then $\mathcal{D} \subset \mathcal{S} \subset \mathcal{E}$, with dense imbeddings. Also $\phi_k \to 0$ in $\mathcal{D}$ (or $\mathcal{S}$) implies $\phi_k \to 0$ in $\mathcal{S}$ and in $\mathcal{E}$ (or in $\mathcal{E}$). Hence for $u\in\mathcal{E}'$ we have $v=u|\mathcal{D}\in\mathcal{D}'$ (and $w=u|\mathcal{S}\in\mathcal{S}'$ in case of $\Omega=\mathbb{R}^n$).

Lemma 5.1. If $u\in\mathcal{E}'$ then the distribution $w=u|\mathcal{D}\in\mathcal{D}'$ has compact support.

Proof. The topology in $\mathcal{E}$ is given by the semi-norms

(5.1) $$\|u\|_{\alpha,K} = \sup \{|u^{(\alpha)}(x)|: x\in K\} \quad , \quad \alpha\in\mathbb{Z}_+^n \; , \; K \text{ compact} .$$

[For detailed discussion c.f. section 6].

Accordingly a functional $u: \mathcal{E} \to \mathbb{C}$ is continuous if and only if we have

(5.2) $$|\langle u,\phi\rangle| \le \sum_{|\alpha|\le k} \|\phi\|_{\alpha,K_0} \quad , \qquad \phi \in \mathcal{E} \quad ,$$

for some integer k and compact set K_0 (Theorem 6.2).

Accordingly we must have $\langle u,\phi\rangle=0$ whenever $\phi=0$ in a neighbourhood of K_0. It follows that supp $w \subset K_0$. Since supp u is closed, by definition, it must be compact, q.e.d.

It is interesting that the converse of lemma 5.1 also is correct. Every distribution $w\in\mathcal{D}'$ with compact support can be uniquely extended, as a linear functional, to be a continuous linear functional over $\mathcal{E}$.

Let $\lambda\in C_0^{\infty}(\Omega)$ be equal to 1 near supp w, so that $\mu=1-\lambda$ vanishes near supp w. For $\phi\in\mathcal{E}(\Omega)$ we define $u\in\mathcal{E}'(\Omega)$ by setting

(5.3) $$\langle u,\phi\rangle = \langle w,\lambda\phi\rangle \quad .$$

It follows that $\langle u,\phi\rangle = \langle w,\phi\rangle$ whenever both u and w are defined, [for $\phi\in\mathcal{D}(\Omega)\subset \mathcal{E}(\Omega)$, since $\langle w,\phi\rangle-\langle u,\phi\rangle=\langle w,\mu\phi\rangle=0$, due to supp $w \cap$ supp $\mu\phi = \emptyset$]. Hence u extends w to $\mathcal{E}$. Moreover, continuity of u is a simple consequence of Leibnitz' formula [$\phi_k \to 0$ in $\mathcal{E}$ implies $\lambda\phi_k \to 0$ in $\mathcal{D}$]. Also, if v is any continuous extension of w it must have compact support - in the sense that for all $\phi\in\mathcal{E}$ vanishing near a given compact set K_0 we get $\langle v,\phi\rangle=0$ (by 5.2) again. The corresponding will be true for the difference u-v. Picking $\lambda_0\in\mathcal{D}(\Omega)$ equal to 1 near K_0 we notice that

$<u - v,\phi> = <u-v,\lambda_0\phi> = 0$, since $\lambda_0\phi\epsilon\mathcal{D}(\Omega)$ [where both functionals agree].

We have proven the result, below.

Theorem 5.2. There is a 1-1 correspondence between $\mathcal{E}'$ and the class of distributions with compact support in $\mathcal{D}'$ given by taking restrictions and continuous extensions, respectively.

The correspondence of theorem 5.2 will be used to identify the elements of $\mathcal{E}'$ with the compactly supported distributions, so that $\mathcal{E}'$ then consists of the distributions with compact support.

Notice also that $\mathcal{E}'\subset\mathcal{S}'$, in case of $\Omega=\mathbb{R}^n$, by virtue of the above identification. We know that $u\epsilon\mathcal{E}'$ induces $v=u|\mathcal{S}\epsilon\mathcal{S}'$. There can be only one extension of v to $\mathcal{E}$, since the extension of $u|\mathcal{D} = v|\mathcal{D}$ to $\mathcal{E}$ was proven to be unique. This imbedds $\mathcal{E}'$ into $\mathcal{S}'$. In other words, the compactly supported distributions all are temperate and in particular have Fourier transforms.

Lemma 5.3. The (inverse) Fourier transform of a distribution u with compact support is a function in $\mathcal{T}\subset\mathcal{E}$. Moreover $u^\wedge=Fu$ and $u^\vee=F^{-1}u$ are entire functions of the n complex variables $x=(x_1,\dots,x_n)$, with value at x given by the formula

$$(5.4)\qquad u^\wedge(x) = (2\pi)^{-n/2}<u,e_x> , \quad u^\vee(x) = (2\pi)^{-n/2}<u,e_{-x}>$$

with the function $e_x\epsilon C^\infty(\mathbb{R}^n)=\mathcal{E}$ defined to take the values

$$(5.5)\qquad e_x(y) = e^{-ix\cdot y}$$

at $y\ \epsilon\ \mathbb{R}^n$. Moreover,we have $u^\wedge\ \epsilon\ \mathcal{T}$, a temperate function.

Proof. For a moment let $v(x)=(2\pi)^{-n/2}<u,e_x>$. Observe that this defines a power series in x which converges in $\mathbb{C}^n$. Indeed, it is evident from common facts in complex variables that we have the power series expansion of the exponential function converging to the function e_x in the space $\mathcal{E}$. In details (for simplicity in only one dimension), introduce the functions $f_k\epsilon\mathcal{E}$ by $f_k(y)=y^k$, $y\epsilon\mathbb{R}^n$, and let $s_x^N = (2\pi)^{-1/2}\sum_{k=0}^{N}(-ix)^k/k!f_k$, then it follows that $e_x-s_x^N\to 0$ as $N\to\infty$ (in $\mathcal{E}$), as defined above, for every given fixed $x\epsilon\mathbb{C}^n$. [This simply is a consequence of the fact that the entire function e_z has its power series expansion uniformly convergent on all compact sets, and that this series may be differentiated infinitely term by term, where each times the differentiated series is a power series which converges uniformly on compact sets.]

This implies that (in one dimension)

$$(5.6)\qquad v(x) = (2\pi)^{-1/2}\sum_{k=0}^{\infty}(-ix)^k/k!<u,f_k> , \quad x\epsilon\mathbb{C}^n .$$

Particularly the right hand side series converges for every $x \in \mathbb{C}$, and v(x) therefore defines an entire analytic function in x.

The above conclusion may be repeated. The expansion (5.6) may be regarded as a series expansion

$$(5.7) \quad v = (2\pi)^{-1/2} \sum_{k=0}^{\infty} (-i)^k / k! \langle u, f_k \rangle f_k \quad ,$$

konvergent in E [that is the difference between v and the partial sums of the series tends to zero in E]. Therefore, if $\phi \in \mathcal{D} \subset E'$ is a testing function, we conclude that

$$(5.8) \quad \langle v, \phi \rangle = (2\pi)^{-1/2} \sum_{k=0}^{\infty} (-i)^k / k! \langle u, f_k \rangle \langle \phi, f_k \rangle \quad .$$

Next look at

$$(5.9) \quad \langle u^{\wedge}, \phi \rangle = \langle u, \phi^{\wedge} \rangle$$

and write

$$(5.10) \quad \phi = (2\pi)^{-1/2} \sum_{k=0}^{\infty} (-i)^k / k! f_k \langle \phi, f_k \rangle$$

where the series again converges in E, as readily confirmed.

Therefore we only have to substitute (5.10) into the right hand side of (5.9) and take the series to the outside. Comparing the result with the right hand side of (4.13) we conclude that $\langle u^{\wedge}, \phi \rangle = \langle v, \phi \rangle$ for all $\phi \in \mathcal{D}$, which amounts to $u^{\wedge} = v$.

This proves the lemma in case of n=1 and for F. The extension to general n and to F^{-1} is straight forward, and will be left to the reader.

Problems: 1) Show that the Dirac distribution (is temperate and) has (inverse) Fourier transform

$$(5.11) \quad \delta^{\wedge}(x) = (2\pi)^{-n/2} \cdot 1 = \delta^{\vee}(x)$$

with '1' denoting the constant function, taking the value 1 everywhere. 2) Prove that the distribution $\langle \delta_z, \phi \rangle = (2\pi)^{n/2} \phi(z)$, $\phi \in \mathcal{S}(\mathbb{R}^n)$ is temperate and has (inverse) Fourier transform given by

$$(5.12) \quad \delta_z^{\wedge}(x) = e^{-ix \cdot z} = e_z(x) \; , \; \delta_z^{\vee}(x) = e_{-z}(x) \quad .$$

3) Calculate the (inverse) Fourier transforms of the derivatives $\delta_z^{(\alpha)}$, as

$$(5.13) \quad \delta_z^{(\alpha)\wedge}(x) = (ix)^{\alpha} e_z(x) \; , \; \delta_z^{(\alpha)\vee}(x) = (-ix)^{\alpha} e_{-z}(x) \; .$$

4) Calculate the (inverse) Fourier transform of a polynomial. In particular show that

$$(5.14) \quad (x^{\alpha})^{\wedge} = (i)^{|\alpha|} \delta_0^{(\alpha)} \; , \; (x^{\alpha})^{\vee} = (-i)^{|\alpha|} \delta_0^{(\alpha)} \quad ,$$

and

(5.15) $P(x)^{\wedge} = P(-D)\delta_0$, $P(x)^{\vee} = P(D)\delta_0$,

5) Obtain $(x^{\alpha}e^{izx})^{\wedge}$, $(x^{\alpha}e^{izx})^{\vee}$, for $z\in\mathbf{R}^n$. 6) Using formulas from [9], regarding integrals involving Bessel functions to calculate the Fourier transform of the distribution defined in 2, problem 7.

6. Order of a distribution and topology in $\mathcal{D}'$, $\mathcal{E}'$ and $\mathcal{S}'$.

It is convenient to have additional criteria available for continuity of a linear functional u over $\mathcal{D}$,$\mathcal{E}$ or $\mathcal{S}$. Accordingly we discuss in the following an estimate as necessary and sufficient condition for continuity, in each of the three spaces.

Theorem 6.1. u over $\mathcal{D}(\Omega)$ is continuous if and only if for every open $\Omega_1 \subset \Omega$, having compact closure K there exist constants c and k with

$$(6.1) \qquad |\langle u,\phi\rangle| \le c \sum_{|\alpha|\le k} \sup_x |\phi^{(\alpha)}(x)| \quad , \ \phi \in \mathcal{D}(\Omega_1) \quad .$$

Theorem 6.2. u over $\mathcal{E}(\Omega)$ is continuous if and only if there exists a compact set K and constants c, k such that

$$(6.2) \qquad |\langle u,\phi\rangle| \le c \sum_{|\alpha|\le k} \sup_{x\in K} |\phi^{(\alpha)}(x)| \quad , \ \phi \in \mathcal{E}(\Omega).$$

Theorem 6.3. u over $\mathcal{S}$ is continuous if and only if there exist constants c and k such that

$$(6.3) \qquad |\langle u,\phi\rangle| \le c \sum_{|\alpha|,|\beta|\le k} \sup_x |x^\alpha \phi^{(\beta)}(x)| \quad , \ \phi \in \mathcal{S} \ .$$

Proofs. For $\mathcal{D}(\Omega)$: It is evident that (6.1) implies continuity of u over $\mathcal{D}$, since the right hand side tends to zero when $\phi=\phi_k$ is substituted where $\phi_k \to 0$ in $\mathcal{D}$. Vice versa, suppose (6.1) is not true for a linear functional u. Then there exists an open Ω_1 with compact closure $K \subset \Omega$ and for every c and k a function for which (6.1) is false. Let us pick c=k=j (any integer > 0) and denote $\phi=\phi_j$, correspondingly. We may normalize ϕ_j according to $\langle u,\phi_j\rangle=1$ and then get

$$(6.4) \qquad j \sum_{|\alpha|\le j} \sup_x |\phi_j^{(\alpha)}(x)| \le 1 \quad , \quad \operatorname{supp} \phi_j \subset K \ , \ j=1,2,\dots, \quad .$$

This implies

$$(6.5) \qquad |\phi_j^{(\alpha)}(x)| \le 1/j \ , \ |\alpha| \le j \quad , \quad x \in \Omega \quad ,$$

or, $\phi_j \to 0$ in $\mathcal{D}(\Omega)$. On the other hand $\langle u,\phi_j\rangle=1$, in contradiction to the requirement that $\langle u,\phi_j\rangle \to 0$. Hence u cannot be continuous, and theorem 6.1 is proven.

For $\mathcal{E}(\Omega)$: Again it is trivial that (6.2) implies continuity of u over $\mathcal{E}$. Vice versa, assume that (6.2) is not correct. Then we may pick an increasing sequence $\{K_j\}$ of compact sets with $\Omega = \bigcup_j K_j$ and set c=k=j (an integer > 0) and then can find $\phi_j \in \mathcal{E}$ such that $\langle u,\phi_j\rangle=1$ while '>' holds in (6.2) with $\phi=\phi_j$. That is, we get

$$(6.6) \qquad j \sum_{|\alpha|\le j} \sup_{K_j} |\phi_j^{(\alpha)}(x)| < 1$$

which implies

(6.7) $\sup_{K_j} |\phi_j^{(\alpha)}(x)| \le 1/j \quad , \quad |\alpha| \le j$.

Every compact subset $K \subset \Omega$ must be contained in K_j for sufficiently large j, by a familiar conclusion. Accordingly (6.7) implies that $\phi_j \to 0$ in $\mathcal{E}$. If u were continuous over $\mathcal{E}$ then $\langle u,\phi_j\rangle \to 0$ follows, which contradicts to $\langle u,\phi_j\rangle = 1$, above. Accordingly u cannot be continuous and theorem 6.2 is proven.

For $\mathcal{S}$: The proof is quite similar to the others, above, and mainly consists of properly negating condition (6.3). This is left to the reader.

As the next subject, let us shortly discuss the problem of topologies in the distribution spaces $\mathcal{D}'$, $\mathcal{E}'$ and $\mathcal{S}'$ since it often will be convenient to have a notion of convergence for a sequence of distributions. It is convenient to do this within the bounds of theory of locally convex spaces.

First of all, there is a 'weak topology' in $\mathcal{D}'$, $\mathcal{E}'$ and $\mathcal{S}'$ each, since in effect they are defined as the adjoint spaces of the locally convex spaces $\mathcal{D}$, $\mathcal{E}$ and $\mathcal{S}$. A sequence $\{u_k\}$ of distributions is said to converge weakly to $u\varepsilon\mathcal{D}'$ if

(6.8) $\langle u_k,\phi\rangle \to \langle u,\phi\rangle \quad , \quad \phi\varepsilon\mathcal{D}$.

Similarly, (with '$\rightharpoonup$' indicating 'weak convergence') ,

(6.9) $u_k \rightharpoonup u$ in $\mathcal{E}' \Longleftrightarrow \langle u_k,\phi\rangle \to \langle u,\phi\rangle \ , \ \phi\varepsilon\mathcal{E}$

and

(6.10) $u_k \rightharpoonup u$ in $\mathcal{S}' \Longleftrightarrow \langle u_k,\phi\rangle \to \langle u,\phi\rangle \ , \ \phi\varepsilon\mathcal{S}$.

It is known (but we shall not prove it in details, c.f. [3]) that $\mathcal{D}'$, $\mathcal{E}'$ and $\mathcal{S}'$ are weakly complete: Suppose, for example, $u_k\varepsilon\mathcal{D}'$ is a Cauchy sequence in this weak sense (i.e. for every $\phi\varepsilon\mathcal{D}$ the sequence $\langle u_k,\phi\rangle$ of complex numbers is Cauchy) then the limits $\lim_{K\to\infty} \langle u_k,\phi\rangle = \langle u,\phi\rangle$ define a continuous linear functional over $\mathcal{D}'$, so that u_k converges weakly to u in $\mathcal{D}'$. Similarly for $\mathcal{E}'$ and $\mathcal{S}'$ again. [This weak completeness essentially is a consequence of the so-called uniform boundedness principle in functional analysis.]

A variety of stronger topologies in $\mathcal{D}'$, $\mathcal{E}'$, $\mathcal{S}'$ is available, which will be of limited importance for us. As adjoint (or topologically dual) spaces each of them also allows a so-called strong topology (for definition and discussion c.f. [1], [2], [7]). Moreover, it is clear that the semi-norms

(6.11) $\mu_{\Omega_1,k}(u) = \sup \{|\langle u,\phi\rangle| : \phi\varepsilon\mathcal{D}, \ \mathrm{supp}\ \phi \subset \Omega_1, \sum_{|\alpha|\le k} \sup_{\Omega} |\phi^{(\alpha)}(x)| \le 1\}$

where k and Ω_1 run over all integers ≥ 0 and all open subsets $\Omega_1 \subset K \subset \Omega$, K compact, respectively, are meaningful for every $u \in \mathcal{D}'$, provided k is sufficiently large, in dependence of K. The smallest integer k such that (6.11) is finite, for a given $u \in \mathcal{D}'$ and Ω_1 will be called the order of u in Ω_1. Generally we shall say that k is an order of u in Ω_1 if $\mu_{\Omega_1,k}(u)$ is finite.

Theorem 6.1 indicates that every $u \in \mathcal{D}'$ has an order over every $\Omega_1 \subset K \subset \Omega$. Let $\mathcal{D}_k'(\Omega_1)$ denote the class of all $u \in \mathcal{D}'$ having k as an order (being 'of order k') in Ω_1. If Ω_j denotes an increasing sequence of open sets $\Omega_j \subset K_j \subset \Omega$, K_j compact, and if $\Omega = \bigcup \Omega_j$, then we clearly get

$$(6.12) \qquad \mathcal{D}' = \bigcap_j \bigcup_k \mathcal{D}_k'(\Omega_j) \quad .$$

For each j the space $\bigcup_k \mathcal{D}_k'(\Omega_j)$ has an inductive limit topology, [defined as the strongest locally convex topology in which all of the injections $\mathcal{D}_k' \to \bigcup \mathcal{D}_k'$ are continuous]. Thus $\mathcal{D}'$ also is a locally convex space with set of semi-norms given as the union of the corresponding collections for $\bigcup_k \mathcal{D}_k'(\Omega_j)$. This offers itself as another 'natural' topology for $\mathcal{D}'$, in particular as it may be seen that the result is independent of the choice of Ω_j. We note that $\mathcal{D}'$ is a Montel space (c.f. [7], [12]). From this it follows that for sequences weak and strong convergence is the same. Also a weakly continuous distribution-valued function of finitely many parameters necessarily is also strongly continuous ([7], Ch. III, §3, Theorem XIII).

Similar considerations apply for $\mathcal{E}'$ and $\mathcal{S}'$, using theorem 6.2 or theorem 6.3, correspondingly. A distribution u in $\mathcal{E}'$ (or $\mathcal{S}'$) has a global order, with respect to $\mathcal{E}$ (or $\mathcal{S}$), defined to be (the smallest) k allowing an estimate (6.2) [or (6.3)]. Matters are slightly simpler, insofar as we have

$$(6.13) \qquad \mathcal{E}' = \bigcup \mathcal{E}_k'$$

and

$$(6.14) \qquad \mathcal{S}' = \bigcup \mathcal{S}_k'$$

with Banach spaces $\mathcal{E}_k'$ and $\mathcal{S}_k'$ defined as the classes of distributions of (global) order k respectively. Accordingly we obtain inductive limit topologies for $\mathcal{E}'$ and $\mathcal{S}'$ ([1], [2]).

We note that these inductive limit topologies coincide with the corresponding strong topologies in $\mathcal{E}'$ and $\mathcal{S}'$. For a proof of this fact in the case of $\mathcal{E}'$ c.f. [7], p.90, in the case of $\mathcal{E}'$, and that proof is easily adapted to the (simpler) case of $\mathcal{S}'$ as well.

In theorem 4.4 we already commented on continuity properties of the various operations introduced for temperate distributions, (without proof). Only the weak topology of $\mathcal{S}'$ was used. Similar results are valid for $\mathcal{E}'$ and $\mathcal{D}'$, as well as for the various stronger topologies, but these are quite numerous and will never be used. It is useful, however, to have available a result like the lemma, below, which, in essence, says that $\mathcal{E}'$ is dense in $\mathcal{S}'$.

Lemma 6.4. If $u\in\mathcal{S}'$ and $\chi\in C_0^\infty(\mathbb{R}^n)$ equals 1 near zero, and $\chi_k\in C_0^\infty$ is defined by $\chi_k(x) = \chi(x/k)$, $k=1,2,3,\dots$, then we have $\chi_k u\in\mathcal{E}'$ and $\chi_k u \to u$ [in weak topology of $\mathcal{S}'$], as $k\to\infty$.

Proof. We get $\langle\chi_k u,\phi\rangle = \langle u,\chi_k\phi\rangle$, $\phi\in\mathcal{S}$ and then only must show that $\omega_k\phi\to 0$ in $\mathcal{S}$, where $\omega_k = 1-\chi_k$. Notice that $\omega_k = 0$ in $|x|\le ck$, where c is a suitable positive constant. Also $\omega_k(x) = \omega(x/k)$, with $\omega = 1-\chi$. Furthermore, $\omega=1$ for large $|x|$ so that derivatives of ω vanish for large $|x|$. Any derivative $\phi_k^{(\alpha)}$ of $\phi_k = \omega_k\phi$ is a linear combination of $k^{-|\beta|}\omega^{(\beta)}(x/k)\phi^{(\gamma)}$, $\beta+\gamma=\alpha$. It follows that

$$(6.15)\qquad (1+|x|)^l|\phi_k^{(\alpha)}(x)| \le c\sum_{|\gamma|\le|\alpha|}\ \sup_{|y|\ge ck}\ (1+|y|)^l|\phi^{(\gamma)}(y)|$$

with the constant c independent of k. As $k\to\infty$ the right hand side will go to zero. This proves the desired convergence $\phi_k\to 0$ in $\mathcal{S}$, q.e.d.

Remark: Using the convolution product of section 8, particularly Lemma 8.6, it is easily verified that even $\mathcal{D}$ is dense in $\mathcal{D}'$ (c.f. 8, problem 6).

Problems: 1) For simplicity let n=1. Prove that every temperate distribution T may be obtained as a finite linear combination of distributions of the form $T=D^k f_k$ with $D = -id/dx$ and some functions $f_k\in L^1_{pol}$. (Hint: Use theorem 6.2.)

2) Show that $\mathcal{S}'$ equals the inductive limit

$$(6.16)\qquad \mathcal{S}' = \bigcup_{k=1}^{\infty} \mathcal{S}'_{k,2}\ ,$$

including its strong topology, with the duals $\mathcal{S}'_{k,2}$ of the Hilbert spaces $\mathcal{S}_{k,2}$, obtained as completion of $\mathcal{S}$ under the norm

$$(6.17)\qquad \|u\|_{k,2} = \{\sum_{|\alpha|,|\beta|\le k}\|x^\alpha u^{(\beta)}\|^2_{L^2(\mathbb{R}^n)}\}^{1/2}.$$

(Hint: Show that $\mathcal{S}_{k-2n}\supset\mathcal{S}_{k,2}\supset\mathcal{S}_{k+2n}$, as a topological inclusion, with $\mathcal{S}_j$ as in (6.14).)

7. Tensor products of distributions.

Let $u \in \mathcal{D}'(\Omega)$, $v \in \mathcal{D}'(\Omega')$ be distributions over $\Omega \subset \mathbb{R}^n$ and $\Omega' \subset \mathbb{R}^m$, respectively. Then a linear functional g^o may be defined over a subspace $\mathcal{D}^o$ of $\mathcal{D}(\Omega\times\Omega')$ by setting

$$\begin{aligned} \mathcal{D}^o &= \mathcal{D}(\Omega) \otimes \mathcal{D}(\Omega') \\ &= \{\phi :\ \phi \in \mathcal{D}(\Omega\times\Omega'), \phi(x,y) = \sum_{j=1}^{N} \phi_j^1(x)\phi_j^2(y), \phi_j^1 \in \mathcal{D}(\Omega), \phi_j^2 \in \mathcal{D}(\Omega')\} \end{aligned} \tag{7.1}$$

and

$$\langle g^o,\phi\rangle = \sum_{j=1}^{N} \langle u,\phi_j^1\rangle\langle v,\phi_j^2\rangle,\quad \phi = \sum_{j=1}^{N}\phi_j^1(x)\phi_j^2(y) \in \mathcal{D}^o \quad . \tag{7.2}$$

Please note that (7.2) is well defined. Suppose we have $\sum_{j=1}^{N}\phi_j^1(x)\phi_j^2(y) = 0$ for all (x,y). It follows that $\sum_{j=1}^{N}\phi_j^1(x)\langle v,\phi_j^2\rangle = 0$ [If the variable x is kept fixed, we obtain a testing function in $\mathcal{D}(\Omega')$, and the above follows by taking the value of v at this testing function]. Again, taking the value of u at

$\lambda(x) = \sum_{j=1}^{N}\phi_j^1(x)\langle v,\phi_j^2\rangle \in \mathcal{D}(\Omega)$ we obtain

$$\sum_{j=1}^{N}\langle u,\phi_j^1\rangle\langle v,\phi_j^2\rangle = 0 \quad , \tag{7.3}$$

which shows that $\langle g^o,\phi\rangle$ in (7.2) is independent of the choice of representation of $\phi\in\mathcal{D}^o$.

It will be seen below, that the space $\mathcal{D}^o$ is a dense subspace of $\mathcal{D}(\Omega\times\Omega')$, and moreover, that g^o admits a unique continuous extension to $\mathcal{D}(\Omega\times\Omega')$. This extension $g \in \mathcal{D}'(\Omega\times\Omega')$ is called the tensor product of u and v and is denoted by $u\otimes v$.

For any $\phi\in\mathcal{D}^o$ and $x\in\Omega$ the restriction of ϕ to the set $\{x\}\times\Omega'$ is a testing function in $\mathcal{D}(\Omega')$, denoted by ψ_x. Clearly we have $\langle v,\psi_x\rangle$ meaningful and get

$$\langle v,\psi_x\rangle = \sum_{j=1}^{N}\phi_j^1(x)\langle v,\phi_j^2\rangle \ , \ x\in\Omega \ . \tag{7.4}$$

It follows that (7.4) defines a function in $\mathcal{D}(\Omega)$, henceforth denoted by $\langle v,\phi\rangle$. Similarly we may define $\langle u,\phi\rangle$, for $u\in\mathcal{D}'(\Omega)$ and $\phi\in\mathcal{D}^o(\Omega\times\Omega')$, which will be a function in $\mathcal{D}(\Omega')$.

We observe that

$$\langle g^o,\phi\rangle = \langle u,\langle v,\phi\rangle\rangle = \langle v,\langle u,\phi\rangle\rangle \quad , \tag{7.5}$$

where each times the outer brackets denote values of functionals at testing functions, while the inner brackets are as defined above.

Lemma 7.1. Let $\phi\in\mathcal{D}(\Omega\times\Omega')$. There exists a sequence $\phi_k\in\mathcal{D}^o$ such that $\phi-\phi_k \to 0$ in $\mathcal{D}(\Omega\times\Omega')$.

Proof. Let us use properties of the Fourier transform, as discussed in section 3 for a convenient construction of the sequence ϕ_k. First, we may regard ϕ as a function in $\mathcal{D}(\mathbb{R}^{n+m}) \subset \mathcal{S}(\mathbb{R}^{n+m})$ by extending it zero outside $\Omega\times\Omega'$. The extended function has its Fourier transform in $\mathcal{S}$, and we may use the Fourier integral in n+m dimensions to write

$$(7.6)\qquad D_x^\alpha D_y^\beta \phi(x,y) = (2\pi)^{-(n+m)/2} \int_{\mathbb{R}^{n+m}} d\xi d\eta\, \xi^\alpha \eta^\beta \phi^\wedge(\xi,\eta)\lambda(x)e^{ix\xi}\mu(y)e^{iy\eta} \quad .$$

Here the functions $\lambda\varepsilon\mathcal{D}(\Omega)$ and $\mu\varepsilon\mathcal{D}(\Omega')$ are chosen such that $\lambda(x)\mu(y) = 1$ near supp ϕ. Denote $\lambda_\xi(x) = \lambda(x)e^{ix\xi}$, $\mu_\eta(y) = \mu(y)e^{iy\eta}$ and observe that $\lambda_\xi\varepsilon\mathcal{D}(\Omega), \mu_\eta\varepsilon\mathcal{D}(\Omega')$ and that (7.6) for vanishing α,β may be written as

$$(7.7)\qquad \phi(x,y) = \int d\xi d\eta\, \omega(\xi,\eta)\lambda_\xi(x)\mu_\eta(y) \ , \ \omega = (2\pi)^{-(n+m)/2}\phi^\wedge \quad .$$

The integrals in (7.6) and (7.7) converge quite well, since the integrand is in $\mathcal{S}$. Thus an approximation of ϕ by suitable sums will be possible, any such approximation being of the form

$$(7.8)\qquad s(x,y) = \sum_j m(K_j)\omega(\xi_j,\eta_j)\lambda_{\xi_j}(x)\mu_{\eta_j}(y) \quad ,$$

with of subsets $K_j \subset \mathbb{R}^{n+m}$, with $(\xi_j,\eta_j)\varepsilon K_j$ and with a finite sum. It follows that $s\varepsilon\mathcal{D}^O$, and we only must confirm that a sequence of such sums can be chosen with the properties of ϕ_k, above.

Certainly ϕ-s has support contained in supp $(\lambda\mu)$, a fixed compact subset of $\Omega\times\Omega'$. Also we have

$$(7.9)\qquad D_x^\alpha D_y^\beta s(x,y) = \sum_j m(K_j)\omega(\xi_j,\eta_j)D_x^\alpha\lambda_{\xi_j}(x)D_y^\beta\mu_{\eta_j}(y) \quad .$$

Introduce a function χ over $\mathbb{R}^{2n+2m}$ by setting

$$(7.10)\qquad \chi(x,y,\xi,\eta) = \omega(\xi_j,\eta_j)\lambda_{\xi_j}(x)\mu_{\eta_j}(y) \quad , \quad (\xi,\eta)\ \varepsilon\ K_j \quad ,$$

and let it be zero elsewhere. Then

$$(7.11)\qquad \begin{aligned} &\phi(x,y) - s(x,y) \\ &= \int d\xi d\eta \left[\omega(\xi,\eta)\lambda_\xi(x)\mu_\eta(y) - \chi(x,y,\xi,\eta)\right] \quad . \end{aligned}$$

We shall have to make the sets K_j mutually disjoint as to have χ well defined. Then these sets have to be chosen in such a way that the L^1-norm in ξ,η of the integrand in (7.11) tends to zero uniformly in x,y. A calculation shows this to be possible, using the fact that $\omega\varepsilon\mathcal{S}$, that the integrand vanishes for (x,y) outside supp μ while the exponential functions also is uniformly continuous over compact subsets.

In details, let the integrand in (7.11) be denoted by $\sigma(x,y,\xi,\eta)$, and write

$$\sigma(x,y,\xi,\eta) = [\omega(\xi,\eta) - \omega(\xi_j,\eta_j)]\lambda_\xi(x)\mu_\eta(y) \tag{7.12}$$
$$+ \omega(\xi_j,\eta_j)[\lambda_\xi(x)-\lambda_{\xi_j}(x)]\mu_\eta(y)+\omega(\xi_j,\eta_j)\lambda_{\xi_j}(x)[\mu_\eta(y)-\mu_{\eta_j}(y)]$$

in the set K_j. The functions λ_ξ and μ_η are uniformly bounded in x,ξ and y,η, resp.. Also each of the families of functions $\lambda_\xi(x)$, $\mu_\eta(y)$, where ξ and η are the variables and x,y the parameters, is equi-continuous, since the exponential functions $e^{ix\xi}$ and $e^{iy\eta}$ are for $x \in \operatorname{supp}\lambda$, and $y \in \operatorname{supp}\mu$, these supports being compact. It follows that by choosing the diameter of K_j small enough we may arrange for

$$|\sigma| < \varepsilon m(K_j) \qquad j=1,\dots,N \tag{7.13}$$

with arbitrary $\varepsilon>0$. Also, we have

$$\sigma(x,y,\xi,\eta) = \chi(x,y,\xi,\eta) \quad \text{for} \quad (\xi,\eta) \notin K_j \quad . \tag{7.14}$$

Accordingly

$$|\phi-s| \le c\int_{K_\infty} |\omega(\xi,\eta)|d\xi d\eta + \varepsilon m(\bigcup K_j) \quad , \tag{7.15}$$

where K_∞ denotes the complement of $\bigcup K_j=K'_\infty$. Using that $\omega\varepsilon S$ we first choose K_∞ to have the first term small. Then we choose the partition $K'_\infty = \bigcup K_j$ as to have the second term small, keeping K_∞ fixed. This makes it clear indeed that a sequence $s=s_k$, $k=1,2,3,\dots$, may be picked such that $\sup|\phi-s_k| \to 0$, as $k\to\infty$.

Similarly we may treat every derivative $D_x^\alpha D_y^\beta\sigma$. One will obtain a representation like (7.12), with λ_ξ and μ_η replaced by $D_x^\alpha\lambda_\xi$ and $D_y^\beta\mu_\eta$. These derivatives are not uniformly bounded anymore, but are of polynomial growth in ξ and η. They are still equi-continuous families in compact subsets. Therefore an estimate like (7.15) will result, with ω replaced by $(1+|\xi|)^{|\alpha|}(1+|\eta|)^{|\beta|}\omega(\xi,\eta)$. This is sufficient for proving that also the derivative $D_x^\alpha D_y^\beta(\phi-s)$ may be made uniformly small by first picking K_∞ and then the partition $K'_\infty = \bigcup K_j$.

The desired sequence ϕ_k may be constructed by setting $\phi_k = s$, where we have arranged K_∞ and $\{K_j\}$ such that

$$|D_x^\alpha D_y^\beta(\phi-\phi_k)| \le 1/k \quad , \quad |\alpha|,|\beta| \le k \quad . \tag{7.16}$$

This completes the proof of lemma 7.1.

Lemma 7.2. Let $\tilde\Omega_1 \subset K \subset \Omega\times\Omega'$, $\tilde\Omega_1$ open, K compact. There exist constants c,k such that

$$(7.17)\qquad |\langle g^{o},\phi\rangle| \le c \sum_{|\alpha|+|\beta|\le k} \sup_{\Omega_1} |D_x^{\alpha}D_y^{\beta}\phi(x,y)| \ ,\ \phi\in\mathcal{D}^{o}\ ,\ \operatorname{supp}\phi\subset\tilde{\Omega}_1 \ .$$

Proof. We will apply formula (7.5) here. First let $\tilde{\Omega}_1 \subset \Omega_1\times\Omega_1' \subset K \subset \Omega$. The distributions u and v both satisfy (6.1) with Ω_1,Ω_1' and suitable c_1,k_1,c_1',k_1'. Let $\mu=\langle v,\phi\rangle\in\mathcal{D}(\Omega)$, and observe that supp $\mu\subset\Omega_1'$, whenever supp $\phi\subset\tilde{\Omega}_1$. Hence (6.1) implies

$$(7.18)\qquad |\langle g^{o},\phi\rangle| \le c_1 \sum_{|\alpha|\le k_1} \sup_x |\mu^{(\alpha)}(x)| \ .$$

However we get $D^{\alpha}\mu = \langle v,D_x^{\alpha}\phi\rangle$, by (7.4), so that another application of (7.1) yields

$$(7.19)\qquad |\mu^{(\alpha)}(x)| \le c_2 \sum_{|\beta|\le k_2} \sup_y |D_x^{\alpha}D_y^{\beta}\phi(x,y)| \ .$$

Then (7.17) follows from (7.18) and (7.19), q.e.d..

It is evident now that lemma 7.2 amounts to the fact that g^{o} is continuous in the relative topology of $\mathcal{D}$ in $\mathcal{D}_o$. Since lemma 7.1 states that $\mathcal{D}_o$ is dense in $\mathcal{D}$, there is a unique continuous extension g of g^{o} to $\mathcal{D}$ which is a distribution in $\mathcal{D}' = \mathcal{D}'(\Omega\times\Omega')$. The complex number $\langle g,\phi\rangle$ for $\phi\in\mathcal{D}(\Omega\times\Omega')$ may be constructed by picking a sequence $\phi_k\in\mathcal{D}^{o}$ with $\phi-\phi_k \to 0$ in $\mathcal{D}$ and defining

$$(7.20)\qquad \langle g,\phi\rangle = \lim_{k\to\infty} \langle g^{o},\phi_k\rangle \ .$$

As mentioned above, the distribution $g\in\mathcal{D}'(\Omega\times\Omega')$ is called the tensor product of $u\in\mathcal{D}'(\Omega)$ and $v\in\mathcal{D}'(\Omega')$, and is denoted by $u\otimes v$.

Lemma 7.3. The two linear maps $\phi \to \langle v,\phi\rangle$ and $\phi \to \langle u,\phi\rangle$, defined from $\mathcal{D}^{o}$ to $\mathcal{D}(\Omega)$ and to $\mathcal{D}(\Omega')$, respectively are continuous with the relative topology of $\mathcal{D}(\Omega\times\Omega')$ in $\mathcal{D}^{o}$.

Proof. Let $\mu=\langle v,\phi\rangle$ again, and recall that $D^{\alpha}\mu = \langle v,D_x^{\alpha}\phi\rangle$ and that (7.19) holds whenever supp $\mu\subset\Omega_1$ (i.e., whenever supp $\phi\subset\tilde{\Omega}_1$). Also that supp $\mu\subset\Omega_1'$ whenever supp $\phi\subset\tilde{\Omega}_1$. This implies continuity of the map $\phi \to \langle v,\phi\rangle$ as stated. Similarly for the other map. Q.E.D..

Following lemma 7.3 we find that $\langle u,\phi\rangle$ and $\langle v,\phi\rangle$ may be extended to supply continuous maps $\mathcal{D}(\Omega\times\Omega') \to \mathcal{D}(\Omega')$ and $\mathcal{D}(\Omega\times\Omega') \to \mathcal{D}(\Omega)$, which will be denoted by the same symbol $\langle u,\phi\rangle$ and $\langle v,\phi\rangle$. Note that also (7.5) now may be extended to g. Using (7.20) and lemma 7.3 we get

$$(7.21)\qquad \langle u\otimes v,\phi\rangle=\langle g,\phi\rangle=\langle u,\langle v,\phi\rangle\rangle=\langle v,\langle u,\phi\rangle\rangle,\ u\in\mathcal{D}'(\Omega),\ v\in\mathcal{D}'(\Omega),\ \phi\in\mathcal{D}(\Omega\times\Omega') \ .$$

Lemma 7.4. Suppose the distribution v is in $L^1_{loc}(\Omega')$. Then we have

$$(7.22)\qquad \langle v,\phi\rangle = \int_{\Omega'} v(y)\phi(x,y)dy$$

with a convergent Lebesgue integral.

Proof. Let $\phi_k \varepsilon \mathcal{D}^0$ converge to ϕ in $\mathcal{D}(\Omega\times\Omega')$. Notice that this implies convergence of the restricted functions $\psi_k^x = \phi|\{x\times\Omega'\}$ to $\psi^x = \phi|\{x\times\Omega'\}$. Therefore the assertion follows from the fact that $\langle v,\phi_k\rangle \to \langle v,\phi\rangle$ by lemma 7.3 and that

$$\langle v,\phi_k\rangle(x) = \langle v,\psi_k^x\rangle \to \langle v,\psi^x\rangle = \int v(y)\phi(x,y)dy$$

as was seen initially [because $\langle v,\phi\rangle$ for $\phi\varepsilon\mathcal{D}^0$ is defined in this way]. Q.E.D..

Lemma 7.5. We have

$$(7.23)\qquad \text{supp } u\otimes v = \text{supp } u \times \text{supp } v \quad .$$

The proof is straight forward, and its discussion is omitted.

Problems: 1) Show that all distributions of the form $\delta^{(j)} \otimes T\varepsilon\mathcal{D}'(\mathbb{R}^2)$, with δ the Dirac measure in $\mathbb{R}$, and any $T\varepsilon\mathcal{D}'(\mathbb{R})$ have their support on the line $\{(x,y):y=0\}$ in $\mathbb{R}^2 = \mathbb{R}\times\mathbb{R}$, with the first and second coordinate denoted by x and y, respectively.
2) Vice versa, show that every distribution $T\varepsilon E'(\mathbb{R}^2)$, with support on the line $y=0$ must be a finite sum $\sum_{j=0}^{N} \delta^{(j)} \otimes T_j$ of distributions as in problem 1. (Hint: Use a technique as in the proof of theorem 6.1. Or else adapt the proof of II, Theorem 1.2.)

8. Convolution product of distributions.

A convolution product $u*v$ for distributions u and v over $\mathbb{R}^n$ was introduced before in (4.22), under the assumption that either u or v has its Fourier transform in $\mathcal{T}$ (the space of temperate functions or multipliers of $\mathcal{S}'$). Formula (4.22) is not entirely satisfactory, since it used the Fourier transform. In particular we shall require the convolution product for a distributions not in $\mathcal{S}'$, thus not allowing a Fourier transform.

Another conventional definition of $u*v$ for distributions u and v, uses the tensor product of section 7 and defines the functional $u*v$ by

$$(8.1) \qquad \langle u*v,\phi\rangle = \langle u\otimes v,\psi\rangle, \psi(x,y)=(2\pi)^{-n/2}\phi(x+y), \ \phi\in\mathcal{D}(\mathbb{R}^n) \quad .$$

Note that ψ as defined is in $\mathcal{E}(\mathbb{R}^{2n})$ but not in $\mathcal{D}(\mathbb{R}^{2n})$. Accordingly (8.1) is meaningful for $u,v\in\mathcal{E}'(\mathbb{R}^n)$, but not necessarily for $u,v\in\mathcal{D}'$. On the other hand the support of ψ always is contained in a 'strip' $|x+y| \le a < \infty$. Suppose the support of $u\otimes v$ has compact intersection with any such strip. Then even a closed neighbourhood N of supp $u\otimes v$ may be constructed with the same property. Construct a $C^\infty(\mathbb{R}^{2n})$ function λ equal to 1 near supp $u\otimes v$ and with support in N, and define

$$(8.2) \qquad \langle u*v,\phi\rangle = \langle u\otimes v,\lambda\psi\rangle \quad , \quad \phi\in\mathcal{D}(\mathbb{R}^n) \quad .$$

In effect the right hand side extends the functional $u\otimes v\in\mathcal{D}'(\mathbb{R}^{2n})$ to a larger class of testing functions again, in a manner similar to that used in section 5 for extending $u\in\mathcal{D}'$ with compact support to all of $\mathcal{E}$. Particularly the extension is independent of λ again, as readily seen. Thus we shall use the notation

$$(8.3) \qquad \langle u\otimes v,\psi\rangle = \langle u\otimes v,\lambda\psi\rangle$$

again whenever convenient and meaningful, but we shall not investigate continuity of the resulting extended functional.

In other words, for a distribution $u\otimes v$ with support of the above property we define $u*v$ by (8.1), with the tensor product's value at ψ defined by (8.3). It is trivial to confirm that (8.1) defines a continuous linear functional over $\mathcal{D}(\mathbb{R}^n)$, using Leibnitz' formula. Thus we have proven the result below.

Theorem 8.1. The convolution product $u*v$ of (8.1) is well defined whenever the sets

$$(8.4) \qquad [\text{supp } u \times \text{supp } v] \cap \{(x,y) : |x+y| \le a\} = K_a \subset \mathbb{R}^{2n}$$

all are compact (for $0\le a<\infty$).

Corollary 8.2. If u and v satisfy (8.4) then

$$(8.5) \quad \text{supp } u*v \subset \text{supp } u + \text{supp } v = \{x+y : x\in\text{supp } u, y\in\text{supp } v\} \quad .$$

Corollary 8.3. The convolution product $u*v$ is well defined by (8.1) whenever $u,v\in\mathcal{D}'$ and at least one of them has compact support. It also is well defined whenever u and $v \in \mathcal{D}'(\mathbb{R}^n)$ both have support in a half-space $x\cdot H\geq 0$, where $H\neq 0$ denotes a vector of $\mathbb{R}^n$, while at least one of them has support in a convex circular cone $\{x\cdot H\geq |x|\}$, $\varepsilon>0$. In the latter case we have supp $u*v\in\{x\cdot H\geq 0\}$ again.

If both u and v have compact support then $u*v$ has compact support. If both supports (of u and v) are in the same convex circular cone then also supp $u*v$ is in such a cone.

Proof. Corollary 8.2 is a direct consequence of (7.23) For corollary 8.3 we notice that the sets $\{|x|\leq b\}$ have compact intersections with $\{|x+y|\leq a\}$ so that (8.4) holds whenever supp u is in $\{|x|\leq a\}$ (i.e. is compact). Similarly, if we assume without loss of generality that $H=(1,0,\dots,0)$, then $x\cdot H=x_1\geq 0$ for all x in supp u $\cup$ supp v implies that $x \in$ supp u, $y\in$ supp v, $|x+y|\leq a$ $0\leq x_1\leq a$, $0\leq y_1\leq a$. Moreover if in addition $x_1\geq\varepsilon|x|$ it follows that $|x|\leq a/\varepsilon$ and hence $|y|\leq ca$, so that the intersection K_a of (8.4) is bounded, hence compact. Thus again theorem 5.1 applies. The assertion about the support of $u*v$ then is readily confirmed, using (8.5), q.e.d.

It is convenient to use distribution integrals and write

$$(8.6)\qquad \langle u*v,\phi\rangle = (2\pi)^{-n/2}\int dxdy\, u(x)v(y)\phi(x+y) \quad .$$

One will find this notation quite suggestive and convenient. In particular one may avoid explicit definition of ψ above. Also the various considerations concerning supports become intuitively evident.

Lemma 8.4. If u and v have compact support then

$$(8.7)\qquad (u*v)^\wedge = u^\wedge\cdot v^\wedge \quad ,$$

with the distribution product at right, so that the two definitions of $u*v$ by (8.1) and by (4.22) agree.

Proof. For an intuition use distribution integrals. $u*v$ has compact support, by corollary 8.2, hence $(u*v)^\wedge(x) = (2\pi)^{-n/2}\langle u*v,e^{-ix\cdot}\rangle$, by (5.4). Accordingly ,

$$(u*v)^{\wedge}(\xi) = (2\pi)^{-n}\int u(x)v(y)\, e^{-i(x+y)\xi}dxdy$$

(8.8)

$$= \int u(x)e^{-ix\xi}dx \int v(y)e^{-iy\xi}dy = \langle u,e^{-i\xi\cdot}\rangle\langle v,e^{-i\xi\cdot}\rangle = u^{\wedge}(\xi)v^{\wedge}(\xi) \quad .$$

The conclusion is 'intuitive', insofar as Fubini's theorem was used for distribution integrals. However one readily confirms that $\langle u\otimes v, e_\xi \otimes e_\xi\rangle = \langle u,e_\xi\rangle\langle v,e_\xi\rangle$ using the definition (7.2) of the tensor product, which gives the corresponding for a product $\phi^1 \otimes \phi^2$ of functions in $\mathcal{D}$, not of functions in $\mathcal{E}$. Q.E.D..

Corollary 8.5. Formula (8.7) remains true for $u\varepsilon\mathcal{E}'$, $v\varepsilon\mathcal{S}'$.

This corollary is a consequence of lemma 6.4 and the continuity of the Fourier transform in weak topology of $\mathcal{S}'$. Also we have to use continuity of the distribution product (theorem 4.4.).

Lemma 8.6. If $u\varepsilon\mathcal{E}$ and $v\varepsilon\mathcal{E}'$ then $u*v\varepsilon\mathcal{E}$; if $u\varepsilon\mathcal{D}'$ and $v\varepsilon\mathcal{D}$ then again $u*v\varepsilon\mathcal{E}$. In either case we have (in terms of distribution integrals)

$$(8.9)\qquad (2\pi)^{n/2}(u*v)(x) = \int dy u(x-y)v(y)dy = \int dy u(y)v(x-y) .$$

Or, in other terms, the value of the C^∞-function $u*v$ at x is given as the value of the distribution u at the testing function $\phi(y) = v(x-y)$ (which is in $\mathcal{E}$ or $\mathcal{D}$, just as needed).

Proof. Suppose we have $u\varepsilon\mathcal{D}'$ and $v\varepsilon\mathcal{D}$, for example. Use (7.21) and (7.22) for

$$(2\pi)^{n/2}\langle u*v,\phi\rangle = \langle u,\langle v,\phi\rangle\rangle = \langle u,\int v(y)\phi(x+y)dy\rangle$$

(8.10)

$$= \langle u,\int v(x-y)\phi(y)dy\rangle = \int dy\langle u,v_y\rangle\phi(y)$$

with the translated testing function $v_y(x) = v(x-y)$.

The integrals in (8.10) exist in the Lebesgue sense, thus the variables substitution used is legitimate. Also for the last step in (8.10) we are using the fact that the integral $\int v(x-y)\phi(y)dy$ converges in the sense of $\mathcal{D}$. That is, a sequence of Riemann sums

$$(8.11)\qquad s(x) = \sum_{j=1}^{N} v(x-y^j)\,\phi(y^j)m(K_j)$$

with the Lebesgue measure $m(K_j)$ of $K_j \subset \mathbb{R}_n$ and suitable sets $K_j \ni x^j$ may be picked such that s-$\int v(x-y)\phi(y)dy \to 0$ in $\mathcal{D}$. This is easily confirmed, in a similar (but easier) way as in the proof of lemma 7.1.

Note that (8.10), in distribution integral terms, amounts to (8.9). It is left to show that the function $\langle u, v_x\rangle$ is infinitely differentiable in x. This is an immediate consequence of the fact that the family $\{v_x\}$ of testing functions (with parameter x) is continuous from $\mathbb{R}^n$ to $\mathcal{D}$ [just check that $x^k \to x$ implies $v_{x^k} - v_x \to 0$ in $\mathcal{D}$].

Finally, the other half of the lemma, referring to E and E' is proven analogously, and is left to the reader. Q.E.D..

Let us offer a version of lemma 8.6 for S and S' as well.

Lemma 8.7. Let $u \in S'$, $v \in S$, then $u * v \in E$ and formula 8.9 holds as well [where now the convolution product is defined by (4.22).].

Proof. We get

$$(8.12) \quad \langle u*v, \phi\rangle = \langle (u^\wedge v^\wedge)^\vee, \phi\rangle = \langle u^\wedge, v^\wedge \phi^\vee\rangle = \langle u, (v^\wedge \phi^\vee)^\wedge\rangle \quad .$$

Now we use that $f^\wedge = f^\vee_-$ $\quad f^\vee = f^\wedge_-$, with the function $f_-(x) = f(-x)$, as follows by a substitution in the Fourier integral. Thus we may continue (8.12) ,

$$= \langle u, (v^\wedge \phi_-^\wedge)^\vee_-\rangle = \langle u, (v * \phi_-)_-\rangle$$

$$= (2\pi)^{-n/2} \langle u, \int v(y)\,\phi(x+y)dy\rangle$$

$$= (2\pi)^{-n/2} \langle u, \int v(y-x)\,\phi(y)dy\rangle$$

$$= (2\pi)^{-n/2} \int \langle u, v^y\rangle \phi(y)dy \quad , \qquad \phi \in S \ ,$$

with $v^y(x) = v(y-x)$ defining a function in S. Again it was used that the Lebesgue integral exists in the sense of convergence in S, as may be easily verified [but will be left as an exercise]. This, and the remark that the family v^y , $y \in \mathbb{R}^n$ is continuous in S together with all its derivatives for x completes the proof.

Finally, let us still come back to convolution operators K_a as introduced in (3.6) and to formal Fourier multipliers a(D). Previously we discussed the relation

$a(D) = K_{a^\vee}$, $a \in T$ (c.f. (4.23)).

Let us here still observe that lemma 8.7 gives the representation

(8.13) $$K_u v(x) = (2\pi)^{-n/2} \int u(x-y)v(y)dy \quad , \qquad v \in S \quad ,$$

of the convolution operator K_u: $S \to E$ defined for $u \in S'$ by

(8.14) $$K_u v = u * v \qquad , \qquad v \in S \quad .$$

Here the integral is a distribution integral.

We shall speak of the <u>distribution kernel</u> $k(x,y) = u(x-y)$, in this connection, where k here, in this special case, denotes a certain family of distributions in S'.

The more general concept of distribution kernel of a linear operator will be introduced and discussed in section 9, below.

<u>Lemma 8.7.</u> Let H and H_1' be defined by

(8.15) $$H = \{u: u \in E \ , \ u=0 \text{ in } x \cdot H \le 0\} \quad ,$$

(8.16) $$H_1' = \{u: u \in D', \ u=0 \text{ in } x \cdot H \le \varepsilon|x| \ , \text{ for some } \varepsilon > 0\} \quad ,$$

with a non-vanishing vector $H \in \mathbb{R}^n$ again, then we get $u*v=w \in H$ for $u \in H$ and $v \in H_1'$.

Moreover supp w is contained in a convex cone $x \cdot H \le \varepsilon'|x|$.

<u>Lemma 8.8.</u> We have the commutative and associative law valid for convolution products as defined in this section as long as the support condition of corollary 8.3 is satisfied for all convolution products occuring in such a formula.

Lemma 8.7 is proven similar to lemmas 8.6 and 8.7. The proof of lemma 8.8 is left to the reader.

<u>Problems:</u> 1) Consider distributions in $D'(\mathbb{R})$ again, with support in $\mathbb{R}^+ = [0,\infty)$. Such a distribution u is said to be of <u>exponential growth</u>, if $e^{-cx}u(x) \in S'$ for some $c \in \mathbb{R}$. For any such u show that $(e^{-\gamma x}u(x))^\wedge = v_\gamma \in S'$, $\gamma > c$ defines a $C^\infty(\mathbb{R})$-function, explicitly given by the formula

(8.17) $$v_\gamma(\xi) = (2\pi)^{-1/2} \langle e^{-cx}u, e^{-(\gamma-c)x-ix\xi}\rangle \quad , \ \xi \in \mathbb{R} \ .$$

Moreover show that the function $w(\xi)=w(\xi+i\eta)=v_{-\eta}(\xi)$ is an analytic function of ξ in $\mathrm{Im}\,\xi<-c$. The function $w(\xi)$ then is called the <u>Laplace transform</u> of u.

2) Calculate explicitly the Laplace transforms of the following functions.

a) e^{px}, $p \in \mathbb{C}$; b) x^m ; c) sin px ; d) cos px ; e) sin px/x (where in each case the function is cut off and replaced by zero for negative x). 3) For two functions u,f, as in problem 1) the convolution product $u*f$ is defined by corollary 8.3. Show that $g = u*f$ again is of exponential growth, and that its Laplace transform equals the product of the Laplace transforms of u and f. 4) Consider distributions in

$\mathcal{D}'(\mathbb{R}^n)$ with support in the half space $\mathbb{R}^n_+ = \{x_n \geq 0\}$. Such distributions are called of exponential growth (in the x_n-variable), if $e^{-cx_n}u(x) \in \mathcal{S}'$. Suppose in addition that $u \in \mathcal{H}_1'$ as defined in (8.16) with the vector $H=(0,\ldots,0,1)$. Show that the Fourier-Laplace transform w of u, defined by

(8.18) $$w(\zeta_1,\ldots,\zeta_n) = (2\pi)^{-n/2} \langle e^{-cx_n}u,\ e^{cx_n - ix\zeta} \rangle$$

is well defined and holomorphic in $\zeta_1,\ldots,\zeta_n$, as $\mathrm{Im}\ \zeta_n > -c$. Similar if u has its support in a half cylinder $x_n \geq 0$, $x_1^2+\ldots+x_{n-1}^2 \leq c_0^2$, with some $c_0 > 0$. 5) Calculate the Fourier-Laplace transform of a convolution product $u * v$ with distributions as in lemma 8.7, assuming that in addition u has support in $x_1^2+\ldots+x_{n-1}^2 \leq c_0^2$. Show that again formula (8.7) holds for the Fourier-Laplace transform instead of the Fourier transform, if u satisfies above conditions. 6) Show that $\mathcal{D}$ is dense in $\mathcal{D}'$, using lemma 6.4 and lemma 8.6, where $v \in \mathcal{D}$ runs through a sequence approximating the Dirac function δ_0.

9. On (Fourier) distribution kernels and kernel multiplication.

In this section we consider continuous linear operators $Q: S \to S'$. For such an operator we introduce the inverse Fourier transform $Q^{\vee} = FQF^{-1}$, which again is acting between S and S'.

Such an operator will be said to have an integral kernel $q(x,y)$ (a Fourier integral kernel $q^{\vee}(\xi,\eta)$) whenever we have

$$(9.1) \qquad \langle Qv,u\rangle = \int q(x,y)u(x)v(y)dxdy \ , \ u,v \in \mathcal{D}(\mathbb{R}^n) \ ,$$

where the right hand side constitutes a Lebesgue integral. In particular we assume that $q \in L^1_{loc}(\mathbb{R}^{2n})$. (Or, for a Fourier kernel, we require $Q^{\vee}$ to have an integral kernel, denoted by $q^{\vee}(\xi,\eta)$). If no confusion is possible we shortly will refer to kernels or Fourier kernels.

It is clear that a kernel exists only under special assumptions on the operator Q. On the other hand it may be proven that a distribution kernel (and a Fourier distribution kernel) will always exist. By this we mean that there will always exist a distribution $q \in \mathcal{D}'(\mathbb{R}^{2n})$, such that (1.1) is satisfied with a distribution integral at right. Moreover we have the result below, often referred to as kernels theorem, c.f. [12].

Theorem 9.1. For every continuous operator $Q: S \to S'$, where S' is taken with its strong topology, there exists a unique distribution $q \in S'(\mathbb{R}^{2n})$ (and a distribution $q^{\vee} \in S'(\mathbb{R}^{2n})$) such that

$$(9.2) \qquad \langle Qv,u\rangle = \langle q,u\otimes v\rangle \ , \ \langle Q^{\vee}v,u\rangle = \langle q^{\vee},u\otimes v\rangle \ , \ u,v \in S(\mathbb{R}^n).$$

Remark. It follows easily that (9.2) defines continuous operators $S \to S'$ for any given pair of distributions $q,q^{\vee} \in S'(\mathbb{R}^{2n})$. Accordingly theorem 9.1 establishes 1-1-correspondences $L(S,S') \leftrightarrow S'(\mathbb{R}^{2n})$, $S=S(\mathbb{R}^n)$, $S'=S'(\mathbb{R}^n)$. In fact these are homeomorphisms, with proper topologies (c.f. [12]).

It is clear that one will attempt a proof of the kernels theorem along the guide lines given by the existence proof for the tensor product of distributions. For $\phi=\phi(x,y) \in S(\mathbb{R}^n)\otimes S(\mathbb{R}^n) \subset S(\mathbb{R}^{2n})$ the functional $\langle q,\phi\rangle$ is at once defined, using (9.2) (Clearly it suffices to look at the existence of q only). One then must provide a continuous extension to $S(\mathbb{R}^{2n})$ of the functional constructed. First we note the Lemma, below.

Lemma 9.2. A linear map $Q: S \to S'$ is continuous if and only if for some integer $k \geq 0$ we have

$$(9.3) \qquad \langle Qv,u\rangle \leq c\,\|u\|_k\,\|v\|_k \quad , \ u,v \in S(\mathbb{R}^n) \ ,$$

where, for the moment, we have denoted (c.f. 6, problem 2)

$$(9.4)\qquad \|w\|_k = \|w\|_{k,2} = \{\sum_{|\alpha|,|\beta|\le k} \|x^\alpha w^{(\beta)}\|^2_{L^2(\mathbb{R}^n)}\}^{1/2} .$$

Proof. It must be verified that the norms $\|\cdot\|_k$ of (9.4) are a system of generating semi-norms for the locally convex topology of S. Also if $S_k = S_{k,2}$ denotes the completion of S under $\|\cdot\|_k$, then the adjoint Hilbert spaces $S'_{k,2}$ are contained in S', by natural identification, (similar as in III, (5.11)). Moreover we have S' as an inductive limit

$$(9.5)\qquad S' = \bigcup_k S'_{k,2}$$

and the topology of S' is the inductive limit topology (c.f. 6, problem 2). It follows that any continuous map Q: $S \to S'$ must map continuously into some S_k. This in turn means that we must have

$$\sup\{|\langle Qu,\psi\rangle| : \|\Psi\|_k \le 1\} \le c\|u\|_1 \quad , \quad u \in S \quad ,$$

with some k,l. Replacing k and l by the larger of the two we get (9.4) for every continuous Q: $S \to S'$. Vice versa it is trivial that any Q: $S \to S'$ satisfying (9.4) must be continuous. Q.E.D.

Continuing with the proof of theorem 9.1, for $\phi = \phi(x,y) \varepsilon S(\mathbb{R}^{2n})$ write

$$(9.6)\qquad \phi(x,y) = \phi_N(x,y)\lambda^N(y) \ , \ \phi_N \varepsilon S(\mathbb{R}^{2n}) \ , \ \lambda(y) = (1+y^2)^{-1/2} .$$

Then express ϕ_N by its Fourier integral, in the y-variables only. One gets

$$(9.7)\qquad \phi(x,y) = \int d\xi \psi_N(x,\xi)(e^{i\xi y}\lambda^N(y)) = \int d\xi \psi_N(x,\xi)\omega_N(y,\xi) ,$$

where the y-Fourier transform $\psi_N(x,\xi)$ of ϕ_N is in $S(\mathbb{R}^{2n})$, and the function $\omega_N(\cdot,\xi)$ is in $S_{k,2}$, as N>k+n, as easily checked. Accordingly one will attempt the definition

$$(9.8)\qquad \begin{aligned} \langle q,\phi\rangle &= \int d\xi \langle q, \psi_N(\cdot,\xi)\otimes\omega_N(\cdot,\xi)\rangle \\ &= \int d\xi \langle Q\psi_N(\cdot,\xi), \omega_N(\cdot,\xi)\rangle \end{aligned} .$$

First of all the integrand is meaningful for N>k+n, since the estimate (9.3), valid for large k, shows that Q can be extended continuously to an operator $S_{k,2} \to S'_{k,2}$. Second the integral is meaningful: We have

$$(9.9)\qquad \|\omega_N(\cdot,\xi)\|_k = O(|\xi|^k) ,$$

by a calculation. Also

$$(9.10)\qquad \|\psi_N(\cdot,\xi)\|_k = O((1+|\xi|)^{-1}) \ , \ 0,1,2,\dots ,$$

since $\psi_N \varepsilon S(\mathbb{R}^{2n})$. Third one verifies that the partial Fourier transform (with respect to y only) even takes $S_l(\mathbb{R}^{2n}) \to S_{l-2n}(\mathbb{R}^{2n})$, as l is sufficiently large. This may be used to show that the functional (9.8) even is meaningful for $\phi\varepsilon S(\mathbb{R}^{2n})$, and its boundedness over the Banach space $S_j(\mathbb{R}^{2n})$ is easily established, as j is sufficiently large. (Refine (9.10), using the norm $\|\phi\|_{j,2}$ as in (9.4), with 2n, for

(9.11) $\quad \|\psi_N(\cdot,\xi)\|_k \le c_{k,j,l} \|\phi\|_{j,2}(1+|\xi|)^{-1}, \; l=1,2,\ldots$

as is easily done). Fourth it is trivial to see that the functional (9.8) extends the functional trivially defined on $S(\mathbb{R}^n)\times S(\mathbb{R}^n)$, starting from (9.2). Thus we now have a continuous functional over some S_l, which also is a distribution, since the injection $S \to S_l$ is continuous.

Finally it must again be shown - similar as in section 7 that $\overset{o}{S} = S(\mathbb{R}^n)\otimes S(\mathbb{R}^n)$ is dense in $S(\mathbb{R}^{2n})$ which establishes uniqueness of q. Q.E.D.

Remark: We were tempted to offer the above discussion of the kernels theorem, which is recognized to be very short and relies on topological facts of the space S' not fully discussed in these notes. Evidently this is an interesting result, especially in the above general form. We also will apply the kernels theorem, but only in certain special cases where the above difficult proof will strongly simplify. For example we will use in chapter IV that operators $\mathfrak{H}_s \to \mathfrak{H}_t$, with the L^2-Sobolev spaces of chapter III, have a Fourier distribution kernel in $\mathcal{D}'(\mathbb{R}^{2n})$. Similarly in IV,8, where a kernel of operators $L^2_0(\mathbb{R}^n) \to L^2_{loc}(\mathbb{R}^n)$ is needed, again only in $\mathcal{D}'(\mathbb{R}^{2n})$ (c.f. problem 2, below).

We now raise the following problem:Observe that for any temperate function $a\varepsilon T(\mathbb{R}^{2n})$ the products aq and aq with the (Fourier)-distribution kernel of Q: $S \to S'$ again are temperate distributions over $\mathbb{R}^{2n}$. By our remark following theorem 9.1 we therefore must have operators P: $S \to S'$ and R: $S \to S'$ such that P has distribution kernel aq, and R has Fourier distribution kernel $aq^{\vee}$. We define P and R as the kernel product and Fourier kernel product of a and Q, respectively, and use the notation

(9.12) $\quad P = a \blacktriangledown Q \quad , \quad R = a \blacktriangle Q \quad .$

Certain simple properties of the kernel products are readily verified.

a) If $c(x,y)=a(x)b(y)$, where $a,b\varepsilon T(\mathbb{R}^n)$, then

(9.13) $\quad c \triangledown Q = a(M)Qb(M) \quad , \quad c \vartriangle Q = a(D)Qb(D) \quad ,$

with the multiplication operator a(M) and the formal Fourier multiplier b(D) as in (4.19).

b) If $c(x,y) = a(x) - a(y)$, with $a\varepsilon T(\mathbb{R}^n)$, then

$$c \triangledown Q = [a(M),Q] = a(M)Q-Qa(M)$$

(9.14)

$$c \blacktriangle Q = [a(D),Q] = a(D)Q-Qa(D) \quad .$$

c) Kernel multiplications commute and are associative. That is, if $a,b\varepsilon T(\mathbb{R}^{2n})$ and if $c = ab$, then we get

(9.15) $$a\triangledown(b\triangledown Q) = b\triangledown(a\triangledown Q) = c\triangledown Q \quad , \quad a\blacktriangle(b\blacktriangle Q) = b\blacktriangle(a\blacktriangle Q) = c\blacktriangle Q \ .$$

But now suppose we have $Q\colon X \to Y$ with spaces $S\subset X$, $S'\supset Y$, and topological imbeddings, so that also $Q\colon S \to S'$. Then we can set up conditions for a such that also $P\colon X \to Y$ (or that $R\colon X \to Y$). A partial answer is given in IV, Lemma 7.4 (c.f. also problem 2, below).

Problems: 1) Obtain the distribution and the Fourier distribution kernel of a differential operator $L = \sum_\alpha a_\alpha(M)D^\alpha$ with $a_\alpha \varepsilon T$. 2) Show that for any bounded operator $A\varepsilon L(\mathfrak{H})$, for the Hilbert space $\mathfrak{H} = L^2(\mathbb{R}^n)$ the kernel product and the Fourier kernel product $a_s \triangledown A$ and $a_s \blacktriangle A$ with the function $a_s(x,y) = (1+|x-y|^2)^{-s/2}$, $s>0$ are bounded operators of $\mathfrak{H}$ again, and that we have a constant c_s such that

(9.16) $$\|a_s\triangledown A\|_{\mathfrak{H}} \le c_s \|A\|_{\mathfrak{H}} \quad , \quad \|a_s\blacktriangle A\|_{\mathfrak{H}} \le c_s \|A\|_{\mathfrak{H}}$$

holds for all $A\varepsilon L(\mathfrak{H})$. 3) Under the assumptions of problem 2), if A even is compact then also $a_s\triangledown A$ and $a_s\blacktriangle A$ are compact.

10. References.

[1] N. Bourbaki, Espaces vectorielles topologiques; Hermann, Paris 1964.

[2] R.E. Edwards, Functional analysis; Holt Rinehart Winstòn, New York, 1965.

[3] I. Gelfand G. Silov; Generalized functions; Acad. Press 1966.

[4] J. Hadamard, Lectures on Cauchys Problem; Dover Classic 1952.

[5] P. Halmos, Measure theory, v. Nostrand, Princeton New York 1950.

[6] L. Hoermander, Linear partial differential equations; Springer New York Goettingen Heidelberg 1963.

[7] Laurent Schwartz, Thêorie des distributions; Paris Hermann 1966.

[8] L.H. Loomis, Abstract harmonic analysis; v. Nostrand New York 1953.

[9] W. Magnus, F. Oberhettinger, R.P. Soni, Formulas and Theorems for the special functions of mathematical physics; Springer Berlin Heidelberg New York 1966.

[10] C.E. Rickart, General theory of Banach algebras; v. Nostrand Princeton New York 1960.

[11] F. Riesz and B.Sz. Nagy, Functional analysis; Ungar Publ. Co. New York 1955.

[12] F. Treves, Topological vector spaces, distributions and kernels; Academic Press New York London 1967.

Chapter II: Distributions with Rational Singularities.

0. Introduction.

In the present chapter we discuss a class of distributions with finite singular support. These are of crucial importance for theory of constant coefficients partial differential equations, and, more generally, for pseudo-differential equations, asymptotic expansions, etc., (c.f. theorem 4.3, theorem 5.2 and 5, problem 2) . Often we assume only one singularity at the origin. For a related discussion the reader is referred to the book of Gelfand and Silov, Vol. I ([3], Chapter I, §3, §4).

The most prominent type will be the homogeneous distributions of (complex) degree τ, in $C^\infty(\mathbb{R}^{n*})$, $\mathbb{R}^{n*}=\mathbb{R}^n-\{0\}$, discussed in section 2. For a function $f\varepsilon C^\infty(\mathbb{R}^{n*})$, homogeneous of degree τ, we have as defining property that $f(\mu x) = \mu^\tau f(x)$, $x\varepsilon\mathbb{R}^n$, $\mu\varepsilon\mathbb{R}$. Correspondingly a natural definition of a homogeneous distribution f of degree τ is the property that

$$
\begin{aligned}
\int f(x)\phi(\mu x)dx &= |\mu|^{-n}\int f(x/\mu)\phi(x)dx \\
&= |\mu|^{-n-\tau}\int f(x)\phi(x)dx, \quad \phi\;\varepsilon\; C_0^\infty(\mathbb{R}^n) = \mathcal{D}\;,\; \mu\;\varepsilon\;\mathbb{R}\;,
\end{aligned}
\tag{0.1}
$$

expressed in terms of distribution integrals. Or with functional terminology we get

$$
\langle f,\phi_\mu\rangle = |\mu|^{-n-\tau}\langle f,\phi\rangle,\quad \phi_\mu(x) = \phi(\mu x)\;,\;\phi\;\varepsilon\;\mathcal{D}\;. \tag{0.1'}
$$

The spaces of all homogeneous $C^\infty(\mathbb{R}^{n*})$-functions (homogeneous distributions with singular support at 0 only) of degree τ will be denoted by H_τ and H'_τ, respectively. One finds that $H'_\tau\subset S'$ so that Fourier transforms of $u\varepsilon H'_\tau$ are well defined.

It is easily seen that 0 must belong to the singular support of $f\varepsilon H'_\tau$, unless f is a form - i.e., a homogeneous polynomial in x, of some integer degree $\tau=N\geq 0$. This distinction of nonnegative integer degrees is parallelled by another distinction of the integers $\tau=-n-l$, $l=0,1,\ldots$. As $\tau>-n$ a function $f\varepsilon H_\tau$ is necessarily in L^1_{pol}, and a distribution in $H'_\tau\subset S'$. Vice versa, for a distribution $u\varepsilon H'$ of any degree τ the restriction to $\mathbb{R}^{n*}$ defines a function in H_τ. This gives a 1-1 correspondence between H_τ and H'_τ, as $\tau>-n$. It will be seen, that there is a similar 1-1-correspondence also for $\tau\leq -n$, except if $\tau=-n-l$, $l=0,1,2,\ldots$ (Lemma 2.2).

On the other hand, if τ is an integer $\leq -n$, then there exist distributions in H'_τ with support at 0 only, like the δ-function, for example, which is in H'_{-n}. In addition, for these degrees certain functions of H_τ cannot be obtained as restrictions of some distribution in H'_τ. For example there is no distribution in H'_{-n} with restriction to $\mathbb{R}^n$ equal to the function $|x|^{-n}$. In other words, for these integer degrees $\tau\leq -n$ the restriction map $H'_\tau \to H_\tau$ is neither injective nor surjective, although it turns out to be Fredholm, of index zero, as we shall see (c.f. Lemma 2.3).

We shall see that the Fourier transform establishes a surjection $H'_\tau \to H'_{-n-\tau}$, for every $\tau\in\mathbb{R}$. (Lemma 4.7.c). In section 5 we also define the Fourier transform of a general homogeneous function in H_τ, even as $\tau\leq -n$. One obtains a map $F: H_\tau \to H''_{-n-\tau}$, where $H''_\sigma = H_\sigma$, except as $\sigma=0,1,\ldots$ or as $\sigma= -n,-n-1,\ldots$. If $\sigma=N\geq 0$ then H''_σ in addition contains certain logarithmic terms. As $\sigma=N\leq -n$, H''_σ is a proper subset of H_σ (c.f. (3.3). Also, Seeley, [7], where the term pseudo homogeneous is used for the elements of H''_σ).

For the purpose of Fourier transform a distribution $f\varepsilon H'_\tau$ has two "singularities", one at 0, the other at infinity. It is natural to separate these by first considering $(\chi f)^\wedge$, and $(\omega f)^\wedge$, with a partition $1=\chi+\omega$, $\omega,\chi\varepsilon C^\infty(\mathbb{R}^n)$, $0\leq\omega,\chi\leq 1$, $\omega=0$, near 0, $\chi=0$ near infinity (for large $|x|$). This leads into the study of two classes called S^0_{ps} and M (Definition 4.1). First $u\varepsilon S^0_{ps}$ equals some function in S, for large $|x|$. We have sing supp $u \subset \{0\}$, and the origin is a <u>rational singularity</u>: We have $x^\alpha u\varepsilon C^k(\mathbb{R}^n)$, as $|\alpha|\geq N(k)$, $k=0,1,\ldots$ (c.f. Definition 1.1.). Second, M consists of the $C^\infty(\mathbb{R}^n)$-functions u with the property that $u^{(\alpha)}(x)=O(|x|^{-k})$ as $|x|\to\infty$, $|\alpha|\geq N(k)$, for $k=0,1,\ldots$.

It turns out that the Fourier transform establishes a bijection $S^0_{ps} \to M$ (Theorem 4.3). Regarding the homogeneous distributions, F therefore simply interchanges the two singularities; we get $(\omega f)\varepsilon M$, $(\chi f)\varepsilon S^0_{ps}$, but $(\omega f)^\wedge\varepsilon S^0_{ps}$, $(\chi f)^\wedge\varepsilon M$.

For $u\varepsilon S^0_{ps}$ the restriction to $\mathbb{R}^{n*}$ defines a <u>function with a rational singularity at 0</u>. The class of functions obtained is called S^0_{ra} (Definition 1.1). Any function u with only rational singularities can be "extended" into a distribution. Such extension is called regularization by Gelfand and Silov; it is simply a matter of extending a linear functional defined over a space with finite codimension to the entire space S (Theorem 1.3). On the other hand there are standard extension procedures of classical analysis, which supply distinguished such extensions, like the <u>Cauchy principal value</u> or the <u>Hadamard finite part</u> p.f.u (c.f. (2.5) and (2.11)). In particular they will supply a homogeneous distribution for a homogeneous function if one exists. Other extensions are important in hyperbolic theory (c.f. 5, problem 3)).

We will consider two kinds of asymptotic expansions. First a distribution (or function) can have an expansion mod D near a rational singularity (section 3). Second, a function in M may have an asymptotic expansion mod $|x|$, at infinity. (Definition 5.2). The asymptotic expansions will have homogeneous distributions or pseudo-homogeneous functions (in the sense of Seeley) as terms. One obtains two classes $S^{0}_{ad} \subset S^{0}_{ps}$ and $M_{as} \subset M$ of distributions (functions) with such asymptotic expansions (at 0 (mod D)or at ∞, (mod $|x|$)), respectively. Again we show that the Fourier transform F provides a bijection F: $S^{0}_{ad} \to M_{as}$ (Theorem 5.2). This has an immediate application for construction of a parametrix of an N-th order elliptic operator (c.f. 5, problem 2)). Similarly it can be used, jointly with the Fourier-Laplace transform of I,8, problems 1) through 5), to solve the hyperbolic Cauchy problem .

1. Rational points and pseudo-functions.

Definition 1.1. Let u denote either a distribution, or a function, not necessarily in L^1_{pol}, but defined and measurable in $\mathbb{R}^n$, with complex values. We shall say that u has a rational point at $z \in \mathbb{R}^n$, if (i) $u \in C^\infty(0<|x-z|<\varepsilon)$ with some $\varepsilon>0$; (ii) for every $k=0,1,2,\ldots$, there exists an integer $N(k)$ such that $(x-z)^\alpha u \in C^k(|x-z|<\varepsilon)$, for all $|\alpha| \geq N(k)$. We define T_{ps} and T_{ra} as the spaces of all distributions and functions, respectively, such that (i) the singular support is finite , (ii) all points of the singular support are rational points; (iii) u equals a function in $T \subset S'$ outside a neighbourhood of its singular support.

It is clear that the function space T_{ra} and the distribution space T_{ps} both extend the space T of temperate functions.

The (translated) Dirac measure δ_z for $z \in \mathbb{R}^n$, is defined as the distribution

$$(1.1) \qquad \langle \delta_z, \phi \rangle = (2\pi)^{n/2} \phi(z) \quad , \quad \phi \in S .$$

Its derivatives $\delta_z^{(\alpha)}$ are given by

$$(1.2) \qquad \langle \delta_z^{(\alpha)}, \phi \rangle = (2\pi)^{n/2} (-1)^{|\alpha|} \phi^{(\alpha)}(z) \quad , \quad \phi \in S ,$$

Note that

$(x-z)^\beta \delta_z^{(\alpha)} = 0$, $|\beta| > |\alpha|$. Accordingly $\delta_z^{(\alpha)} \in T_{ps}$, for every z and α, and, moreover, the distributions

$$(1.3) \qquad w = \sum_{j=1}^{r} P_j(D)\delta_{z^j} \quad , \quad P_j(D) = \sum_{|\alpha| \leq N_j} p_{\alpha,j} D^\alpha , p_{\alpha,j} \in \mathbb{C} ,$$

form a linear subspace of T_{ps}, which is characterized by the fact that its distributions have a finite support, not only a finite singular support. We recall the result, below, first mentioned by L. Schwartz, [6].

Theorem 1.2. A distribution u with finite support is necessarily of the form (1.3).

Proof. Let $\operatorname{supp} u \subset \{0\}$, and let

$$(1.4) \qquad \begin{aligned} &\lambda \in C_0^\infty(\mathbb{R}^n), \ \lambda = 1 \text{ in } |x| \leq 1/2, \ \lambda = 0 \text{ in } |x| \geq 1 , \\ &\lambda_\tau(x) = \lambda(\tau x) . \end{aligned}$$

We get

$$\langle u, \phi \rangle = \langle u, \lambda_\tau \phi \rangle \ , \ \phi \in S \ , \ \tau > 0 .$$

Applying theorem 6.3 we get an integer N such that

$$(1.5) \qquad |\langle u, \phi \rangle| \leq c \sum_{|\alpha| \leq N} \| (\lambda_\tau \phi)^{(\alpha)} \|_C \ , \ \phi \in S \ , \ \tau \geq 1 ,$$

since $\lambda_\pi(x)=0$ for $|x|\geq 1/\tau$, so that $|x^\alpha|\leq 1$ in supp λ_τ.
Consider the space

(1.6) $$S_{z,M} = \{ \phi\varepsilon S : \phi^{(\alpha)}(z) = 0 , |\alpha| \leq M \} .$$

For $\phi\varepsilon S_{0,N}$ we get

(1.7) $$\phi^{(\alpha-\beta)}(x)\lambda_\tau^{(\beta)}(x) = O(\tau^{-N-1+|\alpha|}) ,$$

because of

(1.8) $$\phi^{(\gamma)}(x) = O(|x|^{N+1-|\gamma|}) ,$$

by Taylor's formula, and since $\lambda_\tau^{(\beta)}(x) = \tau^{|\beta|}\lambda^{(\beta)}(\tau x)$ has its support in $|x|\leq 1/\tau$.
We apply Leibnitz' formula (1.23) and get

(1.9) $$\| (\lambda_\tau\phi)^{(\alpha)} \|_C \leq \alpha! \sum_{\beta\leq\alpha} \| \lambda_\tau^{(\beta)}\phi^{(\alpha-\beta)} \|_C \leq c_\alpha \tau^{-N-1+|\alpha|} .$$

Then (5.4) implies $|<u,\phi>| \leq c/\tau \to 0$, $\tau\to\infty$. Or, we conclude that $<u,\phi> = 0$ for $\phi\varepsilon S_{0,N}$. For $\phi\varepsilon S$ we have the unique decomposition

(1.10) $$\phi(x) = \sum_{|\alpha|\leq N} x^\alpha\lambda(x)\phi^{(\alpha)}(0)/\alpha!+\phi_N(x) , \phi_N \varepsilon S_{0,N}.$$

Substituting (1.10) we get

(1.11) $$<u,\phi> = \sum_{|\alpha|\leq N} (i)^{|\alpha|}c_\alpha\phi^{(\alpha)}(0) , c_\alpha = (-i)^{|\alpha|}/\alpha!<u,x^\alpha\lambda> ,$$

hence

$$u = P(D)\delta , P(D) = \sum_{|\alpha|\leq N} c_\alpha D^\alpha .$$

If supp $u\subset\{z\}$, we similarly get $u=P(D)\delta_z$; if in general supp $u = \{z^1,\ldots,z^r\}$, with distinct points z^j, then let

$$u = \sum_{j=1}^{r} u_j , u_j = \lambda_\tau^{z^j} u ,$$

(1.12)
$$\lambda_\tau^z(x) = \lambda_\tau(x-z) , j=1,\ldots,r ,$$

with τ large enough to insure that the balls $|x-z^j|\leq 1/\tau$ are disjoint. It follows that supp $u_j \subset\{z^j\}$, hence $u_j=P_j(D)\delta_{z^j}$, q.e.d.

In the intention to relate the two spaces T_{ps} and T_{ra}, we note that for $u\varepsilon T_{ps}$ there is a unique function $v\varepsilon T_{ra}$, such that $u=v$ in $\mathbb{R}^n$ - sing supp u. Indeed we first get a unique $v\varepsilon C^\infty(\mathbb{R}^n$ - sing supp u) such that

(1.13) $$<u,\phi> = \int v\phi dx , \phi\varepsilon C_0^\infty(\mathbb{R}^n - \text{sing supp } u) .$$

For z ε sing supp u we have $(x-z)^\alpha u\varepsilon C^k(|x-z|<\varepsilon)$. That is $<u,(x-z)^\alpha\phi> = \int w_\alpha\phi dx$, whenever $\phi=0$ outside $|x-z|<\varepsilon$, with a function $w_\alpha\varepsilon C^k(|x-z|<\varepsilon)$, for all $|\alpha|\geq N_j(k)$.

Accordingly, $\int(x-z)^{\alpha}v\phi dx = \int w_{\alpha}\phi dx$, or, $(x-z)^{\alpha}v = w_{\alpha}\varepsilon C^{k}(|x-z|<\varepsilon)$, $|\alpha|\geq N_j(k)$. Hence all points of sing supp v are rational points, and we get $v\varepsilon T_{ra}$, sing supp u $\supset$ sing supp v.

We shall refer to the function $v\varepsilon T_{ra}$ as of the restriction of the distribution u to its C^{∞}-domain.

Vice versa, for any finite subset $\Gamma=\{z^1,\dots,z^r\}\subset\mathbb{R}^n$ and corresponding r-multi-index $\nu=(\nu_1,\dots,\nu_r)\varepsilon\mathbb{Z}_+^r$ let us define

$$S_{\Gamma,\nu} = \bigcap_{j=1}^{r} S_{z^j,\nu_j}$$

(1.14)

$$= \{\phi|\phi\varepsilon S\ ,\ \phi^{(\alpha)}(z_j)=0\ ,\ |\alpha|\leq\nu_j\}\quad .$$

Then, if $v\varepsilon T_{ra}$ is given, with sing supp $v = \Gamma$, and if ν_j is chosen such that

(1.15) $\quad N_j(0)\leq n+\nu_j\quad .$

We conclude that the integral in (1.16), below, exists absolutely and that for $\phi\varepsilon S_{\Gamma,\nu}$ we get

(1.16) $$\int_{\mathbb{R}^n}|v\phi|dx \leq c \sum_{|\gamma|\leq M,\ |\beta|\leq \operatorname{Max} N_j(0)} \|x^{\gamma}\phi^{(\beta)}\|_C \quad ,$$

with suitably large M, depending on the growth of v, for large x. Accordingly, a continuous linear functional

(1.17) $$\langle v,\phi\rangle = \int_{\mathbb{R}^n} v\phi dx \quad ,\quad \phi\ \varepsilon\ S_{\Gamma,\nu}\quad ,$$

is defined, in the relative topology of S.

The subspace $S_{\Gamma,\nu}\subset S$ has finite codimension, and we get the unique decomposition

(1.18) $$\phi(x) = \sum_{j=1}^{r}\sum_{|\alpha|\leq\nu_j}(x-z^j)^{\alpha}\lambda_j(x)\phi^{(\alpha)}(z^j)/\alpha!+\phi_{\Gamma,\nu}\quad ,$$

with $\phi_{\Gamma,\nu}\ \varepsilon\ S_{\Gamma,\nu}$, for every $\phi\varepsilon S$, where $\lambda_j = \lambda_{\tau}^{z^j}$, as in (5.12), with large τ, are kept fixed. For arbitrary complex numbers $c_{\alpha,j}$, $|\alpha|\leq\nu_j$, $j=1,\dots,r$, a distribution v^{ψ}, extending the functional (1.17), is defined by

(1.19) $$\langle v^{\psi},\phi\rangle = \sum_{j=1}^{r}\sum_{|\alpha|\leq\nu_j} c_{\alpha,j}\phi^{(\alpha)}(z^j) + \langle v,\phi_{\Gamma,\nu}\rangle \quad .$$

with $\phi_{\Gamma,\nu}$ uniquely determined by (1.18).

Any such distribution v^{ψ}, with $S_{\Gamma,\nu}$, λ_j as discussed, will be called a pseudo-function (associated to v). The term "pseudo-function" was used before (c.f., for example, [1], [6]) with slightly different meaning.

There are infinitely many pseudo-functions v^{ψ} associated to a given function $v\varepsilon T_{ra}$, and all of them are of the form

(1.20) $$v^{\psi} = v^{\psi'} + w$$

with any particular pseudo-function $v^{\psi'}$ to v, and an arbitrary distribution w of finite support (i.e., of the form (1.3)). Indeed, $v^{\psi} - v^{\psi'}$, for two pseudo-functions v^{ψ} and $v^{\psi'}$ to the same v has its support in sing supp v, hence must be of the form (1.3), by theorem 1.2.

We have proven the result, below.

<u>Theorem 1.3.</u> The space T_{ps} consist precisely of all pseudo-functions associated to the functions of T_{ra}.

<u>Theorem 1.4.</u> The spaces T_{ps} and T_{ra} are invariant under differentiation and multiplication by $a\varepsilon T$. Moreover, if $v\varepsilon T_{ra}$ and $v^{\psi}\varepsilon T_{ps}$ are associated, then so are $v^{(\alpha)}$ and $(v^{\psi})^{(\alpha)}$, and av with $a\cdot v^{\psi}$ associated, respectively.

The proof of theorem 1.4 is a matter of Leibnitz' formula (1.23), which ensures that (i) $x^{\alpha}av$ and $x^{\alpha}v^{(\beta)}$ may be expressed by $x^{\alpha}v$ and $D^{\gamma}(x^{\delta}v)$, and that (ii) $a\cdot$ and D^{α} map $S_{\Gamma',\nu'}$ into $S_{\Gamma,\nu}$ for suitable Γ',ν'.

<u>Problems:</u> 1) The function $f(x) = 1/x$, $0\neq x\varepsilon\mathbb{R}$, has the rational point x=0 and is in T_{ra}. Show that the distribution f^{0} defined by $\langle f^{0},\phi\rangle = \lim_{\varepsilon\to 0,\varepsilon>0}\int_{|x|\geq\varepsilon} f(x)\phi(x)dx =$ $\langle p.v.f,\phi\rangle$, $\phi\varepsilon S$ is a pseudo-function to f, and obtain the family of all pseudo-functions to f in terms of f^{0}. 2) For f as in problem 1) show that two more pseudo-functions to f are defined by $\langle f^{\pm},\phi\rangle = \lim_{\varepsilon\to 0,\varepsilon>0}\int_{\mathbb{R}}\phi(x)dx/(x\pm i\varepsilon)$, $\phi\varepsilon S$. Also show that $f^{0}\neq f^{+}\neq f^{-}\neq f^{0}$. 3) With f as in problem 1) and 2) show that an <u>unsymmetric principal part</u>, like for example $\langle f^{1},\phi\rangle = \lim_{\varepsilon\to 0,\varepsilon>0}\int_{x\notin(-\varepsilon,2\varepsilon)}\phi(x)dx/x$ again defines a pseudo function to f, and obtain it explicitly in terms of f^{0}. 4) With f as in 1) show that f^{0},f^{1},f^{+},f^{-} of 1), 2), 3) all are homogeneous distributions of degree -1, as defined in the introduction. 5) Consider the function $g(x)=|x|^{-n}$, $0\neq x\varepsilon\mathbb{R}^{n}$. Show that $g\varepsilon T_{ra}$, and that it is homogeneous of degree -n, but that the principal value $\langle g^{0},\phi\rangle = \lim_{\varepsilon\to 0,\varepsilon>0}\int\phi(x)dx/|x|^{n}$ does not even define a distribution, hence certainly not a pseudo-function to g. Similarly for $\langle g^{\pm},\phi\rangle = \lim_{\varepsilon\to 0,\varepsilon>0}\int\phi(x)dx/(|x|^{n}\pm i\varepsilon)$.
6) Define pseudo-functions to the functions $h_j(x) = x_j/|x|^{n+1}$, j=1,...,n,

$x=(x_1,\dots,x_n)\in\mathbb{R}^n$. Use the above problems as guide to construct 3 distinct pseudo-functions to h_j, which all are homogeneous distributions. 7) With h_j as in 6) show that, if $\rho(x)$ is any norm on $\mathbb{R}^n$, then $\langle h_j^\rho,\phi\rangle = \lim_{\varepsilon\to 0,\varepsilon>0}\int_{\rho(x)\geq\varepsilon} h_j(x)\phi(x)dx$ defines a pseudo-function to h_j. Show that in general for two different norms one will get different pseudo-functions. (Express h_j^ρ by $h_j^{\rho^0}$ for Euclidean norm ρ^0 and the delta function.) 8) Suppose a pseudo-function $a \in T_{ps}$ with sing supp $a\subset\{0\}$ equals some function in S for $|x|>1$. Show that then the convolution operator $K_a = a*$, defined by I,(3.6) (with I,Theorem 8.1) takes S to S. (Remark: this may be derived either directly, or as a consequence of Theorem 4.3, below). 9) Under the assumptions of problem 8, suppose $\phi,\psi\in\mathcal{D}$ have disjoint supports, and let d = distance{supp ϕ, supp ψ}. Show that for sufficiently large integers N, and $1\leq p\leq\infty$, $p^{-1}+q^{-1}=1$, there exists a constant $c_{N,p}$ such that for all $u,v\in S$ we get, with $(u,v)=\int\bar{u}v dx$,

$$(1.21)\qquad |(K_a(\phi u),(\psi v))| \leq c_{N,p}\, d^{-N}\, \|u\|_{L^p}\|v\|_{L^q} \quad .$$

10) Given a family $\{k(x,z),\ x\in\mathbb{R}^n\}$, of pseudo-functions, in the variable z, with the property that (i) $k(x,z)\in C^\infty(\mathbb{R}^n\times(\mathbb{R}^n-\{0\}))$; (ii) $D_x^\alpha D_z^\beta k(x,z) = O(|z|^{-j})$, as $x,z\in\mathbb{R}^n$, $|z|\geq p>0$, for all α,β, and all j,p, with $\langle z\rangle = (1+|z|^2)^{-\frac{1}{2}}$; (iii) $(D_x^\alpha D_z^\beta(z^\gamma k(x,z)))\in CB(\mathbb{R}^{2n})$, for all α, β, γ such that $|\gamma|\geq N_0(|\beta|)$. Show that

$$(1.22)\qquad (Ku)(x) = \int k(x,x-y)u(y)dy = \langle k(x,.),u(x-.)\rangle \quad ,x\in\mathbb{R}^n\ ,\ u\in S \quad ,$$

with a distribution integral, defined by the expression at right, defines a continuous operator from S to S. 11) Under the assumptions of problem 10 show that the operator K satisfies estimate (1.21) as well: For every (sufficiently large) integer N and every real p, $1\leq p\leq\infty$, these exists a constant $c_{N,p}$ such that (1.21) holds for K instead of K_a, where $c_{N,p}$ may depend on ϕ and ψ, and k, but not on u and v. (Hint for problems 9 and 11: this may be regarded as an application of Schur's Lemma (i.e. I, Lemma 1.1).)

2. Positive homogeneous functions and distributions; finite parts.

Let H_τ, for a given $\tau\epsilon\mathbb{C}$, denote the space of all $C^\infty(\mathbb{R}^n-\{0\})$-functions a, which are (positively) homogeneous of degree τ, that is

(2.1) $a(\mu x) = \mu^\tau a(x)$, $x\neq 0$, $0<\mu<\infty$.

Then $D^\beta x^\alpha a \in H_{\tau+|\alpha|-|\beta|}$, $a\epsilon H_\tau$, and all functions in H_τ, Re $\tau>0$ are in $C^0(|x|<1)$, since

(2.2) $a(x) = |x|^\tau a(x/|x|)$, $x \neq 0$,

implies $\lim_{x\to 0} a(x)=0$. It follows that x=0 is a rational point of every $a\epsilon H_\tau$ which does not vanish identically. Moreover, we get $H_\tau \subset T_{ra}$, with sing supp a $\subset\{0\}$, $a\epsilon H_\tau$, $a\not\equiv 0$.

A distribution $u\epsilon S'$ will be called homogeneous of degree τ, with $\tau\epsilon\mathbb{C}$, if

(2.3) $\langle u,\phi_\mu\rangle = \mu^{-n-\tau}\langle u,\phi\rangle$, $\phi\epsilon S$, $\phi_\mu(x)=\phi(\mu x)$, $0<\mu<\infty$.

We denote by H'_τ the linear space of all homogeneous distributions of degree τ, with singular support in the origin (or void).

It is clear that $H_\tau\subset L^1_{pol}$, whenever Re $\tau>-n$, and we then get $H_\tau = H'_\tau$, via the canonical imbedding I, (4.2). It is natural to ask, for Re $\tau\leq -n$, whether or not some of the pseudo-functions associated to an $a\epsilon H_\tau$, are homogeneous distributions. First we observe that $\langle\delta^{(\alpha)},\phi_\mu\rangle = (-1)^{|\alpha|}\langle\delta,\phi_\mu^{(\alpha)}\rangle = (-\mu)^{|\alpha|}\phi^{(\alpha)}(0) = \mu^{|\alpha|}\langle\delta^{(\alpha)},\phi\rangle$, so that a comparison with (2.3) yields $\delta^{(\alpha)}\epsilon H'_{-n-|\alpha|}$.

Since $H'_\tau\cap H'_{\tilde\tau} = \{0\}$, for all $\tau\neq\tilde\tau$, and since $\delta_z^{(\alpha)} \notin H'_\tau$, unless z=0, $\tau= -n-|\alpha|$, we conclude that a distribution w with finite support (of the form (2.3)) is in H'_τ if and only if

(2.4) $\tau = -n-l$, $w = P(D)\delta$, $P(D) = \sum_{|\alpha|=l} p_\alpha D^\alpha$, $\bar p_\alpha \in \mathbb{C}$;

for some integer l = 0,1,2,... . It follows that there is at most one homogeneous pseudo-function associated to a given $a\epsilon H_\tau$,$a \not\equiv 0$, whenever $\tau' = -n-\tau \neq 0,1,2,\ldots$.

There exists such unique $a^\psi\epsilon H'_\tau$, indeed, which will be denoted by "p.f.a", for every $a\epsilon H_\tau$,$\tau' \neq 0,1,2,\ldots$, and it is defined by

(2.5) $\text{p.f.a} = \Gamma(-\tau')/\Gamma(m-\tau') \sum_{|\alpha|=m} m!/\alpha!\,(x^\alpha a)^{(\alpha)}$,$a \in H_\tau$,

with Eulers gamma-function $\Gamma(t)$, and with any integer m satisfying Re $\tau+n+m>0$. We get $x^\alpha a\epsilon H_{\tau+m}\subset L^1_{pol}$, for $|\alpha|=m$, hence $(x^\alpha u)^{(\alpha)}$ is well defined as a distribution derivative, with $x^\alpha a\epsilon L^1_{pol}\subset S'$ interpreted as a distribution. Since $u\epsilon H'_\tau$ implies $(x^\alpha u)^{(\beta)} \in H_{\tau+|\alpha|-|\beta|}$, it follows that p.f.a, as defined by (2.5), is in H'_τ.

It remains to be shown that p.f.a is independent of m and that it is a pseudo-function to a. However,

$$(2.6)\qquad \sum_{|\alpha|=m} m!/\alpha!x^{\alpha}\phi^{(\alpha)}(x) = r^m\partial^m/\partial r^m\phi(rz)\Big|_{r=|x|,\ z=x/r}$$

so that

$$(2.7)\qquad \begin{aligned} &<\sum_{|\alpha|=m} m!/\alpha!(x^{\alpha}a)^{(\alpha)},\phi> = (-1)^m\int_0^{\infty} r^{\tau+n+m-1}\omega^{(m)}(r)dr,\\ &\omega^{(m)}(r) = d^m/dr^m\int_{|x|=1} a(x)\phi(rx)do_x \ , \end{aligned}$$

with the surface element do_x of the unit sphere $|x|=1$. If $\phi\varepsilon S_{0,m-1}$, as in (1.6), so that $a\phi\varepsilon L^1$, then an m-fold partial integration will carry the right hand side of (2.7) into

$$(2.8)\qquad \Gamma(m-\tau')/\Gamma(-\tau')\int_0^{\infty} r^{\tau+n-1}\omega(r)dr = \int_{\mathbb{R}^n} a\ \phi\ dx\ ,$$

so that indeed p.f.a = a^{ψ}. For similar reason p.f.a is independent of m. Also, p.f.a = a for Re $\tau+n>0$.

The above also motivates the notation <u>finite part of</u> a for the distribution p.f.a (c.f. Hadamard [4]).

<u>Lemma 2.1.</u> For $u\varepsilon H'_{\tau}$,and also for $u\varepsilon H_{\tau}$, we have the identity

$$(2.9)\qquad u = \Gamma(-\tau')/\Gamma(m-\tau')\sum_{|\alpha|=m} m!/\alpha!(x^{\alpha}u)^{(\alpha)}\ ,\quad m=0,1,2,\dots\ ,$$

referred to as the (iterated)(adjoint) <u>Euler formula</u>, whenever $\tau'=-n-\tau \neq 0,1,2,\dots$. Here "(α)" denotes the distribution and function derivative, for $u\varepsilon H'_{\tau}$, and for $u\varepsilon H_{\tau}$, respectively.

<u>Proof.</u> It suffices to prove the distribution identity. For the τ, as specified, we have u=p.f.a, with some $a\varepsilon H_{\tau}$. Then (2.9) follows from substitution of (2.5) into (2.8), using (2.6) and (2.7), and from integration by parts.

The lemma, below, is a consequence of the uniqueness of p.f.a as a homogeneous pseudo-function to a.

<u>Lemma 2.2.</u> If $\tau'=-n-\tau \neq 0,1,2,\dots$, then a → p.f.a defines an isomorphism between the linear spaces H_{τ} and H'_{τ}, and we have

$$(2.10)\qquad \text{p.f.}x^{\alpha}a^{(\beta)} = x^{\alpha}(\text{p.f.a})^{(\beta)}\ ,\quad \alpha,\beta\varepsilon\mathbb{Z}^n_+\ .$$

Next, let us investigate the left-over cases $\tau = -n-l$, $l = 0,1,2,\dots,$. Then we define

$$(2.11)\qquad \text{p.f.a} = (-1)^l\sum_{|\alpha|=l+1}(l+1)/\alpha!(x^{\alpha}\log|x|\ a)^{(\alpha)}\ ,\quad a\varepsilon H_{-n-l}\ .$$

This defines a pseudo-function to a again and we get

$$(2.12)\qquad \langle p.f.a,\phi\rangle = -1/l!\int_0^\infty \log r\ \omega^{(l+1)}(r)\,dr\ ,$$

with notation and derivation, as for (2.7).

This distribution needs not to be in H'_{-n-1}, however. In fact, none of the pseudo-functions to a is in H'_{-n-1}, unless we have

$$(2.13)\qquad q_\alpha = \int_{|x|=1} x^\alpha a\, do_x = 0\ ,\ |\alpha| = 1\ .$$

Indeed, if and only if (2.13) holds, we get

$$(2.14)\qquad \begin{aligned} \langle p.f.a,\phi_\mu\rangle - \mu^l\langle p.f.a,\phi\rangle &= (\mu^l \log\mu)/l!\int_0^\infty \omega^{(l+1)}(r)\,dr \\ &= -(\mu^l\log\mu)/l!\,\omega^{(l)}(0) = -(\mu^l\log\mu)\sum_{|\alpha|=l}\phi^{(\alpha)}(0)q_\alpha/\alpha! = 0\ ,\quad 0<\mu<\infty\ . \end{aligned}$$

If (2.13) holds, we also get the formulas

$$(2.15)\qquad p.f.a = (-1)^l\binom{m}{l}\sum_{|\alpha|=m+1}(m+1)/\alpha!\,(x^\alpha\log|x|\ a)^{(\alpha)}\ ,\ m=l,l+1,\dots,$$

and

$$(2.16)\qquad \langle p.f.a,\phi\rangle = \sum_{|\alpha|=l}\ \lim_{\varepsilon\to0,\varepsilon>0}\int_{|x|\ge\varepsilon} x^\alpha/\alpha!\,a(x)\phi(x)dx\ ,$$

which represents p.f.a as a Cauchy principal value

If (2.13) is not satisfied, however, then the limit in (2.16) will not exist in general. Also, the right hand side of (2.15), for two different integers m, will differ by an expression $P(D)\delta$, $P(D) = \sum_{|\alpha|=l} p_\alpha q_\alpha D^\alpha$, q_α as in (2.13), $p_\alpha\in\mathbb{C}$.

<u>Lemma 2.3.</u> If $\tau' = -n-\tau = l = 0,1,2,\dots,$ then $a \to p.f.a$, with p.f.a defined by (2.11) or (2.15) or (2.16) defines an injection $H''_{-n-1} \to H'_{-n-1}$, where H''_{-n-1} denotes the space of all $a\in H_{-n-1}$ satisfying (2.13). We get

$$(2.17)\qquad H'_{-n-1} = p.f.H''_{-n-1} + \{P(D)\delta : P(x)\in H_l\}\ ,$$

and (2.20) also holds for $a\in H''_{-n-1}$ (but not in general, for $a\in H_{-n-1}$). If we set $H''_\tau = H_\tau$, $\tau'\neq l$, then $a\in H''_{-n-1}$ implies $x^\alpha a^{(\beta)} \in H''_{-n-1+|\alpha|-|\beta|}$.

The proof will not be discussed in details.

<u>Problems.</u> 1) Show that the functions h_j of 1, problem 6) are in H''_{-n}, but that $|x|^{-n}=g(x)$ of 1, problem 5) is not in H''_{-n}. 2) Discuss the proof of the iterated adjoint Euler formula - that is of Lemma 2.1, formula (2.9) - in details. 3) Set up and prove an iterate of the ordinary (not adjoint) Euler formula. 4) Discuss the derivation of the formula (2.6). 5) Using (2.7), for n=1, give a formula for $p.f.|x|^\lambda$, for n=1, as λ is not a negative integer. 6) In the case of problem 5)

show that p.f.x^{λ} also may be defined in two different ways as a limit of a family of complex integrals, similar to $f^{\pm}$ in 1, problem 2). As long as λ is not a negative integer one will get two homogeneous distributions with restrictions to $\mathbb{R}^{n*}$ different. 7) Let $x_+^{\lambda} = x^{\lambda}$ as $x \geq 0$, $= 0$ as $x < 0$. Show that $\langle p.f.x_+^{-3/2}, \phi\rangle = \int_0^{\infty} (\phi(x) - \phi(0))x^{-3/2}dx$. 7) Obtain the codimension of the space H''_{τ} in H_{τ}. 8) Consider again a convolution operator $K_a = a*$,for a pseudo-function a. Show that the operator K_a is L^2-bounded whenever the Fourier transform $a^{\wedge}$ of a is bounded.(Hint:use Parseval's relation). 9) Consider the 'distribution integral operator' K of 1,problem 10,under the assumptions imposed there. Using techniques similar to those in the proof of Theorem 4.3 (in section 4,below) show that we may write

$$(2.18) \qquad Ku(x) = \int d\xi\, \hat{k^2}(x,\xi)\, u^{\wedge}(\xi)\, e^{ix\xi} \quad ,x \in R^n \quad , \quad u \in S \quad ,$$

with $u^{\wedge}$ and $\hat{k^2}$ denoting the Fourier transform of u ,and the Fourier transform of k with respect to the second argument z ,respectively. Moreover,it may be shown that the function $\hat{k^2}(x,z)$ ins in $C^{\infty}(\mathbb{R}^{2n})$,and even in $T(\mathbb{R}^{2n})$. 10) Under the assumptions of problem 9 show that the operator K is $L^2(\mathbb{R}^n)$-bounded if we assume that (i) the support of $\hat{k^2}$ is in some set $K \times \mathbb{R}^n$,with $K \subset \mathbb{R}^n$ compact,and (ii) that all derivatives of $\hat{k^2}(x,z)$ are bounded over $\mathbb{R}^{2n}$. (Hint: One may take the Fourier transform of the equation $Ku = f$,where $u,f \in S$,and then try to apply Schur's Lemma (I,Lemma 1.1)) 11) Suppose , under the assumptions of problem 9 ,that in addition the function $\hat{k^2}(x,z)$ has all of its derivatives bounded (but not necessarily satisfies the assumption (i) of problem 10 , regarding the support of $\hat{k^2}$). Then prove that still the operator K is $L^2(\mathbb{R}^n)$-bounded.(Hint:Use problem 11 of section 1,together with a partition $1 = \sum_{j=1}^{\infty} \phi_j^2$, $x \in \mathbb{R}^n$,where $\phi_j \in \mathcal{D}$ and the supports of ϕ_j cover $\mathbb{R}^n$ .Also apply the preceeding problem,and Schur's Lemma again ,in a discrete form,for infinite series instead of integrals.)(Remark: The technique is suggested by a proof of R.Beals of the so-called Calderon-Vaillancourt Theorem.)

3. Asymptotic expansions modulo D.

It is possible to generalize the concept of finite part to functions in T_{ra} with an asymptotic expansion at each rational point. The definition, below, is partly motivated by the discussion in section 5, mainly theorem 5.2.

Definition 3.1. A function $a\varepsilon T_{ra}$ will be said to have an asymptotic expansion (mod D) at its rational point z=0, and we shall write

$$(3.1) \qquad a \sim \sum_{l=0}^{\infty} a_l \text{ (mod D) , at 0 ,}$$

if there exists a sequence $\sigma_0,\sigma_1,\sigma_2,\ldots$ of distinct complex numbers with non-decreasing real parts, and $\lim_{j\to\infty} \operatorname{Re}\sigma_j = \infty$, and a sequence $a_0,a_1,a_2,\ldots$ of functions $a_j\varepsilon H'''_{\sigma_j}$ such that for k=0,1,2,... there exists an integer N(k) with

$$(3.2) \qquad a - \sum_{l=0}^{N} a_l \varepsilon C^k(|x|<\varepsilon) \quad , \quad N\geq N(k) \;, \quad \varepsilon \text{ small .}$$

Here we were using the notation H'''_τ for the space defined by (3.3), below.

$$(3.3) \qquad \begin{aligned} H'''_\tau &= H''_\tau \;, \quad \tau \neq 0,1,2,\ldots \\ H'''_1 &= H''_1 + \{p_1 \log|x| : p_1(x) = \sum_{|\alpha|=1} p_\alpha x^\alpha ,\; p_\alpha \varepsilon \mathbb{C}\} \;. \end{aligned}$$

At a general rational point z of $a\varepsilon T_{ra}$ we define an asymptotic expansion

$$(3.4) \qquad a \sim \sum_{l=0}^{\infty} a_l \text{ (mod D) , at z ,}$$

by the property of translation invariance. That is, (3.4) means that

$$(3.5) \qquad a^z \sim \sum_{l=0}^{\infty} a_l^z \text{ (mod D) , at 0 ,}$$

with the translated functions $b^z(x) = b(x-z)$.

A rational point with asymptotic expansion (3.4) will be called an asymptotic rational point, or shortly, an asymptotic point, and a function $a\varepsilon T_{ra}$ with only asymptotic points is said to be in T_{as}.

Not all rational points are asymptotic. As examples consider the functions $a,b\varepsilon T_{ra}$ defined by $a(x) = \sin|x|^{-\mu}$, $\mu>0$, $b(x) = \sin\log^2|x|$, which have a non-asymptotic rational point at zero.

Note that, at an asymptotic rational point z, the expansion (3.4) is not unique, but that two such expansions may differ only by a formal power series (with center z).

For a function $a\varepsilon T_{as}$ with singular support $\{z^1,\ldots,z^r\}$, and with asymptotic expansions

(3.6) $a \sim \sum_{l=0}^{\infty} a_l^j \pmod{D}$, at z^j , $j=1,\dots,r$,

let

(3.7) $a = \sum_{j=1}^{r} \lambda_j \sum_{k=0}^{\nu_j} a_k^j + r_\nu$,

with functions λ_j, as in (1.18), and with $\nu=(\nu_1,\dots,\nu_r)$, ν_j large enough to ensure that r_ν, defined by (3.7) is in L^1_{pol}. For any $b\varepsilon L^1_{pol}$ we shall use the notation p.f.b for the distribution assigned to b by I, (4.2).

We define

(3.8) $p.f.a = \sum_{j=1}^{r} \lambda_j \sum_{k=0}^{\nu_j} p\cdot f\cdot a_l^j + r_\nu$,

and note that this definition is independent of the special choice of λ_j and ν_j, subject to the restrictions mentioned.

By T_{ad} we denote the space of all pseudo-functions $u\varepsilon T_{ps}$ with the property that u equals a function $a\varepsilon T_{as}$ in the complement of its singular support.

Then for $u = a^\psi \varepsilon T_{ad}$ we get

(3.9) $u = p.f.a + \sum P_j(D)\delta_{z^j}$,

the sum being taken over all z^j of sing supp u, as follows from (1.20) since p.f.a is a pseudo-function to a.

We obtain an asymptotic expansion

(3.10) $u \sim \sum_{l=0}^{\infty} u_l^j \pmod{D}$, at z^j ,

where $\{u_l^j\}$ is a rearrangement of the homogeneous parts of $P_j(D)$, and the $\{p.f.a_l^j$, $l=0,1,2,\dots\}$ after non-decreasing real parts of their homogeneity degree.

Also we get

(3.11) $u - \sum_{l=0}^{M} u_l^j \varepsilon C^k(|x-z^j|<\varepsilon)$, $N \geq N_0(k)$.

If z=0 is an asymptotic rational point for a, or for $u\varepsilon S'$, then so it is for $x^\alpha a^{(\beta)}$, and $x^\alpha u^{(\beta)}$, respectively, and the asymptotic expansion may be differentiated (or multiplied by x^α) term by term.

By Taylor's formula, and, because $(x-z)^\alpha$ is a linear combination of powers x^γ, $\gamma\leq\alpha$, it follows that a general asymptotic rational point remains invariant under multiplication by $b\varepsilon T$, while the asymptotic expansion of $b\cdot a$ or $b\cdot u$ is a rearrangement of the formal products for increasing real parts of degrees. Especially, the theorem, below, may be verified by rather technical arguments.

Theorem 3.2. If $b \in T$, $u = \text{p.f.}a + w \in T_{ad}$, $a \in T_{as}$, w as in (1.3)(of finite support) then $a \cdot b \in T_{as}$, $b \cdot u \in T_{ad}$.

Moreover,

$$\text{sing supp } (a \cdot b) \subset \text{sing supp } a \quad ,$$

(3.12)

$$\text{sing supp } (a \cdot u) \subset \text{sing supp } u \quad ,$$

and

(3.13) $$b \cdot u = \text{p.f.}(a \cdot b) + b \cdot w \ , \ u^{(\alpha)} = \text{p.f.}a^{(\alpha)} + w^{(\alpha)} \ .$$

Problems: 1) Obtain an asymptotic expansion mod D for a rational function $f(x) = p(x)/q(x)$, $x \in \mathbb{R}$, where p and q are polynomials. (Near every point of sing supp f.) 2) Verify in details that $x=0$ is an asymptotic rational point for $b = x^{\alpha} a^{(\beta)}$, if it is an asymptotic rational point for a. 3) If $f(x) = p(x)/|x|^k$, with $p \in C^{\infty}(\mathbb{R}^n)$ and some integer k, show that $x=0$ is asymptotic and obtain an expansion mod D.

4. Fourier transform of distributions with rational singularities.

Definition 4.1. (i) Let M denote the class of all temperate functions $a\varepsilon T$ such that for $k=0,1,2,\ldots$ there exists $N(k)$ with

$$(4.1)\qquad x^{\alpha}a^{(\beta)}\ \varepsilon\ L^1\ ,\ |\alpha|\ \leq\ k\ ,\ |\beta|\geq N(k)\ .$$

(ii) Let M_{ix} denote the class of all $a\varepsilon T$ of the form

$$(4.2)\qquad a(x)\ =\ \sum_{j=1}^{m} a_j(x)\ e^{iz^jx}\ ,\ a_j\ \varepsilon\ M\ ,\ z^j\ \varepsilon\ \mathbb{R}^n\ .$$

(iii) Let S_{ps} denote the collection of all $u\varepsilon T_{ps}$ (c.f. section 1) which coincide with a function in S, outside a large sphere.

(iv) Let S^{o}_{ps} denote the class of all $u\varepsilon S_{ps}$, with singular support in the origin only.

Examples 4.2. (i) If $a\varepsilon H_{\tau}$, $u\varepsilon H'_{\tau}$, and if

$$(4.3)\qquad \lambda,\omega\varepsilon C_0^{\infty}(\mathbb{R}^n),\ \lambda+\omega=1,\ \lambda\varepsilon C_0^{\infty}(\mathbb{R}^n)\ ,\ \omega=0,\ |x|\leq\varepsilon\ ,\ \varepsilon>0\ ,$$

then λu, $\lambda(\text{p.f.}a)\ \varepsilon\ S^{o}_{ps}$, and ωu, $\omega a\ \varepsilon\ M$.

(ii) If $u\varepsilon T_{ps}$, $b\varepsilon S$, then $bu\varepsilon S_{ps}$.

(iii) Let T^{s}_{ρ} denote the class of all $a\varepsilon C^{\infty}(\mathbb{R}^n)$ such that for all $\alpha\varepsilon \mathbb{Z}^n_+$ we have (with some $s\varepsilon\mathbb{R}$, $\rho>0$)

$$(4.4)\qquad a^{(\alpha)}(x)\ =\ O((1+|x|)^{s-\rho|\alpha|})\ .$$

Then $T^{s}_{\rho}\subset M$. The condition (4.4) essentially is that of Hoermander [5] describing the space $S^{s}_{\rho,0}$ for a function $a(x,\xi)$ independent of the first variable.

For any function $a\varepsilon H_{\tau}$ and ω as in (4.3) we get $\omega a\varepsilon T^{\mathrm{Re}\,\tau}_{-1}$. Also $\sin((1+|x|^2)^{\tau})$ is a function in $T^{0}_{2\tau-1}$, $0<\tau<1/2$.

Theorem 4.3. a) The conditions (i), (ii) and (iii), below, are equivalent:

(i) $a\varepsilon M$; (ii) $a^{\wedge}\varepsilon S^{o}_{ps}$; (iii) $a^{\vee}\varepsilon S^{o}_{ps}$.

b) The conditions (i), (ii) and (iii), below, are equivalent:

(i) $a\varepsilon M_{ix}$; (ii) $a^{\wedge}\varepsilon S_{ps}$; (iii) $a^{\vee}\varepsilon S_{ps}$.

Proof. Let us observe, at first, that the function $b(x)\ =\ a(x)e^{izx}$ $a\varepsilon M$, has its Fourier transform defined by

$$(4.5)\qquad \langle b^{\wedge},\phi\rangle\ =\ \langle a^{\wedge},\phi^{-z}\rangle\quad ,\quad \phi^{-z}(x)\ =\ \phi(x+z)\ .$$

In other words, $b^{\wedge}$ is the translation of the distribution $a^{\wedge}$ along the vector z.

Accordingly, it is sufficient to prove that

(1) $a\in M \Rightarrow a^{\wedge}\in S^{O}_{ps}$; (2) $u\in S^{O}_{p} \Rightarrow u^{\wedge}\in M$.

(1) Suppose $a\in M$, then (4.1) implies $x^{\beta}a^{(\alpha)}\in L^{1}$, or,

(4.6) $D^{\beta}x^{\alpha}a^{\wedge} \in CO$, $|\beta| \le k$, $|\alpha| \ge N(k)$,

using I, (3.28), and I, (4.15). Here D^{β} are distribution derivatives, but from I, lemma 2.4 we conclude that $x^{\alpha}a^{\wedge}\in C^{k}(\mathbb{R}^{n})$, $|\alpha| \ge N(k)$ so that $|x|^{2l}a \in C^{k}(\mathbb{R}^{n})$, $2l \ge N(k)$.

We have $\omega|x|^{-2l}\in T$, with ω as in (4.3). Hence $\omega a^{\wedge} = (\omega|x|^{-2l})|x|^{2l}a\in C^{k}(\mathbb{R}^{n})$, for all k, and all such ω.

It follows that $a\in C^{\infty}(\mathbb{R}^{n}-\{0\})$. By Leibnitz' formula $D^{\alpha}(\omega a^{\wedge})$ is a linear combination of terms $(\omega|x|^{-2l})^{(\beta)}(|x|^{2l}a^{\wedge})^{(\gamma)}$, $\beta\le\alpha$ $\gamma\le\alpha$, hence $a^{\wedge}=O(|x|^{-2l})$, $|x|\ge 1$, l large.

Thus $\omega a^{\wedge}\in S$. Also, 0 is a rational point, as a consequence of (4.6). It follows that $a\in S^{O}$, q.e.d. .

(2) Let $u\in S^{O}_{ps}$, then it a fortiori satisfies the condition

$$\text{sing supp } u \subset \{0\},\ \omega u \in S,\ \ \omega \text{ as in (4.3)},$$

(4.7)

$$D^{\beta}x^{\alpha}u \in L^{1}\ ,\ |\beta| \le k\ ,\ |\alpha| \ge N_{1}(k)\ .$$

Taking the Fourier transform we get

(4.8) $x^{\beta}u^{\wedge(\alpha)} \in CO$, $|\beta| \le k$, $|\alpha| \ge N_{1}(k)$.

In particular, $u^{\wedge(\alpha)} \in CO$, $|\alpha| \ge N_{1}(0)$, which implies that $u^{\wedge}\in C^{\infty}(\mathbb{R}^{n})$. Also (4.8) implies (4.1), with $N(k)=N_{1}(k+n+1)$. Therefore it follows that $u\in M$, q.e.d. .

Lemma 4.4. For $a\in M_{ix}$, of the representation (4.2) we have

(4.9) $\text{sing supp } a = \Gamma_{a} \subset \{z^{1},\dots,z^{m}\}$.

Proof evident.

Corollary 4.5. a) The condition (4.1), describing M, is equivalent to each of the conditions (i), (ii) and (iii), below.

(i) (4.1), with L^{1} replaced by CO ; (ii) (4.1) with "$x^{\beta}a^{(\alpha)}\in L^{1}$" replaced by "$x^{\beta}a^{(\alpha)} = O(1)$, over $\mathbb{R}^{n}$ " ; (iii) $u^{(\alpha)}(x) = O((1+|x|)^{-m_{|\alpha|}})$, with $m_{k} \to \infty$, $\alpha\in\mathbb{Z}^{n}_{+}$.

b) The condition (ii) in definition 1.1, describing a rational point z of a distribution u, is equivalent to each of the conditions (I), (II) or (III), below.

(I) $D^{\beta}(x-z)^{\alpha}u \in CO(|x-z|<\varepsilon)$; (II) $D^{\beta}(x-z)^{\alpha}u \in L^{1}(|x-z|<\varepsilon)$; (III) $D^{\beta}(x-z)^{\alpha}u = O(1)$, near z ; (in each case, above, for $|\beta| \le k$, $|\alpha| \ge N(k)$). (The derivatives are distribution derivatives, each times.)

Proof. Conditions a) (i), or a) (ii), or a) (iii) each imply (4.1), and also, conditions b) (I), or b) (III), or (ii) of definition 1.1 each imply b) (II), using I, lemma 2.4. In our proof of theorem 4.3, part (1) and part (2), we started from (4.1) and (4.7), respectively, and also note that (4,7) is the same as b) (II) for z=0, and u replaced by λu, λ as in (4.3). Use the Fourier inversion formula $u = u^{\wedge\vee}$, and note that the discussion of (1) and (2) in the proof of theorem 4.3 may be amended in such a way that we get a) (ii) for $u^{\wedge} \Rightarrow$ b) (I),(II),(III) for u, and (4.7) for u $\Rightarrow$ a) (i), (ii), (iii), q.e.d. .

Lemma 4.6. a) M and M_{ix} are algebras under "$\cdot$", invariant under $x^{\alpha}\cdot$ and $D^{\alpha}\cdot$.

b) S^{o}_{ps} and S_{ps} are algebras under "$*$", invariant under $x^{\alpha}\cdot$, and $D^{\alpha}\cdot$.

Proof. (a) is a matter of Leibnitz' formula ; (b) then follows from theorem 4.3 and I, (4.22), q.e.d. .

In order to obtain a similar characterization for the Fourier transform of pseudo-functions with asymptotic rational points, than occuring in theorem 4.3 for functions in S_{ps}, we first will seek information about the Fourier transform of homogeneous pseudo-functions.

Lemma 4.7. a) For any $a \in H_{\tau}$ the pseudo-function p.f.a, defined by either (2.16), $\tau' \ne l$, or by (2.11), $\tau' = l$, has its Fourier transform in T_{ps}, with singular support at O only. We get

$$(\text{p.f.}a)^{\wedge} \in H'_{\tau'}\ , \quad \tau' = -n-\tau \ne l,\ l = 0,1,2,\ldots \quad ,$$

(4.10) $$(\text{p.f.}a)^{\wedge} = b(x) + B(x)\log|x| \ , \quad \tau' = -n-\tau = l \quad ,$$

$$b \in H_{l} = H'_{l}\ ,$$

with

(4.11) $$B(x) = (-1)^{l+1}(2\pi)^{-n/2}\int_{|y|=1}(xy)^{l}/l!\ a(y)\,d\sigma_{y} \quad .$$

b) We have

(4.12) $$(Q(D)\delta)^{\wedge} = (2\pi)^{-n/2}Q(x) \quad ,$$

for any polynomial $Q(x)$.

c) The Fourier transform provides a bi-jective isomorphism

(4.13) $$F : H'_{\tau} \to H'_{\tau'}\ , \quad \tau+\tau' = -n \ , \quad \tau \in \mathbb{C} \quad .$$

Proof. Assertion (c) follows from (a) and (b) and (2.17). For (b) we note that $\delta = \lim_{\varepsilon\to\infty}(\varepsilon^{n}\phi_{\varepsilon})$ in weak convergence of S' where $\phi \in C_{0}^{\infty}(\mathbb{R}^{n})$ satisfies $\int \phi\,dx = 1$.

Since F is weakly continuous (I, theorem 4.4 (iv)) we get $\delta^{\wedge} = \lim_{\varepsilon\to 0}\phi_{\varepsilon}^{\wedge}$, that is ,

$$\langle\delta^\wedge,\psi\rangle = \lim_{\varepsilon\to 0}\int\phi^\wedge(\varepsilon x)\psi(x)dx = \phi^\wedge(0)\int\psi dx = (2\pi)^{-n/2}\langle 1,\psi\rangle \quad ,$$

or, $\delta^\wedge = (2\pi)^{-n/2}\cdot 1$.

Formula I, (4.15) then implies (4.12).

To prove (a), let $u=\text{p.f.a}$, $a\varepsilon H_\tau$. With λ,ω as in (4.3) get $\omega u\varepsilon M$, $\lambda u \ \varepsilon \ S^0_{ps}$, so that theorem 4.3 implies $(\omega u)^\wedge \ \varepsilon \ S^0_{ps}$, $(\lambda u)^\wedge \ \varepsilon \ M$, that is, $u^\wedge = (\lambda u)^\wedge + (\omega u)^\wedge \varepsilon \ T_{ps}$. In particular, $u^\wedge \varepsilon C^\infty(\mathbb{R}^n-\{0\})$. If $u\varepsilon H'_\tau$, then

$$\langle u^\wedge,\phi_\mu\rangle = \langle u,(\phi_\mu)^\wedge\rangle = \mu^{-n}\langle u,(\phi^\wedge)_{\mu^{-1}}\rangle = \mu^{-n-\tau'}\langle u^\wedge,\phi\rangle, \text{ so that } u\varepsilon H'_\tau,\ \tau+\tau' = -n.$$

Especially this holds true for $\tau' \neq 0,1,2,\ldots$, by lemma 2.2. For $\tau = -n-l$, $l=0,1,\ldots$, we do not have $u = \text{p.f.a}$ in H'_τ, in general, but we will get (2.14) instead. Substitute $\phi^\wedge$ for ϕ, in (2.14), with $\phi\varepsilon S$, $\phi=0$, near 0. Let f be the restriction of p.f.a to its C^∞-domain $\mathbb{R}^n - \{0\}$. Then (4.12) implies

$$(-1)^{l+1}(2\pi)^{-n/2}(\mu^l\log\mu)\sum_{|\alpha|=l} q_\alpha/\alpha!\langle x^\alpha,\phi\rangle$$

$$= (-1)^{l+1}(\mu^l\log\mu)\sum_{|\alpha|=l} q_\alpha/\alpha!\langle\delta^{(\alpha)},\phi^\wedge\rangle$$

$$= (-\mu^l\log\mu)\sum_{|\alpha|=l}\phi^{(\alpha)}(0)\ q_\alpha/\alpha!$$

$$= \langle \text{p.f.a},(\phi^\wedge)_\mu\rangle - \mu^l\langle\text{p.f.a},\phi^\wedge\rangle$$

$$= \mu^{-n}\langle(\text{p.f.a})^\wedge,\phi_{1/\mu}\rangle - \mu^l\langle(\text{p.f.a})^\wedge,\phi\rangle$$

$$= \langle f_\mu,\phi\rangle - \langle f,\phi\rangle \quad .$$

That is,

$$f(\mu x) = \mu^l f(x) + \mu^l\log\mu\ B(x) \ ,$$

(4.14)

$$x \ \varepsilon\ \mathbb{R}^n - \{0\},\ 0 < \mu < \infty \ ,$$

with B(x) as in (4.11). Define

$$b(x) = f(x) - B(x)\log|x| \ ,$$

then $b\varepsilon C^\infty(\mathbb{R}^n-\{0\})$, and (4.14) implies $b\ \varepsilon\ H_l$.

Accordingly, we get the second formula (4.10), q.e.d. .

Problems: 1) Obtain the Fourier transform of $\text{p.f.}|x|^\lambda$, $x\varepsilon\mathbb{R}^n$, as $\lambda \neq -n-l$, $l=0,1,2,\ldots$. 2) Obtain explicitly the Fourier transform of a general distribution $u\varepsilon H'_\tau$, $\tau \neq -n-l$, $l=0,1,2,\ldots$, expressed by an integral over $|x| = 1$.

5. Fourier equivalence of asymptotic expansions mod D and mod $|x|$.

In the following it will be of use to (temporarily) define the Fourier transform of a function $a \in H_\tau$ (not a distribution) by setting

(5.1) $\quad a^\wedge = \text{r.c.d.}\ (\text{p.f.}a)^\wedge$

with the restriction "r.c.d." to its C^∞-domain $\mathbb{R}^n-\{0\}$, of the distribution $(\text{p.f.}a)^\wedge$. However, for $\tau = -n-1$, we shall regard (5.1) as an equality modulo the homogeneous polynomials of degree 1 only.

With this interpretation of the Fourier transform of a function, and with the space H'''_τ of (3.3) we note that the lemma, below, is evident.

Lemma 5.1. The conditions (i), (ii) and (iii), below, are equivalent.

(i) $a \in H_\tau$; (ii) $a^\wedge \in H'''_{-n-\tau}$; (iii) $a^\vee \in H'''_{-n-\tau}$.

Definition 5.2. (i) A function $a \in T$ will be said to have an asymptotic expansion (mod $|x|$), (at ∞), and we shall write

$$(5.2) \qquad a \sim \sum_{l=0}^{\infty} a_l \ (\text{mod}\ |x|),\ \text{at}\ \infty ,$$

if there exists a sequence $\rho_0, \rho_1, \rho_2, \ldots,$ of distinct complex numbers, with non-increasing real parts, and $\lim_{l\to\infty} \operatorname{Re} \rho_l = -\infty$, and a sequence a_l, $l=0,1,2,\ldots,$ of functions $a_j \in H_{\rho_j}$ such that

$$(5.3) \qquad a - \sum_{l=0}^{M} a_l = O(|x|^{-k}) \ , M \geq N(k),\ |x| \geq 1 .$$

(ii) The space of all functions in M (c.f. definition (4.1)(i)), with an asymptotic expansion, mod $|x|$, at ∞, is denoted by M_{as}.

(iii) The space of all functions $a \in T$ with an expansion

$$(5.4) \qquad a(x) = \sum_{j=1}^{r} a_j(x) e^{iz^j x},\ a_j \in M_{as},\ z^j \in \mathbb{R}^n ,$$

is denoted by M_{iy} .

(iv) We define S_{as} and S_{ad} as the spaces of all functions in T_{as} and distributions in T_{ad}, respectively, which are equal to a function in S, for large $|x|$.

(v) We denote by S^0_{as} and S^0_{ad} the subspaces of S_{as} and S_{ad}, respectively, containing all functions (distributions) with singular support at the origin $z=0$ only.

Theorem 5.2. a) The conditions (i), (ii) and (iii), below, are equivalent.

(i) $a \in M_{as}$; (ii) $a^\wedge \in S^0_{ad}$; (iii) $a^\vee \in S^0_{ad}$.

b) The conditions (i), (ii) and (iii), below, are equivalent.

(i) $a\varepsilon M_{iy}$; (ii) $a^{\wedge}\varepsilon S_{ad}$; (iii) $a^{\vee}\varepsilon S_{ad}$.

c) The Fourier transform (and its conjugate) may be taken "term by term", in the asymptotic expansions (3.10) and (5.2), whenever $u\varepsilon S_{ad}$, or $a\varepsilon M_{iy}$, respectively.

Proof. Again it suffices to establish (1) $a\varepsilon M_{as} \Longrightarrow a^{\wedge}\varepsilon S^{O}_{ad}$, and (2) $u\varepsilon S^{O}_{ad} \Longrightarrow u^{\wedge}\varepsilon M_{as}$.

(1) With the functions ω,λ of (4.3) let

$$r_N = a - \sum_{0}^{N} \text{p.f.}\, a_l \ , \ p_N = \omega r_N \ , \ q_N = \lambda r_N \ ,$$

so that

$$x^{\alpha} a = \sum_{0}^{N} x^{\alpha}\text{p.f.}\, a_l + x^{\alpha} p_N + x^{\alpha} q_N \ ,$$

$$a_l \ \varepsilon \ H_{\rho_l} \ , \ x^{\alpha} p_N = O((1+|x|)^{-k+|\alpha|}), \ x^{\alpha} q_N \ \varepsilon \ S^{O}_{ps} \ .$$

Take the Fourier transform to obtain

$$D^{\alpha}(a^{\wedge} - \sum_{0}^{N} b_l) = (x^{\alpha} p_N)^{\wedge} + (x^{\alpha} q_N)^{\wedge} \ ,$$

$$\text{r.c.d.}\, b_l = a_l^{\wedge} \ \varepsilon \ H'''_{-n-\rho_l} \ , \ (x^{\alpha} p_N)^{\wedge} \ \varepsilon \ CO, \ (x^{\alpha} q_N)^{\wedge} \ \varepsilon \ M \subset C^{\infty}(\mathbb{R}^n) \ ,$$

for sufficiently large N. Accordingly $a^{\wedge} \ \varepsilon \ S^{O}_{as}$, because we also have $a \ \varepsilon \ M$, hence $a^{\wedge} \ \varepsilon \ S^{O}_{ps}$, by theorem 4.3, and its asymptotic expansion (mod D), at O, is obtained by taking the Fourier transform term by term.

(2) Let $u \ \varepsilon \ S^{O}_{ad}$, then $u^{\wedge}\varepsilon M$, by theorem 4.3, and let again $r_N = u - \sum_{l=0}^{N} u_l$, $p_N = \omega r_N \ \varepsilon \ M$,

$$q_N = \lambda r_N \ \varepsilon \ C^{k}_{O}(\mathbb{R}^n) \ , \ \text{for } N \geq N(k) \ ,$$

where u_l are as in (3.10). Write

$$u^{\wedge} = \sum_{l=0}^{N} u_l^{\wedge} + p_N^{\wedge} + q_N^{\wedge} \ , \ \text{r.c.d.}\, u_l^{\wedge} \ \varepsilon \ H_{-n-\rho_l} \ ,$$

$$p_N^{\wedge} \ \varepsilon \ S^{O}_{ps} \ , \ q_N^{\wedge} = O(|x|^{-k}) \ , \ N \geq N(k+n+1).$$

Accordingly, $u^{\wedge}\varepsilon M_{as}$, and again its asymptotic expansion (mod $|x|$), at ∞, is obtained by taking the Fourier transform term by term, q.e.d. .

Observe that the definitions of M_{as}, S_{as} and S^{O}_{ad} may be modified similarly as the definitions of M and S^{O}_{ps}, in corollary 4.5, by replacing C^{O} or L^{1}, by other (local) spaces.

Lemma 5.3. a) M_{as} and M_{iy} are algebras under "$\cdot$", invariant under $x^\alpha\cdot$ and D^α; b) S^o_{ad} and S_{ad} are algebras under "$*$", invariant under $x^\alpha\cdot$ and D^α.

The proof of lemma 5.3 will be omitted.

Let us introduce

$$(5.5)\qquad \|a\|_{\infty,k} = \sum_{|\alpha|=k} \|D^\alpha a\|_\infty,\ \|a\|_\infty = \text{ess sup } \{|a(x)| : |x|\le 1\}$$

as norms on the spaces H_τ, $\tau\in\mathbb{C}$. For H'''_τ define similar norms, denoted by $\|b\|'''_{\infty,k}$, where for $b\in H'''_1$, $b=B'+P\log|x|$, $B'\in H_1$, $P=\sum_{|\alpha|=1} p_\alpha x^\alpha$, we set

$$(5.6)\qquad \|b\|'''_\infty = \inf\{\|b+R\|_\infty : R=\sum_{|\alpha|=1} r_\alpha x^\alpha\} + \|P\|_\infty ,$$

while $\|b\|'''_\infty = \|b\|_\infty$, for all other cases.

Lemma 5.4. For $a\in H_\tau$ and its Fourier transform $a\in H'''_{\tau'}$, as defined in (5.1), $\tau+\tau'=-n$, we get

$$(5.7)\qquad \|a^\wedge\|'''_{\infty,k} \le c\,\|a\|_{\infty,m} ,$$

with constants $c=c_{k,\tau}$ and $m=m_{k,\tau}$, independent of a.

Proof. With the functions ω,λ of (4.3) we get

$$(5.8)\qquad x^\alpha\lambda a\in L^1,\ \Delta^m(x^\alpha\omega a)\in L^1,\ -n<\operatorname{Re}\tau+|\alpha|<-n+2m .$$

Taking Fourier transforms we have

$$(5.9)\qquad \begin{aligned}\|D^\alpha a^\wedge\|_\infty &\le \|D^\alpha(\lambda\,\text{p.f.}\,a)^\wedge\|_{L^\infty} + \||x|^{2m}D^\alpha(\omega a)^\wedge\|_{L^\infty}\\ &\le (2\pi)^{-n/2}(\|x^\alpha\lambda a\|_{L^1} + \|\Delta^m(x^\alpha\omega a)\|_{L^1} \le c\,\|a\|_{\infty,2m} .\end{aligned}$$

If Re $\tau>-n$ we thus get (5.7) with $m>1/2\,(\operatorname{Re}\tau+n+k)$.

Next, if Re $\tau\le -n$, then (5.9) may be used only for $|\alpha|>-n-\operatorname{Re}\tau$. If not $\tau=-n-1$, then Euler's formula yields

$$(5.10)\qquad \|a^\wedge\|_{\infty,k} \le c(k,\tau)\sum_{|\alpha|=k}\|D^\alpha a^\wedge\|_\infty ,$$

which gives (5,7) again, with the same m. Finally let $\tau=-n-1$. Then (5.9) implies

$$(5.11)\qquad \|D^\alpha a^\wedge\|_\infty \le c\,\|a\|_{\infty,2m},\ |\alpha|>1 .$$

Lemma 5.5. For $b\in H'''_1$ we get

$$(5.12)\qquad \|b\|'''_{\infty,1} \le c\sum_{|\alpha|=l+1}\|b^{(\alpha)}\|'''_\infty .$$

Proof. For a function $b\in H_0$, and $|x|=1$ we get $b(x)=b(e)+\int_\Gamma db$, with a suitable countour Γ connecting e and x on $|x|=1$. Accordingly

$$|b(x) - b(e)| \le \pi\sum_1^n \|D_j b\|_\infty \ , \quad |x| = 1 \ ,$$

which may be integrated over $|e| = 1$ for

$$(5.13) \qquad \|b\|_{\infty,0} \le c\Big(\Big|\sum_{|x|=1} b\, do_x\Big| + \sum_{|\alpha|=1} \|D^\alpha b\|_\infty\Big), b \in H_0 \ .$$

Now for

$$b = b' + P \log|x| \in H_1''' \ , \ 1 = 0,1,\ldots, b' \in H_1 \ ,$$

we get the modified Euler formulas

$$(5.14) \qquad 1b = \sum_{j=1}^n x_j b_{x_j} - P \quad , \quad 1P = \sum_{j=1}^n x_j P_{x_j} \quad .$$

For $1=0$ we get

$$(5.15) \qquad \begin{aligned} \|b\|_\infty''' &= \|P\|_\infty + \inf_{\gamma\in\mathbf{C}} \|b+\gamma\|_\infty \\ &\le \Big\|\sum_{j=1}^n x_j b_{x_j}\Big\|_\infty + c\sum_{|\alpha|=1} \|D^\alpha b\|_\infty \le c\sum_{|\alpha|=1} \|D^\alpha b\|_\infty \ , \end{aligned}$$

using (5.13) and (5.14). This amounts to (5.12), for $1=0$. For $1>0$ it follows that

$$(5.16) \qquad \begin{aligned} \|b\|_\infty''' &= \|P\|_\infty + \inf\{\|b+R\|_\infty : \text{ degree } R = 1\} \\ &\le 1/1\ \{\|\sum x_j P_{x_j}\|_\infty + \inf \|\sum x_j (b_{x_j} + R_j)\|_\infty : \text{ degree } R_j = 1-1\} \\ &\le 1/1 \sum_{|\alpha|=1} \|D^\alpha b\|_\infty \quad . \end{aligned}$$

Hence

$$\|b\|_{\infty,1}''' \le c\Big(\|b\|_\infty''' + \sum_{j=1}^n \|D_j b\|_{\infty,1-1}'''\Big) \le c\sum_{|\alpha|=1+1} \|D^\alpha b\|$$

by induction, q.e.d. .

Lemma 5.6. Let $\tau\in\mathbf{C}$, and, with a fixed λ, as in (4.3), let

$$(5.17) \qquad b = u^\wedge - \lambda u^\wedge - \lambda^\wedge * u^\wedge, u = p.f.a \ , \ a \in H_\tau \ .$$

Then we have $b \in S$, and, moreover,

$$(5.18) \qquad \|x^\alpha D^\beta b\|_C \le c\ \|a\|_{\infty,m} \ , \quad a \in H_\tau \ ,$$

with c and m independent of a.

Proof. Write

$$(5.19) \qquad b = \omega u^\wedge - (\lambda u)^\wedge = (\omega u)^\wedge - \lambda u^\wedge \ ,$$

and note that $\omega u^\wedge$, $(\lambda u)^\wedge \in M$, $(\omega u)^\wedge \lambda u^\wedge \in S^0_{ps}$, by theorem 4.3, so that indeed $b \in M \cap S^0_{ps} = S$ follows.

Note the estimates, below, valid for $|\alpha| \le k$, $|\beta| \ge N(k)$.

$$\|x^\alpha D^\beta(\omega u^\wedge)\|_C \le c_1 \|u^\wedge\|'''_{\infty,m_1} \le c\|u\|_{\infty,m} ,$$

(5.20)

$$\|x^\alpha D^\beta(\lambda u)^\wedge\|_C \le c_1 \|D^\alpha x^\beta \lambda u\|_{L^1} \le c\|u\|_{\infty,m} ,$$

which imply (5.18) for $|\alpha| \le k$, $|\beta| \ge N(k)$, by (5.19).
Similarly, using the second formula (5.19), we get (5.18) for $|\beta| \le k$, $|\alpha| \ge N(k)$.
If $|\beta| \ge N(0)$ we get

(5.21) $$\|x^\alpha D^\beta b\|_C \le c(\|D^\beta b\|_C + \sum_{|\gamma|=p} \|x^\gamma D^\beta b\|_C) \le c\|u\|_{\infty,m} ,$$

for arbitrary α (with sufficiently large p).

Finally, (5.18) will follow for arbitrary $\alpha,\beta \in \mathbb{Z}_+^n$, if we can achieve it for $|\beta| \le N(0)$ and $\alpha = 0$. Assume this true for $|\beta| = j \le N(0)$, and let $|\beta| = j-1$, and $e \in \mathbb{R}^n$ and $|e| = 1$ be fixed, then

$$|(D^\beta b)(x)| \le |(D^\beta b)(e)| + c\sum_{|\delta|=1} \|D^{\beta+\delta} b\|_C , \quad |x| \le 1 ,$$

(5.22)

$$\le c(\sum_{|\gamma|=p} \|x^\gamma D^\beta b\|_C + \sum_{|\delta|=1} \|D^{\beta+\delta} b\|_C) \le c\|a\|_{\infty,m} ,$$

and

$$\|D^\beta b\|_C \le c|\sup\{|D^\beta b| : |x| \le 1 + \sum_{|\gamma|=p} \|x^\gamma D^\beta b\|_C)$$

(5.23)

$$\le c\|a\|_{\infty,m} ,$$

q.e.d. .

We mention, finally, that there exist explicit formulas, expressing the Fourier transform a of a homogeneous function a in terms of a distribution integral over $|x|=1$ (c.f. [2]).

Problems: 1) A polynomial $p(x) = \sum_{|\alpha| \le N} p_\alpha x^\alpha$ will be called an elliptic symbol of degree N if its principal part $p_N(x) = \sum_{|\alpha|=N} p_\alpha x^\alpha$ does never vanish at $|x|=1$. Show that (i) for an elliptic symbol the set $\mathcal{N}=\{x \in \mathbb{R}^n : p(x)=0\}$ is compact; (ii) if $\omega(x) \in C^\infty(\mathbb{R}^n)$ equals 1 for large $|x|$ and vanishes near $\mathcal{N}$, then the function $f(x)=\omega(x)/p(x)$ is in $\mathcal{M}_{as}$. 2) A constant coefficient partial differential operator $L=p(D)$ is called elliptic, if $p(x)$ is an elliptic symbol. Show that for an elliptic constant coefficient operator there exists a distribution $F \in \mathcal{S}^0_{ad}$ with the property that $LF=\delta_0+\chi$, with some $\chi \in C^\infty(\mathbb{R}^n)$. (Hint: use problem 1) above, together with theorem 5.2.)

6. References.

[1] R.E. Edwards, Functional analysis; Holt Rinehart Winston, New York 1965.

[2] L. Gårding, Transformations de Fourier des distributions homogènes, Bull. Soc. Math. France 89 (1961), 381-428.

[3] I. Gelfand and G. Silov, Generalized functions; Acad. Press, New York 1966.

[4] J. Hadamard, Lectures on Cauchy's problem; Dover Classic, New York 1952.

[5] L. Hörmander, Pseudo-differential operators and hypo-elliptic equations; Proc. Sympos. Pure Math. Vol. X (Singular integrals) Chicago 1966.

[6] Laurent Schwartz, Théorie des distributions; Paris Hermann 1966.

[7] R.T. Seeley, Topics in pseudo-differential operators; Centro Internazionale Matematico Estivo, Conference at Stresa, 1968.

Chapter III: Finitely Generated Pseudo-Differential Operators.

In this chapter we will consider an algebra of formal expressions, which we denote finitely generated pseudo-differential operators (abbreviated ψdo's). The definition of the algebra, called $\mathring{\Psi\mathcal{O}}$, is simple. It is finitely generated by the algebra L of differential expressions over $\mathbb{R}^n$ with coefficients in $CB^{\infty}(\mathbb{R}^n)$, and the family of formal Fourier multipliers $\Lambda^s = (1-\Delta)^{-s/2} = (1+|D|^2)^{-s/2} = F^{-1}(1+|M|^2)^{-s/2}$, where $CB^{\infty}(\mathbb{R}^n)$ denotes the class of functions in $C^{\infty}(\mathbb{R}^n)$ with all derivatives bounded (section 5).

An element $K \in \mathring{\Psi\mathcal{O}}$ may be regarded as operator $S \to S$, or $S' \to S'$, as a consequence of I,4. In the present chapter, however, we plan a study of these operators acting between L^2-Sobolev spaces $\mathfrak{H}_s$, as s is either finite or infinite. The spaces $\mathfrak{H}_s$, $s \in \mathbb{R}$, are discussed in sections 1 and 2. In particular they are Hilbert spaces of temperate distributions, as s is small, and consist of more and more differentiable functions, as s is large.

Our particular emphasis will be on the Frechet space $\mathfrak{H}_{\infty} = \bigcap_{s\in\mathbb{R}} \mathfrak{H}_s$, a subspace of $CB^{\infty}(\mathbb{R}^n)$, and on the continuous linear operators $L(\mathfrak{H}_{\infty})$. Every $K \in L(\mathfrak{H}_{\infty})$ possesses an order function (section 4), reflecting its topological properties. Operators with the linear order function $s \to s-r$ will be said to be of order r, where $r \in \mathbb{R}$, and the class of all such operators is denoted by $\mathcal{O}(r)$.

It turns out that $\mathring{\Psi\mathcal{O}}$ may be interpreted as a subalgebra of $\mathcal{O}(\infty) = \bigcup_r \mathcal{O}(r)$. Of particular importance for later discussions will be the algebra $\mathring{\Psi\mathcal{O}}_0$ of all $A \in \mathring{\Psi\mathcal{O}}$, having order zero. In particular we focus on the algebra $\mathring{\mathcal{L}} \in \mathring{\Psi\mathcal{O}}_0$, generated by the operators of type $L\Lambda^N$, with $L \in L$ having order ($\leq$) N. The algebras $\mathring{\mathcal{L}}$ and $\mathring{\Psi\mathcal{O}}_0$ may be interpreted as operator subalgebras of $L(\mathfrak{H}_s)$ as well, for every $s \in \mathbb{R}$.

Now it is important that the algebra $\mathcal{O}(0)$ with topology generated by all the operator norms $\|A\|_s$ of the spaces $\mathfrak{H}_s$, $s \in \mathbb{R}$, has a Frechet topology. This is due to an interpolation theorem by Calderon (section 3), allowing an estimate of $\|A\|_s$ by $\|A\|_{[s]}$ and $\|A\|_{[s]+1}$, with the largest integer $[s] \leq s$.

The algebra $\mathcal{O}(\infty)$ has a natural involution $A \to A^*$, called $\mathcal{O}(\infty)$-adjoint, which preserves orders (section 5). The Hilbert space adjoint of $A \in \mathcal{O}(0)$ in some $\mathfrak{H}_s$ is to be distinguished from A^*. They are denoted by $A^{<s>}$.

In section 6 and 7 we deal with an extension of Leibnitz' formula onto products of ψdo's. For a function $a(x) \in C^{\infty}$ and a differential expression p(D) one has Leibnitz' formula in the form

$$(0.1) \qquad p(D)a(M) = \sum_{\theta} (-i)^{|\theta|}/\theta!\ a^{(\theta)}(M)p^{(\theta)}(D) \quad ,$$

where this is a finite sum since the derivatives of the polynomial $p(\xi)$ vanish for sufficiently large θ. If now $a \in CB^{\infty}(\mathbb{R}^n)$, and the polynomial p is replaced by a function in M(as in II, definition 4.1), then the series (0,1) will be infinite. We show that it may be regarded as an asymptotic expansion mod $O(-\infty)$: Subtracting a partial sum $\sum_{|\theta|\leq k}$ from the left hand side in (0.1) will yield an operator of order tending to $-\infty$, as k gets large (theorem 6.4). Such asymptotic expansion can be proven by application of Taylor's formula together with the results on the Fourier transform of distributions with rational singularities in Chapter II. This also is reflected in an asymptotic calculus for more general ψdo's in ΨO (section 7).

In section 8 we discuss a compactness result: The operator $a(M)b(D)$: $\mathfrak{H}_0 \to \mathfrak{H}_0$ is compact whenever $a,b \in CO(\mathbb{R}^n)$. In section 9 this result is extended to general $\mathfrak{H}_s$. There we start considering a subalgebra $\mathring{\mathfrak{A}} \subset \mathring{\mathfrak{L}}$, obtained by chosing the coefficients of the generating differential operators in the algebra

$$(0.2) \qquad A = \{a \in CB^{\infty}(\mathbb{R}^n) : a^{(\alpha)}(x) = o(1), \text{ as } \alpha \neq 0 \text{ and } |x| \to \infty \}.$$

The functions of (0.2) will be said to have oscillation vanishing at ∞. The same notation is used for the functions of the closure $CM(\mathbb{R}^n)$ of A in $CB(\mathbb{R})$, under sup norm.

Section 9 and 10 develop some results preparatory to a Banach algebra study attempted in chapter IV. The closure of $\mathring{\mathfrak{A}}$ under operator norm of any $\mathfrak{H}_s$, $s \in \mathbb{R}$, will be seen to be a C^*-subalgebra of $L(\mathfrak{H}_s)$, containing the compact ideal $K(\mathfrak{H}_s)$ and having compact commutators. In fact all algebras $\mathfrak{A}_s$ obtained in this way will prove canonically isomorphic. A key of these future investigations will be an improved asymptotic expansion of the operator $C_s = \Lambda^{-s}a(M)\Lambda^s - a(M)$, as $s \in \mathbb{R}$, $a \in A$. This operator will prove compact, in every $\mathfrak{H}_s$, in particular.

It is clear that the operators Λ^s ,$s \geqslant 0$, may be written as (singular) convolution operator with kernel being the well known Bessel potential. Similarly all the finitely generated ψdo's prove to be (singular) integro-differential operators.In chapter IV we shall derive certain necessary and sufficient conditions for operators in our algebras to be normally solvable (Fredholm).This will be the principal motivation of the studied undertaken in the present chapter.

1. Definition of L^2-Sobolev spaces

We introduce the spaces $\mathfrak{H}_s$ of temperate distributions over $\mathbb{R}^n$ by

(1.1) $$\mathfrak{H}_s = \left\{ u \in \mathcal{S}' : u^\wedge \in L^1_{loc} , \int_{\mathbb{R}^n} |u^\wedge(\xi)|^2 (1+\xi^2)^s d\xi < \infty \right\} ,$$

where s is any real number and where "^" denotes the Fourier transform (I,3). The space $\mathfrak{H}_s$ will be called <u>the (L^2-) Sobolev space (of order s) (over $\mathbb{R}^n$)</u>. Theory of these spaces is simple, as compared to the case of L^p or more general domains $\Omega \subset \mathbb{R}^n$, but it is all we require at the moment.

First of all we notice that the space

(1.2) $$F\mathfrak{H}_s = \{ u \in \mathcal{S}' : u = v^\wedge , v \in \mathfrak{H}_s \}$$

coincides with $L^2(\mathbb{R}^n, d\mu_s)$ with the measure $d\mu_s = (1+x^2)^s dx$. Recall that $F: \mathcal{S}' \to \mathcal{S}'$ is an isomorphism of the vector space $\mathcal{S}'$ onto itself. Also $L^2(\mathbb{R}^n, d\mu_s)$ is a Hilbert space, with inner product

(1.3) $$(u,v)^\sim_s = \int_{\mathbb{R}^n} \bar{u}(x) v(x) (1+x^2)^s dx .$$

Accordingly the Hilbert space structure can be taken over into $\mathfrak{H}_s$, and we have proven:

<u>Theorem 1.1.</u> $\mathfrak{H}_s$ is a Hilbert space of temperate distributions with

(1.4) $$(u,v)_s = \int_{\mathbb{R}^n} \overline{u^\wedge(\xi)}\, v^\wedge(\xi)\, d\mu_s , \qquad \|u\|_s^2 = \int_{\mathbb{R}^n} |u^\wedge(\xi)|^2 d\mu_s$$

as inner product and norm.

<u>Corollary 1.1'.</u> We have

(1.5) $$\mathfrak{H}_0 = \mathfrak{H} = L^2(\mathbb{R}^n, dx) .$$

<u>Proof.</u> Using I, lemma 4.5 (i.e., Parseval's relation) we conclude that

(1.6) $$\int_{\mathbb{R}^n} |u^\wedge|^2 d\mu_0 = \int_{\mathbb{R}^n} |u^\wedge|^2 dx = \int_{\mathbb{R}^n} |u|^2 dx$$

which implies the corollary.

<u>Corollary 1.1''.</u> $\{ \mathfrak{H}_s , -\infty < s < \infty \}$ constitutes a decreasing family of linear subspaces of $\mathcal{S}'$. That is,

(1.7) $$\mathfrak{H}_s \subset \mathfrak{H}_t \quad \text{as} \quad s > t .$$

Moreover, the norm $\|u\|_s$ increases as s increases, i.e.

1.8) $$u \in \mathfrak{H}_s , \quad s > t \quad \text{implies} \quad \|u\|_s \geq \|u\|_t .$$

Proof. This follows because of

(1.9) $$d\mu_s/d\mu_t = (1+x^2)^{s-t} \geq 1 , \quad \text{for all } x \in \mathbb{R}^n \text{ and } s \geq t.$$

Remark. It is clear nor that $\mathcal{G}_s$, $s \geq 0$, contains functions only, and no proper distributions, since already the elements of $\mathcal{G}_0$ are L^2-functions. In fact, the class $\mathcal{G}_s$ gets more and more restricted as s increases because of the increase of the measure $d\mu_s$. This is reflected in the result below.

Theorem 1.2. (Sobolev's lemma). We have

(1.10) $$\mathcal{G}_s \subset CO(\mathbb{R}^n) \quad \text{for} \quad s > n/2 ,$$

with the space $CO(\mathbb{R}^n)$ of bounded continuous functions with limit zero at ∞ , as defined in I,1.

Proof. Let $s > n/2$ and let $u \in \mathcal{G}_s$. We get $u^\wedge \in L^1_{loc}$ and $\int |u^\wedge|^2 (1+\xi^2)^s d\xi < \infty$ so that (by Schwarz' inequality)

(1.11) $$\int_{\mathbb{R}^n} |u^\wedge| d\xi \leq \left\{ \int |u^\wedge|^2 (1+\xi^2)^s d\xi \int d\xi/(1+\xi^2)^s \right\}^{1/2} = c_s \|u\|_s$$

where $c_s = \left[\int (1+\xi^2)^{-s} d\xi \right]^{1/2} < \infty$, due to $s > n/2$. Accordingly we get $u^\wedge \in L^1(\mathbb{R}^n)$ and thus $u = u^{\wedge\vee} \in CO(\mathbb{R}^n)$, by I, theorem 3.6, q.e.d.

Corollary 1.2'. For $s > n/2$ and $u \in \mathcal{G}_s$ we have

(1.12) $$\|u\|_{L^\infty(\mathbb{R}^n)} \leq (2\pi)^{-n/2} c_s \|u\|_s ,$$

with c_s as above. Moreover ,

(1.13) $$|u(x)-u(y)| \leq \delta_s(|x-y|) \|u\|_s , \quad x,y \in \mathbb{R}^n ,$$

with the continuous function

(1.14) $$\delta_s(t) = 2(2\pi)^{-n/2} \left\{ \int_{\mathbb{R}^n} \sin^2 \xi_1 t/2 \, d\mu_{-s}(\xi) \right\}^{1/2}$$

Proof. Formula (1.12) is evident from the above and I,(3.30). Regarding (1.13) we write [x)]

(1.15) $$\begin{aligned} |u(x) - u(y)| &= \left| \int u^\wedge(\xi)(e^{ix\xi} - e^{iy\xi}) đ\xi \right| \\ &\leq \left\{ \int |e^{ix\xi} - e^{iy\xi}|^2 đ\mu_{-s}(\xi) \int |u^\wedge|^2 đ\mu_s \right\}^{1/2} \leq \delta_s(|x-y|) \|u\|_s \end{aligned}$$

x) Here and in the following we denote $đx = (2\pi)^{-n/2} dx$, and similarly $đ\mu = (2\pi)^{-n/2} d\mu$ for any measure over $\mathbb{R}^n$.

with

$$\delta_s(|x-y|) = (2\pi)^{-n/2}\left\{\int |e^{ix\xi} - e^{iy\xi}|^2 \, d\mu_{-s}(\xi)\right\}^{1/2}$$

(1.16)

$$= 2(2\pi)^{-n/2}\left\{\int \sin^2 (x-y)\xi/2 \, d\mu_{-s}(\xi)\right\}^{1/2} ,$$

where the right hand side is a function of $|x-y|$ only because the integral does not change if x-y is substituted by O(x-y) with an orthogonal matrix O. Thus we may set $x-y = |x-y|(1,0,\dots,0)$ which brings (1.16) onto a form corresponding to (1.14), q.e.d.

Corollary 1.2''. The injection map $\mathfrak{H}_s \to CO(\mathbb{R}^n)$ of (1.10) is a compact operator from the Hilbert space $\mathfrak{H}_s$ to the Banach space $CO(\mathbb{R}^n)$ (with the norm of L^∞).

Proof. A bounded sequence of functions in $\mathfrak{H}_s$ is bounded and equicontinuous over (the 1-point compactification of) $\mathbb{R}^n$, and thus must contain a uniformly convergent subsequence, by the theorem of Arzela-Ascoli, q.e.d.

Corollary 1.2'''. Let $s > n/2 + k$ for some nonnegative integer k then

$$\text{(1.17)} \qquad \mathfrak{H}_s \subset CO^k(\mathbb{R}^n)$$

and the injection operator is compact again. Here

$$\text{(1.18)} \qquad CO^k(\mathbb{R}^n) = \{ \ u \in \mathcal{S}' : u^{(\alpha)} \in CO(\mathbb{R}^n) \ \text{ for all } |\alpha| \le k \ \}$$

where $u^{(\alpha)}$ denotes the distribution derivative of order α of u, and where $CO^k(\mathbb{R}^n)$ is regarded as a Banach space under the norm

$$\text{(1.19)} \qquad \|u\|_{CO^k} = \sum_{|\alpha|\le k} \|u^{(\alpha)}\|_{L^\infty} .$$

The proof of Corollary 1.2''' is evident after the lemma, below.

Lemma 1.3. Let $u \in \mathfrak{H}_s$ then we have $u^{(\alpha)} \in \mathfrak{H}_{s-|\alpha|}$ for all multiindices α, where $u^{(\alpha)}$ denotes the αth distribution derivative of u. Moreover we have $\|u^{(\alpha)}\|_{s-|\alpha|} \le \|u\|_s$.

Proof. Recall that for $u \in \mathcal{S}'$ we have (I,(4.15))

$$\text{(1.20)} \qquad (D^\alpha u)^\wedge(\xi) = \xi^\alpha u^\wedge(\xi) ,$$

For $u \in \mathfrak{H}_s$ we thus have $u^{(\alpha)\wedge} = i^{|\alpha|}(D^\alpha u)^\wedge \in L^1_{loc}$ and then get

$$\text{(1.21)} \qquad \int |u^{(\alpha)\wedge}|^2 \, d\mu_{s-|\alpha|} = \int |u^\wedge|^2 \, \xi^{2\alpha}(1+\xi^2)^{-|\alpha|} \, d\mu_s \le \|u\|_s^2$$

because one easily verifies that

$$\text{(1.22)} \qquad \xi^{2\alpha} \le (1+\xi^2)^{|\alpha|} .$$

Remark. Please note that I, lemma 2.4 and Corollary 1.2 may be used to prove that the space $CO^k(\mathbb{R}^n)$ of (1.18) is a subspace of $C^k(\mathbb{R}^n)$.

Next we define the two locally convex spaces

$$(1.23)\qquad \mathfrak{H}_\infty = \bigcap_{s\,\varepsilon\,\mathbb{R}} \mathfrak{H}_s\,, \qquad \mathfrak{H}_{-\infty} = \bigcup_{s\,\varepsilon\,\mathbb{R}} \mathfrak{H}_s\,.$$

The space $\mathfrak{H}_\infty$ carries the locally convex topology induced by the system $\{\|u\|_s : s\,\varepsilon\,\mathbb{R}\}$ of all $\mathfrak{H}_s$-norms. It is a Frechet space since the monotony (i.e. (1.8)) of norms allows to select a countable subset of norms, for example $\{\|u\|_k : k=0,1,\dots\}$.

On the other hand the space $\mathfrak{H}_{-\infty}$ is equipped with the inductive limit topology (strongest locally convex topology leaving all the injections $\mathfrak{H}_s \to \mathfrak{H}_{-\infty}$ continuous).

As a trivial consequence of theorem 1.2 we have:

Lemma 1.4. $\mathfrak{H}_\infty$ consists of C^∞-functions. In fact, we get

$$(1.24)\qquad \mathfrak{H}_\infty \subset CO^\infty(\mathbb{R}^n) = \{\ u\,\varepsilon\, C^\infty(\mathbb{R}^n) : u^{(\alpha)}\,\varepsilon\, CO(\mathbb{R}^n)\ \}\,.$$

We also find the following variant of Sobolev's Lemma useful.

Lemma 1.5. If $s>1/2$, then for $u\,\varepsilon\,\mathfrak{H}_s \subset L^2(\mathbb{R}^n)$ the assignment

$$(1.25)\qquad x_n \to u(\cdot,x_n)$$

defines a bounded continuous map from $\mathbb{R} = \{-\infty<x_n<\infty\}$ to $L^2(\mathbb{R}^{n-1})$.

Proof. For $s>1/2$ and $u\,\varepsilon\,\mathfrak{H}_s$ we have $u^\wedge \varepsilon L^2(\mathbb{R}^n, d\mu_s)$, by (1.1), which gives $v_{\xi_n} = u^\wedge(\cdot,\xi_n)\ \varepsilon\ L^2(\mathbb{R}^{n-1})$ for almost all $\xi_n\,\varepsilon\,\mathbb{R}$ and

$$(1.26)\qquad \int \|v_{\xi_n}\|^2_{L^2(\mathbb{R}^{n-1})} (1+\xi_n^2)^s d\xi_n \le \|u\|_s^2 < \infty\,,$$

by Fubini's theorem. Accordingly

$$(1.27)\qquad \int \|v_{\xi_n}\| d\xi_n \le \{\int d\xi_n (1+\xi_n^2)^{-s} \int \|v_{\xi_n}\|^2_{L^2(\mathbb{R}^{n-1})} (1+\xi_n)^2 d\xi_n\}^{1/2} < \infty\,,$$

as an amendment of (1.11). Using Parseval's relation in n-1 dimensions one concludes that (1.25) defines a bounded function $\mathbb{R} \to L^2(\mathbb{R}^{n-1})$ given in the form

$$(1.28)\qquad u(\cdot,x_n) = \int e^{ix_n\xi_n} v^\vee_{\xi_n} \,d\!\!\!{}^-\xi_n\,.$$

Similar to (1.13), by (1.15), we get

$$(1.29)\qquad \|u(\cdot,x_n)-u(\cdot,y_n)\|_{L^2(\mathbb{R}^{n-1})} \le \delta(|x_n-y_n|)\,\|u\|_s\,,\quad x_n,y_n\varepsilon\mathbb{R},\ \text{q.e.d.}$$

Problems. 1) Show that $E' \subset \mathfrak{H}_{-\infty}$. 2) Show that $S \subset \mathfrak{H}_\infty$, and that this is a proper inclusion. 3) Investigate compactness of the Sobolev imbedding $\mathfrak{H}_\infty \to CO^\infty(\mathbb{R}^n)$ (For compactness of operators between Frechet spaces c.f. IV,3). 4) Show that the function of (1.25) even maps $\mathfrak{H}_s$ into $CB(\mathbb{R}, \mathfrak{H}_t(\mathbb{R}^{n-1}))$, for every $t < s - 1/2$.

2. Some more simple properties of Sobolev spaces.

Let us recall the notion of formal Fourier multiplier, as defined in I, section 4: For a function $a \in T$ (i.e. for a multiplier of S [all derivatives are bounded by powers of $1+|x|$]) we introduced the linear operator $a(D): S' \to S'$ by the sequence

$$(2.1) \qquad S' \underset{F}{\to} S' \underset{u \to au}{\to} S' \underset{\bar{F}}{\to} S' .$$

That is, if the multiplication operator $u \to au$ is denoted by $a(M)$ we get

$$(2.2) \qquad a(D) = F^{-1} a(M) F .$$

Note that the functions $\lambda^s(x) = (1+x^2)^{-s/2}$ are in T, so that the formal Fourier multipliers

$$(2.3) \qquad \lambda^s(D) = (1-\Delta)^{-s/2} , \quad \Delta = -|D|^2 = \sum_{j=1}^{n} \partial^2/\partial x_j^2$$

are well defined.

Lemma 2.1. We have

$$(2.4) \qquad \mathfrak{H}_s = \left\{ u \in S' : (1-\Delta)^{s/2} u \in L^2(\mathbb{R}^n) \right\}$$

and

$$(2.5) \qquad \|u\|_s = \|(1-\Delta)^{s/2} u\|_{L^2(\mathbb{R}^n)} .$$

Proof. Using Parseval's relation (I, lemma 4.5) we get

$$(2.6) \qquad (1+\xi^2)^{s/2} u^\wedge \in L^2 \text{ if and only if } ((1+\xi^2)^{s/2} u^\wedge)^\vee = (1-\Delta)^{s/2} u \in L^2 ,$$

and, in fact, both functions have the same L^2-norm. Thus (2.4) follows from (1.1), q.e.d.

Lemma 2.2. Let $k=0,1,2,\ldots$ be a non-negative integer, then we get $u \in \mathfrak{H}_k$ if and only if all distribution derivatives $u^{(\alpha)}$, $|\alpha| \leq k$, are in $L^2(\mathbb{R}^n)$. Moreover the Sobolev norm $\|u\|_k$ is equivalent to the norm

$$(2.7) \qquad \|u\|^{\bullet}_k = \left\{ \sum_{|\alpha| \leq k} \|u^{(\alpha)}\|^2_{L^2(\mathbb{R}^n)} \right\}^{1/2} .$$

We have (with some number p_k depending only on k, not on u)

$$(2.8) \qquad \|u\|^{\bullet}_k \leq \|u\|_k \leq p_k \|u\|^{\bullet}_k .$$

Proof. It follows from Lemma 1.3 that $u \in \mathfrak{H}_k$ has all $u^{(\alpha)}$, $|\alpha| \leq k$, in $\mathfrak{H}_0 = L^2$. Vice versa, if the latter is the case, we must have

$$(2.9) \qquad \|u^{(\alpha)}\|_{L^2} = \|(u^{(\alpha)})^\wedge\|_{L^2} = \|\xi^\alpha u^\wedge\|_{L^2} < \infty , \quad |\alpha| \leq k .$$

Accordingly

$$(2.10) \qquad \|u\|_k^2 = \int |u^\wedge|^2 (1+\xi^2)^k d\xi = \sum_{|\alpha|\le k} c_\alpha \|\xi^\alpha u^\wedge\|^2 < \infty ,$$

with the numbers $c_\alpha \geq 1$ determined by the multinomial expansion

$$(2.11) \qquad (1 + x^2)^k = \sum_{|\alpha|\le k} c_\alpha x^{2\alpha} , \quad x \in \mathbb{R}^n .$$

In particular (2.8) follows with $p_k = \operatorname{Max}_{|\alpha|\le k} c_\alpha^{1/2}$, q.e.d.

The notion of inner product, commonly introduced for $u,v \in \mathfrak{H} = \mathfrak{H}_0 = L^2(\mathbb{R}^n)$ by

$$(2.12) \qquad (u,v) = (u,v)_0 = \int \bar{u}\, v\, dx = \int \bar{u}^\wedge v^\wedge d\xi$$

may be extended by setting

$$(2.13) \qquad (u,v) = \int \bar{u}^\wedge v^\wedge d\xi \qquad \text{for } u \in \mathfrak{H}_s , \quad v \in \mathfrak{H}_{-s} .$$

It is evident that the integral in (2.13) is meaningful, and that we have Schwarz' inequality in the form

$$(2.14) \qquad |(u,v)| \le \|u\|_s \cdot \|v\|_{-s} , \quad u \in \mathfrak{H}_s , \quad v \in \mathfrak{H}_{-s} .$$

Also this shows that the bounded linear functionals over $\mathfrak{H}_s$, i.e. the elements of the adjoint space $\mathfrak{H}_s^*$ are naturally identified with the elements of $\mathfrak{H}_{-s}$: Every bounded linear functional over $\mathfrak{H}_s$ is uniquely of the form

$$(2.15) \qquad l(u) = (v,u) , \quad u \in \mathfrak{H}_s ,$$

with some unique $v \in \mathfrak{H}_{-s}$.

Note that this representation of $\mathfrak{H}_s^*$ is different from the usual representation of the dual of the Hilbert space $\mathfrak{H}_s$ by $\mathfrak{H}_s$ itself, involving the inner product $(u,v)_s$, not (u,v) above.

Note also that we get

$$(2.16) \qquad \|u\|_{-s} = \sup \left\{ |(u,v)| : v \in \mathfrak{H}_s, \ \|v\|_s = 1 \right\} .$$

Problems.

1) Show that $C_0^\infty(\mathbb{R}^n)$ is dense in every $\mathfrak{H}_s$, including $s = \infty$.

2) For a nonnegative integer k the space $\mathfrak{H}_k$ is equal to the class of all functions $u \in L^2(\mathbb{R}^n)$ with the property that there exists a sequence $\phi_j \in C_0^\infty(\mathbb{R}^n)$ such that $\phi_j \to u$ in $L^2(\mathbb{R}^n)$ while for each multi-index α with $|\alpha| \le k$ the sequence of derivatives $_j^{(\alpha)}$, j=1,2,... is Cauchy in $L^2(\mathbb{R}^n)$. [Prove this fact; it is interesting because it amounts to a distribution free characterization of $\mathfrak{H}_k$.]

3) The limit of the Cauchy sequence $\phi_j^{(\alpha)}$ of problem 2 is independent of the choice of the sequence ϕ_j. This limit is often referred to as the strong $L^2(\mathbb{R}^n)$-derivative of u, and of order α.

3. $\mathfrak{H}_s$ - bounded operators.

Our main interest in this section aims toward boundedness of multipliers and formal Fourier multipliers as operators from $\mathfrak{H}_s$ to $\mathfrak{H}_s$ [or, more generally from $\mathfrak{H}_s$ to $\mathfrak{H}_t$].

First of all boundedness of formal Fourier multipliers is trivial. We have seen in section 1 that $F\mathfrak{H}_s = L^2(\mathbb{R}^n, d\mu_s)$. In fact the Fourier transform acts as an isometry between $\mathfrak{H}_s$ and $L^2(\mathbb{R}^n, d\mu_s)$. Accordingly for $a \in T$ we get

$$(3.1) \qquad \|a(D)u\|_t = \|(1+D^2)^{t/2} a(D)u\|_{L^2} = \|a(D)(1+D^2)^{t/2}u\|_{L^2},$$

using Lemma 2.1, and that formal Fourier multipliers commute. Also

$$(3.2) \qquad \|a(D)v\|_{L^2} = \|av^\wedge\|_{L^2} \le c_{st}\|(1+\xi^2)^{(s-t)/2}v^\wedge\|_{L^2}$$

$$= \|(1-\Delta)^{(s-t)/2}v\|_{L^2}$$

whenever

$$(3.3) \qquad c_{st} = \sup\left\{ |a(x)| \; (1+x^2)^{-(s-t)/2} : x \in \mathbb{R}^n \right\} \le \infty$$

Then (3.1) and (3.2) imply that

$$(3.4) \qquad \|a(D)u\|_t \le c_{st}\|u\|_s .$$

Lemma 3.1. A formal Fourier multiplier $a(D)$, $a \in T$ constitutes a bounded operator from $\mathfrak{H}_s$ to $\mathfrak{H}_t$ if and only if the constant c_{st} of (3.3) above is finite. Moreover we get c_{st} equal to the operator norm. That is,

$$(3.5) \qquad \sup\left\{ \|a(D)u\|_t : \|u\|_s \le 1 \right\} = c_{st} .$$

Proof. In view of (3.4) it suffices to show that a sequence $u_k \in \mathfrak{H}_s$ can be constructed with $\|u_k\|_s = 1$ and $\|a(D)u_k\|_t \to c_{st}$, as $k \to \infty$. Such a sequence is readily obtained by requiring that the Fourier transform $u_k^\wedge$ has its support in a set of $\mathbb{R}^n$ where one has

$$(3.6) \qquad |a(x)| \; (1+x^2)^{(t-s)/2} \ge c_{st} - 1/k ,$$

q.e.d.

Next we ask the question for a class of multipliers leaving $\mathfrak{H}_s$ invariant. The Lemma , below, is merely a curiosity for us, not a matter of vital importance.

Lemma 3.2. Let $s > n/2$, then $\mathfrak{H}_s$ is a Banach algebra under pointwise multiplication. That is, $u,v \in \mathfrak{H}_s$ implies $uv \in \mathfrak{H}_s$ and we have

$$(3.7) \qquad \|uv\|_s \le c_s\|u\|_s\|v\|_s$$

with a constant c_s independent of u,v.

The proof below seems due to T. Kato (unpublished) : Notice that (by I, (3.26), for $u,v \in S$)

$$(3.8) \qquad (uv)^\wedge(\xi) = \int u^\wedge(\xi-\eta)v^\wedge(\eta)đ\eta \ .$$

Let us introduce the functions

$$(3.9) \qquad u^\wedge(1+\xi^2)^{s/2} = U \ , \quad v^\wedge(1+\xi^2)^{s/2} = V \ , \quad \lambda(\xi) = (1+\xi^2)^{-1/2} \ .$$

Substituting this into (3.8) and using (1.4) we conclude that

$$(3.10) \qquad \begin{aligned} \|uv\|_s^2 &= \int |(uv)^\wedge|^2 \, d\mu_s \\ &= (2\pi)^{-n} \int \lambda^{-2s}(\xi)d\xi \, \left|\int U(\xi-\eta)V(\eta)\lambda^s(\xi-\eta)\lambda^s(\eta) \, d\eta \right|^2 . \end{aligned}$$

Applying Schwarz' inequality we may continue estimate (3.10) as follows:

$$\leq (2\pi)^{-n} \int d\xi \int d\eta \, |U(\xi-\eta)|^2 |V(\eta)|^2 \rho(\xi)$$

with *)

$$(3.11) \qquad \rho(\xi) = \lambda^{-2s}(\xi) \int d\eta \, \lambda^{2s}(\xi-\eta)\lambda^{2s}(\eta) < \infty \qquad \text{as} \quad s > n/2 \ .$$

A simple calculation shows that $\rho(\xi)$ is bounded over $\mathbb{R}^n$, and thus we get

$$(3.12) \qquad \|uv\|_s \leq c_s\left\{\int d\xi \, d\eta \, |U(\xi)|^2|V(\eta)|^2\right\}^{1/2} = c_s\|u\|_s\|v\|_s \ ,$$

q.e.d.

The next following result is of fundamental importance for much of our discussion in this and the following chapter.

<u>Lemma 3.3.</u> Let $a \in CB^\infty(\mathbb{R}^n)$ then $a(M)$ [the multiplication operator induced] is a bounded operator of every Hilbert space $\mathfrak{H}_s$. Here $CB^\infty(\mathbb{R}^n)$ denotes the collection of all $C^\infty(\mathbb{R}^n)$-functions which are bounded with all their derivatives.

<u>Proof.</u> First we note that the assertion is evident for the case of $s = k \geq 0$ being an integer. In that case it is sufficient to show that

$$(3.13) \qquad \|(a(M)u)^{(\alpha)}\|_{L^2} \leq c_\alpha\|u\|_k \ , \quad u \in \mathfrak{H}_s \ , \quad |\alpha| \leq k \ ,$$

in view of the equivalence (2.8). But by Leibnitz formula I, (1.23) we may estimate

$$(3.14) \qquad |(au)^{(\alpha)}(x)|^2 \leq c_\alpha\|a\|^2_{C^k(\mathbb{R}^n)} \sum_{|\beta|\leq|\alpha|} |u^{(\beta)}(x)|^2 \ ,$$

which implies (3.13), if we integrate.

---

*) By Fubini's Theorem we get $\int d\xi \int d\eta |U^2(\xi-\eta)V^2(\eta)| = \int d\xi \int d\eta |U(\xi)|^2|V(\eta)|^2 < \infty$

Similarly we conclude the statement of the lemma for $s = -k < 0$, using the duality between $\mathfrak{H}_k$ and $\mathfrak{H}_{-k}$: For $u, \phi \in C_0^\infty(\mathbb{R}^n)$ we have

(3.15) $$|(a(M)u,\phi)| = |(u,\bar{a}(M)\phi)| \leq \|u\|_{-k} \|\bar{a}(M)\phi\|_k \leq c_k\|u\|_{-k}\|\phi\|_k$$

using the boundedness of $\bar{a}(M)$ over $\mathfrak{H}_k$. Then (2.16) implies that

$$\|a(M)u\|_{-k} = \sup\left\{ |(a(M)u,\phi)| : \|\phi\|_s \leq 1 \right\} \leq c_k\|u\|_{-k} .$$

It follows that $a(M)$ is trivially a bounded operator over $\mathfrak{H}_k$, for each integer k.

The same boundedness for general $\mathfrak{H}_s$ seems less trivial. One way of proving this is discussed in Cordes-Herman [3], using a commutator relation between $a(M)$ and the operators $\Lambda^s = (1-\Delta)^{-s/2} = (1+D^2)^{-s/2}$ [lemma 6 in [3]] which is stated for the purposes there under the stronger assumption that the derivatives of $a(x)$ vanish at infinity. However the same technique, employing an asymptotic expansion like (18) in [3] (c.f. also section 6 below) allows us to prove L^2-boundedness of $\Lambda^s a(M)\Lambda^{-s}$ for all $s \in \mathbb{R}$ if only $a \in CB^\infty(\mathbb{R}^n)$.]

Let us discuss another way of proving this, by means of an interpolation theorem, which seems due to A. Calderon (c.f. Seeley |11|, Theorem 5).

Theorem 3.4. Suppose a linear map $A: \mathcal{S} \to \mathcal{S}'$ satisfies the boundedness relations (for some pair $s < t$ of real numbers) $Au \in \mathfrak{H}_t$, $u \in \mathcal{S}$, and

(3.16) $$\|Au\|_s \leq c_s \|u\|_s \quad \text{and} \quad \|Au\|_t \leq c_t\|u\|_t , \quad u \in \mathcal{S}.$$

Then we have

(3.17) $$\|Au\|_r \leq c_r \|u\|_r , \quad s \leq r \leq t , \quad u \in \mathcal{S} ,$$

where we may choose

(3.18) $$c_r = c_s^{(t-r)/(t-s)} \cdot c_t^{(r-s)/(t-s)} .$$

It is trivial that $\mathcal{S}$ is dense in every $\mathfrak{H}_s$. Therefore since we know that $a(M)$ for $a \in CB^\infty(\mathbb{R}^n)$ is bounded over $\mathfrak{H}_k$ for each integer $k = 0,\pm 1,\pm 2,\ldots$ theorem 3.4 implies lemma 3.3 for all s. (Just apply it for $s=k$, $t=k+1$). There will be another important application of theorem 3.4 later on.

Proof of theorem 3.4. We shall have to use the Phragmen-Lindeloef principle of complex variables theory, which says that a bounded and holomorphic function $f(z)$ over the strip $0 \leq \text{Re } z \leq 1$ must satisfy the estimate

(3.19) $$|f(x+iy)| \leq M_0^{1-x} M_1^x , \quad y \in \mathbb{R}, \quad 0 \leq x \leq 1 ,$$

with

(3.20) $$M_j = \sup\left\{ |f(j+iy)| : y \in \mathbb{R}\right\}, \quad j = 0,1 .$$

(C.f. J. Marsden, [10], p. 338).

Let us introduce the function (for fixed $u,v \in S$)

$$(3.21) \qquad f(z) = < A \Lambda^{s(z)} u , \Lambda^{-s(z)} v >$$

with $A \Lambda^{s(z)} u \in S'$, $\Lambda^{-s(z)} v \in S$ and the operator $\Lambda = (1-\Delta)^{-1/2}$ as above, and with $s(z) = (1-z)s+zt$. Clearly the relations (3.16) imply continuity of A from $\mathfrak{H}_s$ to $\mathfrak{H}_s$, thus certainly from S to S' (use weak topology in S'). Accordingly one may show that the function f(z) above is well defined for all z and that its complex derivative for z exists everywhere [because $d/dz(\Lambda^{\pm s(z)} u)$ exists in the convergence of S, by continuity of $F: S \to S$ and $F^{-1}: S \to S$, and using that $d/dz((1+x^2)^{s(z)} u)$ exists in the convergence of S.] Moreover the function is bounded in the strip $0 \le \text{Re } z \le 1$. Indeed, we may write

$$(3.22) \qquad f(z) = < \Lambda^{-s} A \Lambda^{s} u(z), v(z) >$$

where $A_s = \Lambda^{-s} A \Lambda^{s}$ is L^2-bounded (as a consequence of (3.16)) and with $u(z) = \Lambda^{s(z)-s} u$, $v = \Lambda^{s-s(z)} v$ being L^2-bounded in the above strip, so that $|f(z)| \le c_s \|u(z)\|_{L^2} \|v(z)\|_{L^2} \le c$. Respectively, on the lines $\text{Re } z = 0;1$, we get $|f(z)| \le c_s \|u\|_0 \|v\|_0$, $|f(z)| \le c_t \|u\|_0 \|v\|_0$, due to $\|\Lambda^{i\sigma}\|_{L^2} = 1$.

Accordingly the Phragmen - Lindeloef principle gives

$$(3.23) \qquad |<A \Lambda^{s(z)} u, \Lambda^{-s(z)} v>| \le c_r \|u\|_0 \|v\|_0 , \quad r = (1-x)s+xt ,$$

with c_r as in (3.18). In particular for z on the real axis this amounts to

$$(3.24) \qquad |< \Lambda^{-r} A \Lambda^{r} u, v >| \le c_r \|u\|_0 \|v\|_0 , \quad u,v \in S .$$

Accordingly the temperate distribution $\Lambda^{-r} A \Lambda^{r} u$ defines a bounded linear functional over $L^2 = \mathfrak{H}_0$ and therefore must be a function in $L^2 = \mathfrak{H}_0$, by the Frechet-Riesz theorem. Moreover we have

$$(3.25) \qquad \| \Lambda^{-r} A \Lambda^{r} u\|_0 \le c_r \|u\|_0 , \quad u \in S .$$

Letting $\Lambda^{r} u = v$ in (3.24) we get (3.17), due to (2.5), q.e.d.

Problems.

1) Write out in details the proof of the fact that f(z) above is bounded and analytic in the strip $0 \le \text{Re } z \le 1$.

2) Show that the equivalence (2.7) of the L^2-Sobolev norms $\|u\|_k$ and $\|u\dot{\|}_k$ has its counterpart for L^p-Sobolev norms, with $\|u\|_{s,p}$ as above and $\|u\dot{\|}_{k,p} = \sum_{|\alpha| \le k} \|u^{(\alpha)}\|_{L^p(\mathbb{R}^n)}$, provided that $1 < p < \infty$.
[Hint: Use that the operators Λ, $D_j \Lambda$ are L^p-bounded singular integral operators].

3) Discuss validity of theorem 3.4 (and lemma 3.3) for L^p-Sobolev norms defined by

$$(3.26) \qquad \|u\|_{s,p} = \|(1-\Delta)^{s/2}u\|_{L^p(\mathbb{R}^n)} \; .$$

4) The L^p-boundedness of $S_j = D_j\Lambda$, above, can be derived from the $L^p(\mathbb{R})$-boundedness of the (one dimensional) Hilbert transform $u \to Hu$ with

$$(Hu)(x) = \lim_{\varepsilon\to 0} \int_{|x-y|\geq\varepsilon} u(y)dy/(x-y)$$

using that S_j is a singular convolution operator with odd kernel (by II,Theorem 5.2. Can you reconstruct this three line proof ?(Hint: For an odd pseudo-function $a \in T^0_{ps}$ we have (with a distribution integral) $a(x) = a(-x)$,

$$K_a u(x) = \int a(y)u(x-y)dy = (8\pi)^{-\frac{1}{2}} \int_{|y^0|=1} dS_{y^0} \int_0^\infty a(\rho y^0)((u(x-\rho y^0)-u(x+\rho y^0))\rho^{n-1}\, d\rho \; .$$

In the case of S_j we must chose $a(x) = D_j((1+|\xi|^2)^{-\frac{1}{2}})^{\vee}(x)$,which may be calculated in terms of modified Hankel functions (c.f. I,3,problem 3). Then the inner integral may be linked to the Hilbert transform,while,for the outer integral we require a triangle inequality only.) (Remark: The proof seems due to Calderon and Zygmund.)

5) Consider again a <u>distribution integral operator</u> of the form discussed in II,1,problem 10 . Show that such an operator is in $L(\mathfrak{H}_s)$,for every $s \in \mathbb{R}$,if only the assumptions of II,2,problem 11 are satisfied,i.e.,that the Fourier transform $k^{\wedge 2}(x,\xi)$ with respect to z has all (x,ξ)-derivatives of arbitrary order bounded over $\mathbb{R}^{2n}$.(Hint: The proof of II,2,problem 10 may be repeated with slight amendment.)

6) Under the assumptions of problem 5 above,if the condition that all derivatives of $k^{\wedge 2}(x,\xi)$ are bounded is replaced by the condition that all these derivatives are $O(\langle\xi\rangle^r)$,with some r independent of differentiation, then the operator K is in $L(\mathfrak{H}_s, \mathfrak{H}_{s-r})$,for every s .

4. Operators of $L(\mathfrak{H}_\infty)$ and their true order function.

We have introduced the spaces $\mathfrak{H}_\infty$ and $\mathfrak{H}_{-\infty}$ in (1.23), and there we shortly discussed their topologies. In the following it will prove helpful to introduce certain classes of operators in $L(\mathfrak{H}_\infty)$, denoted by $O(r)$, where r is an extended real number. The operators in $O(r)$ will be said to have order r.

Lemma 4.1. A linear operator $T: \mathfrak{H}_\infty \to \mathfrak{H}_\infty$ is continuous if and only if for every $s \in \mathbb{R}$ there exists $t(s) \in \mathbb{R}$ such that (with some c_s)

$$(4.1) \qquad \|Tu\|_s \le c_s \|u\|_{t(s)} \;, \quad u \in \mathfrak{H}_\infty \;, \quad s \in \mathbb{R} .$$

Proof. Since $\|u\|_s$ increases as s increases a neighbourhood base of the origin will be given by the collection of balls

$$(4.2) \qquad \{\|u\|_s < \varepsilon : \varepsilon > 0, \quad s \in \mathbb{R}\} .$$

Hence T is continuous if and only if for every ball $B_{s,\varepsilon} = \{\|u\|_s < \varepsilon\}$ there exists a ball $B_{t,\delta}$ such that $Tu \in B_{s,\varepsilon}$ whenever $u \in B_{t,\delta}$. It follows that (4.1) holds for $t(s) = t$ and $c_s = \varepsilon/\delta$, q.e.d.

Lemma 4.2. The space $\mathfrak{H}_\infty$ is dense in every $\mathfrak{H}_s$, $s \in \mathbb{R}$.

Proof. We know from section 1 that $F\mathfrak{H}_s = L^2(\mathbb{R}^n, d\mu_s)$ and thus $F\mathfrak{H}_\infty$ is the intersection of all these L^2-spaces over $\mathbb{R}^n$ with different measures $d\mu_s$. By definition of norm in $\mathfrak{H}_s$ the operator $F: \mathfrak{H}_s \to L^2(\mathbb{R}^n, d\mu_s)$ acts as an isometry. Clearly the intersection $\bigcap_{s\in\mathbb{R}} L^2(\mathbb{R}^n, d\mu_s)$ contains the space $C_0^\infty(\mathbb{R}^n)$ which is dense in $L^2(\mathbb{R}^n, d\mu_s)$ for every s. Accordingly $\mathfrak{H}_\infty$ must be dense in every $\mathfrak{H}_s$, q.e.d.

For a given operator $T \in L(\mathfrak{H}_\infty)$ a function $t = t(s)$, $s \in \mathbb{R}$ satisfying (4.1) will be called an order function for T. Every continuous operator of $\mathfrak{H}_\infty$ has infinitely many order functions, because if an arbitrary function $t_1(s)$ satisfies $t_1(s) \ge t(s)$ then it also is an order function, due to $\|u\|_{t(s)} \le \|u\|_{t_1(s)}$. We may look at the infimum $\tau(s) = \inf t(s)$ taken for fixed s and all order functions t(s). The function $\tau(s)$ sometime is called the true order function, although it needs not to be an order function.

Lemma 4.3. The function $\tau(s)$ takes extended real values; it is always either finite or $-\infty$, but never $+\infty$. It is defined for all real s. It is nondecreasing, continuous and convex from below. If $\tau(s)$ is not constantly equal to $-\infty$ or to $r\in\mathbb{R}$ then it increases to $+\infty$.

Proof. Let t(s) be an order function and let $s_0 \in \mathbb{R}$. Note that

$$(4.3) \qquad \|Tu\|_s \le \|Tu\|_{s_0} \le c_{s_0} \|u\|_{t(s_0)}, \quad u \in \mathfrak{H}_\infty, \quad s \le s_0,$$

which shows that the function $t_1(s) = t(s_0)$, $s \le s_0$, $t_1(s) = t(s)$, $s \ge s_0$ also defines an order function. This shows that indeed $\tau(s)$ must be non-decreasing, due to $\inf t(s) \le \inf t_1(s) = \inf t(s_0)$, $s \le s_0$. Now the convexity from below of $\tau(s)$ is a consequence of (a trivial extension of) theorem 3.4, and all other statements of the lemma follow trivially, q.e.d.

Although the true order function does not have to be an order function, it is evident that there always exist order functions which are monotone and convex. In particular we may assume $t(s)$ strictly increasing so that the inverse function $s(t)$ is defined.

Since $\mathfrak{H}_\infty$ is dense in $\mathfrak{H}_s$, by lemma 4.2, there is a unique continuous extension T_t: $\mathfrak{H}_t \to \mathfrak{H}_{s(t)}$ of T, in view of estimate (4.1). The function $t(s)$ increases from some asymptotic value $t_{-\infty} = t(-\infty) \ge -\infty$ to $+\infty$, in view of its assumed convexity. Therefore the inverse function $s(t)$ is defined in $t_{-\infty} < t < \infty$. For any such t an extension T_t will result and these extensions will coincide wherever they are jointly defined.

Let us now assume that

$$(4.4) \qquad \lim_{s\to-\infty} \tau(s) = -\infty,$$

Then we may assume that $t_{-\infty} = -\infty$, and the extension T_t exist for all $t \in \mathbb{R}$. We may introduce an operator $T_{-\infty}$: $\mathfrak{H}_{-\infty} \to \mathfrak{H}_{-\infty}$ by setting $T_{-\infty}u = T_t u$ in $\mathfrak{H}_t$. This will be an operator of $L(\mathfrak{H}_{-\infty})$ since its restrictions T_t to $\mathfrak{H}_t$ are continuous. In fact these restrictions map boundedly into $\mathfrak{H}_{s(t)}$ where $s(t)$, defined above as inverse function is increasing and convex from above, and takes $\mathbb{R}$ continuously onto itself.

Next let us consider a linear map $S \in L(\mathfrak{H}_{-\infty})$. Following [2], p. 14 the continuity of S amounts to the requirement that all the restrictions $S_t = S|\mathfrak{H}_t$ are continuous as maps $\mathfrak{H}_t \to \mathfrak{H}_{-\infty}$. Specifically, if S_t maps boundedly into some $\mathfrak{H}_{s(t)}$, for all $t \in \mathbb{R}$, then S: $\mathfrak{H}_{-\infty} \to \mathfrak{H}_{-\infty}$ is continuous. Or, in other words, continuity of S results if for each $t \in \mathbb{R}$ there exists an $s(t)$ such that

$$(4.5) \qquad \|Su\|_{s(t)} \le c_t \|u\|_t, \quad u \in \mathfrak{H}_t, \quad t \in \mathbb{R}.$$

By a dual chain of conclusions we next introduce the supremum $\sigma(t) = \sup s(t)$, again a <u>true order function</u>, but not necessarily an order function. Again it follows that $\sigma(t)$ is extended real-valued non-decreasing convex from above, may assume ∞ but not $-\infty$. Suppose the condition (dual to (4.4)) holds that

$$(4.6) \qquad \lim_{t\to+\infty} \sigma(t) = +\infty.$$

Then if $S_\infty = S|\mathfrak{H}_\infty$, it follows that S_∞ maps into $\mathfrak{H}_\infty$ since for $u \in \mathfrak{H}_\infty$ and $t \in \mathbb{R}$ we get $S_\infty u = S_t u \in \mathfrak{H}_{s(t)}$, so that $S_\infty u \in \mathfrak{H}_s$ for all s, due to $s(t) \to \infty$. We have proven:

Theorem 4.4. The assignment $T \to T_{-\infty}$ and the reverse assignment $S \to S_\infty$ as introduced above define a 1-1-correspondence between the operators in $L(\mathfrak{H}_\infty)$ and in $L(\mathfrak{H}_{-\infty})$ with true order function satisfying (4.4) and (4.6), respectively.

Problems:

1) A differential operator D^α has true order function $s \to s+|\alpha|$ [or $t \to t-|\alpha|$]. The operator $a(M)$ of multiplication by the function $a \in CB^\infty(\mathbb{R}^n)$ has order function $s \to s$ [where $CB^\infty(\mathbb{R}^n) = \{u: u \in C^\infty(\mathbb{R}^n),\ u^{(\alpha)} = O(1),\ \alpha \in \mathbb{Z}^n_+\}$].

2) Obtain the true order function of the operator $C = \phi \rangle\langle \psi$ defined by $Cu = \phi(\overline{\psi},u)_0$, $u \in \mathfrak{H}_\infty$, with the inner product of $\mathfrak{H}_0$ where $\phi \in \mathfrak{H}_\infty$ but $\psi \in \mathfrak{H}_0$ and $\psi \notin \mathfrak{H}_s$ for any $s > 0$.

3) Let T be an integral operator defined by

$$Tu(x) = \int_{\mathbb{R}^n} T(x,y)u(y)dy, \quad u \in \mathfrak{H}_\infty,$$

where $T(x,y)$ denotes a function in $S(\mathbb{R}^{2n})$. Show that $T \in L(\mathfrak{H}_\infty)$ and that its true order function equals the constant $-\infty$.

4)*) Can you construct an operator in $L(\mathfrak{H}_\infty)$ which has a finite and strictly convex true order function ?

5. The spaces $\mathcal{O}(r)$, $-\infty \le r \le +\infty$ and the algebras $\mathcal{L}$, $\psi\mathring{\mathcal{O}}$.

An operator $T \in L(\mathfrak{H}_\infty)$ having the order function $s \to s+r$ (where $r \in \mathbb{R}$ is constant) will be said to have order r. For such operator the true order function certainly satisfies (4.4). Therefore T admits natural extensions to every $\mathfrak{H}_s$ (mapping from $\mathfrak{H}_s$ to $\mathfrak{H}_{s-r}$) and to $\mathfrak{H}_{-\infty}$ (to $\mathfrak{H}_{\infty}$)). The class of operators in $L(\mathfrak{H}_\infty)$ with a given order $r \in \mathbb{R}$ is denoted by $\mathcal{O}(r)$. Its operators may be likewise regarded as operators $\mathfrak{H}_{-\infty} \to \mathfrak{H}_{-\infty}$ or from $\mathfrak{H}_s$ to $\mathfrak{H}_{s-r}$. We frequently will neglect to distinguish between these various operators arising from each other by a unique extension or restriction procedure. We also define

$$\mathcal{O}(\infty) = \bigcup_{r\in\mathbb{R}} \mathcal{O}(r) , \quad \mathcal{O}(-\infty) = \bigcap_{r\in\mathbb{R}} \mathcal{O}(r) , \tag{5.1}$$

and note that evidently $\mathcal{O}(\infty)$ is an algebra and that $\mathcal{O}(-\infty)$ is a 2-sided ideal of $\mathcal{O}(\infty)$. In fact the elements of $\mathcal{O}(-\infty)$ are characterized by the property of having true order function equal to the constant $-\infty$. Therefore we shall say that these operators have order $-\infty$. More generally we get

Lemma 5.1. $A \in \mathcal{O}(r)$, $B \in \mathcal{O}(p)$ implies

$$A+B \in \mathcal{O}(\mathrm{Max}\,(r,p)) \quad \text{and} \quad AB \in \mathcal{O}(r+p) . \tag{5.2}$$

Lemma 5.2. We have $T \in \mathcal{O}(-\infty)$ if and only if for every $s,t \in \mathbb{R}$ there exists a constant $c_{s,t}$ such that

$$\|Tu\|_s \le c_{s,t}\|u\|_t , \quad u \in \mathfrak{H}_\infty . \tag{5.3}$$

The proofs are evident.

We also note that $\mathcal{O}(0)$ is an algebra, by (5.2), and that again $\mathcal{O}(-\infty)$ is a two-sided ideal of $\mathcal{O}(0)$.

It is easy to construct some concrete operators in $\mathcal{O}(\infty)$, the most important example being that of a differential operator $L: \mathfrak{H}_\infty \to \mathfrak{H}_\infty$ with coefficients in $CB^\infty(\mathbb{R}^n)$. Indeed let L be defined by

$$Lu = \sum_{|\alpha|\le N} a_\alpha(M)D^\alpha u , \quad u \in \mathfrak{H}_\infty, \quad [a \in CB^\infty(\mathbb{R}^n)] , \tag{5.4}$$

then we have $L \in \mathcal{O}(N)$, as an immediate consequence of Lemma 3.3 and Lemma 1.3. As an immediate consequence of Leibnitz' formula I, (1.23) we note that the collection $\mathcal{L}$ of all such differential operators of arbitrary N and arbitrary coefficients in CB^∞ forms an algebra, and a subalgebra of $\mathcal{O}(\infty)$.

On the other hand the algebra $\mathcal{L}$ above contains few operators of order zero. In fact it may be seen that the only operators of order zero are the multiplications a(M).

Our attention will focus on the operators of L, since we are out to derive properties of differential operators. However in this context it proves advantageous to introduce a larger subalgebra of $\mathcal{O}(0)$ (or $\mathcal{O}(\infty)$). Notice in this respect that the operator

$$(5.5) \qquad \Lambda^s = (1-\Delta)^{-s/2} = (1+D^2)^{-s/2}, \quad s \in \mathbb{R},$$

has order (and true order) $-s$. Let us therefore introduce the algebra $\mathring{\mathcal{L}} \subset \mathcal{O}(0)$ generated by the operators

$$(5.6) \qquad A = L\,\Lambda^N$$

with L of the form (5.4). Clearly all operators (5.6) are in $\mathcal{O}(0)$. By the terminology 'algebra generated by' we mean the smallest subalgebra of $\mathcal{O}(0)$ containing the elements (5.6). That is, $\mathring{\mathcal{L}}$ consists of all sums of finite products of elements of the form (5.6). On the other hand let us introduce a large algebra containing L and $\mathring{\mathcal{L}}$, above, by letting $\Psi\mathring{\mathcal{O}}$ denote the algebra generated by L and the operators Λ^s, $s \in \mathbb{R}$.

We note that $\mathcal{O}(0)$ is endowed with a natural locally convex topology: The generating collection of semi-norms is given by

$$(5.7) \qquad \{ \|T\|_s : s \in \mathbb{R} \},$$

with the operator norms

$$(5.8) \qquad \|T\|_s = \sup \{ \|Tu\|_s : \|u\|_s \le 1 \}.$$

Theorem 5.3. $\mathcal{O}(0)$ with the above topology is a Frechet space.

Proof. First of all, it is a consequence of theorem 3.4 that the collection

$$(5.9) \qquad \{ \|T\|_k : k = 0, \pm 1, \pm 2, \ldots \}$$

generates the same topology on $\mathcal{O}(0)$, because for $s = \tau k + (1-\tau)(k+1)$ we get

$$(5.10) \qquad \|T\|_s \le \|T\|_k^{\tau} \|T\|_{k+1}^{(1-\tau)},$$

which implies that every ball $\{\|T\|_s < \varepsilon\}$ contains an intersection of two suitable balls for s being an integer.

Next, if $\{T_j\}$ is a Cauchy sequence, it follows that there exists a limit in every norm $\|\cdot\|_k$ (by AII, p. 1.3), and it is easily seen that this limit must be independent of k, which proves completeness and establishes the theorem.

Let us recall that the elements of $\mathcal{O}(0)$ have a natural interpretation as operators in $L(\mathfrak{H}_s)$ for $-\infty \le s \le +\infty$. Accordingly we may take closure of $\mathring{\mathcal{L}}$ in any such $L(\mathfrak{H}_s)$, as well as in $\mathcal{O}(0)$. Also we shall introduce a 'closure' of the algebra $\mathring{\Psi\mathcal{O}}$ later on. The operators in $\mathring{\mathcal{L}}$ and in $\mathring{\Psi\mathcal{O}}$ shall be called pseudo-differential operators, or, shortly, ψ do-s, following a recent convention. The same notation also shall be used for the elements of the various closures mentioned above. In cases where we wish to distinguish $\mathring{\Psi\mathcal{O}}$ from its closures we shall use the term finitely generated ψdo for the operators of $\mathring{\Psi\mathcal{O}}$.

Returning to the algebra $\mathcal{O}(\infty)$ of operators with an order, let us next discuss the matter of taking adjoints of elements in $\mathcal{O}(\infty)$. Already in (2.13) we introduced the form $(u,v) = \int \overline{u^\wedge} v^\wedge dx$ which was seen to be meaningful for $u \in \mathfrak{H}_s$, $v \in \mathfrak{H}_{-s}$ and it was noted that (u,v) gives a natural identification between the spaces $\mathfrak{H}_s^*$ and $\mathfrak{H}_{-s}$, which is different from the usual identification of $\mathfrak{H}_s^*$ with $\mathfrak{H}_s$, based on the inner product $(u,v)_s$.

This matter is important for a natural definition of the adjoint of an operator in $\mathcal{O}(\infty)$. Let $A \in \mathcal{O}(r)$ for some r, then there is a continuous extension A_s to all of $\mathfrak{H}_s$, for every $s \in \mathbb{R}$, where $A_s : \mathfrak{H}_s \to \mathfrak{H}_{s-r}$ is a bounded operator, having a (Hilbert space) adjoint

(5.11) $$A_s^* : \mathfrak{H}_{-s+r} \to \mathfrak{H}_{-s}$$

assuming the identification $\mathfrak{H}_s^* = \mathfrak{H}_{-s}$ and $\mathfrak{H}_{s-r}^* = \mathfrak{H}_{-s+r}$. Now it is readily verified that A_s is a restriction of A_t, as $s \geq t$, in the sense that $A_s u = A_t u$, $u \in \mathfrak{H}_s$. Also we have

(5.12) $$\begin{aligned} (A_s u,v) &= (u,A_s^* v), \quad u \in \mathfrak{H}_s, \quad v \in \mathfrak{H}_{-s+r}, \\ (A_t u,v) &= (u,A_t^* v), \quad u \in \mathfrak{H}_t, \quad v \in \mathfrak{H}_{-t+r}. \end{aligned}$$

Accordingly for $u \in \mathfrak{H}_s$ and $v \in \mathfrak{H}_{-t+r}$ we get

(5.13) $$(u,A_s^* v) = (u,A_t^* v)$$

which shows that the two distributions $A_s^* v$ and $A_t^* v$ must coincide. Or, in other words, we have $A_s^* \supset A_t^*$ (A_s^* is an extension of A_t^*, whenever $s > t$.)

Let A^* denote the restriction of A_s^* to $\mathfrak{H}_\infty$. Then evidently A^* is independent of s, and the boundedness of the Hilbert space adjoint (5.11) implies

(5.14) $$\| A^* u \|_{-s} \leq c_s \| u \|_{-s+r}, \quad u \in \mathfrak{H}_\infty, \quad s \in \mathbb{R}.$$

We may replace $-s$ by s in (5.14) and then conclude that $A^* \in \mathcal{O}(r)$, and that

(5.15) $$(A^*)_s = (A_{-s+r})^*, \quad s \in \mathbb{R}.$$

It is natural to refer to the operator A^* as of the adjoint (or $\mathcal{O}(\infty)$-adjoint) of $A \in \mathcal{O}(\infty)$. In that respect it also may be noted that A^* is independent of the choice of r: For $A \in O(r) \cap O(\rho)$ we obtain the same A^*, using r or ρ in the above procedure, because evidently we get

(5.16) $$(Au,v) = (u,A^* v), \quad u,v \in \mathfrak{H}_\infty$$

which determines the operator A^* uniquely.

Also (5.16) offers a convenient approach to an explicit calculation of A^*, since for $u,v \in \mathcal{G}_\infty \subset \mathcal{G}_0 = L^2(\mathbb{R}^n)$ we get

$$(5.17) \qquad (u,v) = (u,v)_0 = \int \bar{u}v\,dx ,$$

so that A^* must coincide with the formal adjoint of A in case where A is a differential operator, for example.

Problems:

1) Show that the operator $L \in \mathcal{O}(N)$ defined by

$$(5.18) \qquad Lu(x) = \sum_{|\alpha|\le N} a_\alpha(x)\, u^{(\alpha)}(x) , \quad u \in \mathcal{G}_\infty ,$$

where the coefficients a_α are in $CB^\infty(\mathbb{R}^n)$, has its adjoint L^* given by the formally adjoint differential expression $u \to \sum_{|\alpha|\le N} (-1)^{|\alpha|}(\bar{a}\,u)^{(\alpha)}$.

2) For $a \in M$ (as in I, definition 6.1) show that the formal Fourier multiplier $a(D)$ is in $\mathcal{O}(\infty)$, and calculate its ($\mathcal{O}(\infty)$-) adjoint.

3) For a function $a \in M$ show that the true order (infimum of all orders) of $a^{(\alpha)}(D)$ tends to $-\infty$, as $|\alpha| \to \infty$.

4) Let K be an integral operator $(Ku)(x) = \int_{\mathbb{R}^n} k(x,y)u(y)dy$ with kernel $k \in S(\mathbb{R}^{2n})$. Show that $K \in \mathcal{O}(-\infty)$ and calculate K^*.

5) Show that a distribution integral operator of the form K in II,problem 10 is an operator of order 0 ,in the above sense ,if only the assumptions of 3,problem 5 are satisfied.

6) Show that K ,as in problem 5 , is of order r if only the family k(x,z) is in S^0_{ad} ,in the variable z ,for every $x \in R^n$, with lowest homogeneity exponent σ_0 being $\ge -r-n$,and with a uniformity in x ,for $x \in R^n$. That is,we have (i) $z^\alpha D_z^\beta k(x,z) = O(1)$,as $x,z \in R^n$, $|z| \le 1$,and (ii) $D_z^\alpha z^\beta k(x,z)$ $\in CB(R^n \times \{|z| \le 1\})$,as $|\beta| \ge N_0(|\alpha|)$, and (iii) an asymptotic expansion

$$(5.18) \quad k(x,z) - \sum_{l=0}^{N} k_l(x,z) \in CB^j(R^n \times \{|z| \le 1\}) , \text{ as } N > N_0(j) ,$$

with distributions $k_l \in \mathcal{D}'(R^{2n})$, $k_l \in CB^\infty(R^n \times \{|z| = 1\})$, $k_l(x,.) \in H''_{\sigma_l}$,$\sigma_l \uparrow \infty$, for every $x \in R^n$.(Hint: Reduce this to the preceeding problem,or to problem 6 of section 3.)

6. A Leibnitz formula for finitely generated ψdo's.

Let us first recall Taylors formula with integral remainder, in the following form,

Lemma 6.1. For $a \in C^{N+1}(\mathbb{R}^n)$ we have

$$a(x) = \sum_{|\theta| \le N} a^{(\theta)}(y)/\theta!\ (x-y)^{\theta} + r_N(x,y) \tag{6.1}$$

where

$$\begin{aligned} r_N(x,y) &= \sum_{|\theta|=N+1} (N+1)/\theta!\ (x-y)^{\theta}\ r_{\theta,N}(x,y) , \\ r_{\theta,N}(x,y) &= \int_0^1 t^N dt\ a^{(\theta)}(\tau x+ty) , \quad t+\tau=1 . \end{aligned} \tag{6.2}$$

Proof. We introduce

$$\phi(\tau) = a(y + \tau(x-y)) , \quad 0 \le \tau \le 1 , \tag{6.3}$$

and apply Taylors formula in one variable to ϕ:

$$\begin{aligned} \phi(1) &= \sum_{k=0}^{N} \phi^{(k)}(0)/k! + \rho_N \\ \rho_N &= 1/N! \int_0^1 (1-\tau)^N \phi^{(N+1)}(\tau)d\tau . \end{aligned} \tag{6.4}$$

Let us verify (6.4) by induction: it is trivial for N=0, and an integration by parts implies that

$$\begin{aligned} \rho_N &= \int_0^1 (1-\tau)^{N-1}/(N-1)!\ \phi^{(N)}(\tau)d\tau + (1-\tau)^N/N!\ \phi^{(N)}(\tau)\Big]_0^1 \\ &= \rho_{N-1} - \phi^{(N)}(0)/N! = \phi(1) - \sum_{k=0}^{N} \phi^{(k)}(0)/k! \end{aligned} \tag{6.5}$$

using (6.4) for N-1 as an induction hypothesis. Notice that $\phi(1) = a(x)$, $\phi(0) = a(y)$, $\phi'(0) = (x-y)\ \nabla a(y) = \sum_{|\theta|=1} (x-y)^{\theta}\ a^{(\theta)}(y)$, and more generally

$$\phi^{(k)}(\tau) = \sum_{|\theta|=k} (x-y)^{\theta}\ a^{(\theta)}(y + \tau(x-y))k!/\theta! \tag{6.6}$$

as follows by induction, using the multi-nomial formula

$$(\mu_1 + \mu_2 + \dots + \mu_n)^k = \sum_{|\theta|=k} k!/\theta!\ \mu^{\theta} . \tag{6.7}$$

Substituting (6.6) into (6.4) yields (6.1) and (6.2). In particular,

$$\rho_N = (N+1) \sum_{|\theta|=N+1} (x-y)^{\theta}/\theta! \int_0^1 (1-\tau)^N a^{(\tau)}((1-\tau)y + \tau x)d\tau , \tag{6.8}$$

which becomes (6.2) if we still substitute the integration variable by $1-\tau$, q.e.d.

Lemma 6.1 implies a corresponding expansion of the operator $a(M)b(D)$, as in the lemma, below.

Lemma 6.2. Let us assume that $a,b \in M$ (with the space M of II,5). Then we have the formula

$$(6.9)\qquad a(M)b(D) = \sum_{|\theta|\le N} i^{|\theta|}/\theta!\; b^{(\theta)}(D)a^{(\theta)}(M) + R_N$$

as N is sufficiently large, where $R_N: S \to S$ is the integral operator

$$(6.10)\qquad R_N u(x) = \int_{\mathbb{R}^n} r_N(x,y)u(y)dy\;,\qquad u \in S\;,$$

with kernel

$$(6.11)\qquad r_N(x,y) = \sum_{|\theta|=N+1} (N+1)\; i^{|\theta|}/\theta! \int_0^1 t^N dt\; a^{(\theta)}(\tau x + ty)b^{(\theta)\vee}(x-y)\;.$$

Proof. Using formula I, (4.27) we write (c.f. also Lemma 4.7)

$$(6.12)\qquad b(D)u(x) = \int b^{\vee}(x-y)u(y)đy\;,$$

with the right hand side signifying a distribution integral (that is, the value of the linear functional $b^{\vee} \in S'$ at the testing function $v(y) = u(x-y) \in S$). However, since $b \in M$, by assumption, its inverse Fourier transform $b^{\vee}$ is in S^{O}_{ps} (c.f. II, def. 4.1). That is, $b^{\vee}$ has its singular support at the origin only; it equals a function in S for $|x| \ge 1$ and it is a pseudo-function, or,

$$(6.13)\qquad x^{\theta}b^{\vee} = i^{|\theta|}b^{(\theta)\vee} \in C^{k}(\mathbb{R}^{n}),\qquad |\theta| \ge N_O(k)\;.$$

Accordingly, with Taylors formula above we get

$$(6.14)\qquad \begin{aligned} a(M)b(D)u(x) &= \int_{\mathbb{R}^n} b^{\vee}(x-y)a(x)u(y)đy \\ &= \sum_{|\theta|\le N} 1/\theta! \int_{\mathbb{R}^n} (x-y)^{\theta}b^{\vee}(x-y)a^{(\theta)}(y)u(y)đy \\ &\quad + \sum_{|\theta|=N+1} (N+1)/\theta! \int_{\mathbb{R}^n} (x-y)^{\theta}b^{\vee}(x-y)r_{N,\theta}(x-y)u(y)đy. \end{aligned}$$

In (6.14) we introduce (6.13). This will bring the terms of the first sum onto the form

$$(6.15)\qquad \int_{\mathbb{R}^n} i^{|\theta|}b^{(\theta)\vee}(x-y)a^{(\theta)}(y)u(y)đy = i^{|\theta|}b^{(\theta)}(D)a^{(\theta)}(M)u(x)\;.$$

In other words the first sum goes into the sum of (6.9). On the other hand, the terms of the second sum will contain ordinary Lebesgue (or improper Riemann-) integrals whenever N gets sufficiently large, because $x^{\theta}b^{\vee}$ then will be a function, by (6.13). It also is evident that this second sum is the desired integral operator (6.10) with kernel (6.11), q.e.d.

Notice that the remainder term R_N in (6.9) will become zero for sufficiently large N whenever at least one of the functions a or b is a polynomial. Then formula (6.9) will be identical with the adjoint Leibnitz formula I, (1.23). (If b is a polynomial, then b(D) will be the corresponding differential operator with constant coefficients. Accordingly, theorem 6.4 below may be regarded as an extension of Leibnitz' formula to formal Fourier multipliers.

Definition 6.3. Let $A \in \mathcal{O}(\infty)$ and a sequence $A_j \in \mathcal{O}(r_j)$, $j=1,2,\ldots$, be given, where we assume that

$$\lim_{j\to\infty} r_j = -\infty \quad \text{(monotonically)} \quad . \tag{6.16}$$

We shall say that A admits the asymptotic expansion

$$A \sim \sum_{j=0}^{\infty} A_j \quad (\mathrm{mod}\ \mathcal{O}(-\infty)) \tag{6.17}$$

if we have

$$A - \sum_{j=0}^{N} A_j \in \mathcal{O}(r_{N+1}) \qquad N=0,1,2,\ldots \quad . \tag{6.18}$$

Theorem 6.4. Let $a \in CB^{\infty}(\mathbb{R}^n)$, $b \in M$, then we have the asymptotic expansions

$$\begin{aligned} a(M)b(D) &\sim \sum_{l=0}^{\infty} \sum_{|\theta|=l} i^{|\theta|}/\theta!\, b^{(\theta)}(D)a_{(\theta)}(M) \quad (\mathrm{mod}\ \mathcal{O}(-\infty)), \\ b(D)a(M) &\sim \sum_{l=0}^{\infty} \sum_{|\theta|=l} (-i)^{|\theta|}/\theta!\, a^{(\theta)}(M)b_{(\theta)}(D) \quad (\mathrm{mod}\ \mathcal{O}(-\infty)). \end{aligned} \tag{6.19}$$

Proof. For $u,v \in \mathcal{G}_\infty$ and the form of (2.13) we get

$$(u,a(M)b(D)v) = (\bar{b}(D)\bar{a}(M)u,v) \quad , \tag{6.20}$$

as readily verified (c.f. also problem 2 in section 5). Also for $A \in \mathcal{O}(r)$ we have $A^* \in \mathcal{O}(r)$, with the $\mathcal{O}(\infty)$-adjoint A^* of A. The proposition below is therefore evident.

Lemma 6.5. We have the asymptotic expansion (6.17) if and only if

$$A^* \sim \sum_{j=0}^{\infty} A_j^* \quad (\mathrm{mod}\ \mathcal{O}(-\infty)) \ . \tag{6.21}$$

Accordingly, since both classes M and CB^{∞} are invariant under taking complex conjugates, in the proof of theorem 6.4 it is sufficient to verify the first relation (6.19), since the second one then follows by taking $\mathcal{O}(\infty)$-adjoints.

Now, regarding the first relation (6.19) we will use Lemma 6.2, of course. If $a \in CB^{\infty}$ then we get $a_{(\theta)}(M) \in \mathcal{O}(0)$ for all θ, using Lemma 3.3. Also we conclude that $b^{(\theta)}(D) \in \mathcal{O}(r_j)$ for all $|\theta| \le j$, where $r_j \to -\infty$ (c.f. problem 3 of section 5). Accordingly, the asymptotic expansion (6.19) is meaningful; it meets the general requirement of the definition. To prove the theorem we must look at the remainder term R_N of (6.9). All we have to show is the proposition below.

Lemma 6.6. For $s \in \mathbb{R}$ and $k=1,2,\ldots$ there exists an $N_0(s,k)$ such that

$$(6.22) \qquad \| R_N u \|_{s+k} \le c_s \| u \|_s \,, \quad u \in \mathcal{S}, \quad N \ge N_0(s,k) \;.$$

Indeed, if A and A_l, $l=0,1,2,\ldots$ are chosen as the left hand side and the inner sums at right of $(6.19)_1$, respectively, then lemma 6.6 implies that

$$(6.23) \qquad A - \sum_{l=0}^{N} A_l = \sum_{l=N+1}^{M} A_l + R_M \,, \quad M \ge N_0(s,k)$$

maps continuously from $\mathcal{G}_s$ to $\mathcal{G}_{s+\sigma}$, with

$$(6.24) \qquad \sigma = \mathrm{Min}\,[k, -r_{N+1},\ldots,-r_M] = \mathrm{Min}\,[k,-r_{N+1}] \;.$$

Chosing k (that is M) large enough the minimum at right equals $-r_{N+1}$ which is independent of s. This shows that the left hand side of (6.23) must be of order r_{N+1}, as required for the expansion (6.19).

Next we reduce the assertion still further: For lemma 6.6 it suffices to prove the proposition, below.

Lemma 6.7. For every non-negative integer k there exists an $N_1(k)$ such that for $N \ge N_1(k)$ the operator $(1-\Delta)^k R_N (1-\Delta)^k$ defined for functions in $\mathcal{S}$, is $L^2(\mathbb{R}^n)$-bounded. That is,

$$(6.25) \qquad \| (1-\Delta)^k R_N (1-\Delta)^k u \|_{L^2(\mathbb{R}^n)} \le c_k \| u \|_{L^2(\mathbb{R}^n)} \,, \quad u \in \mathcal{S}.$$

Indeed, in (6.25) we may introduce $v = (1-\Delta)^k u$, $u = \Lambda^{2k} v$. It is clear that the class of all such v describes the space $\mathcal{S}$ again. Using (1.8) and (2.5) we conclude that

$$(6.26) \qquad \| R_N v \|_{2k} \le c_k \| v \|_{-2k} \,, \quad v \in \mathcal{S},$$

which yields

$$(6.27) \qquad \| R_N v \|_s \le c_k \| v \|_t \,, \quad v \in \mathcal{S} \,, \quad s \le 2k, \quad t \ge -2k \,,$$

and thus certainly implies (6.22) since k can be made arbitrarily large.

Proof of Lemma 6.7. We have the explicit form (6.11) for the kernel $r_N(x,y)$ of R_N and recall that (6.13) holds. It follows that $r_N \in C^j(\mathbb{R}^{2n})$ whenever N is sufficiently large. Also with fixed x, $r_N(x,y)$ is equal to a function in $\mathcal{S}$ for large $|y|$. (In particular we note that a with all its derivatives is bounded.). It follows that we have $w = Wu = (1-\Delta)^k R_N (1-\Delta)^k u$ given by

$$(6.28) \qquad w(x) = W(x,y)u(y)dy$$

with

$$(6.29) \qquad W(x,y) = (1-\Delta_x)^k (1-\Delta_y)^k \, r_N(x,y) \,,$$

because (i) the left differentiation $(1-\Delta)^k$ can be pulled trough the integral to land as $(1-\Delta_x)^k$ on the kernel r_N. Also, (ii) the right differentiation $(1-\Delta_y)^k$ may be thrown to the kernel again, by a partial integration with no boundary terms. In all of this we must chose N large enough as to have $b^{(\theta)} \varepsilon C^{4k}$ at zero, for all $|\theta| = N+1$.

It now follows that the function W(x,y) satisfies the conditions I, (1.24) of I, lemma 1.1, so that indeed the operator W above is L^2-bounded. In fact, the function W(x,y) of (6.29) is a linear combination of terms of the form

$$(6.30) \qquad \int_0^1 t^p \tau^q dt\, a^{(\alpha)}(\tau x+ty)(b^{(\theta)\vee})^{(\beta)}(x-y), \quad |\alpha|,\ |\beta| \geq N+1, \quad |\beta| \geq 0, \quad p,q = 0,1,2,\ldots ,$$

as readily confirmed by induction. For each term (6.30) the integrals I, (1.24) are $= O(\|(b^{(\theta)\vee})^{(\beta)}\|_{L^1(\mathbb{R}^n)})$, which is finite as N is sufficiently large.

Accordingly lemma 6.7 and therefore theorem 6.4 are established.

We shall adopt the convention of writing the two asymptotic expansions (6.19) in the simplified form

$$(6.31) \qquad \begin{aligned} a(M)b(D) &\sim \sum_\theta i^{|\theta|}/\theta!\, b^{(\theta)}(D)a^{(\theta)}(M) \quad (\mathrm{mod}\ \mathcal{O}(-\infty)) , \\ b(D)a(M) &\sim \sum_\theta (-i)^{|\theta|}/\theta!\, a^{(\theta)}(M)b^{(\theta)}(D) \quad (\mathrm{mod}\ \mathcal{O}(-\infty)) . \end{aligned}$$

They shall be referred to as the two generalized Leibnitz' formulas.

The corollary , below, will be useful for our discussion in section 9.

Corollary 6.8. Under the assumptions of Theorem 6.4, if in addition there exists an integer k_0 such that all derivatives $a^{(\theta)}$ for $|\theta| \geq k_0$ tend to zero as $|x| \to \infty$, then the operators

$$(6.32) \qquad V = (1-\Delta)^k R_N , \quad Z = R_N(1-\Delta)^k$$

are compact (from $L^2(\mathbb{R}^n)$ to $L^2(\mathbb{R}^n)$) for every fixed k and sufficiently large $N \geq N_2(k)$.

Proof. From the above discussion around the operator (6.28) we know that the operators V and Z are integral operators with kernel in C^{2k} as soon as N gets sufficiently large. In fact, the kernel V(x,y) of V, for example, must be a linear combination of terms of the form

$$(6.33) \qquad \int_0^1 \tau^{|\alpha|} dt\, a^{(\alpha+\theta)}(\tau x+ty)(b^{(\theta)\vee})^{(\beta)}(x-y), \quad |\alpha|,\ |\beta| \leq 2k, \ |\theta|=N+1 .$$

If N is chosen very large in comparison to k we will get $(b^{(\theta)\vee})^{(\beta)} \varepsilon L^2$ because at ∞, $b^{(\theta)}$ equals a function in $\mathcal{S}$ so its derivatives are in L^2 there, while for large θ, $b^{(\theta)\vee}$ is C^{2k}, so certainly is L^2_{loc}. It follows that each expression (6.33) is a Schmidt kernel, whenever we assume for a moment that $a \varepsilon C^\infty$ has compact support. Indeed, let us denote (6.33) by $p(x,y)$, then

$$\iint dxdy\, |p(x,y)|^2 \le \sup\left\{ \iint dxdy\, |a^{(\alpha+\theta)}(\tau x+ty)|^2 |(b^{(\theta)\vee})^{(\beta)}(x-y)|^2 \right\}, \tag{6.34}$$

the supremum being taken over $t+\tau = 1,\ 0 \le t,\tau \le 1$.
In (6.34) at right introduce new integration variables $\tau x+ty = \xi,\ x-y = \eta$. Observing that $\partial(\xi,\eta)/\partial(x,y) = 1$ we calculate the right hand side as

$$\| a^{(\theta+\alpha)} \|^2_{L^2} \, \| (b^{(\theta)\vee})^{(\beta)} \|^2_{L^2} < \infty \tag{6.35}$$

[In fact, the integral under the sup of (6.34) proves independent of t and τ which only makes (6.34) correct.]

Accordingly the operator V is compact, as stated, provided that a has compact support.

Under the more general assumption of the lemma we at least will get the kernel v a linear combination of expressions

$$\int_0^1 \tau^\mu\, q(\tau x+ty)\phi(x-y)\,dt, \quad q \in CO(\mathbb{R}^n), \quad \phi \in L^1 \cap L^2 . \tag{6.36}$$

In that case let us approximate q by a sequence q_j of functions with compact support. Then the kernel (6.36) with q replaced by q_j will be Schmidt and thus will define a compact operator on L^2, as follows by a conclusion similar to the above one. Moreover, it follows from I, Lemma 1.1 that the operator with kernel

$$\int_0^1 \tau^\mu (q - q_j)(\tau x+ty)\phi(x-y)dt \tag{6.37}$$

goes to zero in L^2-operator norm, as $j \to \infty$. Since the compact ideal $K(L^2)$ is norm closed we thus conclude that all the operators with kernels (6.36) are in $K(L^2)$, so that indeed $V \in K(L^2)$. The proof for compactness of Z proceeds similar, and will be left as an exercise.

Problems.

1) We may introduce an order concept different from the one introduced in section 5 by saying that $C \in L(\mathfrak{H}_\infty)$ has compact order r if it maps compactly from $\mathfrak{H}_s$ to $\mathfrak{H}_{s-r}$, for every $s \in \mathbb{R}$. Show that the classes $\mathcal{Q}(r)$ or operators of compact order r together with $\mathcal{Q}(\infty) = \bigcup \mathcal{Q}(r)$ and $\mathcal{Q}(-\infty) = \bigcap \mathcal{Q}(r)$ satisfy the equivalent of Lemma 5.1 again, and propose an equivalent of Lemma 5.2.

2) Let $C \in \mathcal{Q}(r)$, $C_j \in \mathcal{Q}(r_j)$ where $r_j \searrow -\infty$. Define the asymptotic expansion $C \sim \sum_j C_j \pmod{\mathcal{Q}(-\infty)}$ to amount to

$$C - \sum_{j=0}^{N} C_j \in \mathcal{Q}(r_{N+1}),\ N = 0,1,2,\dots . \tag{6.38}$$

Show that for $a \in CB^\infty(\mathbb{R}^n)$ with $a^{(\alpha)} \in CO^\infty(\mathbb{R}^n)$ for $\alpha \ne 0$ we can state the two asymptotic expansions (6.31) mod $\mathcal{Q}(-\infty)$ rather than only mod $\mathcal{O}(-\infty)$.

3) Let $a \in \mathcal{M}$ and let $\lambda(x) = (1+x^2)^{-1/2}$. Show that we have the asymptotic expansion (for all $t \in \mathbb{R}$)

$$\lambda^t(M)a(D)\lambda^{-t}(M) \sim \sum_\theta i^{|\theta|}/\theta!\, a^{(\theta)}(D)\mu_\theta(M) \qquad (\mathrm{mod}\ \mathcal{O}(-\infty)) \tag{6.39}$$

with $\mu_{(\theta)} = (\lambda^t)^{(\theta)}\lambda^{-t}$.

7. Calculus of ψdo-s.

In this section we only obtain some very formal extensions of the results of section 6, notably of the generalized Leibnitz formulas. Also we want to establish a normal form for finitely generated ψdo-s. Or rather, there will be several such normal forms.

Let the function $a \in C^\infty(\mathbb{R}^{3n})$ be of the special form

$$a(x,y,\xi) = \sum_{j=1}^{N} b_j(x)c_j(y)d_j(\xi) \quad , \qquad b_j, c_j, d_j \in T \tag{7.1}$$

(Recall that T as defined in I,4 is the space of multipliers of S).
Then we define the linear operator $A: S \to S$ by setting

$$A = \sum_{j=1}^{N} b_j(M)d_j(D)c_j(M) \quad . \tag{7.2}$$

This operator will be denoted by $A = a(M_l,M_r,D)$ referring to the fact that the factors involving the variable x and y have been put to the left and to the right of the differentiations $d_j(D)$, respectively.

If only the variable x (or only y) is present (i.e. if the b_j (or the c_j) are all constant so that we have $a = a(x,\xi)$ (or $a = a(y,\xi)$), then we shall write the operator A as $A = a(M_l,D)$, or as $A = a(M_r,D)$, respectively.

Using that $d_j(D) = F^{-1}d_j(M)F$, where F is the Fourier transform one may easily express the three operators $a(M_l,M_r,D)$, $a(M_l,D)$ and $a(M_r,D)$ (with respective choices of a) as

$$a(M_l,M_r,D)u(x) = \int d\!\!\!{}^-\xi \int d\!\!\!{}^-y \; a(x,y,\xi)\; e^{i\xi(x-y)}u(y)dy, \qquad u \in S , \tag{7.3}$$

and

$$a(M_l,D)u(x) = \int d\!\!\!{}^-\xi \;\; e^{ix\xi}\, a(x,\xi)u^\wedge(\xi) \; , \qquad u \in S \; , \tag{7.4}$$

and

$$(a(M_r,D)u)^\wedge(\xi) = \int d\!\!\!{}^-y \;\; e^{-iy\xi}\, a(y,\xi)u(y)dy \; , \quad u \in S \; . \tag{7.5}$$

Conventionally since about 1965 calculus of ψdo-s starts with discussion of the above 3 types of operators $S \to S$ in case where the function a is not a finite sum of products as in (7.1), but is taken from a more general class varying with the author (c.f. Hoermander [5], Friedrichs [4], Kohn-Nirenberg [6], Kumano-go [7]). In contrast our discussion of more general ψdo-s will not depart from the above, but will use a more abstract extension principle, to be discussed later on (Chapter IV).

If an operator $A: S \to S$ can be written in the form $A = a(M_l,M_r,D)$ with some function a of the form (7.1) then we shall call the function $a(x,y,\xi)$ a <u>formal symbol</u> of A. The differential operators of L (as in section 5) certainly all have formal symbols. Notice that any such differential operator was written in the form

(7.6) $$L = \sum_{|\alpha|\leq N} a_\alpha(M)D^\alpha \quad , \quad a \in CB^\infty(\mathbb{R}^n)$$

Introducing the function

(7.7) $$l(x,\xi) = \sum_{|\alpha|\leq N} a_\alpha(x)\xi^\alpha$$

it is evident that $L = l(M_l,D)$. However, the same differential operator L also may be written in the form

(7.8) $$L = \sum_{|\alpha|\leq N} D^\alpha b_\alpha(M) \quad ,$$

with different coefficients $b_\alpha \in CB^\infty$. Each b_α is a combination of derivatives of the a_β, and therefore is in CB^∞ indeed. The transformation between (7.6) and (7.8) is a matter of Leibnitz' formula, of course. Then we also have

(7.9) $$L = l_1(M_r,D) \quad ,$$

as readily seen. This shows that in general there are many formal symbols for a given differential operator in L, although there is only one symbol independent of y (or of x).

There are general ways and means to provide the finitely generated ψdo-s in $\mathring{\Psi O}$ with formal symbols but we shall not be interested in this task. However, let us ask the question for a normal form again, which will be answered by the result below.

Theorem 7.1. Every operator $A \in \mathring{\Psi O}$ admits an asymptotic expansion of the form

(7.10) $$A \sim \sum_{j=0}^{\infty} A_j \qquad (\text{mod } O(-\infty))$$

where

(7.11) $$A_j = a_j(M_l,D) = L_j \Lambda^{\sigma_j} \quad , \qquad L_j \in L \; .$$

For the proof we need the lemma, below.

Lemma 7.2. Let

(7.12) $$A \sim \sum_{l=0}^{\infty} A_j \qquad (\text{mod } O(-\infty))$$

with $A_l \in O(r_l)$, $r_l \searrow \infty$, and let

(7.13) $$A_l \sim \sum_{j=0}^{\infty} A_{lj} \qquad (\text{mod } O(-\infty)) \; ,$$

where $A_{lj} \in O(r_{lj})$, $r_{l0} = r_l$, $r_{lj} \searrow -\infty$, as $j \to \infty$, l fixed. Then, if ρ_k,

$k=0,1,2,\ldots$ is any reordering of the collection $\{r_{lj}: l,j = 0,1,\ldots\}$ which forms a non-increasing sequence, and if we denote $B_k = A_{lj}$ with corresponding integers l,j and k, we have

$$(7.14) \qquad A \sim \sum_{j=0}^{\infty} B_k \qquad (\text{mod } \mathcal{O}(-\infty)) \ .$$

Proof. Under the assumption of the lemma there can be only finitely many r_{lj} above $-p$, for $p=1,2,\ldots$. Therefore at least one reordering as proposed exists. For any ρ_k let r_l be chosen such that $r_l < \rho_k$. Then all terms $B_0, B_1,\ldots,B_k$ must coincide with a corresponding $A_{l'j'}$, $l' \le l$, so that

$$(7.15) \qquad A - \sum_{m=0}^{k} B_m = (A - \sum_{l'=0}^{l} A_{l'}) + \sum_{l'=0}^{l} \sum_{j'=0}^{j} (A_{l'} - A_{l'j'}) + \sum B_q \ .$$

Here j has been chosen such that $r_{l'j} < \rho_k$, $l'=0,1,\ldots,l$. The terms in the last sum are the B_q, $q > k$ which coincide with one of the $A_{l'j'}$, $l' \le l$, $j' \le j$. There are only finitely many such terms and all of them are in $\mathcal{O}(\rho_{k+1})$. The other terms at right have that order too, by construction, so that we indeed get $A - \sum_{m=0}^{k} B_m \in \mathcal{O}(\rho_{k+1})$, which implies (7.13), q.e.d.

Proof of theorem 7.1. This proof proceeds by induction. It suffices to show that a finite product of generators has the asymptotic expansion required. This certainly is correct for one factor only (we of course admit the case that all but finitely many terms of the asymptotic expansion vanish). Suppose all products of less than N factors $F_j = a(M)$, or $= D^\alpha$, or $= \Lambda^s$ admit the desired asymptotic expansion. Then consider a product

$$(7.16) \qquad P = F_1F_2\ldots F_N = \tilde{P}F_N \ , \quad \tilde{P} \sim \sum_{l=0}^{\infty} P_l \quad (\text{mod } \mathcal{O}(-\infty)) .$$

Trivially we get

$$(7.17) \qquad P \sim \sum_{l=0}^{\infty} P_l F_N \quad (\text{mod } \mathcal{O}(-\infty)) .$$

Now, if $F_N = \Lambda^s$ or $= D^\alpha$, then all summands $P_l F_N$ are of the desired form (7.11) and we are done. On the other hand, if $F_N = a(M)$ we must apply our generalized Leibnitz formula $(6.31)_2$ (or in details, $(6.19)_2$) together with lemma 7.2 and then again will obtain the desired asymptotic expansion for P, q.e.d.

Theorem 7.3. (Calculus of finitely generated ψdo-s)

1) Let $A = a(M_1,M_r,D)$ with a as in (7.1), $b_j,c_j \in CB^\infty$, $d_j \in M$. Then we get the asymptotic expansions

$$(7.18) \qquad A \sim \sum_\theta i^{|\theta|}/\theta!(D_y^\theta D_\xi^\theta a)(M_1,M_1,D) \sim \sum_\theta (-i)^{|\theta|}/\theta!(D_x^\theta D_\xi^\theta a)(M_r,M_r,D)$$

$$(\mathrm{mod}\ \mathcal{O}(-\infty))\ .$$

2) Let $A = a(M_1,D)$, a as in (7.1), $c_j = 1$, $b_j \in CB^\infty$, $d_j \in M$. Then the $\mathcal{O}(\infty)$-adjoint A^* has the asymptotic expansion

$$(7.19) \qquad A^* \sim \sum_\theta (-i)^{|\theta|}/\theta!(D_x^\theta D_\xi^\theta \bar{a})(M_1,D) \quad (\mathrm{mod}\ \mathcal{O}(-\infty))\ .$$

3) Given $a(x,\xi)$ and $b(x,\xi)$ as under (2), and let $A = a(M_1,D)$, $B = b(M_r,D)$, $C = AB$. We have $C = c(M_1,M_r,D)$, $c(x,y,\xi) = a(x,\xi)b(y,\xi)$, and C admits the asymptotic expansion

$$(7.20) \qquad C \sim \sum_\theta (-i)^{|\theta|}/\theta!(D_\xi^\theta(aD_x^\theta b))(M_1,D) \quad (\mathrm{mod}\ \mathcal{O}(-\infty))\ .$$

The proof of Theorem 7.3 is completely formal. One simply must apply the generalized Leibnitz formulas of section 6 to interchange the corresponding differentiations and multiplications. This is left to the reader.

8. A compactness result

First we recall the operator $\Lambda = (1+D^2)^{-1/2} = (1-\Delta)^{-1/2} = F^{-1}(1+M^2)^{-1/2}F$, and its powers $\Lambda^s = (1+D^2)^{-s/2}$, and prove the following proposition.

Lemma 8.1. Let $\varepsilon > 0$, and let $a \in CO(\mathbb{R}^n)$ (c.f. I,1), then the operators $\Lambda^\varepsilon a(M)$ and $a(M)\Lambda^\varepsilon$ extend to compact operators of $\mathfrak{H}_0 = L^2(\mathbb{R}^n)$ to itself.

Proof. We will make use of the fact that the formal Fourier multiplier Λ^ε can be written as a (singular) convolution operator. If we let

(8.1) $$\lambda(x) = (1+x^2)^{-1/2} ,$$

then clearly $\lambda \in T$, and [by I, (4.23) and lemma 8.7] we get

(8.2) $$\Lambda^\varepsilon u(x) = \lambda^\varepsilon(D)u(x) = (\lambda^{\varepsilon\vee} * u)(x) = \int \lambda^{\varepsilon\vee}(x-y)u(y)dy , \quad u \in S ,$$

with a distribution integral, at right.

However, the function $\lambda^{\varepsilon\vee}$ actually will be seen to be in $L^1(\mathbb{R}^n)$ so that the integral in (8.2) turns out to be a Lebesgue integral. In fact, it is well known that

(8.3) $$\lambda^{\varepsilon\vee}(x) = 2^{1-\varepsilon/2}/\Gamma(\varepsilon/2)|x|^{\varepsilon/2-n/2} K_{\varepsilon/2-n/2}(|x|)$$

with the Euler Gamma function $\Gamma(t)$ and the modified Hankel function $K_\nu(t)$. This function is known as a Bessel potential (c.f. [1]). Formula (8.3) may be derived by applying well known formulas on integrals involving Bessel functions (c.f. [9]) together with the fact, that the Fourier transform of a radially symmetric function coincides with the Hankel transform (c.f. I, 3, problem 2).

With the above knowledge it is possible to apply the Hankel asymptotic expansion for K_ν at ∞ and the power series expansion of K_ν near zero (c.f. [9]) which gives an exponential decay of λ^ε at ∞ and a behaviour like $|x|^{\varepsilon-n}$ near zero, as $0 < \varepsilon$ is small. Both facts imply that $\lambda^{\varepsilon\vee} \in L^1(\mathbb{R}^n)$ indeed.

While the above is interesting information on the operators Λ^ε we shall by no means have to depend on the necessary heavy use of theory of special functions for our lemma. Rather we observe that almost the same can be concluded from our results in II, 5 on the Fourier transform of distributions with an isolated singularity. Indeed, we get (using the binomial series expansion)

(8.4) $$\lambda^\varepsilon(x) = |x|^{-\varepsilon}(1+1/x^2)^{-\varepsilon/2} = \sum_{j=0}^{\infty} \binom{-\varepsilon/2}{j} |x|^{-\varepsilon-2j}, \quad |x| > 1,$$

and it is evident that (8.4) also may be interpreted as an asymptotic expansion of λ^ε at ∞ (mod $|x|$) (c.f. II, (5.2)). Accordingly, since $\lambda^\varepsilon \in M$ is easily verified, it follows that $\lambda^\varepsilon \in M_{as}$ (c.f. II, Def. 5.2). Then, applying II, Theorem 5.2

we find that $\lambda^{\varepsilon\vee} \varepsilon S^{O}_{ad}$, with corresponding asymptotic expansion

(8.5) $$\lambda^{\varepsilon\vee} \sim \sum_{l=0}^{\infty} b_l^{\vee} \quad (\text{mod } D\,) \text{ at } 0 \;,$$

where

(8.6) $$b_l = \binom{-\varepsilon/2}{l} |x|^{-\varepsilon-2l} .$$

Assuming that $0 < \varepsilon < 1$, as to avoid the possible formal difficulties arising from the task of extending a homogeneous function of degree $-n-k$, $k=0,1,\ldots$, to a homogeneous distribution, (c.f. II, lemma 2.2 and II, lemma 2.3), it is evident that the b_l are (extend to) homogeneous distributions of degree $-\varepsilon-2l$, so that their Fourier transforms are homogeneous of degree $-n+\varepsilon+2l$. In fact, they must be $C^{\infty}(\mathbb{R}^n-\{0\})$ by II, Lemma 4.7. Also, since the b_l are radially symmetric, so must be the $b_l^{\vee}$. Since all homogeneity degrees $-n+\varepsilon+2l$ are larger than $-n$ the $b_l^{\vee}$ must be functions of $|x|$ only. In fact we must get

(8.7) $$b_l^{\vee}(x) = c_l |x|^{\varepsilon-n+2l}$$

with certain constants c_l.

As a function in S^{O}_{ad}, $\lambda^{\varepsilon\vee}$ behaves like a function in S for $|x| \geq 1$. It is C^{∞}, except at zero. So again it follows that $\lambda^{\varepsilon\vee} \varepsilon L^1(\mathbb{R}^n)$. Accordingly we may write

(8.8) $$a(M)\Lambda^{\varepsilon}u(x) = \int a(x)\lambda^{\varepsilon\vee}(x-y)u(y)\,dy$$

with an ordinary Lebesgue integral at right.

To prove the compactness stated, let us first assume that the function a has compact support. Also introduce the function

(8.9) $$\mu_N(x) = \text{Min } \{N, \text{ Max } [-N, \lambda^{\varepsilon\vee}(x)] \}$$

In that respect notice that $\lambda^{\varepsilon\vee}$ as inverse Fourier transform of a real-valued even function is real-valued. The operator

(8.10) $$Q_N u(x) = \int a(x)\mu_N(x-y)u(y)\,dy \quad , \; u \; \varepsilon \; S \; ,$$

has finite Schmidt norm and therefore is compact from $\mathfrak{H}_0$ to $\mathfrak{H}_0$. One calculates the Schmidt norm as

(8.11)
$$\begin{aligned} |||Q_N||| &= (2\pi)^{-n/2}\left\{\iint dxdy\, |a(x)|^2|\mu_N(x-y)|^2\right\}^{1/2} \\ &= (2\pi)^{-n/2}\|a\|_{L^2}\|\mu_N\|_{L^2} < \infty \end{aligned} .$$

Keeping $a \; \varepsilon \; C_0(\mathbb{R}^n)$ fixed we next let N tend to ∞ and observe that

(8.12) $$\| a(M)\Lambda^{\varepsilon} - Q_N \|_0 \longrightarrow 0 \;,$$

by I, Lemma 1.1, because the two constants I, (1.24) tend to zero, due to

$$\|\lambda^{\varepsilon} - \mu_N\|_{L^1} \leq \int_{|\lambda^{\varepsilon\vee}(x)|\geq N} |\lambda^{\varepsilon\vee}(x)|dx \longrightarrow 0, \quad N \to \infty .$$

Since the compact ideal $K(\mathfrak{H}_0)$ is closed in $L(\mathfrak{H}_0)$ (the Banach algebra of bounded operators over $\mathfrak{H}_0$) we conclude that $a(M)\Lambda^{\varepsilon} \in K(\mathfrak{H}_0)$ whenever $a \in C_0(\mathbb{R}^n)$.

Finally if now $a \in CO(\mathbb{R}^n)$ then a sequence $a_k \in C_0$ may be constructed converging uniformly over $\mathbb{R}^n$ to a. It follows that

$$\|a(M)\Lambda^{\varepsilon} - a_k(M)\Lambda^{\varepsilon}\|_{\mathfrak{H}_0} \leq \|a(M) - a_k(M)\|_{\mathfrak{H}_0}\|\Lambda^{\varepsilon}\|_{\mathfrak{H}_0}$$

$$(8.13) \qquad \leq \|a-a_k\|_{L^\infty}\|\lambda^{\varepsilon}\|_{L^\infty} = \|a-a_k\|_{L^\infty} \to 0, \quad k \to \infty .$$

Again since the compact ideal $K(\mathfrak{H}_0)$ is norm closed we conclude that $a(M)\Lambda^{\varepsilon} \in K(\mathfrak{H}_0)$ for all $a \in CO(\mathbb{R}^n)$. Also, the operator $\Lambda^{\varepsilon}a(M)$ evidently may be treated exactly in the same way, after we have seen that Λ^{ε} admits the integral representation (8.2) with a kernel in L^1. Hence our Lemma is established.

Problems:

1) Verify the two results used above, that, for $a \in CB(\mathbb{R}^n)$,

$$(8.14) \qquad \|a(M)\|_{L^2(\mathbb{R}^n)} = \|a(D)\|_{L^2(\mathbb{R}^n)} = \|a\|_{L^\infty} = \sup_{x\in\mathbb{R}^n}\{|a(x)|\}.$$

2) Lemma 8.1 may be proven in a completely different way, avoiding the theory of I,5 and I,6 as follows.

   a) Verify that it suffices to show the result for $\varepsilon=1$ (Hint: Use Stone - Weierstrass to approximate the continuous function $\lambda^{\varepsilon}(x)$ by polynomials in $\lambda(x)$.).

   b) For $u \in C_0^\infty(\mathbb{R}^n)$ we get

$$(8.15) \qquad \|\Lambda^{-1}u\|^2_{L^2} = \|u\|^2_{L^2} + \sum_{j=1}^{n} \|u_{x_j}\|^2$$

   c) Use Fourier series for functions in a cube $Q_N = \{|x_j| < N, j=1,\dots,n\}$, with large fixed N to prove that for each integer M there exist finitely many functions ϕ_j such that

$$(8.16) \qquad \|\Lambda^{-1}u\| > M\|u\| \quad , \quad u \in C_0^\infty(Q_N)$$

   for each $u \in S$ orthogonal (in the sense of $\mathfrak{H}_0$) to all ϕ_j.

   d) Apply Rellich's criterion to prove compactness of $a(M)\Lambda^{\varepsilon}$ for functions $a \in C_0^\infty$. (c.f. AI, Criterion 4.1).

9. Commutator relations between multiplications and differentiations.

In this section we will start choosing our multipliers from the algebra A of functions defined by

(9.1) $$A = \{ \quad a \in C^{\infty}(\mathbb{R}^n) : \ a = O(1), \ a^{(\alpha)} = o(1) \ \text{ as } \ |x| \to \infty, \ \alpha \neq 0 \},$$

Clearly $CO^{\infty}(\mathbb{R}^n) \subset A \subset CB^{\infty}(\mathbb{R}^n)$, and CO^{∞} is an ideal of A. All operators $a(M)$ with $a \in A$ are of order zero, as we know. Here we first want to prove the result below.

Lemma 9.1. Let $\lambda^s(x) = (1+x^2)^{-s/2}$, as before, and let $\Lambda^s = \lambda^s(D)$. For each $a \in A$ and $s,t \in \mathbb{R}$ we have

(9.2) $$C_s = \Lambda^{-s} a(M) \Lambda^s - a(M) \in K(\mathfrak{H}_t) .$$

More precisely the operator $S \to S$ at left of (9.2) is bounded from $\mathfrak{H}_t$ to $\mathfrak{H}_t$ for every $t \in \mathbb{R}$, and the continuous extension to $\mathfrak{H}_t$ is a compact operator of $\mathfrak{H}_t$.

Proof. We already know that the left hand side of (9.2) is in $O(0)$, using the fact that $a \in O(0)$ and that $\Lambda^s \in O(-s)$. Let us first prove the compactness of C_s for t=0. This is an easy consequence of Lemma 6.2 and Corollary 6.8. Indeed write

(9.3) $$a(M)\Lambda^s = \Lambda^s a(M) + \sum_{1 \le |\theta| \le N} i^{|\theta|}/\theta! \ \lambda^{s(\theta)}(D) a^{(\theta)}(M) + R_N$$

with R_N as in Lemma 6.2. Note that

(9.4) $$\lambda^{s(\theta)}(x) = \lambda^s(x)\mu_\theta(x) , \quad \mu_\theta \in CO(\mathbb{R}^n), \quad \theta \neq 0 ,$$

as follows by induction and straight forward differentiation.

Accordingly,

(9.5) $$\Lambda^{-s} a(M)\Lambda^s - a(M) = \sum_{1 \le |\theta| \le N} i^{|\theta|}/\theta! \ \mu_\theta(D) a^{(\theta)}(M) + \Lambda^{-s} R_N .$$

The sum at right of (9.5) is in $K(\mathfrak{H}_0)$, by Lemma 8.1, using our above assumptions. On the other hand let us pick an integer $k \ge 0$ with $-s+2k \ge 0$ and write

(9.6) $$\Lambda^{-s} R_N = \Lambda^{2k-s}(1-\Delta)^k R_N = \Lambda^{2k-s} V_N$$

with $V_N = V$ as in Corollary 6.8. The operator Λ^{2k-s} is L^2-bounded and V_N is compact as N is sufficiently large. Therefore the operator C_s is in $K(\mathfrak{H}_0)$. Finally to remove the restriction t=0 we note the Lemma, below.

Lemma 9.2. Λ^s is (extends to) an isometry $\mathfrak{H}_0 \to \mathfrak{H}_s$, and we then have

(9.7) $$L(\mathfrak{H}_s) = \Lambda^s L(\mathfrak{H}_0)\Lambda^{-s} , \quad K(\mathfrak{H}_s) = \Lambda^s K(\mathfrak{H}_0) \Lambda^{-s} .$$

That is, every operator in $L(\mathfrak{H}_s)$ can be written as $A = \Lambda^s B \Lambda^{-s}$, with a unique $B \in L(\mathfrak{H}_0)$, and for every $B \in L(\mathfrak{H}_0)$ the operator A above is well defined and in $L(\mathfrak{H}_s)$ (and similar with K instead of L).

The proof is evident, in view of lemma 2.1.

Coming back now to the proof of Lemma 9.1 for general t, we notice that

$$\Lambda^{-t}C_s\Lambda^t = \Lambda^{-(s+t)}a(M)\Lambda^{s+t} - \Lambda^{-t}a(M)\Lambda^t$$

(9.8)

$$= C_{s+t} - C_t \in K(\mathfrak{H}_0) ,$$

so that Lemma 9.2 yields $C_s \in \Lambda^t K(\mathfrak{H}_0)\Lambda^{-t} = K(\mathfrak{H}_t)$, q.e.d.

Lemma 9.3. Let $a \in CO^\infty(\mathbb{R}^n)$, $\varepsilon > 0$, then

(9.9) $$a(M)\Lambda^\varepsilon \in K(\mathfrak{H}_t), \quad \Lambda^\varepsilon a(M) \in K(\mathfrak{H}_t), \quad t\in\mathbb{R} .$$

Proof. Applying Lemma 8.1 we have the operators (9.9) in $K(\mathfrak{H}_0)$. Also we get

(9.10) $$\Lambda^{-t}a(M)\Lambda^\varepsilon\Lambda^t = C_t\Lambda^\varepsilon + a(M)\Lambda^\varepsilon \in K(\mathfrak{H}_0) ,$$

by Lemma 9.1. Thus Lemma 9.2 will supply the statement for the first operator, and the treatment of the second operator is similar, q.e.d.

Lemma 9.4. Let $a \in A$, and let us introduce the operators $S_j \in O(0)$ by

(9.11) $$S_j = D_j\Lambda \quad , j=1,\ldots,n .$$

For every $t \in \mathbb{R}$ we have

(9.12) $$[a(M),\Lambda] \in K(\mathfrak{H}_t), \quad [a(M),S_j] \in K(\mathfrak{H}_t), \quad j=1,\ldots,n .$$

Proof. The compactness of the first commutator is trivial, because (9.2) for s=1 implies

(9.13) $$[a(M),\Lambda] = \Lambda C_1 \in K(\mathfrak{H}_t) .$$

On the other hand,

(9.14) $$[a(M),S_j] = [a(M),D_j]\Lambda + D_j\Lambda C_1$$

with C_1 as in (9.2) for s=1 again.

But $[a(M),D_j] = i\, a_{x_j}(M)$, with the partial derivative $a_{x_j} = \partial a/\partial x_j \in CO^\infty(\mathbb{R}^n)$ again. Hence the first term at right is in $K(\mathfrak{H}_t)$, by Lemma 9.1. Also the second term is in $K(\mathfrak{H}_t)$ since C_1 is, by Lemma 9.1 and due to $D_j\Lambda = S_j \in O(0)$, q.e.d.

Finally, in this section, let us introduce the $\mathfrak{H}_s$-adjoint $A^{<s>}$ of an operator in $L(\mathfrak{H}_s)$. This adjoint is different from the $O(\infty)$-adjoint A^* of A, which we defined in section 5. It is defined as the Hilbert space adjoint of $A \in L(\mathfrak{H}_s)$ with respect to the inner product $(u,v)_s$, and therefore it is an operator in

$L(\mathcal{H}_s)$ again. Note that

(9.15) $$(u,v)_s = (\Lambda^{-2s}u,Av) = (A^*\Lambda^{-2s}u,v) = (\Lambda^{2s}A^*\Lambda^{-2s}u,v)_s$$

for $u,v \in \mathcal{H}_\infty$ and $A \in \mathcal{O}(0)$, which shows that

(9.16) $$A^{<s>} = \Lambda^{2s}A^*\Lambda^{-2s} \quad \text{on } \mathcal{H}_\infty ,$$

for $A \in \mathcal{O}(0)$, $s \in \mathbb{R}$. We summarize:

Lemma 9.5. For $A \in \mathcal{O}(0)$ we get $A^{<s>} \in \mathcal{O}(0)$ (rather its restriction to $\mathcal{H}_\infty$ defines an operator in $\mathcal{O}(0)$), and $A^{<s>}$ is explicitly given by (9.16).

Corollary 9.6. For $a \in A$ and $s,t \in \mathbb{R}$ we have

(9.17) $$a(M)^{<s>} - \bar{a}(M) \in K(\mathcal{H}_t) .$$

Also we have

(9.18) $$a(D)^{<s>} = a(D) .$$

Proof: The first is an immediate consequence of (9.2) for s replaced by $-2s$, using that $a(M)^* = \bar{a}(M)$. The second follows because formal Fourier multipliers commute, using (9.16).

Problems:

1) Show that relation (9.5) may be improved to obtain an asymptotic expansion

(9.19) $$\Lambda^{-s}a(M)\Lambda^{s}-a(M) = \sum_{1\le|\theta|} i^{|\theta|}/\theta!\ \mu_\theta(D)a^{(\theta)}(M) \quad (\text{mod } \mathcal{Q}(-\infty)).$$

10. The algebra $\mathring{\mathcal{A}}$ of finitely generated ψdo-s with zero oscillation at ∞.

In this section we are preparing some additional auxiliary results involving the operator algebra $\mathring{\mathcal{A}}$ generated by

(10.1) $$a(M) : a \in A , \Lambda , S_1,\dots,S_n .$$

Speaking precisely, $\mathring{\mathcal{A}}$ is defined to be the smallest sub-algebra of $\mathcal{O}(0)$ containing all operators (10.1).

Clearly $\mathring{\mathcal{A}}$ has a unit element, the identity operator of $\mathcal{O}(0)$, or the operator $e(M)$ with the function $e(x) = 1$, $x \in \mathbb{R}^n$. Usually we shall regard $\mathring{\mathcal{A}}$ as an algebra of continuous linear operators $\mathcal{H}_\infty \to \mathcal{H}_\infty$. However, since any such operator has a unique continuous extension mapping $\mathcal{H}_s \to \mathcal{H}_s$ where s is fixed, $-\infty \leq s \leq \infty$, we may just as well regard $\mathring{\mathcal{A}}$ as a sub-algebra of $L(\mathcal{H}_s)$ for any given fixed s. We shall not distinguish in notation between $A \in \mathring{\mathcal{A}}$ and its various continuous extensions mentioned. Similarly we shall not distinguish between the various interpretations of $\mathring{\mathcal{A}}$ mentioned. Note that these operators could just as well be interpreted to map from $\mathcal{S}$ to $\mathcal{S}$, or from $\mathcal{S}'$ to $\mathcal{S}'$, or to map between L^p-spaces or L^p-Sobolev spaces, etc.

It perhaps is important to mention this point, since in general an operator is considered well defined only after a precise definition of its domain. Here we are considering ψdo-s just about like formal differential operators. A formal principle is given by such an operator, to map functions or distributions of a very general class into other functions or distributions, but the precise fixing of the domain of definition is changed from time to time, without changing notation.

Lemma 10.1. For any $s \in \mathbb{R}$, given fixed, let $u \in \mathcal{H}_s$, $u \neq 0$ be given fixed. Then for every $\varepsilon > 0$ and non-negative integer k and every $v \in C_0^\infty(\mathbb{R}^n)$ there exists an operator $A \in \mathring{\mathcal{A}}$ such that

(10.2) $$\| v - Au \|_k \leq \varepsilon .$$

For the proof we will need the proposition below.

Lemma 10.2. The operator Λ^2 may be written as an integral operator with positive kernel.

Proof. It already was seen earlier (proof of Lemma 8.1) that Λ^2 may be written as an integral operator with kernel in L^1. In fact the kernel is the inverse Fourier transform of $\lambda^2(x) = (1+x^2)^{-1}$ up to a multiple of 2π. But the function $\mu = \lambda^{2\vee}$ must be radially symmetric - i.e., a function of $|\xi|$ only, since $\mu(O\xi) = \mu(\xi)$ for every orthogonal matrix O, as readily seen with an integral substitution. Accordingly it suffices to look at $\phi(\rho) = \mu(0,\dots,0,\rho)$ only, for $\rho > 0$. Notice that

$$(10.3) \qquad \phi(\rho) = \int d\tilde{x} \int dx_n \, e^{ix_n\rho}/(1+\tilde{x}^2+x_n^2)$$

where we have introduced $\tilde{x} = (x_1,\dots,x_{n-1})$, $\sigma = (1+\tilde{x}^2)^{1/2}$. The inner integral is easily calculated, using complex techniques together with a decomposition by partial fractions. Thus one obtains

$$(10.4) \qquad \phi(\rho) = (2\pi)^{-n/2}\pi \int d\tilde{x}/\sigma \; e^{-\sigma\rho} = (2\pi)^{-n/2}\pi\omega_{n-1}\int_0^\infty \frac{d\tau\; \tau^{n-2} e^{-\rho(1+\tau^2)^{1/2}}}{(1+\tau^2)^{1/2}} .$$

The existence of all integrals involved is undisputed, assuming that $\rho > 0$. We leave it to the reader to verify that this indeed gives the desired Fourier transform (i.e., that (8.2) for $\varepsilon=2$ holds with)

$$(10.5) \qquad \lambda^{2\vee}(x) = (2\pi)^{-n/2}\pi\omega_{n-1}\int_0^\infty d\tau\; \tau^{n-2} e^{-|x|(1+\tau^2)^{1/2}}/(1+\tau^2)^{1/2}$$

(where ω_{n-1} denotes the area of the unit sphere $|\tilde{x}| = 1$). Note in that respect that the n-dimensional Fourier integral for the function λ^2 is divergent, as n is larger than 1. On the other hand, (10.5) shows that indeed $\mu = \lambda^{2\vee} > 0$ for all x, q.e.d.

Remark. Lemma 10.2 is evident if we accept the representation (8.3) of λ^ε by modified Hankel functions. In fact, it then follows at once that every power Λ^ε $\varepsilon > 0$ is an integral operator with positive kernel. Also, positivity of $\lambda^{2\vee}$ is an easy consequence of the maximum principle for the equation $\Delta u = u$.

Proof of Lemma 10.1. Suppose $0 \neq u \in \mathfrak{H}_s$ is given, then let $w = \Lambda^t u$ and observe that w is continuous and not identically zero if only we let $t > -s+n/2$, by Sobolev's lemma, since $w \in \mathfrak{H}_{s+t}$, due to $\Lambda^t \in \mathcal{O}(-t)$. In fact, by letting $t > -s+1+n/2$ we can make $w \in C^1(\mathbb{R}^n)$.

Since w is continuous and not identically zero it must be bounded away from zero in some disk $|x-x^0| \leq \delta$. Moreover, we may assume that all values of w in that disk are in the complex halfplane $\mathrm{Re}\,(e^{i\theta}w) \geq \gamma > 0$, with suitable constants θ, γ,δ. Then let $a \in C_0^\infty(\mathbb{R}^n)$ be non-negative, and let it have support in that disk. Also let $\int a dx \neq 0$. Conclude that $z = \Lambda^2 a(M)w = \Lambda^2 a(M)\Lambda^t u$ is never zero, because all the values of z(x) for $x\in\mathbb{R}^n$, must be in the halfplane $\mathrm{Re}\,(e^{i\theta}z) > 0$, using Lemma 10.2. Also we get $a(M)w \in C_0^1(\mathbb{R}^n) \subset \mathfrak{H}_1$ (using Lemma 2.2.). Accordingly $z \in \mathfrak{H}_{1+2}$, provided only that $t > -s+1+n/2$. Next we note that $z \in CO^k(\mathbb{R}^n)$ whenever $1+2 > k+n/2$, using again Sobolev's Lemma. Accordingly the function $b(x) = v(x)/z(x)$ is well defined and in $C_0^k(\mathbb{R}^n)$. It follows that

$$(10.6) \qquad v = b(M)\Lambda^2 a(M)\Lambda^t u \; .$$

If we still replace b(x) by a suitable C_0^∞-function c(x) approximating b in C^k-norm then Lemma 10.1 becomes evident: We must choose the operator $A \in \mathfrak{A}$ as

(10.7) $$A = c(M)\Lambda^2 a(M)\Lambda^t \quad .$$

Corollary 10.3. The operator A of Lemma 10.1 may be chosen in the smaller algebra with multiplications in $C_0^\infty(\mathbb{R}^n)$ only.
The proof is evident after (10.7) .

Next let us introduce the class

(10.8) $$CM(\mathbb{R}^n) = \{ a \in C(\mathbb{R}^n) : a = O(1),\ cm_x(a) \to 0, \text{ as } |x| \to \infty\}.$$

Here we were using the local oscillation $cm_x(a) = cm_{x,1}(a)$ of the continuous function a(x) where for $\varepsilon > 0$ we define

(10.9) $$cm_{x,\varepsilon}(a) = \text{Max } \{|a(x+h)-a(x)|: h \in \mathbb{R}^n, |h| < \varepsilon \} .$$

Thus we may say that $CM(\mathbb{R}^n)$ is the class of continuous functions over $\mathbb{R}^n$ with local oscillation vanishing at infinity.

Notice that the condition $cm_x(a) \to 0$, as $|x| \to \infty$, is equivalent to $cm_{x,\varepsilon}(a) \to 0$ as $|x| \to \infty$, where $\varepsilon > 0$ may be chosen arbitrary fixed. Indeed, the number $cm_{x,\varepsilon}(a)$ decreases for fixed x and a, as ε decreases, and, on the other hand, we get

(10.10) $$cm_{x,s}(a) \le \sum_{l=1}^{M} cm_{x+h^l,\varepsilon}(a)$$

where the vectors h^l are chosen such that the balls with center h^l and radius ε cover the ball with center 0 and radius s.

Lemma 10.4. $CM(\mathbb{R}^n)$ is an algebra, and it is the closure of the algebra A defined in (9.1), with respect to the sup norm over $\mathbb{R}^n$.

Proof. It is sufficient to prove the second statement, since we know that A is an algebra. Clearly CM contains A since we get

(10.11) $$|a(x+h)-a(x)| = \left|\int_0^1 d/dt\ (a(x+th))dt\right| \le |h| \sup \{|\nabla a(y)|: |x-y|\le h\}$$

so that $cm_x(a) \to 0$ whenever $a^{(\alpha)} \to 0$ for $|\alpha| = 1$, as $|x| \to \infty$. Vice versa to approximate $a \in CM$ by a sequence in A we use the well known regularization technique (c.f. [8], for example). Let $\phi \in C_0^\infty(\mathbb{R}^n)$ have support in $|x| \le 1$, let it be nonnegative, let

(10.12) $$\int_{\mathbb{R}^n} \phi\, dx = 1 ,$$

and let us define ϕ_ε, $\varepsilon > 0$, by $\phi_\varepsilon(x) = \varepsilon^{-n}\phi(x/\varepsilon)$. Then for $a \in CM$ define

$a_\varepsilon = \phi_\varepsilon * a$, which is explicitely given by the convolution integral

(10.13) $$a_\varepsilon(x) = \int \phi_\varepsilon(x-y)a(y)đy \quad .$$

Evidently a_ε is in $C^\infty(\mathbb{R}^n)$, and we get

(10.14) $$a_\varepsilon^{(\alpha)}(x) = \int \phi_\varepsilon^{(\alpha)}(x-y)a(y)đy \quad ,$$

because the integral is over the compact ball $|x-y| \leq \varepsilon$ only, outside of which the integrand vanishes. All partial derivatives of the integrand are continuous in x and y jointly.

For $\alpha = 0$ we get $|a_\varepsilon(x)| \leq \sup\{|a(x)|: x \in \mathbb{R}^n\} < \infty$, since $\int đy\, |\phi_\varepsilon(y)| = \int \phi\, đx = 1$ by assumption. This gives boundedness of a_ε uniformly in $x \in \mathbb{R}^n$, $0 < \varepsilon \leq 1$. On the other hand, for $\alpha \neq 0$ we may write

(10.15) $$a_\varepsilon^{(\alpha)}(x) = \int \phi_\varepsilon^{(\alpha)}(x-y)\ (a(y)-a(x))đy \quad ,$$

because $\int \phi_\varepsilon^{(\alpha)}(x-y)đy = 0$, due to the fact that ϕ vanishes for large x. From (10.15) we estimate

(10.16) $$|a_\varepsilon^{(\alpha)}(x)| \leq \int đy\, |\phi_\varepsilon^{(\alpha)}(y)| \quad \sup\{|a(y)-a(x)|: \ |y-x| \leq \varepsilon\}$$

which amounts to $|a_\varepsilon^{(\alpha)}(x)| \leq c_\alpha(\varepsilon)\ cm_{x,\varepsilon}(a)$ and shows that $a_\varepsilon \in A$ for fixed $\varepsilon > 0$.

Finally, as $\varepsilon \to 0$ we get

(10.17) $$|a(x) - a_\varepsilon(x)| = |\int đy\ \phi_\varepsilon(x-y)(a(x)-a(y))| \leq cm_{x,\varepsilon}(a) \to 0 \quad .$$

The convergence is uniform over $x \in \mathbb{R}^n$, as follows for example from Dini's Lemma, since $cm_{x,\varepsilon}$ for fixed ε and a is continuous in x and goes to zero monotonically as $\varepsilon \to 0$ and due to $cm_{x,\varepsilon} \to 0$ as $|x| \to \infty$. This establishes Lemma 10.4.

Finally, in this section we prove the following converse to Lemma 8.1 .

Lemma 10.5. Let $a,b \in CM(\mathbb{R}^n)$, as defined in (10.8). If the operator

(10.18) $$A = a(M)b(D)$$

is compact from $\mathfrak{H}_0$ to $\mathfrak{H}_0$ then either one of the functions a,b vanishes identically, or we have $a,b \in CO(\mathbb{R}^n)$.

Remark. Strictly speaking the formal Fourier multiplier b(D) is defined only if $b \in T$ which is not always the case for $b \in CM$. Note however that $b(D) = F^{-1}b(M)F$ is indeed well defined as an operator in $L(\mathfrak{H}_0)$, due to boundedness of b and the fact that $F|\mathfrak{H}_0$ is a unitary operator $\mathfrak{H}_0 \to \mathfrak{H}_0$. We thus shall use that notation with this slightly different meaning.

<u>Proof of Lemma 10.5.</u> It suffices to prove that A is not compact if $a \notin CO$ and $b \not\equiv 0$, because we get

(10.19) $$FAF^{-1} = c(D)b(M) \ , \quad c(x) = a(-x) \ .$$

Hence, if $a \not\equiv 0$ and $b \notin CO$ then $\bar{b}(M)\bar{c}(D) = (FAF^{-1})^*$ cannot be compact which gives the full statement of the lemma.

If $a \notin CO$ we may chose a sequence x^j, $|x^j| \to \infty$ such that $|a(x^j)| \geq 2p > 0$. Since the oscillation of a vanishes at ∞ we get

(10.20) $$|a(x)| \geq p \quad \text{in} \quad |x-x^j| \leq 1$$

for sufficiently large j. Also, by thinning out the sequence x^j, i.e. omitting some of the points, we can obtain a subsequence $|x^{j_l} - x^{j_{l-1}}| > 2$. By this construction we have reached a sequence (called x^j again) such that

(10.21) $$|a(x)| \geq p \quad \text{in} \quad |x-x^j| \leq 1, \quad |x^j-x^l| > 2, \ j \neq l, \quad j,l=1,2,\ldots \ .$$

Also, since $b \not\equiv 0$ we can find $u \in \mathcal{G}_0$ such that $b(D)u \neq 0$. It is known that Fourier multipliers are translation invariant. That is we get

(10.22) $$T_z b(D) = b(D) T_z \qquad ,$$

for every translation operator $T_z\colon \mathcal{G}_0 \to \mathcal{G}_0$, defined by $T_z u(x) = u(x-z)$, $x \in \mathbb{R}^n$, where $z \in \mathbb{R}^n$ is a given fixed vector. Translation operators T_z are unitary maps of $\mathcal{G}_0$, since T_{-z} inverts T_z and an integral substitution shows that $\|T_z u\|_0 = \|u\|_0$, $u \in \mathcal{G}_0$, $z \in \mathbb{R}^n$. Since $b(D)u \neq 0$ there must be some ball $\mathcal{B}_z$ of radius 1 such that $\int_{\mathcal{B}_z} |b(D)u|^2 dx = \gamma > 0$. Then it follows that

(10.23) $$\int_{|x-x^1| \leq 1} |b(D)u|^2 dx = \gamma > 0$$

if the original u is replaced by $T_{x^1-z} u \in \mathcal{G}_0$.

It follows that a C_0^∞-function ϕ^1 with support in $\mathcal{B}_{x^1}$ can be constructed such that $\phi^1(M)b(D)u \neq 0$. Using (10.21) one therefore concludes that

(10.24) $$v^1 = \phi^1(M)Au = a(M)\phi^1(M)b(D)u \neq 0 \ .$$

Also, let $V_l = T_{x^l-x^1}$ and let $w^l = V_l\phi^1(M)b(D)u = \phi^l(M)b(D)u^l$ with $\phi^l(x) = \phi^1(x-x^l+x^1)$ having support in $\mathcal{B}_{x^l}$, and with $u^l = V_l u$. Since the V_l are unitary it follows that u^l and $f^l = a(M)w^l$ satisfy

(10.25) $$\|u^l\|_0 = \|u\|_0, \quad \|f^l\| \geq p\|w^l\| = p\|\phi^1(M)b(D)u\|_0 > 0,$$

since the w^l have support in $\mathcal{B}_{x^l}$ where $a(M)$ is $\geq p$.

Note also that the f^l are mutually orthogonal in $\mathfrak{H}_0$, since their supports are disjoint (i.e. $\operatorname{supp} f^l \subset B_{x^l}$, $B_{x^l} \cap B_{x^j} = \emptyset$, as $j \neq l$, due to (10.21)). Therefore the sequence f^l cannot contain a convergent subsequence. However then the sequence Au^l cannot contain a convergent subsequence as well. For suppose $Au^{l_k} \to g$ in $\mathfrak{H}_0 = L^2(\mathbb{R}^n)$, then

$$\|f^{l_k}\|_0^2 = \|\phi^{l_k}(M)Au^{l_k}\|_0^2 \leq 2\|\phi^{l_k}(M)(Au^{l_k}-g)\|_0^2 + 2\|\phi^{l_k}g\|_0^2 \tag{10.26}$$

$$\leq c\|Au^{l_k}-g\|_0 + c\int_{B_{x^l}} |g(x)|^2 dx \to 0$$

which is a contradiction. This proves the Lemma.

Problems:

1) Prove Lemma 10.2 by means of the maximum principle for the Helmholtz equation $\Delta u = u$.

2) Prove that we have $a(M)b(D) \in K_0$ whenever $a, b \in CO(\mathbb{R}^n)$.

References

[1] N. Aronszajn and K.T. Smith, Theory of Bessel potentials. Part I; Ann. Institute Fourier, Grenoble 11 (1961), 385-475.

[2] N.Bourbaki, Espaces vectorielles topologiques; Hermann, Paris 1964.

[3] H.O. Cordes and E. Herman, Gelfand theory of pseudo-differential operators, Amer. J. Math. 90 (1968) 681-717.

[4] K.O. Friedrichs, Pseudo-differential operators: an introduction; Courant Institute NYU, Lecture Notes New York 1970.

[5] L. Hoermander, Pseudo-differential operators and hypo-elliptic equations; Proc. Symp. Pure Math. Vol X (Singular integrals) Chicago 1966.

[6] J.J. Kohn and L. Nirenberg, An algebra of pseudo-differential operators; Comm. Pure Appl. Math. 18 (1965) 501-517.

[7] H. Kumano-go, Pseudo-differential operators and the uniqueness of the Cauchy problem; Comm. Pure Appl. Math. 23 (1969) 73-129.

[8] L. Schwartz, Théorie des distributions; Hermann Paris 1966.

[9] W. Magnus and F. Oberhettinger, Special functions; Springer Berlin 1948.

[10] J. Marsden, Basic complex analysis; Freeman, San Francisco 1973.

[11] R.T. Seeley, Topics in Pseudo-differential operators, Centro Internazionale Matematico Estivo, Conference at Stresa; 1968.

Chapter IV: On the Laplace Comparison Algebras for $L^2(\mathbb{R}^n)$.

In the present chapter we come to the point of our efforts. If we take norm closure in $L(\mathfrak{H}_s)$ of the algebra $\mathring{\mathfrak{A}}$, of III, section 10, - that is, basically of the algebra of all zero order finitely generated ψdo's with coefficients having zero oscillation at ∞ then a C^* - subalgebra $\mathfrak{A}_s$ of $L(\mathfrak{H}_s)$ is generated which contains the compact ideal $K(\mathfrak{H}_s)$. Moreover, $\mathfrak{A}_s$ has compact commutators, so that $\mathfrak{A}_s / K(\mathfrak{H}_s)$ is a commutative C^*-algebra with unit. By the Gel'fand-Naïmark theorem therefore it must be isometrically isomorphic to a function algebra $C(\mathbb{M})$ with some compact Hausdorff space $\mathbb{M}$ (c.f. AII, Theorem 4.1).

It will be proven that the symbol space $\mathbb{M}$ is (homeomorphic to) the boundary of $\mathbb{R}^n \times \mathbb{R}^n = \mathbb{R}^{2n}$ in a certain compactification $\mathbb{P}^n \times \mathbb{B}^n$ of (x,ξ)-space $\mathbb{R}^{2n}$. The symbols of the generators $a(M)$ and $b(D)$, of the set III, (10.1), simply are the restrictions to $\mathbb{M}$ of the continuous extensions of the functions $a(x)$ and $b(\xi)$ to $\mathbb{P}^n \times \mathbb{B}^n$. Here the symbol of an operator $A \in \mathfrak{A}_s$ is defined to be the continuous function $a = \sigma_A$ associated to the co-set $A^{\vee} = \{A + K(\mathfrak{H}_s)\}$ by the Gel'fand-Naïmark isomophism. All above results are discussed in sections 1 and 2. The algebras $\mathfrak{A}_s$ will be called the Laplace comparison algebras of $L^2(\mathbb{R}^n)$ (or, of $\mathfrak{H}_s$).

These facts gain their importance by the fact that the C^*-algebras $\mathfrak{A}_s$ are Fredholm closed, by Appendix AI,5 (c.f. also Lemma 2.5). Recalling AI, theorem 4.8 it therefore follows that an operator $A \in \mathfrak{A}_s$ is Fredholm if and only if its symbol σ_A does never vanish on $\mathbb{M}$. Notice that the symbol of a finitely generated ψdo in $\mathring{\mathfrak{A}}$ is readily available, so that a useful criterion results.

This in effect may be regarded the root of all theory of elliptic differential operators, because for a differential operator

$$(0.1) \qquad L = \sum_{|\alpha| \leq N} a_\alpha(M) D^\alpha \quad , \quad a_\alpha \in A \, ,$$

over $\mathbb{R}^n$ with coefficients having zero oscillation at infinity one may introduce the operator $A = L\Lambda^N \in \mathring{\mathfrak{A}}$. Clearly then $L: \mathfrak{H}_s \to \mathfrak{H}_{s-N}$ will be Fredholm if and only if A is Fredholm, since $\Lambda^N: \mathfrak{H}_{s-N} \to \mathfrak{H}_s$ acts as an isometry. Hence a necessary and sufficient condition for the differential equation $Lu = f$, $u \in \mathfrak{H}_s$, $f \in \mathfrak{H}_{s-N}$ to be normally solvable is that the symbol quotient $\sigma_A = \tau_L^N$ does never vanish at $\mathbb{M}$ - that is for either $|x| = \infty$ or $|\xi| = \infty$ or both.

We distinguish the following subspaces of the symbol space $\mathbb{M}$. The principal symbol space $\mathbb{M}_p$ is the collection of all $(x,\xi) \in \mathbb{M}$ with $|\xi| = \infty$. Its interior $\mathring{\mathbb{M}}_p = W = \{(x,\xi): |x| < \infty \, , \, |\xi| = \infty\}$ is called the wave front space. Also its complement $\mathbb{M}_s = \mathbb{M} - \mathbb{M}_p = \{(x,\xi): |x| = \infty, \, |\xi| < \infty\}$ will be called the secondary symbol space. An operator $K = A\Lambda^{-r}$, $r \in \mathbb{R}$, $A \in \mathfrak{A}_s$, will be called elliptic if its symbol quotient of order r, defined as $\tau_K = \sigma_A$ is $\neq 0$ on the principal symbol space $\mathbb{M}_p$. If this is even true on all of $\mathbb{M}$, then K is called md-elliptic. (The notation hints at the symmetry of the algebra $\mathfrak{A}_s$ and its symbol space with respect to multiplications "M" and differentiations "D".) For a differential operator K of order N "elliptic" means uniformly elliptic over $\mathbb{R}^n$ in the conventional sense.

One should mention that the compactification $\mathbb{P}^n$ of $\mathbb{R}^n$ used, being the maximal ideal space of the algebra $CM(\mathbb{R}^n)$, has properties similar to that of a Stone-Čech compactification, and therefore is not very intuitive. However the non-vanishing of a symbol on $\mathbb{M}$ or parts of it is easy to check, at least for finitely generated ψdo's. Indeed such symbols are naturally extended into all of $\mathbb{P}^n \times \mathbb{B}^n$. Thus one will have $\sigma \neq 0$ on $\mathbb{M}$ if and only if the extension to $\mathbb{P}^n \times \mathbb{B}^n$ is bounded away from zero in $\mathbb{R}^n \times \mathbb{R}^n$, as $|x| + |\xi|$ is sufficiently large.

The symbol of a differential operator L of positive order in general will not be defined on all of $\mathbb{M}$, but only on the secondary symbol space, where $\sigma_\Lambda \neq 0$. There we define backwardly

$$\text{(0.2)} \qquad \sigma_L = \sigma_A/\sigma_{\Lambda^N} = \tau_L^N/\sigma_{\Lambda^N}, \quad (x,\xi) \in \mathbb{M}_s .$$

It proves that the secondary symbol is responsible for the essential spectrum of the operator L, considered an unbounded operator of $\mathfrak{H}_s$ with domain $\mathfrak{H}_{s+N}$ (dense in $\mathfrak{H}_s$) (c.f. 2, problem 10).

In sections 3 and 4 we consider the algebra $\mathfrak{A}_\infty$ obtained by closing the algebra $\mathring{\mathfrak{A}} \in \mathcal{O}(0)$ under the Frechet topology of $\mathcal{O}(0)$. Clearly, $\mathfrak{A}_\infty \subset \mathfrak{A}_s$, for every s. Therefore an operator $A \in \mathfrak{A}_\infty$ has a symbol in every $\mathfrak{A}_s$. But one finds this symbol to be independent of s, as $A \in \mathfrak{A}_\infty$. Therefore the notation σ_A is used for it again. The question for existence of a specific Fredholm inverse, called Green inverse is investigated in section 4. A green inverse K of an operator $L \in \mathcal{O}(\infty)$ is defined by the property that both 1-AB, 1-BA are operators of finite rank, continuous from $\mathfrak{H}_{-\infty}$ to $\mathfrak{H}_\infty$ (c.f. (4.1); for the definition of ordinary Fredholm inverse c.f. AI, 2). We prove that $A \in \mathfrak{A}_\infty$ admits a Green inverse of order 0 if and only if its symbol σ_A does never vanish on $\mathbb{M}$ (that is if it is md-elliptic). Similarly, a differential operator L of order N admits a Green inverse of order -N if and only if its symbol quotient τ_L^N of order N does never vanish (i.e. again, if and only if it is md-elliptic).

In sections 5 through 8 we attack the problem of constructing operators in $\mathfrak{A}_\infty$ to a given symbol $a = \sigma_A \in C(\mathbb{M})$. Clearly, this would be a trivial question for an algebra $\mathfrak{A}_s$, for finite s, because then every continuous function $a \in C(\mathbb{M})$ is a symbol, as we know. We express the conjecture that the same is true also for $\mathfrak{A}_\infty$. However at present we will prove the existence of $A \in \mathfrak{A}_\infty$ with symbol a only under the much stronger condition that all the x-derivatives of a exist as functions in $C(\mathbb{M})$ and in norm convergence of $C(\mathbb{M})$. That is the successive difference quotients (with respect to x and with $x+h = x$ as $|x| = \infty$) converge uniformly over $\mathbb{M}$ to a function in $C(\mathbb{M})$. (Theorem 7.1). At the same time we show that again a Leibnitz formula of the asymptotic kind of chapter III holds for $Ab(D)$ and $b(D)A$, for any such operator with symbol a, and $b \in \mathcal{M}$ (c.f. II,4).

This result will become significant later on (chapter V), where we will use it for proving a Gårding inequality for strongly elliptic ψdo's. This will establish the connection to "conventional" tecniques of apriori estimates in theory of partial differential equations. In particular it will enable us to solve the Dirichlet problem at least for strongly elliptic 2m-th order equations in domains with smooth boundary, not necessarily compact. The proof of theorem 7.1 again uses the results of chapter II together with an operation called Fourier kernel multiplication in I,9. Also we will have to solve commutator equations like $[D_j, X] \pm iX = F$.

We have attached the final two sections because they in effect show that every bounded operator of $\mathfrak{H}_0 = L^2(\mathbb{R}^n)$ with the property that all successive commutators $[D_{j_1}, [D_{j_2}, \ldots, [D_{j_r}, [M_{l_1}, [M_{l_2}, \ldots, [M_{l_s}, A] \ldots\ldots]$ are bounded operators of $\mathfrak{H}_0$, necessarily must be a ψdo of the form III, (7.19), with (formal) symbol $a \in CB^\infty(\mathbb{R}^n)$.*) (There is no complete proof, but all the formalism is there).

We notice that all applications to classical analysis and differential equations of our abstract results in this chapter are stated in form of problems, at the end of various sections. Our algebra $\Psi\mathcal{D}$, for example, contains a large variety of differential operators and of singular integro-differential operators as well which were the focus of earlier investigations.

Perhaps we should note explicitly that the extension of most results obtained to systems of ψdo's is a simple matter. It can be accomplished by an abstract investigation of the tensor products $L(\mathbb{C}^n) \otimes \mathfrak{A}_s$ and $L(\mathbb{C}^n) \otimes \mathcal{O}(\infty)$, etc. In fact this will be discussed in V,8 and V,9. for a different purpose, even for certain $\infty\times\infty$-systems.

---

*)For a detailed discussion,including proofs,c.f. Cordes,Manuscripta Math.28(1979), p.51 .A very similar result was published earlier by R. Beals,Duke Math. Journal 44 (1977),p.45.

1. The Laplace comparison algebras

We recall the algebra $\mathring{\mathfrak{A}}$ of finitely generated ψdo's with oscillation vanishing at $x = \infty$, as defined in III, section 10. In particular it was pointed out that $\mathring{\mathfrak{A}}$ may be regarded as subalgebra of either $\mathcal{O}(0)$ or of $L(\mathfrak{H}_s)$, $s\in\mathbb{R}$. Now $\mathcal{O}(0)$ is a Frechet space and the $L(\mathfrak{H}_s)$ are Banach spaces under operator norm. We may take closure of $\mathring{\mathfrak{A}}$ in $\mathcal{O}(0)$ to obtain a complete topological algebra denoted by $\mathfrak{A}_\infty$. Similarly the closure of $\mathring{\mathfrak{A}}$ in $L(\mathfrak{H}_s)$ will be a Banach sub-algebra of $L(\mathfrak{H}_s)$ for every $s\in\mathbb{R}$, denoted by $\mathfrak{A}_s$. $\mathfrak{A}_s$, $-\infty<s\leq\infty$ will be called the Laplace comparison algebras of $\mathbb{R}^n$. Specifically we shall speak of the L^2-comparison algebra - meaning $\mathfrak{A}_0$, or the $\mathfrak{H}_s$-comparison algebra $\mathfrak{A}_s$, $s\in\mathbb{R}$, or the $\mathcal{O}(0)$ comparison algebra, meaning $\mathfrak{A}_\infty$. The notation is motivated by our later technique using operators of $\mathfrak{A}_s$ to compare a general ψdo with a power of the Laplace operator - or rather a power of $(1-\Delta)$.

At first we will be focusing on the algebras $\mathfrak{A}_s$, $s \neq \infty$, because they are Banach algebras and therefore accessible to a well developed theory. Later on (section 3 f.) the information gained will be used to analyze $\mathfrak{A}_\infty$ which for many respects might be seen the real focus of our interest.

<u>Theorem 1.1.</u> $\mathfrak{A}_s$, for finite s, is a C^*-subalgebra of $L(\mathfrak{H}_s)$, with the involution given by the Hilbert space adjoint $A \to A^{<s>}$, (c.f. III,9). Moreover, $\mathfrak{A}_s$ contains the identity operator and all of the compact ideal $K_s = K(\mathfrak{H}_s)$ of $\mathfrak{H}_s$ and its commutator is contained in K_s.

<u>Proof.</u> We use Corollary 4.1.10 of [7] saying that any adjoint invariant Banach subalgebra $\mathfrak{A}$ of $L(\mathfrak{H})$ which is irreducible and contains a non-vanishing compact operator must contain the entire compact ideal. Here $\mathfrak{H}$ is any Hilbert space. An explicit proof, independent of all discussions in [7] is easy. For its basic ideas c.f. the proof of Lemma 3.3, below. Indeed it is an evident consequence of III, Lemma 10.1 that $\mathfrak{A}_s$ is irreducible. For if $u \in \mathfrak{H}_s$, $u \neq 0$, and if $v \in C_0^\infty(\mathbb{R}^n)$ the lemma implies that there exists a sequence $A_l \in \mathring{\mathfrak{A}}$ such that $\|v-A_l u\|_k \to 0$, $l \to \infty$. We may chose $k > s$ and then conclude that $\|v-A_l u\|_s \to 0$, $l \to \infty$. Since C_0^∞ is dense in $\mathfrak{H}_s$ (c.f. III, 2, problem 1) it follows that $\mathring{\mathfrak{A}}u$ is dense in $\mathfrak{H}_s$ so that indeed $\mathfrak{A}_s$ is irreducible. Also $\mathring{\mathfrak{A}}$ contains $C = [\Lambda,\lambda(M)] \in K_0$, and $C\neq 0$.

Next it certainly is evident that $\mathfrak{A}_0$ is a C^*-algebra because the generators Λ, S_j are self-adjoint while we get $a(M)^{<0>} = a(M)^* = \bar{a}(M)$, where $\bar{a} \in A$. Accordingly $\mathfrak{A}_0$ contains K_0. All remaining statements are trivial now, as far as $\mathfrak{A}_0$ is concerned, in view of III, Lemma 9.4. In particular the identity operator can be written as $I = e(M)$ with $e(x) = 1$ for all x, $e \in A$.

Next let us look at $\mathfrak{A}_s$ for general $s\in\mathbb{R}$. Again the generators Λ, S_j are self-adjoint, also as operators $\mathfrak{H}_s \to \mathfrak{H}_s$. On the other hand for $a \in A$ we now have the asymptotic expansion

(1.1) $$a(M)^{<s>} = \Lambda^{2s}\bar{a}(M)\Lambda^{-2s} \sim \sum_\theta i^{|\theta|}/\theta!\ \mu_\theta(D)a^{(\theta)}(M) \quad \text{mod } \mathcal{Q}(-\infty)$$

with

(1.2) $$\mu_\theta(x) = \lambda^{-2s(\theta)}(x)/\lambda^{-2s}(x) .$$

(c.f. III, 9, problem 1).

It is readily verified that $\mu_\theta(D)$ is a combination of the generators Λ, S_j. Accordingly we conclude that

(1.3) $$R_N = a(M)^{<s>} - Q_N \in \mathcal{Q}(-M)$$

for every $M = 0,1,2,\ldots$ and sufficiently large $N > N_3(M)$, where

(1.4) $$Q_N = \sum_{0\le|\theta|\le N} i^{|\theta|}/\theta!\ \mu_\theta(D)a^{(\theta)}(M) \in \mathring{\mathfrak{A}} .$$

For the remainder of this proof assume $s > 0$. The reader will easily see how the procedure must be modified in case $s < 0$. We then conclude that $V_N = \Lambda^{-s}R_N\Lambda^{-s} \in K_0 = K(\mathfrak{H}_0)$ as N is sufficiently large. Hence V_N may be approximated in $L(\mathfrak{H}_0)$ by a sequence $V_N^1 \in \mathring{\mathfrak{A}}$, such that $\|V_N - V_N^1\|_0 \to 0$, $1 \to \infty$. As we know this gives

(1.5) $$\|\Lambda^s V_N \Lambda^{-s} - \Lambda^s V_N^1 \Lambda^{-s}\|_s = \|R_N\Lambda^{-2s} - \Lambda^s V_N^1 \Lambda^{-2}\|_s \to 0,\ 1 \to \infty,$$

which implies that

(1.6) $$\|R_N - \Lambda^s V_N^1 \Lambda^s\|_{-s} \to 0, \quad \text{as } 1 \to \infty.$$

Also note that $\Lambda^s \in \mathfrak{A}_t$ for all t, because the binomial expansion

(1.7) $$\Lambda^s = (1-(1-\Lambda))^s = \sum_{j=0}^\infty (-1)^j \binom{s}{j} (1-\Lambda)^j$$

converges in every $L(\mathfrak{H}_t)$, as $s > 0$. Accordingly $\Lambda^s V_N^1 \Lambda^s \in \mathfrak{A}_s$ and it follows that $R_N \in \mathfrak{A}_s$, or that $a(M)^{<s>} \in \mathfrak{A}_s$, which shows that $\mathfrak{A}_s$ for general s is a C^*-algebra,

Again the remainder of the theorem now is trivial, in view of III, Lemma 9.4. q.e.d.

It is an immediate consequence of Theorem 1.1 that the quotient algebra $\mathfrak{A}_s/K_s$ is well defined, and is a commutative C^*-algebra with unit. (c.f. AII, theorem 8.1). Then we may apply the Gel'fand-Naimark theorem (AII, theorem 7.7) to conclude that

(1.8) $$\check{\mathfrak{A}}_s = \mathfrak{A}_s/K_s \approx C(\mathbb{M}) .$$

That is, the quotient algebra $\check{\mathfrak{A}}_s$ is isometrically isomorphic to an algebra $C(\mathbb{M})$ of all continuous complex-valued functions over a compact Hausdorff space $\mathbb{M}$, equipped with the sup norm, In details, there exists a 1-1-correspondence

$A^{\vee} \to \phi_{A^{\vee}}$ between the cosets $A^{\vee} = \{A + K_s\}$ of the elements in $\mathfrak{A}_s$ and the functions $\phi = \phi(m)$, $m\epsilon\mathbb{M}$, leaving the algebraic operations invariant and such that

$$(1.9) \qquad \|\tilde{A}\| = \inf \{\|A+C\| : C \in K_s\} = \sup \{|\phi_A(m)|: m\epsilon\mathbb{M}\} = \|\phi_{A^{\vee}}\|_{C(\mathbb{M})} .$$

Now it will be important, in the following, to find an explicit description of the space $\mathbb{M}$ - which will be referred to as the symbol space of our algebra $\mathfrak{A}_s$. Likewise for an operator $A \in \mathfrak{A}_s$ the continuous function $\phi_{A^{\vee}}$ over $\mathbb{M}$ associated to the coset A of A mod K_s will be called the symbol of A (with respect to $\mathfrak{A}_s$), and will be denoted by σ_A.

Let us first observe that the algebra $\mathfrak{A}_0$ contains two commutative C^*-subalgebras, namely the algebras $\mathfrak{A}^+$ and $\mathfrak{A}^{\#}$ generated by the multiplications a(M) with $a \in A$, and by the formal Fourier multipliers $\Lambda, S_1,\dots, S_n$, respectively. (c.f. III (10.1)) In fact, for our following investigations of the space $\mathbb{M}$ it is sufficient to consider the case of $s = 0$ only. Moreover theorem 1.2 below makes it evident that $\mathbb{M}$ is independent of s.

Theorem 1.2. We have

$$(1.10) \qquad \mathfrak{A}_s = \Lambda^s \mathfrak{A}_0 \Lambda^{-s} = \{ A \in L(\mathfrak{H}_s) : A = \Lambda^s B \Lambda^{-s}, \ B \in \mathfrak{A}_0\} .$$

Proof. We have learned before that $\Lambda^s: \mathfrak{H}_0 \to \mathfrak{H}_s$ acts as an isomtry between the two spaces so that conjugation with Λ^s takes $L(\mathfrak{H}_0)$ onto $L(\mathfrak{H}_s)$ isometrically and isomorphically (III, Lemma 9.2). Specifically this implies that a sequence $A_j \in L(\mathfrak{H}_0)$ converges (in operator norm) if and only if the sequence $B_j = \Lambda^s A_j \Lambda^{-s}$ converges in $L(\mathfrak{H}_s)$. But for any generator A of $\mathfrak{A}_0$, as listed in III, (10.1), we get $\Lambda^s A \Lambda^{-s} \in \mathfrak{A}_s$. Indeed, for a Fourier multiplier this is trivial, since Λ^s commutes with it, while for a multiplication a(M) we may apply III, Lemma 9.1 and Theorem 1.1 above for the same conclusion. Accordingly, if $\mathring{\mathfrak{A}}$ is regarded as subalgebra of $L(\mathfrak{H}_0)$ then $\Lambda^s \mathring{\mathfrak{A}} \Lambda^{-s} \subset \mathfrak{A}_s$ and therefore $\Lambda^s \mathfrak{A}_0 \Lambda^{-s} \subset \mathfrak{A}_s$, by taking closure. Similarly one concludes $\Lambda^{-s} \mathfrak{A}_s \Lambda^s \subset \mathfrak{A}_0$ which implies $\mathfrak{A}_s \subset \Lambda^s \mathfrak{A}_0 \Lambda^{-s} \subset \mathfrak{A}_s$, or the desired relation (1.10), q.e.d.

Lemma 1.3. The algebra $\mathfrak{A}^+$ defined as closure of $\{a(M): a \in A\}$ is explicitely given by

$$(1.11) \qquad \mathfrak{A}^+ = \{a(M): \ a \in CM(\mathbb{R}^n) \} ,$$

with the function algebra CM of III, (10.8). Moreover the correspondence $a \leftrightarrow a(M)$ defines an isometry $CM(\mathbb{R}^n) \leftrightarrow \mathfrak{A}^+$.

Proof. In view of III, Lemma 10.4 it is sufficient to consider the last statement. On the other hand for $a \in CM$ (or even for $a \in L^{\infty}(\mathbb{R}^n)$) it is evident that $\|a(M)\| \le \|a\|_{L^\infty}$, with the L^2-operator norm at left. Also to prove the reverse inequality one merely must chose a function in L^2 with support very close to those points of $\mathbb{R}^n$, where $|a|$ comes close to $\|a\|_{L^\infty} = \|a\|_C$, assuming that a is continuous, q.e.d.

Let us introduce the function

$$(1.12) \qquad y = s(x) = x/(1+x^2)^{1/2}, \quad x \in \mathbb{R}^n.$$

Evidently $s\colon \mathbb{R}^n \to \overset{\circ}{B}{}^n = \{y\colon\ y \in \mathbb{R}^n,\ |y| < 1\}$ constitutes a homeomorphism between $\mathbb{R}^n$ and the open unit ball $\overset{\circ}{B}{}^n$ in $\mathbb{R}^n$. In fact the inverse function s^{-1} is explicitely given by

$$(1.13) \qquad x = s^{-1}(y) = t(y) = y/(1-y^2)^{1/2},$$

as is easily calculated.

We define the class $CS(\mathbb{R}^n)$ by

$$(1.14) \qquad CS(\mathbb{R}^n) = \{ a \in C(\mathbb{R}^n) ; a \circ s^{-1} \in C(B^n) \}$$

where $B^n = \{ y \in \mathbb{R}^n ; |y| \le 1 \}$ is the closure of $\overset{\circ}{B}{}^n$.
More precisely $CS(\mathbb{R}^n)$ consists of all bounded continuous functions a over $\mathbb{R}^n$ such that the composed function $b(y) = a(t(y))$ (which is defined and continuous over $\overset{\circ}{B}{}^n$) admits a continuous extension to the closure B^n of $\overset{\circ}{B}{}^n$.

Evidently $CS(\mathbb{R}^n)$ is a closed sub-algebra of the algebra $CB(\mathbb{R}^n)$ of all continuous and bounded functions over $\mathbb{R}^n$. In fact we also get

$$(1.15) \qquad \begin{aligned} &CL(\mathbb{R}^n) \subset CS(\mathbb{R}^n) \subset CM(\mathbb{R}^n) \subset CB(\mathbb{R}^n), \\ &CL(\mathbb{R}^n) = CO(\mathbb{R}^n) + \{\gamma\cdot 1\colon\ \gamma \in \mathbb{C}\}, \end{aligned}$$

with proper inclusions throughout. The functions of $CS(\mathbb{R}^n)$ are characterized by the property that for every $z \in \mathbb{R}^n$, $|z| = 1$, the limit $\lim_{\rho\to\infty} a(\rho z)$ exists, and uniformly so in z, over all of the unit sphere.

<u>Lemma 1.4.</u> The algebra $\mathfrak{A}^{\#}$ above, defined as closure of the algebra generated by $\Lambda, S_1,\dots,S_n$ in $L(\mathfrak{H}_0)$ is explicitely given by

$$(1.16) \qquad \mathfrak{A}^{\#} = \{a(D)\colon\ a \in CS(\mathbb{R}^n)\}.$$

Moreover the correspondence $a \leftrightarrow a(D)$ defines an isometry $CS(\mathbb{R}^n) \leftrightarrow \mathfrak{A}^{\#}$. Also the correspondence $b = a\circ t \leftrightarrow a(D)$ is an isometry $C(B^n) \leftrightarrow \mathfrak{A}^{\#}$.

<u>Proof.</u> Let $\overset{\circ}{\mathfrak{A}}{}^{\#}$ be the finitely generated algebra with generators Λ, S_j, and observe that $\overset{\circ}{\mathfrak{A}}{}^{\#}$ is in 1-1-correspondence with the finitely generated function algebra $\overset{\circ}{A}{}^{\#}$ with generators $\lambda(x)$, $s_j(x)$, as defined earlier. Also we get (as in

Lemma 1.3)

$$(1.17)\qquad \|a(D)\| = \|F^{-1}a(M)F\| = \|a(M)\| = \|a\|_{L^\infty} = \|b\|_{L^\infty},$$

with L^2-operator norms, and with $b(y) = a(t(y))$.

For any of the generators $a = \lambda, s_j$ we get $b = (1-y^2)^{1/2}$, and $b = y_j$, respectively, and it follows that $b \in C(B^n)$, in each case. Moreover, these n+1 functions strongly separate the points of B^n (Already the functions $1, y_1, \ldots, y_n$ do) and they are self-adjoint functions. Therefore the Stone Weierstrass theorem implies that the closure of $\mathcal{A}$ in $C(B^n)$ equals $C(B^n)$. This and (1.17) imply the statement, q.e.d.

Let us denote the maximal ideal spaces of the commutative C^*-algebras listed in (1.15) by $\mathbb{S}^n$, $\mathbb{B}^n$, $\mathbb{P}^n$ and $\mathbb{Q}^n$, respectively.

Lemma 1.5. The spaces $\mathbb{S}^n, \mathbb{B}^n, \mathbb{P}^n, \mathbb{Q}^n$ are compactifications of $\mathbb{R}^n$, each of them containing $\mathbb{R}^n$ as an open dense subset, and each of them with the property that the elements of the corresponding function algebra CO, CS, CM and CB admit a continuous extension to the compactification, respectively. $\mathbb{S}^n$ is the 1-point compactification of $\mathbb{R}^n$, $\mathbb{Q}^n$ is the Stone Cech compactification of $\mathbb{R}^n$; $\mathbb{B}^n$ is homeomorphic to the ball B^n above, where the homeomorphism is explicitely given by the function (1.12) - the components s_j of which are in CS and therfore extend continuously to $\mathbb{B}^n$.

Proof. It is evident that

$$(1.18)\qquad CL(\mathbb{R}^n) \cong C(\mathbb{S}^n) \quad , \quad CS(\mathbb{R}^n) \cong C(B^n)$$

where the isometry between the algebras is given by continuous extension or restriction to an open dense subset correspondingly. Also, the space $\mathbb{R}^n$ is canonically imbedded in each of the four spaces, since $m_{x^0} = \{a: a(x^0) = 0\}$ is a maximal ideal of the function algebra, for each $x^0 \in \mathbb{R}^n$. Let $\mathring{m}$ be a general maximal ideal, of CM, for example, and let a neighbourhood

$$(1.19)\qquad N_{\mathring{m},a_1,\ldots,a_N,\varepsilon} = \{|\phi_{a_j}(m) - \phi_{a_j}(\mathring{m})| < \varepsilon,\ j=1,\ldots,N\}$$

be considered. We have $\phi_a(m_x) = a(x)$ for $x \in \mathbb{R}^n$. Therefore the function $\phi_a(m)$ constitutes a continuous extension of $a \in CM$ to $\mathbb{P}^n$. Suppose the above neighbourhood does not contain any point of $\mathbb{R}^n$. Write $\mu_j = \phi_{a_j}(\mathring{m})$, and observe that

$$(1.20)\qquad \mathbb{R}^n \cap N_{\mathring{m},a_1,\ldots,a_N,\varepsilon} = \{x: |a_j(x) - \mu_j| < \varepsilon,\ j=1,\ldots,N\}.$$

If this set is empty then it follows that the function

$$(1.21)\qquad f(x) = \sum_{j=1}^{N} |a_j(x) - \mu_j|^2$$

is bounded away from zero - by ε^2. This function is in $CM(\mathbb{R}^n)$, and even is in the maximal ideal $\mathring{m}$, since $\phi_f(\mathring{m}) = 0$. However, f would be an invertible element in a maximal ideal, which is a contradiction. This proves that $\mathbb{R}^n$ is dense in $\mathbb{P}^n$, and similarly in $\mathbb{Q}^n$, $\mathbb{S}^n$, $\mathbb{B}^n$.

Next we look at the associated dual maps

$$\mathbb{Q}^n \to \mathbb{P}^n \to \mathbb{B}^n \to \mathbb{S}^n \tag{1.22}$$

of the injection isomorphisms defined by the inclusions (1.15). From AII, Corollary 7.14 and AII, Proposition 5.8 it follows that each of these maps is surjective. Moreover, the subset $\mathbb{R}^n$ is left invariant, in each case, and so is its complement: For example, if $m \in \mathbb{P}^n - \mathbb{R}^n$ would be mapped onto a point $x^0 \in \mathbb{R}^n \subset \mathbb{S}^n$, then if $m_{x^0} = \{a \in CM : a(x^0) = 0\}$ we get $m \cap CL = m_{x^0} \cap CL$. Since $m \neq m_{x^0}$ we can choose $f \in m$, $f \notin m_{x^0}$, i.e. $f(x^0) \neq 0$. Then if $g \in CL$, $g(x^0) \neq 0$ we conclude that $fg \in m \cap CL$, but $fg \notin m_{x^0}$, which is a contradiction. Now it is trivial that $\mathbb{R}^n$ is an open subset of $\mathbb{S}^n$ - or also of $\mathbb{B}^n$. Since the inverse image of $\mathbb{R}^n$ in $\mathbb{P}^n$ under the associated dual map $\mathbb{P}^n \to \mathbb{S}^n$ is $\mathbb{R}^n$, we find that $\mathbb{R}^n$ is an open subset of $\mathbb{P}^n$ as well (the associated dual map is continuous.). Similarly in $\mathbb{Q}^n$. Finally, it is well known that $\mathbb{Q}^n$ is the Stone-Čech compactification of $\mathbb{R}^n$, since it is one of the characteristica of the Stone-Čech compactification that every bounded continuous function over $\mathbb{R}^n$ admits a continuous extension to it (c.f. [13] Theorem 3.2.11), q.e.d.

The above compactifications have been prepared for the statement of the theorem below.

Theorem 1.6. The symbol space $\mathbb{M}$ of $\mathfrak{A}_0$ as defined above is homeomorphic to the compact subset of the cartesian product $\mathbb{P}^n \times \mathbb{B}^n$ given by

$$\{(x,\xi): \ x \in \mathbb{P}^n, \ \xi \in \mathbb{B}^n, \ |x| + |\xi| = \infty\}, \tag{1.23}$$

where we extend the definition of $|x|$, $|\xi|$, by setting

$$|x| = \infty \text{ on } \mathbb{P}^n - \mathbb{R}^n, \qquad |\xi| = \infty \quad \text{on } \mathbb{B}^n - \mathbb{R}^n \quad . \tag{1.24}$$

Moreover the homeomorphism may be chosen such that for $m \sim (x,\xi)$ we get the following formulas for symbols of generators of the algebra $\mathfrak{A}_0$.

$$\sigma_{a(M)}(x,\xi) = a(x) \ , \qquad \sigma_{b(D)}(x,\xi) = b(\xi) \ , \tag{1.25}$$

for $a \in CM(\mathbb{R}^n)$, $b \in CS(\mathbb{R}^n)$, It is understood here, that a and b have been extended continuously onto $\mathbb{P}^n$ and $\mathbb{B}^n$, respectively, before they are used at right of (1.25).

Problems. In the following let $\mathcal{L}_s$ be the C^*-algebra generated by norm closing in $L(\mathcal{H}_s)$ the algebra generated by

(1.26) $s_j(M)$, $s_j(D)$, $j=1,\ldots,n$,

with $s_j(x) = x_j(1+x^2)^{-1/2}$.

1) Show that $\mathcal{L}_s$ contains $K(\mathfrak{H}_s)$, and that $\mathcal{L}_s/K(\mathfrak{H}_s)$ is commutative.

2) Show that $\mathcal{L}_s$ is invariant under translations and rotations (That is we get $U^{-1}\mathcal{L}_s U = \mathcal{L}_s$, for all unitary operators of the form $Uu(x) = u(x+h)$, with $h \in \mathbb{R}^n$ and of the form $Uu(x) = u(Ox)$ with an orthogonal $n\times n$-matrix O, having real coefficients). 3) Using Lemma 2.1 show that $\mathcal{L}_o^{\vee} = \mathcal{L}_o/K(\mathfrak{H}_o) \cong C(\mathbb{M}_{\mathcal{L}_o})$, with the maximal ideal space $\mathbb{M}_{\mathcal{L}_o}$ being a compact subset of $\mathbb{M}' = \mathbb{B}^n\times\mathbb{B}^n - \mathbb{R}^n\times\mathbb{R}^n$. 4) Using problem 2) and properties of the associate dual map show that $\mathbb{M}_{\mathcal{L}_o} = \mathbb{M}'$, where the symbols of the generators are explicitly given by

(1.27) $s_j(M) \to s_j(x)$, $s_j(D) \to s_j(\xi)$, $j=1,\ldots,n$, $(x,\xi) \in \mathbb{M}'$

5) Obtain the result of problem 4) for $\mathcal{L}_s$, $s \in \mathbb{R}$.

2. Herman's Lemma and the proof of theorem 1.6.

For the proof of theorem 1.6 we require the following result, the importance of which was first realized by E. Herman [10].

Lemma 2.1. Let a C^*-operator algebra $\mathfrak{A} \subset L(\mathfrak{H})$ ($\mathfrak{H}$ a Hilbert space) contain the compact ideal K of $\mathfrak{H}$ and let it contain two commutative C^*-subalgebras $\mathfrak{A}_1, \mathfrak{A}_2$ with unit. Assume that $\mathfrak{A}$ is generated by the elements of K, $\mathfrak{A}_1$ and $\mathfrak{A}_2$ and assume the commutator of $\mathfrak{A}$ to be in K. Let τ_j' be the associate dual map of the homomorphism

$$(2.1)\qquad \mathfrak{A}_j \to \mathfrak{A} \to \mathfrak{A}/K ,$$

(by injection and projection). Then a homeomorphism $\iota: \mathbb{M} \to \mathbb{M}_1 \times \mathbb{M}_2$ from the maximal ideal space $\mathbb{M}$ of $\mathfrak{A}/K$ onto a compact subset of the cartesian product $\mathbb{M}_1 \times \mathbb{M}_2$ of the maximal ideal spaces $\mathbb{M}_j$ of $\mathfrak{A}_j$ is defined by setting

$$(2.2)\qquad \iota(m) = (\tau_1'(m), \tau_2'(m)) \quad , \quad m \in \mathbb{M} .$$

Proof. It is trivial that ι is a continuous map $\mathbb{M} \to \mathbb{M}_1 \times \mathbb{M}_2$. Since $\mathbb{M}$ is compact the image $\iota(\mathbb{M})$ will be compact, and for the continuity of the inverse map it suffices to prove that ι is 1-1. Suppose therefore that we have $\iota(m) = \iota(\tilde{m})$. This means that $\tau_j'(m) = \tau_j'(\tilde{m})$, $j=1,2$. Using the basic property of the associate dual map we get

$$(2.3)\qquad \phi_{A_j^\vee}(m) = \phi^j_{A_j}(\tau_j'(m)) = \phi^j_{A_j}(\tau_j'(\tilde{m})) = \phi_{A_j^\vee}(\tilde{m}) , \quad A_j \in \mathfrak{A}_j ,$$

where $A_j^\vee$ denotes the coset of A_j mod K. However, since the A_j and the compact operators generate $\mathfrak{A}$ we conclude that $\phi_{A^\vee}(m) = \phi_{A^\vee}(\tilde{m})$ for all $A \in \mathfrak{A}$ which implies $m = \tilde{m}$, q.e.d.

Proof of theorem 1.6. We may at once apply Herman's Lemma to our algebra $\mathfrak{A}_0$, where the two generating commutative subalgebras are given by $\mathfrak{A}^+$ and $\mathfrak{A}^\#$ as defined in section 1. Accordingly $\mathbb{M}$, the symbol space of the L^2-Laplace comparison algebra $\mathfrak{A}_0$ must be homeomorphic to a compact subset of $\mathbb{P}^n \times \mathbb{B}^n$, and the homeomorphism is explicitly given by (2.2), which at once implies (1.25), by a conclusion as used in (2.3). All we have to prove then is that the compact subset is precisely given by (1.23). Let us for the moment denote the set (1.23) by $\tilde{\mathbb{M}}$. First we show that $\mathbb{M} \subset \tilde{\mathbb{M}}$: Indeed, we have proven that for $a \in CM$, $b \in CS \subset CM$ the operator $C = a(M)b(D)$ is compact if and only if either one of the functions vanishes identically or both are in CO. We certainly can chose $a,b \in CO$ such that both functions do not vanish in all of $\mathbb{R}^n$. Now the above operator C is compact and we calculate from (1.25) that

$$(2.4)\qquad \sigma_C(x,\xi) = a(x)b(\xi) \quad , \quad (x,\xi) \in \mathbb{M} .$$

Since C is compact we must have $\sigma_C(x,\xi) = 0$ for all $(x,\xi) \in \mathbb{M}$. But the product (2.4) vanishes only if either $|x| = \infty$ or $|\xi| = \infty$ or both. Accordingly it follows that $\mathbb{M} \subset \tilde{\mathbb{M}}$.

Suppose the above is a proper inclusion, so that $\tilde{\mathbb{M}} - \mathbb{M}$ contains at least one point (x^0, ξ^0). One may construct neighbourhoods N_{x^0}, N_{ξ^0} such that the product $N_{x^0} \times N_{\xi^0}$ is disjoint from $\mathbb{M}$ and then functions f,g over the Hausdorff spaces $\mathbb{P}^n$ and $\mathbb{B}^n$ such that $\operatorname{supp} f \subset N_{x^0}$, $\operatorname{supp} g \subset N_{\xi^0}$ while $f(x^0) = g(\xi^0) = 1$. The restrictions of f and g to $\mathbb{R}^n$ are both not identically zero, since $\mathbb{R}^n$ is dense. At least one of these functions cannot be in CO, since not both x^0, ξ^0 are finite. Thus $C = f(M)g(D)$ cannot be compact but its symbol vanishes identically since $f(x)\, g(\xi)$ vanishes on $\mathbb{M}$, by construction. Thus we get a contradiction and Theorem 1.6 is established.

<u>Theorem 2.2.</u> The symbol space of $\mathfrak{A}_s$ is homeomorphic to $\mathbb{M}$, as in (1.23) again, for all $s \in \mathbb{R}$.

<u>Proof.</u> This simply is a consequence of the fact that $\mathfrak{A}_s$ and $\mathfrak{A}_0$ are isometrically isomorphic, by the isometry $\Lambda^s: \mathfrak{H}_0 \to \mathfrak{H}_s$. The associate dual map of the isomorphism $\mathfrak{A}_0 \to \mathfrak{A}_s$ defined by conjugation with Λ^s defines a homeomorphism between the symbol spaces of $\mathfrak{A}_s$ and $\mathfrak{A}_0$. Since the symbol space of $\mathfrak{A}_0$ is homeomorphic to the set (1.23), by theorem 1.6, the statement results, q.e.d.

Let us chose the map ι of (2.2) again to identify the set (1.23) with the symbol space of $\mathfrak{A}_0$, and then use the associate dual map mentioned above to identify the symbol spaces of $\mathfrak{A}_0$ and $\mathfrak{A}_s$. Once and for ever we shall use these identifications. Then we have the Corollary below.

<u>Corollary 2.3.</u> Let $A \in \mathfrak{A}_s$. Then the symbol of A -with respect to the algebra $\mathfrak{A}_s$- is given by the function

$$(2.5) \qquad \sigma_{\Lambda^{-s} A \Lambda^{+s}}(x,\xi), \qquad (x,\xi) \in \mathbb{M}.$$

<u>Theorem 2.4.</u> An operator $A \in \mathfrak{A}_s$ is Fredholm if and only if its symbol does never vanish over $\mathbb{M}$.

<u>Proof.</u> Since $\mathfrak{A}_s$ is a C^*-algebra it is Fredholm closed, by AII, Corollary 7.16. Also since $\mathfrak{A}_s$ contains the Riesz ideal $K(\mathfrak{H}_s)$ it also contains all inverses modulo $K(\mathfrak{H}_s)$ in $L(\mathfrak{H}_s)$ of its elements (AI, Lemma 5.7). Accordingly $A \in \mathfrak{A}_s$ admits an inverse mod $K(\mathfrak{H}_s)$ in $L(\mathfrak{H}_s)$ if and only if its coset $A^\vee = \{A + K(\mathfrak{H}_s)\} \simeq \sigma_A$ is invertible in $\mathfrak{A}_s/K(\mathfrak{H}_s) \simeq C(\mathbb{M})$. That is, by AI, theorem 4.8, we have $A \in \mathfrak{A}_s$ Fredholm if and only if $\sigma_A \neq 0$ on $\mathbb{M}$, q.e.d.

Problems.

1) Consider the singular integral operator (involving a distribution integral)

$$(2.7)\qquad u \to Ku(x) = a(x)u(x) + \int_{\mathbb{R}^n} k(x,x-y)u(y)dy$$

where we have $a \in A$, and where the distribution $k \in \mathcal{D}'(\mathbb{R}^{2n})$ also is in $C^\infty(\mathbb{R}^n \times \mathbb{R}^{n*})$. For each fixed x let $k(x,.) \in S^0_{ad}$, with lowest exponent $\sigma_0 \geq -n$ (c.f. II,5), and uniformly so over $\mathbb{P}^n$. [That is (i) $z^\alpha D_z^\beta k(x,z) = O(1)$ as $x,z \in \mathbb{R}^n$, $|z| > 1$, and (ii) $D_z^\alpha z^\beta a(x,z) \in C(\mathbb{P}^n \times \{|z| \leq 1\})$ as $|\beta| \geq N(|\alpha|)$, and (iii) an asymptotic expansion

$$(2.8)\qquad k(x,z) \sim \sum_{l=0}^{\infty} k_l(x,z) \pmod{D_z} \text{ (at 0) uniformly over } \mathbb{P}^n,$$

amounting to (a) $k_l \in C^\infty(\mathbb{P}^n \times \mathbb{R}^{n*})$, (b) $k_l \in \mathcal{D}'(\mathbb{R}^{2n})$, $k_l(x,\cdot) \in C^\infty(\mathbb{P}^n, \mathcal{D}'(\mathbb{R}^n))$ and (c) $k_l(x,.) \in H'''_{\sigma_l}$ ($\sigma_l \uparrow +\infty$, $\sigma_0 \geq -n$) in z, and (d)

$$(2.9)\qquad k(x,z) - \sum_{l=0}^{N} k_l(x,z) \in C^k(\mathbb{P}^n \times \mathbb{R}^n) \text{ as } N > N_0(k).]$$

(Note that in the above $C^k(\mathbb{P}^n, \mathcal{D}'(\mathbb{R}^n))$ and $C^k(\mathbb{P}^n \times \mathbb{R}^n)$ refers to the functions with derivatives over $\mathbb{R}^n$ (and $\mathbb{R}^n \times \mathbb{R}^n$, resp.)) having continuous extensions to $\mathbb{P}^n$ (or $\mathbb{P}^n \times \mathbb{R}^n$), resp.).)

Prove that the singular integral operator (2.7) is in $\mathfrak{A}_0$. (Note that the distribution integral may be interpreted as a Cauchy principal value.) (Hint: The L^2-boundedness follows from III,5, problem 6. In oder to construct an approximation from $\mathfrak{A}^0$ one must invoke Stone-Weierstrass, and the compactness of $\mathbb{P}^n$.)

2) Obtain the symbol of K, as in (2.7). (Hint: In essence the symbol should be the restriction to $\mathbb{M}$ of the sum of a(x) and $k^{\wedge}(x,\xi)$, with "$^\wedge$" denoting the Fourier transform with respect to the z-variable.) Thus apply Theorem 2.4 to obtain a necessary and sufficient condition for the singular integral operator K to be Fredholm, as a bounded operator from $L^2(\mathbb{R}^n)$ to itself.

3) *) Discuss $\mathfrak{H}_s$-theory of the singular integral operator K of (2.7): Show, that this operator, under the conditions stated in Problem 1) is in $\mathfrak{A}_s$, for every s; obtain its s-symbol and state a necessary and sufficient criterion for K to be Fredholm from $\mathfrak{H}_s$ to $\mathfrak{H}_s$.

4) *) Again for the operator (2.7), if $\sigma_0 < -n$, discuss definition of the operator by means of a finite part integral (c.f. II, 2 and II, 5) rather than by Cauchy principal value. Again set up conditions insuring the redefined operator to be Fredholm from $\mathfrak{H}_s$, to $\mathfrak{H}_0$. (Hint: s should be equal to $-(\sigma_0+n)$; the operator $L\Lambda^s$ should be in every $\mathfrak{H}_s$, as again should follow from III,5, problem 6 and the Stone-Weierstrass Theorem.)

5) Show that a differential expression

$$(2.10) \qquad L = \sum_{|\alpha| \le N} a_\alpha(x) D^\alpha \quad , \quad a_\alpha \in CM(\mathbb{R}^n),$$

defines a Fredholm operator from $\mathfrak{H}_N$ to $\mathfrak{H}_0$ if and only if it is md-elliptic [Here we call an differential expression md-elliptic if it (i) is uniformly elliptic over $\mathbb{R}^n$: $\sum_{|\alpha|=N} a_\alpha(x)\xi^\alpha \neq 0$, as $x,\xi \in \mathbb{R}^n$, $|\xi|=1$ and if in addition (ii) we have it m-elliptic:

$$\left|\sum_{|\alpha|\le N} a_\alpha(x)\xi^\alpha\right| \ge p(1+\xi^2)^{N/2} \quad \text{as} \quad x,\xi \in \mathbb{R}^n,\ |x| \ge a \ ,$$

for positive p and sufficiently large a, both independent of x,ξ].

6) Obtain a result corresponding to that of problem 5 for linear integrodifferential operators, of the general form

$$(2.11) \qquad L = \sum_{|\alpha|\le N} K_\alpha D^\alpha$$

with (singular) integral operators K_α of the form (2.7) satisfying the assumptions of problem 1.

7) Obtain $\mathfrak{H}_s$-results of the kind stated in problems 5 and 6. For example, if the coefficients a of the operator (2.10) are in A, rather than only in $CM(\mathbb{R}^n)$, show that md-ellipticity of L is necessary and sufficient for L to be Fredholm $\mathfrak{H}_s \to \mathfrak{H}_{s-N}$, for any $s \in \mathbb{R}$.

8) Consider the operator L of (2.10) as an unbounded linear operator of the Hilbert space $\mathfrak{H}_0 = L^2(\mathbb{R}^n)$ to itself with domain dom $L = \mathfrak{H}_N \subset \mathfrak{H}_0$. Show that this operator is a closed unbounded Fredholm operator if and only if L is md-elliptic.

9) Formulate results of the kind stated in problem 8 for general $\mathfrak{H}_s$, and for integro-differential operators as well as for operators defined by finite part integral, as introduced in the preceeding problems.

10) Under the assumptions of problem 8 let L of (2.10) be uniformly elliptic over $\mathbb{R}^n$. Show that the Fredholm spectrum of L in the domain dom $L = \mathfrak{H}_N$ coincides with the set of values of the symbol

$$(2.12) \qquad \sigma_L(x,\xi) = \sum_{|\alpha|\le N} a_\alpha(x)\xi^\alpha$$

on the secondary symbol space (introduction). (The Fredholm spectrum is defined to be the set of complex z such that $L-z$ is not Fredholm).

3. The algebra $\mathfrak{A}_\infty$, and the algebra $\Psi\mathcal{D}$ of ψdo-s.

We already have defined the algebra $\mathfrak{A}_\infty \subset \mathcal{O}(0)$ as the closure of $\mathring{\mathfrak{A}}$ in the Frechet topology of $\mathcal{O}(0)$. An operator $A \in \mathcal{O}(0)$ therefore is in $\mathfrak{A}_\infty$ if and only if there exists a sequence $A_j \in \mathring{\mathfrak{A}}$ such that

$$(3.1) \qquad \| A - A_j \|_s \to 0, \quad j \to \infty, \quad s \in \mathbb{R}.$$

From (3.1) it follows that $\mathfrak{A}_\infty$ may be interpreted as a subalgebra of every $\mathfrak{A}_s$, $s \in \mathbb{R}$. Accordingly for $A \in \mathfrak{A}_\infty$ we have a symbol defined with respect to every $\mathfrak{H}_s$.

However we also notice that the symbol is the same function over $\mathbf{M}$ for each $s \in \mathbb{R}$. Indeed, this first of all is true for the generators $a(M)$ and $b(D)$ of $\mathring{\mathfrak{A}}$, in view of (2.5), since we get

$$(3.2) \qquad \Lambda^{-s} a(M)\, \Lambda^s = a(M) + C, \qquad \Lambda^{-s} b(D)\, \Lambda^s = b(D), \quad C \in K_0,$$

whenever $a \in A$, or $b = \lambda, s_j$ (c.f. III, Lemma 9.1). We have $\sigma_C = 0$ since C is compact in $L(\mathfrak{H}_0)$. Therefore

$$(3.3) \qquad \sigma_{\Lambda^{-s} a(M) \Lambda^s} = \sigma_{a(M)}, \qquad \sigma_{\Lambda^{-s} b(D)\, \Lambda^s} = \sigma_{b(D)} .$$

Accordingly we get

$$(3.4) \qquad \sigma_{\Lambda^{-s} A\, \Lambda^s} = \sigma_A , \qquad A \in \mathring{\mathfrak{A}} .$$

For a general $A \in \mathfrak{A}_\infty$ we obtain a sequence $A_j \in \mathring{\mathfrak{A}}$ such that (3.1) holds. However we get

$$(3.5) \qquad \| \Lambda^{-s} A\, \Lambda^s - \Lambda^{-s} A_j\, \Lambda^s \|_0 = \| A - A_j \|_s \to 0, \quad \text{as } j \to \infty.$$

It follows that

$$(3.6) \qquad \sigma_A = \lim_{j\to\infty} \sigma_{A_j} , \qquad \sigma_{\Lambda^{-s} A\, \Lambda^s} = \lim_{j\to\infty} \sigma_{\Lambda^{-s} A_j\, \Lambda^s} ,$$

in uniform convergence over $\mathbf{M}$, by (1.9). Since the right hand sides are equal, by (3.4), it follows that

$$(3.7) \qquad \sigma_{\Lambda^{-s} A\, \Lambda^s} = \sigma_A , \qquad A \in \mathfrak{A}_\infty .$$

Or, indeed, we have shown that the symbol of an operator $A \in \mathfrak{A}_\infty$ is independent of s.

It is therefore natural to speak about <u>the</u> symbol σ_A of an operator in $\mathfrak{A}_\infty$. This symbol always is a continuous function over the space $\mathbf{M}$ of (1.23). On the other hand we are not quite well prepared so far to decide about the question what continuous functions over $\mathbf{M}$ are symbols of operators in $\mathfrak{A}_\infty$, except that, of course we can say that any continuous function $a(x,\xi)$ which can be represented as a sum of products

(3.8) $$\sum_{j=1}^{N} b_j(x)c_j(\xi), \quad b_j \in A,\ c_j = \xi^{\alpha}\lambda^{|\alpha|+m_j}(\xi)\ ,$$

is the symbol of an operator in $\mathring{\mathfrak{A}}$, not only in $\mathfrak{A}_\infty$.

In the next following sections we are going to look into this <u>lifting problem</u>.

Here we only remark that the above discussion may also be used with little amendement to prove the lemma, below.

<u>Lemma 3.1.</u> For $A \in \mathfrak{A}_\infty$ we have

(3.9) $$\Lambda^{-s}A\,\Lambda^{s} - A \in K_t\ , \quad s,t \in \mathbb{R}\,.$$

Next let us investigate some classes of compact operators in $L(\mathfrak{H}_\infty)$. First of all, it seems that the concept of compact operator between Frechet spaces is not a uniform one throughout the literature. The following three properties have been used to define compact operators (All of them agree on Hilbert spaces or reflexive Banach spaces).

(R) There exists a neighbourhood of 0 which is transformed into a relatively compact set.

(B) Every bounded set is transformed into a relatively compact set

(H) Any weak Cauchy sequence is transformed into a convergent sequence .

For a discussion of the relation between (R), (B) and (H) for various classes of topological vector space see Edwards [8], p. 618. For the purpose of this chapter we will call an operator $C \in L(\mathfrak{H}_\infty)$ <u>compact</u> if it satisfies (B) or equivalently (H). Note that the compact operators of $\mathfrak{H}_\infty$ thus defined form a two-sided ideal of $L(\mathfrak{H}_\infty)$. This class of operators will be denoted by $K_\infty = K(\mathfrak{H}_\infty)$.

In the frame of our general discussion, during the preceeding sections, we will become interested in the classes

(3.10) $$\mathcal{L}_s = \{C \in \mathcal{O}(s):\ C \in K(\mathfrak{H}_r, \mathfrak{H}_{r-s}),\ r\in\mathbb{R}\}\ .$$

Clearly the operators of $\mathcal{L}_s$ are in K_∞. Also it is evident that

(3.11) $$\mathcal{L}_s = \Lambda^{-s}\mathcal{L}_0\ , \qquad \Lambda^{s}\mathcal{L}_t = \mathcal{L}_{t-s}\ , \qquad s,t \in \mathbb{R}\,,$$

because the operator Λ^s takes $\mathcal{O}(t)$ onto $\mathcal{O}(t-s)$, as we have seen, and also this operator is continuous $\mathfrak{H}_r \to \mathfrak{H}_{r+s}$, so that its product with $C \in K(\mathfrak{H}_r, \mathfrak{H}_{r-t})$ is in $K(\mathfrak{H}_r, \mathfrak{H}_{r-(t-s)})$. Accordingly we mainly shall focus on the class $\mathcal{L}_0$ which is a closed two-sided ideal of the algebra $\mathcal{O}(0)$, evidently. In particular the closedness of $\mathcal{L}_0$ in the topology of $\mathcal{O}(0)$ is an immediate consequence of the closedness of the ideals $K(\mathfrak{H}_r, \mathfrak{H}_r) = K_r$ in norm convergence of $L(\mathfrak{H}_r)$.

We are not going to investigate in details the relationship between $\mathcal{L}_s$ and K_∞, nor between the classes of operators $\mathfrak{H}_\infty \to \mathfrak{H}_\infty$ satisfying (R), or (H), or

(B) above, although this might prove to be of some interest of curiosity to the reader. (R) is due to F. Riesz [12], (H) is due to Hilbert [11] (in case of special types of spaces each).

Our interest in the class $\mathcal{L}_0$ is motivates in part by the fact that it implicitly occurred in all of the compactness results of III, 9, for example. Lemma 9.1 (of III) for example says that we get

$$(3.12)\qquad \Lambda^{-s}a(M)\,\Lambda^{s} - a(M) \in \mathcal{L}_0, \quad \text{as } a \in A.$$

Similarly for III, Lemma 9.3 and III, Corollary 9.6. Also we shall see, later on (section 4), that we have $A \in \mathfrak{A}_\infty$ invertible modulo the closed two-sided ideal $\mathcal{L}_0$ if and only if its symbol σ_A does never vanish on $\mathbb{M}$. This result would be a natural extension of Theorem 2.4 for the algebra $\mathfrak{A}_\infty$. On the other hand it is possible to prove a much stronger form of the above result involving invertibility modulo an ideal of operators of finite rank. Such inverses will be called Green-inverses, later on (section 4), and we postpone the entire discussion of this question to that section. As a preparation for this discussion we now want to discuss this kind of operator of finite rank.

We define a <u>dyad of order $-\infty$</u> (or an $\mathcal{O}(-\infty)$-dyad) to be a linear operator $C: \mathfrak{H}_\infty \to \mathfrak{H}_\infty$ of the special form

$$(3.13)\qquad C = \sum_{j=1}^{N} \phi_j{>}{<}\psi_j \quad , \quad \phi_j, \psi_j \in \mathfrak{H}_\infty,$$

where we set

$$(3.14)\qquad (\phi{>}{<}\psi)u = (\bar{\phi},u)\psi \ , \quad u \in \mathfrak{H}_{-\infty}.$$

Clearly an $\mathcal{O}(-\infty)$-dyad is in $L(\mathfrak{H}_{-\infty},\mathfrak{H}_\infty)$ and its image is the linear span of the N-vectors ϕ_j, $j=1,\dots,N$. It therefore is of finite rank, and its adjoint is given by

$$(3.15)\qquad C^{*} = \sum_{j=1}^{N} \bar{\psi}_j{>}{<}\bar{\phi}_j \ .$$

In (3.14) we of course were using the generalized inner product of III, (2.13). The notation is chosen in analogy of that of Dirac [6]. In fact it will be convenient to also introduce the bi-linear form

$$(3.16)\qquad \langle u,v\rangle = (\bar{u},v) \ , \quad u \in \mathfrak{H}_s,\ v \in \mathfrak{H}_{-s}, \quad s \in \mathbb{R} \cup \{\infty\} \ .$$

Then we get (3.14) in the form

$$(3.17)\qquad Cu = \sum_{j=1}^{N} \phi_j \langle \psi_j, u\rangle \quad , \quad u \in \mathfrak{H}_{-\infty} \ .$$

Let us notice that the collection $\mathcal{F}_\infty$ of all $\mathfrak{H}_\infty$-dyads forms a 2-sided ideal of each of the algebras $\mathcal{O}(\infty)$, $\mathcal{O}(0)$ and $\mathcal{O}(-\infty)$. Also $\mathcal{F}_\infty$ is contained in every $\mathcal{O}(r)$, with the proper interpretation of its operators, and we trivially get $\mathcal{F}_\infty \subset \mathcal{L}_s$ $s \in \mathbb{R}$.

Lemma 3.2. The ideal $\mathcal{L}_0$ is the closure of $\mathcal{F}_\infty$ in $\mathcal{O}(O)$.

Proof. Let $F_m = 1 - E_{1/m}$, $m=1,2,\ldots$, where E_μ is the spectral family of the self-adjoint operator $\Lambda \in L(\mathcal{G}_0)$ [In fact we notice that $F_m = \chi_{B_m}(D)$ with the characteristic function χ_{B_m} of $B_m = \{|x| \le (m^2-1)^{1/2}\}$, so that no spectral theory has to be involved]. We have $\lim F_m = 1$ in every $\mathcal{G}_s$ with strong operator convergence. Hence if $C \in \mathcal{L}_0 \subset K_s$ (for every s) then we get $\lim_{m\to\infty} F_m C F_m = C$ in norm convergence of every $\mathcal{G}_s$. It follows in particular that

$$(3.18) \qquad \lim_{m\to\infty} F_m C F_m = C \quad \text{in } \mathcal{O}(O) .$$

Hence the Lemma is proven if only we can show that $C_m = F_m C F_m$ can be approximated arbitrarily by dyads in $\mathcal{F}_\infty$.

Let us look for a moment at the operators $C_m^\vee = F C_m F^{-1}$. Clearly,

$$(3.19) \qquad C_m^\vee = \chi(M) C^\vee \chi(M) , \quad \chi = \chi_{B_m} , \quad C^\vee = FCF^{-1} .$$

Recall that $F\mathcal{G}_s = L^2(\mathbb{R}^n, d\mu_s)$, so that $C^\vee$ and $C_m^\vee$ may be regarded as operators between these L^2-spaces. In fact, (3.19) shows that $C_m^\vee$ is naturally identified with an operator of $L^2(B_m)$. Also on B_m all the measures $d\mu_s$ and hence all the corresponding L^2-norms are equivalent.

It is known that the ideal of operators of finite rank is dense (in norm convergence) in the compact ideal of any Hilbert space (c.f. AI,Proposition 5.3).This may be applied to the Hilbert space $L^2(B_m, dx) = \tilde{\mathcal{R}}_m$. $\tilde{\mathcal{R}}_m$ is naturally imbedded in $F\mathcal{G}_\infty = \bigcap_s L^2(\mathbb{R}^n, d\mu_s)$, the imbedding being achieved by extending the functions of $\tilde{\mathcal{R}}_m$ zero outside B_m. With this imbedding it is noticed that $C_m^\vee$ takes $\tilde{\mathcal{R}}_m$ to itself and is uniquely determined by its restriction to $\tilde{\mathcal{R}}_m$ (giving the above-mentionned identification). Let this restriction $\tilde{C}_m^\vee : \tilde{\mathcal{R}}_m \to \tilde{\mathcal{R}}_m$ be approximated in operator norm of $\tilde{\mathcal{R}}_m$ by a sequence

$$(3.20) \qquad \tilde{C}_{mk}^\vee = \sum_{j=1}^{N_{km}} \phi^\wedge_{jmk} >< \omega^\wedge_{jmk} , \quad \phi^\wedge_{jmk}, \omega^\wedge_{jmk} \in \tilde{\mathcal{R}}_m ,$$

so that we get

$$(3.21) \qquad \| \tilde{C}_m^\vee - \tilde{C}_{mk}^\vee \|_{L^2(B_m)} \to 0, \quad \text{as } k \to \infty .$$

Again extend the functions $\phi^\wedge, \omega^\wedge$ zero outside B_m and take the inverse Fourier transform ϕ_{jmk}, ψ_{jmk} of $\phi^\wedge_{jmk}$ and $\psi^\wedge_{jmk}(x) = \omega^\wedge_{jmk}(-x)$, extended to $\mathbb{R}^n$, as described. It follows that $\phi_{jmk}, \psi_{jmk} \in \mathcal{G}_\infty$, by construction. Moreover we get

$$(3.22) \qquad C_{mk} = F^{-1} C_{mk}^\vee F = \sum_{j=1}^{N_{km}} \phi_{jmk} >< \psi_{jmk}$$

where

$$(3.23) \qquad C_{mk}^\vee u = \sum_{j=1}^{N_{mk}} \phi_{jmk} (\bar{\omega}_{jmk}, u) , \quad u \in L^2(\mathbb{R}^n, d\mu_s) ,$$

with the extended functions. Also it is evident that

(3.24) $$\lim_{k\to\infty} C_{mk} = C_m \quad \text{in } \mathcal{O}(0) ,$$

because we get

(3.25) $$\lim_{k\to\infty} \check{C}_{mk} = \check{C}_m \quad \text{in } L(L^2(\mathbb{R}^n, d\mu_s)) ,$$

for each s, using relation (3.21) and the equivalence of $L^2(\mathbb{R}^n, d\mu_s)$-norms over $\tilde{\mathcal{R}}_m$. Thus the lemma now follows from (3.18) and (3.24), q.e.d.

Lemma 3.3. The algebra $\mathfrak{A}_\infty$ as defined above (to be the closure of $\mathring{\mathfrak{A}}$ in $\mathcal{O}(0)$) contains $\mathcal{L}_0$ and $\mathcal{F}_\infty$.

Proof. In view of Lemma 3.2 it is sufficient to show that $\mathfrak{A}_\infty$ contains all $\mathcal{O}(-\infty)$-dyads. This in turn may be proven in a manner entirely parallel to the proof of Theorem 1.1. Therefore we only sketch the proof. Again we use III, Lemma 10.1 to show that $\mathfrak{A}_\infty$ contains all of $\mathcal{F}_\infty$ if only it contains one dyad $0 \neq C_0 = f\rangle\langle g$, $f,g \in \mathfrak{H}_\infty$. Indeed the same conclusion works to show that in every neighbourhood of $\phi \in \mathfrak{H}_\infty$ (which must contain some ball $\|u-\phi\|_k \leq \varepsilon$) there are points of the form Af, with $A \in \mathring{\mathfrak{A}}$, which gives this assertion. Thus it is left to show that $\mathfrak{A}_\infty$ contains C_0 as described. But $\mathfrak{A}_\infty$ contains nontrivial operators of $\mathcal{L}_0$, as for example the commutator $C_1 = [\lambda(M), \Lambda]$ which is nonzero and in every K_s hence in $\mathcal{L}_0 = \bigcap_s K_s$, by Lemma 9.4 (of Ch. III). Then the operator

(3.26) $$C_2 = C_1^* C_1 = C_1^{\langle 0\rangle} C_1$$

is compact and self-adjoint in $L(\mathfrak{H}_0)$ and also is compact in every $L(\mathfrak{H}_s)$ and is in $\mathcal{L}_0$, and is not zero. One obtains an orthonormal base of eigenfunctions to C_2, denoted by ϕ_j, $j=1,2,\ldots$ satisfying $C_2\phi_j = \lambda_j\phi_j$. For any eigenvalue λ_j the corresponding projection to the eigenspace is given by a complex integral

(3.27) $$P_j = -(2\pi i)^{-1} \oint_{|\mu-\lambda_j|=\varepsilon} (C_2-\mu)^{-1} d\mu$$

with sufficient small ε.

Conclude that all ϕ_j with $\lambda_j \neq 0$ must in fact be in $\mathfrak{H}_\infty$: We get $\phi_j \in \mathfrak{H}_0$ and

(3.28) $$\phi_j = \lambda_j^{-m} C_2^m \phi_j \in \mathfrak{H}_{4m}, \quad m = 1,2,\ldots ,$$

because C_2 is readily seen to be of order-4. The operator C_2 is compact in every $\mathfrak{H}_s$ and the above shows that its eigenvalues must be independent of s, also that all its eigenfunctions to eigenvalues different from zero must be in $\mathfrak{H}_\infty$. Hence the resolvent integral in (3.27) is norm convergent in every $L(\mathfrak{H}_s)$, because the resolvent $(C_2-\mu)^{-1}$ is continuous for all complex μ except the eigenvalues [there is no spectrum in any $\mathfrak{H}_s$ except 0 and the λ_j since $C_2 \in K_s$ for all s] . It follows that the integral defining P_j even converges in $\mathcal{O}(0)$ and hence we get $P_j \varepsilon \mathfrak{A}_\infty$ [In fact the resolvent $R(\mu) = (C_2-\mu)^{-1}$ is in $\mathfrak{A}_0$ wherever it exists, since $\mathfrak{A}_0$ is a C^*-algebra which contains all inverses of its elements (AII, Corollary 7.14. etc.). Also we get $\Lambda^{-s}R(\mu)\Lambda^s - R(\mu) \varepsilon K_0$ for all $s\varepsilon\mathbb{R}$ by Lemma 3.1. Thus Lemma 5.3 (of this chapter) may be applied to conclude that $R(\mu) \varepsilon \mathfrak{A}_\infty$. Thus we indeed also get $P_j \varepsilon \mathfrak{A}_\infty$].

Now if there is one non-vanishing eigenvalue of C_2 which has multiplicity one then we have the desired C_0. At any rate all those eigenvalues must have finite multiplicity, say P_j has multiplicity N. Then

$$(3.29) \qquad P = P_j = \sum_{l=1}^{N} \psi_l >< \overline{\psi}_l$$

with an orthonormal system $\{\psi_l\}$ of eigenfunctions to $\lambda = \lambda_j$. For an arbitrary operator $A \varepsilon \mathfrak{A}_\infty$ we get

$$(3.30) \qquad PAP = \sum_{l,k=1}^{N} a_{lk}\, \psi_l >< \psi_k$$

with certain complex constants a_{lk}. If $N > 1$ one may find an $A \varepsilon \mathfrak{A}_\infty$ such that we do not have $a_{lk} = \mu\delta_{lk}$ with a complex μ. For example A may be chosen such that $A\psi_1$ is close to ψ_2 so that $a_{21} = (\psi_2, A\psi_1) \neq 0$.

It follows that for some complex μ_0 the operator $PAP - \mu_0 P = B$ must have rank less than N, but still ≥ 1. The above construction then may be repeated, starting with B instead of C_1. Or somewhat simpler, one directly observes that the resolvent $S(\mu) = (PAP-\mu)^{-1}$ is in $\mathfrak{A}_\infty$ again -just use Lemma 5.3 again. But this resolvent may be written as

$$(3.31) \qquad S(\mu) = PT(\mu)P + \mu^{-1}(1-P)$$

where $T(\mu)$ denotes the resolvent of the restriction $A|$ im P, an operator im T → im T. Or, with reference to the orthonormal basis ψ_l of im P we get

$$(3.32) \qquad T(\mu) = ((a_{jk} - \mu\delta_{jk}))^{-1} \quad .$$

From (3.31) it follows that $PT(\mu)P \varepsilon \mathfrak{A}_\infty$. Thus we may take a complex curve integral again to conclude that the projections of the decomposition of identity belonging to this operator of finite rank all are in $\mathfrak{A}_\infty$. By construction there exists one such projection of rank ≥ 1, but $<N$. Call it Q. Conclude that $Q^*Q \varepsilon \mathfrak{A}_\infty$ and repeat the same resolvent construction to show that the (<0>-orthogonal) projections of the decomposition of identity of Q^*Q again are in $\mathfrak{A}_\infty$. Among these there must be one <0>-orthogonal projection with rank ≥ 1, but $<N$. Call it R, and notice that R satisfies exactly the same assumptions than P but has rank less than N. Hence we must arrive at such a projection in $\mathfrak{A}_\infty$ of rank 1 after a finite number of repetitions of the above construction (At most N-1 times). This proves Lemma 3.3.

Let us finitely remark that during the above proof of Lemma 3.3 we also have proven the following results (which should be put on record as a specific theorem).

<u>Theorem 3.4.</u> The algebra $\mathfrak{A}_\infty$ is adjoint closed: If $A \in \mathfrak{A}_\infty$ then the $\mathcal{O}(\infty)$-adjoint A^* (as defined in III,5) also is in $\mathfrak{A}_\infty$. Moreover, if $A \in \mathfrak{A}_\infty$ is invertible in any $\mathfrak{A}_s$ then its inverse is in $\mathfrak{A}_\infty$.

The proof is implicitely contained in the proof of Lemma 3.3. It entirely rests on Lemma 5.3.

<u>Corollary 3.5.</u> If $A \in \mathfrak{A}_\infty$ then $\Lambda^{-s} A \Lambda^{s} \in \mathfrak{A}_\infty$ for every $s \in \mathbb{R}$.

<u>Proof.</u> Apply Lemma 3.1 to show that $\Lambda^{-s} A \Lambda^{s} = A + C$ with $C \in \mathcal{L}_0$. Then apply Lemma 3.3 to conclude that $C \in \mathcal{L}_0 \subset \mathfrak{A}_\infty$, q.e.d.

Let us introduce a subalgebra $\overset{\circ}{\Psi\mathcal{D}}$ of $\overset{\circ}{\Psi\mathcal{O}}$ (as in III,5) by replacing $CB^\infty(\mathbb{R}^n)$ in (5.4) by A and then using the new class L as before for generation of $\overset{\circ}{\Psi\mathcal{D}}$, together with the powers of Λ. We then extend $\overset{\circ}{\Psi\mathcal{D}}$ to obtain $\Psi\mathcal{D}$, below, a natural algebra for singular elliptic theory.

We define $\Psi\mathcal{D}$ to consist of all operators $K = A\Lambda^{-s}$, with $A \in \mathfrak{A}_\infty$ and $s \in \mathbb{R}$. Notice that we also get $K = \Lambda^{-s} A_1$, with $A_1 = \Lambda^{s} A \Lambda^{-s} \in \mathfrak{A}_\infty$, by Corollary 3.5.

<u>Lemma 3.6.</u> The class $\Psi\mathcal{D}$ is an algebra containing $\overset{\circ}{\Psi\mathcal{D}}$.

<u>Proof.</u> The elements of $\overset{\circ}{\Psi\mathcal{D}}$ are finite sums of finite products of $A_j \in \mathfrak{A}_\infty$ and Λ^{s_j}. Typically such a product will have the form

$$(3.33)\qquad \begin{aligned} &A_1 \Lambda^{s_1} A_2 \Lambda^{s_2} \ldots A_N \Lambda^{s_N} = \\ &A_1 (\Lambda^{s_1} A_2 \Lambda^{-s_1})(\Lambda^{s_1+s_2} A_3 \Lambda^{-s_1-s_2}) \ldots \Lambda^{s_1+s_2+\ldots+s_N} \end{aligned}$$

Evidently the right hand side is of the form $A_1' A_2' \ldots A_N' \Lambda^{-s} = A \Lambda^{-s}$, using Corollary 3.5 again, and the fact that $\mathfrak{A}_\infty$ is an algebra. Hence we indeed get $\overset{\circ}{\Psi\mathcal{D}} \subset \Psi\mathcal{D}$. For $K = A\Lambda^{-s}$ and $L = B\Lambda^{-t}$ assume without loss that $s \le t$ and get $K + L = (A\Lambda^{t-s} + B)\Lambda^{-t} = C\Lambda^{-t}$ with $C = A\Lambda^{t-s} + B \in \mathfrak{A}_\infty$, so that $K+L \in \Psi\mathcal{D}$. Also $KL = A(\Lambda^{-s} B \Lambda^{s})\Lambda^{-s-t} \in \Psi\mathcal{D}$ again, so that indeed $\Psi\mathcal{D}$ is an algebra.

We still distinguish the classes

$$(3.34)\qquad \Psi\mathcal{D}_s = \{K = A\Lambda^{-s} : A \in \mathfrak{A}_\infty\} = \mathfrak{A}_\infty \Lambda^{-s} = \Lambda^{-s}\mathfrak{A}_\infty ,$$

for a specific $s \in \mathbb{R}$, so that $\Psi\mathcal{D} = \bigcup_s \Psi\mathcal{D}_s$ is a graded algebra [Notice that $\Psi\mathcal{D}_s \subset \mathcal{O}(s)$ so that each element of $\Psi\mathcal{D}$ has some order s.]

Problems.

1) Let $L = \sum_{|\alpha|\leq N} a_\alpha(M)D^\alpha$, with coefficients in A, and assume that L is formally self-adjoint:

$$(3.35)\qquad \tilde{L} = \sum_{|\alpha|\leq N} D^\alpha \bar{a}_\alpha(M) = L$$

(i.e. the two differential expressions of L and $\tilde{L}$ coincide coefficientwise after using Leibnitz formula to write

$$(3.36)\qquad \tilde{L} = \sum_{|\alpha|\ N} \tilde{a}_\alpha(M)D^\alpha ,$$

with $\tilde{a}_\alpha \in A$ composed of derivatives of the a_α.)

Assume also that L is uniformly elliptic over $\mathbb{R}^n$ (c.f. 2, problem 5). Prove (i) the minimal operator L_0, defined as the unbounded hermitian linear operator of $L^2(\mathbb{R}^n)$ with domain dom $L_0 = C_0^\infty(\mathbb{R}^n)$, is essentially self-adjoint [its closure is self-adjoint]. Also (ii) that the closure L of L_0 has domain $\mathfrak{H}_N \subset L^2(\mathbb{R}^n) = \mathfrak{H}_0$. Also (iii) prove that the limit spectrum of the self-adjoint operator L is given by the set

$$(3.37)\qquad \tilde{\sigma}(L) = \bigcap\{\sigma^a:\ a \geq 0\}$$

where

$$(3.38)\qquad \sigma^a = \{\sum_{|\alpha|\leq N} a_\alpha(x)\xi^\alpha:\ x,\xi \in \mathbb{R}^n,\ |x| \geq a\}^{\text{closure}}$$

[The limit spectrum of L is defined to be the collection of all points in the spectrum of L which are not isolated pointeigenvalues of finite multiplicity].

2) Prove a generalization of problem 1 in case where L is a ψdo of order $s > 0$ in $\Psi\mathcal{D}_s$ which is uniformly elliptic over $\mathbb{R}^n$ and such that for at least one $\lambda_0 \in \mathbb{C}$ the operator $L-\lambda_0\Lambda^{-s}$ is md-elliptic (c.f. 2, pbm.5), as follows.
(i) The concepts of weak solutions and strong solutions of the equation Lu = f coincide [that is, we have $\tilde{L}_0^{\ *} = L_1^{**} = L$, with the minimal operator $\tilde{L}_0$ of the adjoint expression $\tilde{L}$, and the maximal operator L_1, defined for all $u \in C^\infty(\mathbb{R}^n)$ with $Lu \in L^2(\mathbb{R}^n)$ by setting $L_1u = Lu$, and with the operator L in the domain $\mathfrak{H}_N$, defined by means of strong L^2-derivatives (c.f. III,2) .
(ii) The Fredholm spectrum of the unbounded operator L [that is the collection of all points $\mu \in \mathbb{C}$ such that $L-\mu$ is not a Fredholm operator] coincides with the set $\tilde{\sigma}(L)$ defined in (3.37) and (3.38).

4. Fredholm theory in the algebra $\mathfrak{A}_\infty$; Green Inverse.

Two operators $A,B \in \mathcal{O}(\infty)$ will be called Green inverses of each other if we have

(4.1) $$1 - AB \in \mathfrak{F}_\infty, \quad 1 - BA \in \mathfrak{F}_\infty ,$$

with the ideal of $\mathcal{O}(-\infty)$-dyads, as defined in section 3.

The object of this section is the discussion of existence of a Green inverse for operators in $\Psi\mathcal{O}$ as defined in section 3.

Recall that an operator in $\Psi\mathcal{O}$ must be in some $\Psi\mathcal{O}_s$, and then will have the form $K = A\Lambda^{-s} = \Lambda^{-s}A_1$, with $A,A_1 \in \mathfrak{A}_\infty$. Notice that A and A_1 differ by an element of $\mathfrak{C}_0$ and therefore have the same symbol $\sigma_A = \sigma_{A_1}$. This symbol will be called the symbol quotient of order s of the operator $K \in \Psi\mathcal{O}$, and will be denoted by $\tau^s_K = \tau^s_K(x,\xi)$. In other words, we get

(4.2) $$\tau^s_K(x,\xi) = \sigma_{K\Lambda^s}(x,\xi) = \sigma_{\Lambda^s K}(x,\xi) .$$

Theorem 4.1. An operator $K \in \Psi\mathcal{D}_s$ admits a Green inverse of order -s if and only if its s-symbol quotient τ^s_K does never vanish on $\mathbb{M}$. Moreover, any such Green inverse also is in $\Psi\mathcal{D}_{-s}$.

Proof. Let L be a Green-inverse of $K \in \Psi\mathcal{D}_s$, and let $L \in \mathcal{O}(-s)$. We can write $K = A\Lambda^{-s}$, $L = \Lambda^s B$, with $A \in \mathfrak{A}_\infty$, $B \in \mathcal{O}(0)$. Then we get

(4.3) $$KL = AB = 1 + F, \quad LK = \Lambda^s BA\Lambda^{-s} = 1 + F', \quad F,F' \in \mathfrak{F}_\infty .$$

Or, equivalently, (with $G = \Lambda^{-s}F'\Lambda^s$)

(4.4) $$AB = 1 + F, \quad BA = 1 + G, \quad F,G \in \mathfrak{F}_\infty .$$

Moreover we get

(4.5) $$\tau^s_K = \sigma_A , \quad \tau^{-s}_L = \sigma_B .$$

This shows that it suffices to prove the theorem for the case $s = 0$. Or, in other words, we must show that an operator $A \in \mathfrak{A}_\infty$ admits a Green inverse of order 0 if and only if σ_A does never vanish, and also that any such Green inverse is in $\mathfrak{A}_\infty$ again.

Regarding the latter fact, suppose that $A \in \mathfrak{A}_\infty$ and $B \in \mathcal{O}(0)$ satisfy (4.4). We clearly have $B \in \mathfrak{A}_0$ because it is a Fredholm inverse of $A \in \mathfrak{A}_0$ and $\mathfrak{A}_0$ contains all Fredholm inverses of its elements because it is a C^*-algebra with unit containing the compact ideal, hence is Fredholm closed. (c.f. AI, 5 or AII, Corollary 7.16). Also we get

(4.6) $$A\Lambda^{-s}B\Lambda^s = 1 + C, \quad \Lambda^{-s}B\Lambda^s A = 1 + C', \quad C,C' \in K_0 ,$$

as follows from the fact that $\Lambda^s A \Lambda^{-s} = A + C_1$, $C_1 \in \mathcal{L}_0$, due to $A \in \mathfrak{A}_\infty$. (Also we must apply (4.4)). But (4.6) and (4.4) imply that both operators B and $B_s = \Lambda^{-s} B \Lambda^s$ are inverses modulo K_0 of the operator $A \in \mathfrak{A}_0$. Thus it follows that $B - B_s \in K_0$, and Lemma 5.3 may be applied to conclude that $B \in \mathfrak{A}_\infty$.

Next we observe that a Green inverse certainly is a Fredholm inverse of the operator $A \in \mathfrak{A}_\infty$ in the algebra $\mathfrak{A}_0$. In other words, if A admits a Green inverse, then it (or more precisely its continuous extension to $\mathfrak{H}_0$) must be Fredholm in the algebra $L(\mathfrak{H}_0)$. Accordingly we then must have $\sigma_A(x,\xi) \neq 0$ on $\mathbf{M}$.

Finally assume that $A \in \mathfrak{A}_\infty$ has its symbol never vanishing on $\mathbf{M}$. Let the continuous extensions to $\mathfrak{H}_s$ (which are in $\mathfrak{A}_s$, as we have seen) be denoted by A_s again. All the A_s have the same symbol σ_A which never vanishes on $\mathbf{M}$. Therefore these operators all are Fredholm (in $L(\mathfrak{H}_s)$).

Now we use the property of $\Lambda^s: \mathfrak{H}_0 \to \mathfrak{H}_s$ being an isometry and conclude that $\Lambda^{-s} A_s \Lambda^s : \mathfrak{H}_0 \to \mathfrak{H}_0$ also is a Fredholm operator, with the same Fredholm index as A_s. However, we get

$$(4.7) \qquad \Lambda^{-s} A_s \Lambda^s = A_0 + C_0, \qquad C_0 \in K_0,$$

as a consequence of Lemma 3.1. This shows that $\Lambda^{-s} A_s \Lambda^s$ and A_0 must have the same Fredholm index, and thus we get $\text{ind } A_s = \text{ind } A_0$ constant and independent of s for all $s \in \mathbb{R}$.

Next we easily confirm that (c.f. also AI, 3)

$$(4.8) \qquad \text{ind } A_s = \dim \ker A_s - \dim \ker (A^*)_{-s}$$

Actually we know that $\mathfrak{H}_{-s}$ is the adjoint space of $\mathfrak{H}_s$ under the duality implied by III, (2.13), and that $(A^*)_{-s}$, (with the $\mathcal{O}(\infty)$-adjoint A^* of $A \in \mathcal{O}(0)$) is the Banach space adjoint of A_s. Therefore (4.8) is evident (In particular the co-dimension of the image of A_s which is a closed subspace of $\mathfrak{H}_s$ is equal to the dimension of its orthogonal complement - namely the kernel of the Banach space adjoint). (For details about $\mathcal{O}(\infty)$-adjoints c.f. III, section 5).

It is trivial that $\ker A_s$ can only increase as s decreases, since A_s is the continuous extension of A, defined in $\mathfrak{H}_\infty$, to $\mathfrak{H}_s$. Similarly $\ker (A^*)_t$ is non-decreasing, as t decreases. Therefore, by (4.8), the constant function $\text{ind } A_s$ is the sum of the two non-increasing functions $\dim \ker A_s$ and $(- \dim \ker (A^*)_{-s})$. It follows that the two summands each must be constant. In other words, we get $\ker A_s$ and $\ker (A^*)_{-s}$ each independent of s. In particular we get

$$(4.9) \qquad N = \ker A_s \subset \mathfrak{H}_\infty, \qquad N^* = \ker (A^*)_{-s} \subset \mathfrak{H}_\infty .$$

Evidently N is the nullspace of A, and we shall see that N^* is a complement of im A, with $A: \mathfrak{H}_\infty \to \mathfrak{H}_\infty$. First of all we get $(A^*)_0 = A_0^{\langle 0 \rangle}$, so that in $\mathfrak{H}_0$ we have the direct decompositions

(4.10) $$\mathfrak{h}_0 = \mathcal{N}^* \oplus \text{im}\, A_0, \qquad \mathfrak{h}_0 = \mathcal{N} \oplus \text{im}\, A_0^{<0>},$$

which both are orthogonal decompositions. Define a special Fredholm inverse of A_0 by setting

(4.11) $$B_0 = 0 \text{ in } \mathcal{N}^*, \qquad B_0 = (A_0 |\, \text{im}\, A_0^{<0>})^{-1} \text{ on } \text{im}\, A_0 .$$

This is a well defined operator of $L(\mathfrak{h}_0)$, because A_0 is bijective between the complement of its nulspace and its image.

Conclude that

(4.12) $$A_0 B_0 = 1 - P_{\mathcal{N}^*}, \qquad B_0 A_0 = 1 - P_{\mathcal{N}},$$

with $P_{\mathcal{T}}$ the orthogonal projection (in $\mathfrak{h}_0$) onto the closed subspace $\mathcal{T}$. If $\{\phi_j\}$ denotes an orthonormal base of $\mathcal{N} \subset \mathfrak{h}_\infty$ (with respect to $(u,v)_0$) then we find that

(4.13) $$P_{\mathcal{N}} = \sum_j \phi_j >< \bar{\phi}_j ,$$

so that $P_{\mathcal{N}}$ (and similarly $P_{\mathcal{N}^*}$) is an $O(-\infty)$-dyad.

Now let us prove that $\mathcal{N}^*$ is a complement of $\text{im}\, A_s$ in $\mathfrak{h}_s$. Notice that we have $\dim \mathcal{N}^* = \text{codim im}\, A_s$, because of (4.9) and the fact that $(A^*)_{-s}$ is the Banach space adjoint of A_s [and also because we know A_s to be Fredholm and thus $\text{im}\, A_s$ to be closed.] . Since N^* has finite dimension it must be a complement of $\text{im}\, A_s$ provided that we can show that $\text{im}\, A_s \cap \mathcal{N}^* = \{0\}$. Let $v \in \text{im}\, A_s \cap \mathcal{N}^*$. We know that $v \in \mathfrak{h}_\infty \subset \mathfrak{h}_{-s}$ so that (v,v) is meaningful. But we must have $(v,v) = 0$, because N^* is the ortho-complement of $\text{im}\, A_s$ in the duality III, (2.13). This may be written as $(v,v)_0 = 0$ and thus implies $v = 0$. Thus indeed $\mathcal{N}^*$ is a complement of $\text{im}\, A_s$ in $\mathfrak{h}_s$, for every $s \in \mathbb{R}$. In other words every $u \in \mathfrak{h}_\infty$ admits a unique decomposition

(4.14) $$u = v_s + w_s, \qquad v_s \in \mathcal{N}^*, \qquad w_s \in \text{im}\, A_s .$$

But we have $\text{im}\, A_s \subset \text{im}\, A_t$, as $s > t$. Accordingly the decomposition must be independent of s; v_s and w_s must be constant in s. It follows that $w = w_s = A_s \phi_s$, where the ϕ_s is uniquely determined if we require $(\phi_s, z) = 0$ for $z \in \mathcal{N}$. Again we conclude that ϕ_s must be independent of s, because $A_s \phi_s = A_t \phi_t$ and $s < t$ implies $A_t \phi_s = A_t \phi_t$, and therefore $\phi_s = \phi_t$. Hence we finally get $\phi_s = \phi \in \mathfrak{h}_\infty$, and the decomposition (4.14) can be written as

(4.15) $$u = v + w, \qquad v \in \mathcal{N}^*, \qquad w = A \ \ \varepsilon\ \text{im}\, A .$$

As a conclusion we have that N^* also is a complement of $\text{im}\, A$ in $\mathfrak{h}_\infty$. Also the projections $P_{\mathcal{N}}$ and $P_{\mathcal{N}^*}$ are $O(-\infty)$-dyads, and thus are bounded in every $_s$. Hence we may define

(4.16) $$B_s = 0 \text{ in } \mathcal{N}^*, \text{ and } \quad B_s = (A_s |\, \ker\, (P_{\mathcal{N}})_s)^{-1} \text{ on } \text{im}\, A_s$$

and then get $B_s \in L(\mathfrak{H}_s)$ and

(4.17) $$A_s B_s = 1 - (P_{N^*})_s , \qquad B_s A_s = 1 - (P_N)_s .$$

Moreover it is clear that B_s extends B_t, as $t < s$ and that the restriction B of B_s to $\mathfrak{H}_\infty$ is in $\mathcal{O}(0)$, and has extension B_s to $\mathfrak{H}_s$. Thus we get

(4.18) $$AB = 1 - P_{N^*} , \qquad BA = 1 - P_N .$$

Since P_{N^*} and P_N are $\mathcal{O}(-\infty)$-dyads this means that B and A are Green inverses of each other. This proves existence of B. It then is an immediate consequence of Lemma 5.3, in the next-following section, that every Green inverse B of A is in $\mathfrak{A}_\infty$.

Problems.

1) Let $L = \sum_{|\alpha|\le N} a_\alpha(M)D^\alpha$ have coefficients $a_\alpha \in A$, and let it be md-elliptic (c.f. 2, problem 5). Show that then L has a Green inverse of order -N, and, vice versa that existence of a Green inverse of order -N for an L with coefficients in A implies that L must be md-elliptic.

2) For a differential operator L as in problem 1) show that every Green inverse of order -N may be written as an integral operator

(4.19) $$Gu(x) = \int g(x,y)u(y)dy$$

with kernel $g(x^0,\cdot) \in L^1(\mathbb{R}^n)$, $g(\cdot,y^0) \in L^1(\mathbb{R}^n)$, as $x^0, y^0 \in \mathbb{R}^n$.

3) *) Show that $g(x,y)$ of problem 2) is C^∞ for all $x,y \in \mathbb{R}^n$, except at $x = y$, and that $g(x,y) = O(|x-y|^{-n+N})$ as $|x-y| \to 0$, whenever $N < n$. (Obtain a similar estimate for general N and n). Hint: Use the Frechet - Riesz - Theorem.

4) Prove a global regularity theorem of the following type: Let $K \in \Psi\mathcal{D}_s$, and let $K - \mu\Lambda^{-\tau}$ be md-elliptic for some $\mu \in \mathbb{C}$, $\tau < s$, $s \ge 0$. Then we have

(4.20) $$Ku = f , \quad u \in \mathfrak{H}_{-\infty} , \quad f \in \mathfrak{H}_t$$

only if $u \in \mathfrak{H}_{t+s}$. Specifically, if $f \in \mathfrak{H}_\infty$ then all solutions of equation (4.20) necessarily are in $\mathfrak{H}_\infty$.

5) Let $A \in \mathfrak{A}_\infty$. Prove that, if λ is an isolated point eigenvalue of A having finite multiplicity then every eigenfunction of A_s to this eigenvalue must be in $\mathfrak{H}_\infty$. (here A_s denotes the canonical extension of A to $\mathfrak{H}_s$ (in $L(\mathfrak{H}_s)$), and s is any real.)

5. Some more auxiliary results.

First of all let us consider the action of certain simple invertible operators of $L(\mathfrak{H}_s)$ on our algebras. For example we look at the operators $e^{it\cdot D} = F^{-1}e^{it\cdot M}F$, with $t = (t_1,\ldots,t_n)$, $t\cdot D = \sum_1^n t_jD_j$. Clearly these are unitary operators $\mathfrak{H}_s \to \mathfrak{H}_s$ for every finite s, since formal Fourier multipliers commute and e^{itx} has norm 1 so that multiplication by it defines an isometry. We observe that

$$(5.1)\qquad e^{itD}b(D)e^{-itD} = b(D), \qquad e^{itD}a(M)e^{-itD} = a(M+t) .$$

The first of these relations is trivial. As to the second we note that

$$(5.2)\qquad \begin{aligned} Fe^{itD}u(\xi) &= e^{it\xi}u^{\wedge}(\xi) = \int đx e^{ix\xi}u(x)e^{it\xi} = \int đx e^{-i(x-t)\xi}u(x) \\ &= \int đx e^{-ix\xi}\, u(x+t) \quad . \end{aligned}$$

Taking the inverse Fourier transform we find that

$$(5.3)\qquad e^{itD}u(x) = u(x+t) \quad ,$$

which implies (5.1) and also shows that e^{itD} simply is the translation operator mapping u onto u_t with $u_t(x) = u(x+t)$.

Clearly the collection $\{e^{itD}\colon\ t \in \mathbb{R}^n\}$ is a group of unitary operators over each $\mathfrak{H}_s$, $s\in\mathbb{R}$, which is isomorphic to the group $\mathbb{R}^n$ with vector addition as group operation.

Lemma 5.1. The conjugation $A \to A_t = e^{itD}Ae^{-itD}$ defines $*$-automorphisms of each of the algebras $\mathfrak{A}_s$, $s\in\mathbb{R}$, (onto itself), as well as of the two generating subalgebras $\mathfrak{A}^+$, $\mathfrak{A}^\#$ of $\mathfrak{A}_0$, as defined in (1.11) and (1.16). In fact, we have

$$(5.4)\qquad a(M)_t = a(M+t), \qquad b(D)_t = b(D), \quad a \in CM(\mathbb{R}^n), \quad b \in CS(\mathbb{R}^n).$$

Proof. It suffices to look at the case $s = 0$, since e^{itD} commutes with powers of Λ. On the other hand that case is trivial, since in effect we already verified (5.4), this being a direct conseqeunce of (5.1). Since $\mathfrak{A}_0$ is generated by $\mathfrak{A}^+$ and $\mathfrak{A}^\#$ the invariance of $\mathfrak{A}_0$ under the conjugation listed is a consequence of the invariance of $\mathfrak{A}^+$ and $\mathfrak{A}^*$, q.e.d.

We now may ask about the associate dual maps of the automophisms of Lemma 5.1, which must be homeomorphisms of the corresponding maximal ideal spaces denoted by $\tau\colon \mathbf{M} \to \mathbf{M}$, and $\tau^+\colon \mathbf{P}^n \to \mathbf{P}^n$, and $\tau^\#\colon \mathbf{B}^n \to \mathbf{B}^n$, respectively (c.f. AII, 5).

Lemma 5.2. We have $\tau^{\#}$ = identity map in $\mathbb{B}^n$, while τ^+ is the tranlation $x \to x+t$, mapping $\mathbb{P}^n$ to itself, where one defines $x+t = x$ as $|x| = \infty$. Moreover τ is the restriction of the cartesian product $\tau^+ \times \tau$ to $\mathbf{M}$, so that we have

$$(5.5) \qquad \tau((x,\xi)) = (x+t,\xi) , \quad (x,\xi) \in \mathbf{M} .$$

Also for $A \in \mathfrak{O}_0$ we get

$$(5.6) \qquad \sigma_{A_t}(x,\xi) = \sigma_A(x+t,\xi) , \quad (x,\xi) \in \mathbf{M} .$$

Proof. To verify the description of τ^+ and $\tau^{\#}$ just consider the action of the automorphism on the maximal ideals $\{a(M): a(x^0) = 0\}$ or $\{b(D): b(\xi^0) = 0\}$. While the second kind of ideal is invariant under conjugation with e^{-itD}, the first will go into $\{a(M): 0 = a(x^0-t)\}$. The associate dual map turns out to be the image of the maximal ideal under the inverse automorphism so that the proper formulas for $\tau^{\#}$ and τ^+ result. For the description of τ we only must recall that our interpretation of the space $\mathbf{M}$ has been achieved via the homeomorphism ι of (2.2) constructed with Herman's Lemma, which uses the associate dual maps of the two canonical maps $\mathfrak{O}^+ \to \mathfrak{O}_0/K_0$ and $\mathfrak{O}^{\#} \to \mathfrak{O}_0/K_0$. Accordingly the map τ must be as described. In particular it must follow that the compact set $\mathbf{M} \subset \mathbb{P}^n \times \mathbb{B}^n$ must be left invariant by $\tau^+ \times \tau$, which is indeed evident. Finally (5.6) is an immediate consequence of (5.5), by (AII, 5), q.e.d.

Next it may be observed that there exist largely parallel results for another group of unitary operators of $\mathfrak{z}_0$, namely the multiplications e^{itM}, $t \in \mathbb{R}^n$ by the function e^{itx}. It is readily seen that these operators correspond to translations $\xi \to \xi-t$ in the phase space. Conjugation with e^{itM} also leaves $\mathfrak{O}^{\#}$, $\mathfrak{O}^+$ and $\mathfrak{O}_0$ invariant and the corresponding associate dual maps will shift the ξ-variable but leave the x-variable fixed.

Also of potential interest is the fact that $\mathfrak{O}^+$, $\mathfrak{O}^{\#}$ and $\mathfrak{O}_0$ remain invariant under conjugation with general linear substitution operators

$$(5.7) \qquad G_g u = u_g , \quad u_g(x) = u(g(x)), \quad g(x) = Px+t ,$$

with $t \in \mathbb{R}^n$ and a real invertible $n\times n$-matrix P. In that case, for t=0, we get

$$(5.8) \qquad G_g a(M) G_g^{-1} = a_g(M) , \quad G_g b(D) G_g^{-1} = b_q(D), \quad q(x) = P^{-1^T} x = Qx .$$

as readily verified. This fixes the corresponding associate dual map

$$(5.9) \qquad \tau_g((x,\xi)) = (g(x), q(\xi))$$

where the corresponding maps $\mathbb{R}^n \to \mathbb{R}^n$ are to be extended to $\mathbb{P}^n$ or $\mathbb{B}^n$ continuously. We shall not use these above maps in the following, except in certain secondary applications (although the reader might find it interesting to play with these larger groups leaving $\mathfrak{O}_0$ invariant, after studying our results in the next-following sections, which are based on the translation group only. Therefore we leave the detailed verification of the above to the reader.)

Finally in this section we will prove a result characterizing $\mathfrak{A}_\infty$ as a subalgebra of $\mathfrak{A}_0$. It is interesting that Lemma 5.3 below establishes identity of $\mathfrak{A}_\infty$ with an algebra studied by Taylor [14] in his Ph.D.-thesis under the autors supervision.

Lemma 5.3. The algebra $\mathfrak{A}_\infty$, as a subalgebra of $\mathfrak{A}_0$ is identical with the class of all $A \in \mathfrak{A}_0$ such that for all $s \in \mathbb{R}$ we have $\Lambda^s A \Lambda^{-s}$ defined in a dense subspace of $\mathfrak{H}_0$ and extending continuously to a bounded operator A_s such that

(5.10) $$A_s - A = \Lambda^s A \Lambda^{-s} - A = C_s \in K_0 , \quad s \in \mathbb{R} .$$

[we shall denote A_s by $\Lambda^s A \Lambda^{-s}$ again, in agreement with our general convention of sometimes leaving the domain of an operator undetermined].

Proof. An operator $A \in \mathfrak{A}_0$ satisfying (5.10) must necessarily be in $O(0)$, since $\|A\|_s = \|\Lambda^{-s} A \Lambda^s\|_0 \leq \|A\|_0 + \|C_{-s}\|_0 < \infty$. Also Lemma 3.1 implies that every $A \in \mathfrak{A}_\infty$ satisfies the condition (5.10) above. For $A \in \mathfrak{A}_0$ satisfying (5.10) we then will construct a sequence of operators in $\mathfrak{A}_\infty$ converging to A in $\mathfrak{A}_\infty$. We start with a sequence $A_k \in \mathfrak{S}$ converging to A in $\mathfrak{A}_0$: $\|A - A_k\|_0 \to 0$, as $k \to \infty$. Also let $\chi \in C_0^\infty(\mathbb{R}^n)$, $\chi = 1$ near 0, $0 \leq \chi \leq 1$, let $\chi_m(x) = \chi(x/m)$, $m=1,2,\ldots$, and let $F_m = \chi_m(D)$.

Then it is trivial that we have

(5.11) $$\|F_m A_k F_m - F_m A F_m\|_l \to 0 , \quad k \to \infty ,$$

for each fixed $m = 1,2,\ldots$ and $l = 0,\pm1,\ldots$, since $(\chi_m \lambda^s)(D) = F_m \Lambda^s \in L(\mathfrak{H}_0)$ for fixed m and s. Now (5.11) implies that $F_m A F_m \in \mathfrak{A}_\infty$, $m = 1,2,\ldots$. Introduce

(5.12) $$A_{km} = F_m A F_m + (1-F_m) A_k F_m + A_k (1-F_m) ,$$

which clearly is in $\mathfrak{A}_\infty$. Now we use (5.10) -which is valid for $(A-A_k)$, using (3.9) for

(5.13) $$\begin{aligned}\Lambda^{-l}(A-A_{km})\Lambda^l &= \Lambda^{-l}(1-F_m)(A-A_k)F_m \Lambda^l + \Lambda^{-l}(A-A_k)(1-F_m)\Lambda^l \\ &= (1-F_m)(A-A_k)F_m + (A-A_k)(1-F_m) + C_{kl} - F_m C_{kl} F_m ,\end{aligned}$$

with certain operators $C_{kl} \in K_0$, being independent of m. For $N = 1,2,\ldots$ first chose $k = k_N$ such that $\|A-A_k\|_0 \leq 1/4N$. Then chose $m = m_N$ large to ensure that

(5.14) $$\|C_{k_N l} - F_{m_N} C_{k_N l} F_{m_N}\|_0 \leq 1/4N, \quad l = 0, \pm1, \ldots \pm N .$$

It follows that

(5.15) $$\| A - A_{k_N m_N} \|_1 \leq 1/N , \qquad |l| \leq N ,$$

so that indeed

(5.16) $$\lim_{N\to\infty} A_{k_N m_N} = A \quad \text{in } \mathfrak{A}_\infty \quad .$$

Note that (5.14) can be achieved due to the fact that $F_m \to 1$ in strong convergence of $L(\mathfrak{z}_0)$ while C_{kl} are compact so that $F_m C_{kl} \to C_{kl}$ and $C_{kl} F_m \to C_{kl}$ in norm convergence, as $m \to \infty$, by a well known theorem , q.e.d.

Problems.

1) Prove that for the conclusion of Lemma 5.3 it is sufficient to assume that (5.10) only holds for all $s = k = 0, \pm 1, \pm 2, \ldots$.

6. Commuting ψdo-s with differentiations.

In this section we are going to consider multiple commutators between differentiations D_j and pseudo-differential operators of order zero. It is rather evident that for an arbitrary operator in the algebra $\mathfrak{A}_0$, for example the commutator $[D_j,A]$ needs not to be well defined in the proper sense, if D_j is regarded as an unbounded operator of $\mathfrak{H}_0$ with domain $\mathfrak{H}_1$, because in general A will not take $\mathfrak{H}_1$ to itself [just look at sufficiently irregular compact operators which all are in the algebra $\mathfrak{A}_0$.] On the other hand, for $A \in \mathfrak{A}_\infty$ we indeed get the commutator $[D_j,A] = D_jA - AD_j$ defined as an unbounded operator of $\mathfrak{H}_0$ with domain $\mathfrak{H}_1$ indeed.

We pose the question under what conditions $[D_j,A]$ is not only defined in $\mathfrak{H}_1$ but is even a bounded operator -thus has its closure or continuous extension in $L(\mathfrak{H}_0)$. In that respect, if we look at the algebra of finitely generated ψdo-s then we find this always true: we get

$$(6.1) \qquad [D_j,a(M)] = -ia_{x_j}(M) , \qquad [D_j,b(D)] = 0 ,$$

which shows that not only (for the generators of $\mathring{\mathfrak{A}}$) we have $[D_j,A] \in L(\mathfrak{H}_0)$, but even we get $[D_j,A] \in \mathring{\mathfrak{A}}$. By algebraic extension this must be true for all of $\mathring{\mathfrak{A}}$.

Theorem 6.1, below, will be giving a condition for such a result for more general operators in $\mathfrak{A}_0$. In fact we will be considering repeated commutators, and to this extent introduce the notation

$$(6.2) \qquad \begin{aligned} (\mathrm{ad}\, D_j)A &= D_jA - AD_j = [D_j,A] , \\ (\mathrm{ad}\, D)^\alpha &= \prod_{j=1}^{n} (\mathrm{ad}\, D_j)^{\alpha_j} . \end{aligned}$$

In order to interpret (6.2) concretely and as something more than purely formal we assume $A \in L(\mathfrak{H}_\infty, \mathfrak{H}_{-\infty})$ and interpret the commutator $[D_j,A]$ as an operator in $L(\mathfrak{H}_\infty, \mathfrak{H}_{-\infty})$ again, where D_j actually is used to denote two different operators, either $D_j\colon \mathfrak{H}_\infty \to \mathfrak{H}_\infty$ or $D_j\colon \mathfrak{H}_{-\infty} \to \mathfrak{H}_{-\infty}$, in the two products AD_j and D_jA, respectively.

Next we introduce a decomposition of the symbol space

$$(6.3) \qquad \mathbf{M} = \mathbf{M}_p \cup \mathbf{M}_s$$

with the <u>principal symbol space</u> $\mathbf{M}_p$ and the <u>secondary symbol space</u> $\mathbf{M}_s$, defined (as subsets of $\mathbf{M}$) by

$$(6.4) \qquad \mathbf{M}_p = \{(x,\xi)\colon |\xi| = \infty\} , \qquad \mathbf{M}_s = \{(x,\xi)\colon |\xi| < \infty\} .$$

Also we distinguish interior $\mathring{\mathbf{M}}_p = \mathbf{W}$ and boundary $\partial\mathbf{M}_p$ of $\mathbf{M}_p$, the sets

$$(6.5) \qquad \partial\mathbf{M}_p = \{(x,\xi)\colon |x| = |\xi| = \infty\} , \qquad \mathring{\mathbf{M}}_p = \{(x,\xi)\colon |x| < \infty\}.$$

We shall refer to $\mathbb{W}$ as to the <u>wave front space</u>, at some later occasions, for reasons to be discussed.

We will require derivatives for functions defined on $\mathbb{M}$, in the sequel. At first this will be only x-derivatives. For $k=0,1,\dots,\infty$ (including) we introduce the class

(6.6) $$C^{[k]}(\mathbb{M}) = \{a: \ a \in C(\mathbb{M}), \ a_{(\alpha)} \in C(\mathbb{M}), \ |\alpha| \le k \}$$

with the modified norm $[\alpha] = \operatorname{Max}_{j=1}^{n} \alpha_j$ for multi-indices, where

(6.7) $$a_{(\alpha)}(x,\xi) = i^{|\alpha|} D_x^\alpha a(x,\xi) \quad \text{on } \mathring{\mathbb{M}}_p, \quad = 0 \text{ on } \partial\mathbb{M}_p \cup \mathbb{M}_s .$$

Note that a is globally differentiable in the following classical sense: The difference quotients

(6.8) $$\nabla_j a(x,\xi) = (a(x+e_j h,\xi) - a(x,\xi))/h , \quad e_j = (\delta_{1j},\dots,\delta_{nj})$$

converge uniformly to $a_{(e_j)}$ over all of $\mathbb{M}$, as $h \neq 0$, $h \to 0$, provided that we define $x+e_j h = x$ whenever $|x| = \infty$ (i.e., on $\mathbb{M}_s \cup \partial\mathbb{M}_p$), as is natural when adding a finite and an infinite quantity. Similarly for higher derivatives.

Now we can formulate the first result of this section.

<u>Theorem 6.1.</u> Let $k \le \infty$, $a \in C^{[k]}(\mathbb{M})$. Then there exists a collection $\{A_{(\alpha)}\}$ with α ranging over all multi-indices $|\alpha| \le k$ such that

(6.9) $$A_{(\alpha)} \in \mathcal{O}\!\!\mathcal{l}_0 , \quad \sigma_{A_{(\alpha)}} = a_{(\alpha)}$$

and that for $A = A_{(0)}$ we get

(6.10) $$i^{|\alpha|}(\mathrm{ad}\, D)^\alpha A = A_{(\alpha)} ,$$

again for all multi-indices $[\alpha] \le k$.

We prepare the proof of theorem 6.1 in a series of Lemmata.

<u>Lemma 6.2.</u> For $A \in \mathcal{O}\!\!\mathcal{l}_0$ with symbol a let us define

(6.11) $$B_j = \int_0^\infty e^{-iD_j t} A e^{iD_j t} e^{-t} dt , \quad j = 1,\dots,n .$$

Then (i) these integrals converge as improper Riemann integrals in norm convergence of $L(\mathcal{H}_0)$, and therefore we have $B_j \in \mathcal{O}\!\!\mathcal{l}_0$. Also (ii) the symbol $b_j = \sigma_{B_j}$ has its first derivative $b_{j(e_j)}$ continuous and is explicitely given by the formula (with $e_j = (\delta_{1,j}, \dots ,\delta_{nj})$)

(6.12) $$b_j = \int_0^\infty a(x-te_j,\xi)e^{-t}dt \quad (\text{with } x-te_j = x \text{ on } \mathbb{M}_s \cup \partial\mathbb{M}_p) .$$

In fact (iii) the function b_j is uniquely determined by the property of being in

$C(\mathbb{M})$, together with its derivative $b_{(e_j)}$, and solving the ordinary differential equation

(6.13) $$b_{j(e_j)} + b_j = a \quad .$$

Furthermore (iv) the commutator $[D_j, B_j]$ is not only in $L(\mathfrak{g}_1, \mathfrak{g}_{-1})$ but also is in $L(\mathfrak{g}_0)$, and, in fact, is in $\mathfrak{A}_0$ and explicitely given by

(6.14) $$i[D_j, B_j] + B_j = A \ .$$

Proof. Let the integral in (6.11) be denoted by $I(t)$, for a moment. Let us look at a Riemann sum for the integral $\int_0^T$, $T < \infty$. That is,

(6.15) $$S = S_P = \sum_{j=1}^{N} I(t_j^*)\Delta t_j \ , \qquad \Delta t_j = t_j - t_{j-1}, \quad t_{j-1} \le t_j^* \le t_j,$$

defined for a partition

(6.16) $$P: \quad 0 = t_0 < t_1 < \ldots\ldots < t_N = T$$

with norm

(6.17) $$\delta_P = \operatorname*{Max}_{j=1}^{N} (t_j - t_{j-1}) \quad .$$

For existence of the proper Riemann integral $\int_0^T$ we must show that

(6.18) $$\| S_P - S_{P'} \|_0 < \varepsilon \ , \text{ as } \quad \delta_P \ , \ \delta_{P'} < \delta_0(\varepsilon) \ .$$

Now if A equals any of the generators $a(M)$ of $b(D)$ of the algebra $\mathfrak{A}_0$ (c.f. III, (10.1)), then it follows that the integrand $I(t)$ is norm continuous in $L(\mathfrak{g}_0)$. Indeed we know from Lemma 5.1 that

(6.19) $$e^{-iD_j t}\, a(M)\, e^{+iD_j t} = a(M - e_j t)$$

which certainly is norm continuous since $a \in A$ is uniformly continuous over $\mathbb{R}^n$, evidently. Also since Fourier multipliers commute we get $e^t I(t) = b(D) =$ constant in case of $A = b(D)$, so again it is norm continuous. Accordingly $I(t)$ will be continuous for any operator A in the finitely generated algebra $\mathring{\mathfrak{A}}$.

It is evident that (6.18) holds whenever the integrand is (uniformly) continuous; one then just may repeat the usual proof for complex- or real-valued functions. Also, for $A \in \mathfrak{A}_0$, we get

(6.20) $$\| S \|_0 \le \sup\{\| I(t) \|_0 : \ 0 \le t \le T\} \le \| A \|_0$$

whatever the choice of the partition P may be.

Let $\varepsilon > 0$ be given, then pick $A_1 \in \mathring{\mathfrak{A}}$ with

(6.21) $$\| A - A_1 \|_0 < \varepsilon / 3T \quad .$$

Then write

$$\|s_p(A) - s_{p'}(A)\|_0 \le \|s_p(A_1) - s_{p'}(A_1)\|_0 + \|s_p(A-A_1)\|_0 + \|s_{p'}(A-A_1)\|_0 . \tag{6.22}$$

Using (6.20) and (6.21) it follows that the last two terms at right are less than $\varepsilon/3$ each. Then, keeping A_1 fixed we may chose $\delta_0(\varepsilon)$ small enough to insure that the first term also is $< \varepsilon/3$ whenever $\delta_p, \delta_{p'} < \delta_0(\varepsilon)$, since for $A_1 \varepsilon \mathring{\mathfrak{A}}$ the integrand I(t) is continuous. This proves existence of the proper Riemann integral $\int_0^T$. Then the existence of the improper integral $\int_0^\infty$ is evident, since the integrand has the convergent majorant $\|A\|_0 e^{-t}$. This settles (i).

Next we observe that the projection $\mathfrak{A}_0 \to \mathfrak{A}_0/K_0 \simeq C(\mathbf{M})$ is continuous. Thus existence of the integral in norm convergence of $\mathfrak{A}_0$ allows us to take symbols under the integral sign. Using Lemma 5.1 we get the integrand to be an operator in $\mathfrak{A}_0$ for every fixed t. Moreover from formula (5.6) we get

$$\sigma_{I(t)} = a(x - te_j, \xi)\, e^{-t} , \tag{6.23}$$

which implies (6.12).

Introducing $x_j - t$ as a new integration variable in (6.12) we get

$$b_j(x,\xi) = e^{-x_j} \int_{-\infty}^{x_j} a((\tilde{x},\tau),\xi)\, e^{\tau}\, d\tau , \tag{6.24}$$

with

$$(\tilde{x},\tau) = (x_1,\dots,x_{j-1},\tau,x_{j+1},\dots,x_n) . \tag{6.25}$$

From (6.24) it follows at once that $\partial b_j/\partial x_j$ exists for $|x| < \infty$, and that $b = b_j$ satisfies the differential equation (6.13) over $\mathring{\mathbf{M}}_p$. On the other hand, over $\partial\mathbf{M}_p \cup \mathbf{M}_s$ we get $x - te_j = x$, so that the integral may be calculated to equal $a(x,\xi)$. Also since $a \in C(\mathbf{M})$ is a uniform limit of symbols of generators - evidently - we get the restriction $a' = a|\mathbf{M}_p$ in $CM(\mathbb{R}^n)$ for every fixed $\xi \in \partial\mathbb{B}^n$ and uniformly so over $\partial\mathbb{B}^n$. That is, we have

$$\lim_{|x|\to\infty} cm_x(a') = 0, \qquad cm_x(a') = \sup_{|h|\le 1} |a'(x+h,\xi) - a'(x,\xi)| , \tag{6.26}$$

uniformly for $\xi \in \partial\mathbb{B}^n$.

This fact may be used to show that b_j is continuous at $\mathbf{M}_p$. For (6.26) amounts to the fact that $a(y,\xi)$ is nearly constant in y along larger and larger balls around x, as $|x|$ gets large. Since on the other hand

$$\left|\int_T^\infty a(x-te_j,\xi)e^{-t}dt\right| \le e^{-T}\|a\|_{C(\mathbf{M})} \tag{6.27}$$

one indeed gets

(6.28) $$\left|\int_0^\infty (a(x-te_j,\xi) - a(x,\xi))e^{-t}dt\right| \leq e^{-T}\|a\|_{C(\mathbb{M})} + cm_x(a')$$

which shows that (uniformly in ξ) we have

(6.29) $$\lim_{|x|\to\infty} (b_j(x,\xi) - a(x,\xi)) = 0 , \qquad \xi \in \partial\mathbb{B}^n .$$

Accordingly $b_{j(e_j)} \in C(\mathbb{M})$ and (6.13) holds over all of $\mathbb{M}$.

Suppose next that any function $b \in C(\mathbb{M})$ has $b_{(e_j)}$ defined and in $C(\mathbb{M})$ and that (6.13) holds for b instead for b_j. Let $c = b-b_j$ which satisfies $\partial c/\partial x_j + c = 0$ over $\mathring{\mathbb{M}}_p$. It follows at once that

(6.30) $$c(x,\xi) = c_0(\tilde{x},\xi)\, e^{-x_j}$$

with a certain c_0 independent of x_j. If we let $x_j \to -\infty$ it follows at once that c_0 must vanish identically since certainly $c \in C(\mathbb{M})$ is bounded, while $e^{-x_j} \to \infty$, as $x_j \to -\infty$. Thus we conclude $c = 0$ on $\mathring{\mathbb{M}}_p$. On the other hand the differential equation (6.13) implies that $b = a$ on $\mathbb{M}_s$, so we get $b = b_j$ everywhere on $\mathbb{M}$. This proves (ii) and (iii).

Proof of (iv). We consider the commutator as an operator in $L(\mathfrak{H}_1,\mathfrak{H}_{-1})$ as discussed. Now the left and right multiplication by D_j each constitute continuous maps $L(\mathfrak{H}_0) \to L(\mathfrak{H}_1,\mathfrak{H}_{-1})$, evidently. Using (i) it follows that in the expression arising from the substitution of (6.11) into the commutator $[D_j,B_j]$ the operator D_j may be taken under the integral sign. This yields

$$[D_j,B_j] = \int_0^\infty (D_j e^{-iD_jt} A\, e^{iD_jt} - e^{-iD_jt} AD_j e^{iD_jt})\, e^{-t}dt$$

(6.31) $$= i\int_0^\infty d/dt\, (e^{-iD_jt} Ae^{iD_jt})\, e^{-t}dt$$

$$= i\, e^{-iD_jt} A\, e^{iD_jt}\, e^{-t}\Big|_0^\infty + i\int_0^\infty e^{-iD_jt} Ae^{iD_jt} e^{-t}dt$$

$$= -iA + iB_j .$$

In (6.31) the first two integrals are norm convergent as improper Riemann integrals in $L(\mathfrak{H}_1,\mathfrak{H}_{-1})$. Also the derivative d/dt exists in norm convergence of $L(\mathfrak{H}_1,\mathfrak{H}_{-1})$. However, the partial integration carried out brings us back to integrals which also are norm convergent in $L(\mathfrak{H}_0)$. Thus (iv) and specifically (6.14) follow at once and the Lemma is established.

Proof of Theorem 6.1. For $k < \infty$ this proof is now almost evident. Let us introduce $a_k = \tilde{D}^{\alpha^k} a$, with the unique largest multi-index $\alpha^k = (k,k,\ldots,k,k)$ in the norm $[\alpha]$, and with

(6.32) $$\tilde{D}_j = \partial/\partial x_j + 1 , \qquad \tilde{D}^{\alpha} = \Pi D_j^{\alpha_j} .$$

Then pick any operator $A_{\{\alpha^k\}} \in \mathfrak{A}_0$ with symbol a_k. Let $E_j : \mathfrak{A}_0 \to \mathfrak{A}_0$ denote the operation $A \to B_j$ of (6.11). From (6.20) it follows at once that E_j is continuous and contracts the norm of $\mathfrak{A}_0$; its operator norm is ≤ 1. Moreover, it is clear that the operators E_j all commute:

(6.33) $$E_j E_l = E_l E_j , \qquad j,l = 1,\ldots,n .$$

Then it is natural to define (with $E^{\beta} = \Pi E_j^{\beta_j}$)

(6.34) $$A_{\{\alpha\}} = E^{(\alpha^k - \alpha)} A_{\{\alpha^k\}}$$

These are well defined operators of $\mathfrak{A}_0$ which must have symbol

(6.35) $$\sigma_{A_{\{\alpha\}}} = \tilde{D}^{\alpha} a ,$$

in view of the uniqueness contained in Lemma 6.2.

Then it is clear by induction that the operators $A_{(\alpha)}$ can be defined recursively:

(6.36) $$A_{(0)} = A_{\{0\}} , \qquad A_{(e_j)} = A_{\{e_j\}} - A_{\{0\}}$$
$$A_{(e_j+e_l)} = A_{\{e_j+e_l\}} - A_{\{e_j\}} - A_{\{e_l\}} + A_{\{0\}}, \qquad \text{etc.}$$

Also the commutator relations (6.10) follow trivially.

After settling the case of a finite k we notice that for a function $a \in C^{[\infty]}$ we will obtain a sequence of selections $A_{(\alpha),k}$ and $A_k = A_{(0),k}$, $k=0,1,2,\ldots$, $[\alpha] \leq k$ such that (6.9) and (6.10) holds as far as the operators are defined. It is clear also that

(6.37) $$A_k - A_l = C_{kl} \in K_0 , \qquad A_{(\alpha),k} - A_{(\alpha),l} \in K_0$$

for all k,l and α such that the terms are defined. Specifically,

(6.38) $$A_{k+1} - \chi_k(D) C_{k+1,k} \chi_k(D) = A_k + C_{k+1,k} - \chi_k(D) C_{k+1,k} \chi_k(D)$$

with a suitable function $\chi_k \in C_0^{\infty}$, nonnegative and equal to 1 near 0. Introducing the notation F_{k+1} and G_{k+1} for the corrective compact operators at left and right in (6.38) , respectively, we find that

(6.39) $$A_k - \sum_{l=1}^{k} F_l = A_m + \sum_{l=m+1}^{k} G_l - \sum_{l=1}^{m} F_l$$

where we have used induction. Let us denote the left hand side of (6.39) by $\tilde{A}_k$. Then it is clear that $(\mathrm{ad}\ D)^{\alpha}\tilde{A}_k \in \mathcal{O}\!\!\iota_0$, $[\alpha] \le k$ again and that $\tilde{A}_k$ satisfies the assumptions of Theorem 6.1 for k again, because the compact operators F_l do not influence the symbol and also may carry as many commutators D_j as desired, in view of the factors $\chi_j(D)$ which neutralize all possible factors D_j, since $\chi_l(D)D_j$ are bounded. On the other hand a proper choice of the functions χ_j will result in

$$(6.40)\qquad \|G_j\| \le \varepsilon_j \quad ,$$

where ε_j may be prescribed arbitrarily. For example, if we chose

$$(6.41)\qquad \chi_j(x) = \chi(\tau_j x)$$

with a given fixed χ independent of j meeting the assumptions then it can be observed that

$$(6.42)\qquad \lim_{\tau\to 0} \chi(\tau D) = 1$$

in strong operator convergence of $L(\mathcal{G}_0)$ - not in norm convergence, of course. By a well known result going back to Hilbert [11] this implies norm convergence of $\chi(\tau D)C$ to C for any compact operator C (Similarly $C\chi(\tau D) \to C$ in norm). This indeed implies that τ_j may be chosen to satisfy (6.40).

Chosing $\varepsilon_j < 2^{-j}$ then will make the series $\sum G_j$ convergent and suggests that we define

$$(6.43)\qquad A = A_0 + \sum_l G_l \quad .$$

This operator is not satisfactory however for the infinite version of Theorem 6.1, unless we impose the additional condition that also

$$(6.44)\qquad (\mathrm{ad}\ D)^{\alpha}\tilde{A}_k \to (\mathrm{ad}\ D)^{\alpha} A, \qquad k\to\infty\ , \quad \text{in } L(\mathcal{G}_0)\ .$$

However for $k \ge [\alpha] = m$ we get from (6.39) that

$$(6.45)\qquad (\mathrm{ad}\ D)^{\alpha}\tilde{A}_k = (\mathrm{ad}\ D)^{\alpha}\Big(A_m - \sum_{l=1}^{m} F_l + \sum_{l=m+1}^{k} G_l\Big)\ .$$

We therefore must be concerned about the convergence in $L(\mathcal{G}_0)$ of

$$(6.46)\qquad \sum_{l=m+1}^{\infty}(\mathrm{ad}\ D)^{\alpha}G_l = \sum_{l=m+1}^{\infty}\big((\mathrm{ad}\ D)^{\alpha}C_{l,l-1} - \chi_l(D)(\mathrm{ad}\ D)^{\alpha}C_{l,l-1}\chi_l(D)\big)\ .$$

First of all the $(\mathrm{ad}\ D)^{\alpha}C_{l,l-1}$ are in $L(\mathcal{G}_0)$, as follows from (6.37), due to $l \ge [\alpha]$. Therefore we may repeat the above conclusion for each individual fixed α, replacing (6.40) by the corresponding estimate

$$(6.47)\qquad \|(\mathrm{ad}\ D)^{\alpha}G_j\| \le 2^{-j}\ , \qquad j \ge [\alpha]\ .$$

Notice that (6.47) for all α and j with $j \geq [\alpha]$ amounts to finitely many conditions to be imposed on the corresponding τ_j - one for each $[\alpha] \leq j$. Accordingly a choice is possible enforcing convergence of $(\mathrm{ad}\ D)^{\alpha}\tilde{A}_k$ in $L(\mathcal{J}_0)$ for every α, as $k \to \infty$. Then we may set

$$(6.48) \qquad A = \lim_{k\to\infty} \tilde{A}_k \quad , \quad A_{(\alpha)} = \lim_{k\to\infty} i^{|\alpha|}(\mathrm{ad}\ D)^{\alpha}\tilde{A}$$

and then obtain a collection of operators for Theorem 6.1 with $k = \infty$. Q.E.D.

7. Symbols of operators in $\mathcal{O}_\infty$, and a Leibnitz formula.

We are going to prove the following result.

Theorem 7.1. Let $a \in C^{[\infty]}(M)$ as defined in the preceeding section. Then there exists an operator $A \in \mathcal{O}_\infty$ with $\sigma_A = a$. Moreover $A \in \mathcal{O}_\infty$ with $\sigma_A = a$ can be selected together with other operators $A_{(\alpha)} \in \mathcal{O}_\infty$, $A_{(0)} = A$ having symbol $a_{(\alpha)} = \sigma_{A_{(\alpha)}}$ such that for every function $b \in M$ (c.f. I, 6) we get the asymptotic expansions

$$(7.1)\qquad \begin{aligned} b(D)A &\sim \sum_\theta i^{-|\theta|}/\theta!\; A_{(\theta)} b^{(\theta)}(D) \quad \text{mod } \mathcal{Q}(-\infty) ,\\ Ab(D) &\sim \sum_\theta i^{|\theta|}/\theta!\; b^{(\theta)}(D)\; A_{(\theta)} \quad \text{mod } \mathcal{Q}(-\infty) . \end{aligned}$$

To prepare the proof let us first apply Taylors formula to the function $b(\xi)$. Recalling Fourier kernel multiplication (I,9) we conclude that (with $\zeta = \xi - \eta$)

$$(7.2)\qquad b(D)A = b(\xi)\blacktriangle A = \sum_{\theta \le N} b^{(\theta)}(\eta)/\theta!\; \zeta^\theta \blacktriangle A + r_N \blacktriangle A$$

with r_N as in III, (6.2). Since we get

$$(7.3)\qquad \zeta^\theta \blacktriangle A = (\text{ad } D)^\theta A = i^{-|\theta|} A_{(\theta)} ,$$

relation (7.2) will lead into an asymptotic formula resembling the first formula (7.1) if only we can handle the remainder term.

In details the remainder can be formally written as

$$(7.4)\qquad R_N = \int_0^1 \tau^N d\tau \sum_{|\theta|=N+1} (N+1)/\theta!\; i^{-|\theta|} b^{(\theta)}(t\xi + \tau\eta) \blacktriangle A_{(\theta)}$$

using (7.3) again. Next we will control the kernel multiplication by using the Fourier integral representation

$$(7.5)\qquad b^{(\theta)}(t\xi + \tau\eta) = \int đ\kappa\; b^{(\theta)\vee}(\kappa)\; e^{i\kappa\xi t}\, e^{i\kappa\eta\tau}$$

in view of (7.20) of Lemma 7.4, below. Accordingly we get

$$(7.6)\qquad b^{(\theta)}(t\xi + \tau\eta)\blacktriangle A_{(\theta)} = \int đ\kappa\; b^{(\theta)\vee}(\kappa)\; e^{i\kappa D} A_{(\theta),-\kappa\tau} = J$$

with the translated operator

$$(7.7)\qquad P_\kappa = e^{iD\kappa}\, P\, e^{-iD\kappa} .$$

Next, we recall that the function $c(\kappa) = b^{(\theta)\vee}(\kappa) = i^{\pm|\theta|}\, b^\vee(\kappa)\kappa^\theta$ is in S^0_{ps} (c.f. II, 3); it equals a function in S for $|\kappa| \ge 1$ and is C^k at 0 (its only singularity) as $|\theta|$ is sufficiently large, for every k. One will write

(7.8) $$e^{i\xi\kappa} = (1+\xi^2)^{-1}(1-\Delta_\kappa)^1 e^{i\xi\kappa}$$

and then formally arrive at

(7.9) $$J = \Lambda^{-1}\int d\kappa\; e^{i\kappa D}(1-\Delta_\kappa)^1 (c(\kappa)A_{(\theta),-\kappa\tau})$$

There will be no trouble in making this a rigorous formula with a norm convergent integral (in the norm of $L(\mathfrak{H}_0)$) after the Lemma, below.

Lemma 7.2. For a given $a \in C^{[\infty]}(\mathbb{M})$ let $A, A_{(\alpha)}$ be a collection of operators in A_0, as obtained from the construction of Theorem 6.1. Then we have

(7.10) $$A_t = e^{iDt} A e^{-iDt} \in C^\infty(\mathbb{R}^n, \mathfrak{A}_0) .$$

Moreover the derivatives are explicitely given as

(7.11) $$i^{|\alpha|}D_t^\alpha A_t = A_{(\alpha),t} = e^{iDt}A_{(\alpha)}e^{-iDt} .$$

Proof. It suffices to discuss the existence of only the first derivatives, by induction. By Lemma 6.2 we get (for $|\alpha| = 1$)

(7.12) $$A = \int_0^\infty d\tau\; e^{-\tau}(A_{(\alpha)} + A)_{-\tau\alpha}$$

which yields (with $\alpha = e_j$, $\tilde{t} = (t_1,\ldots,t_{j-1},0,t_{j+1},\ldots,t_n)$)

(7.13) $$A_t = \int_0^\infty d\tau\; e^{-\tau}(A_{(\alpha)} + A)_{t-\tau\alpha} = e^{-t_j}\int_{-\infty}^{t_j} e^{+\tau}(A_{(\alpha)} + A)_{\tilde{t}+\alpha\tau} .$$

This shows at once that the derivative $\partial A_t/\partial t_j$ exists in $\mathfrak{A}_0$ and is continuous on $\mathbb{R}^n$. Moreover one calculates that

(7.14) $$iD_tA_t = -A_t + A_t + A_{(\alpha),t} = A_{(\alpha),t} ,$$

q.e.d.

Returning to the proof of Theorem 7.1 it now is found that the expression J of (7.9) is a linear combination of expressions

(7.15) $$\Lambda^{+21}\int d\kappa\; e^{i\kappa D}c^{(\alpha)}(\kappa)\, A_{(\theta+\beta),-\kappa\tau}\tau^{|\beta|} , \quad |\alpha| + |\beta| \le 21 .$$

Here $c^{(\alpha)} \in L^1$ for sufficiently large N, by construction. Therefore it follows that the terms (7.15) all are of the form $\Lambda^{21}P$ with an operator $P \in \mathfrak{A}_0$. In fact, this process now may be repeated, creating a power Λ^{21} on the other side: Write the operator P as

(7.16) $$\int d\kappa\; c^{(\alpha)}(\kappa)\, A_{(\theta+\beta),+\kappa t}\, e^{i\kappa D} ,$$

where again we may use (7.8) and integrate by parts, use Lemma 7.2, etc. As a conclusion we find the following.

Lemma 7.3. For every integer $l = 0,1,2,\ldots$ there exists $N_0(l)$ such that for $N \geq N_0(l)$ the remainder

$$(7.17)\qquad R_N = b(D)A - \sum_{|\theta|\leq N} i^{-|\theta|}/\theta!\; A_{(\theta)} b^{(\theta)}(D)$$

can be written in the form

$$(7.18)\qquad R_N = \Lambda^{2l} \varrho_{N,l}\, \Lambda^{2l}\;, \qquad \varrho_{N,l} \in \mathfrak{A}_0\;.$$

Next, in the proof of Theorem 7.1 we use Lemma 5.3 and Lemma 7.3 for the special function $b(\xi) = \lambda^s(\xi) = (1+\xi^2)^{-s/2}$. Similarly as in III, (9.5) one concludes that

$$(7.19)\qquad \Lambda^s A\, \Lambda^{-s} - A = \sum_{1\leq|\theta|\leq N} i^{-|\theta|}/\theta!\; A_{(\theta)}\mu_\theta(D) + \Lambda^{2l}\varrho_{N,l}\, \Lambda^{2l-s}\;.$$

Looking at the symbols of operators occuring it is clear that the first sum is compact, because $a_{(\theta)} = 0$ at $|x| = \infty$ and $\mu_\theta = 0$ at $|\xi| = \infty$. Also the remainder term proves to be compact, because $\varrho_{N,l}$ is composed of integrals involving an integrand with the factor $A_{(\gamma),s}\Lambda^\varepsilon \in K_0$, for the above reason. Thus it follows that the assumption of Lemma 5.3 is satisfied and we get $A \in \mathfrak{A}_\infty$, as stated in Theorem 7.1.

Now the operators $A_{(\theta)}$ must be in $\mathfrak{A}_\infty$ for exactly the same reason: One will get Lemma 7.3 for $A_{(\alpha)}$ instead of A with $A_{(\theta)}$ replaced by $A_{(\alpha+\theta)}$, etc. But we have $\mathfrak{A}_\infty \subset \mathcal{O}(0)$ so that we also get $A_{(\theta)} \in \mathcal{O}(0)$. Thereafter the asymptotic expansion $(7.1)_1$ makes sense formally because we now see that the terms are in $\mathcal{O}(\infty)$ with order tending to $-\infty$. Furthermore the asymptotic expansion (i.e. the proper estimate of the remainder in $\mathcal{O}(\infty)$) then is an immediate consequence of Lemma 7.3 again, using the same technique as earlier in the proof of III, Theorem 6.4, q.e.d.

For our discussion of the Fourier kernel product (7.4) we were using the lemma below.

Lemma 7.4. Let $Q \in L(\mathcal{G}_s, \mathcal{G}_t)$, for a given fixed finite pair (s,t). Also assume that the function $a \in C^\infty(\mathbb{R}^{2n})$ allows an integral representation (with some domain $\Omega \subset \mathbb{R}^m$, for simplicity)

$$(7.20)\qquad a(\xi,\eta) = \int_\Omega \overline{\lambda}(\xi,\tau)\mu(\eta,\tau)d\kappa(\tau)$$

with a bounded measure $d\kappa$ over Ω and with bounded continuous functions λ, μ over $\Omega\times\mathbb{R}^n$, having also bounded continuous derivatives $D_\xi^\alpha\lambda$ and $D_\eta^\beta\mu$ (over $\Omega\times\mathbb{R}^n$). Then the Fourier kernel product $R = a\text{▴}Q$ is well defined and is in $L(\mathcal{G}_s, \mathcal{G}_t)$ again.

Proof. It follows that (for $u,v \in \mathcal{D}(\mathbb{R}^n)$ we get)

(7.21) $$\langle aq, \overline{u}\otimes v\rangle = \langle q, a(\overline{u}\otimes v)\rangle = \int_\Omega d\kappa(\tau) \ \langle q, \overline{f}_\tau \otimes g_\tau\rangle$$

with

(7.22) $$f_\tau(\xi) = u(\xi)\lambda(\xi,\tau), \qquad g_\tau(\xi) = v(\xi)\mu(\xi,\tau) .$$

Using the properties of u and v (to be in C_0^∞) and those of λ,μ one concludes that the integral composing $a(\overline{u}\otimes v)$ (as in (7.20)) converges in $\mathcal{D}(\mathbb{R}^{2n})$ and thus may be pulled out of the distribution. Accordingly we get

(7.23) $$\begin{aligned} |\langle aq, \overline{u}\otimes v\rangle| &= \left|\int_\Omega d\kappa(\tau)(f_\tau, \varrho\, g_\tau)\right| \\ &\leq \left(\int_\Omega d\kappa\right) \sup_{\tau,\xi} \{|\lambda|,|\mu|\}^2 \|u^\vee\|_{-t} \|v^\vee\|_s \|\varrho\|_{s,t} . \end{aligned}$$

Using the Frechet Riesz theorem one concludes that there exists a bounded operator $P: \mathfrak{H}_s \to \mathfrak{H}_t$ such that

(7.24) $$\langle aq, \overline{u}\otimes v\rangle = (u^\vee, Pv^\vee) = (u, \overset{\vee}{P} v), \qquad u,v \in C_0^\infty(\mathbb{R}^n),$$

q.e.d.

8. More smoothness conditions for the symbol.

We have observed earlier that the condition $a \in C^{[k]}$ or $a \in C^{[\infty]}$ does not amount to any restriction at all besides continuity for the behaviour of $a(x,\xi)$ over the secondary symbol space $|\xi| < \infty$. For reason of symmetry we get a dual of Theorem 6.1 which refers to commutators

(8.1) $$(\mathrm{ad}\, M)^{\alpha} = (\mathrm{ad}\, M_j)^{\alpha_j} , \quad \mathrm{ad}\, M_j\, A = M_j\, A - A\, M_j \ ,$$

with the multiplication operator $(M_j u)(x) = x_j u(x)$. The operator $(\mathrm{ad}\, M)^{\alpha}$ is regarded to map from $\mathcal{H}_{\infty} = \bigcap_s L^2(\mathbb{R}^n, d\mu_s)$ to $\bigcup_s L^2(\mathbb{R}^n, d\mu_s) = \mathcal{H}_{-\infty}$, with the measures $d\mu_s = (1+x^2)^s dx$, as before in III, 1.

We define ξ-derivatives

(8.2) $$a^{(\alpha)} = i^{|\alpha|} D_{\xi}^{\alpha} a \ \text{ on } \mathbb{M}_s \ , \quad = 0 \ \text{ on } \mathbb{M}_p$$

and denote the collection of $a \in C(\mathbb{M})$ with $a^{(\alpha)} \in C(\mathbb{M})$ for all $[\alpha] \leq k$ by $C^{<k>}(\mathbb{M})$. The proof of Theorem 8.1, below is completely parallel to the proof of Theorem 6.1, and therefore will not be presented.

Theorem 8.1. Let $k \leq \infty$, and let $a \in C^{<k>}(\mathbb{M})$. There exists a collection $A = A^{(0)} \in \mathfrak{A}_0$, $A^{(\alpha)} \in \mathfrak{A}_0$, $[\alpha] \leq k$, such that

(8.3) $$\sigma_{A^{(\alpha)}} = a^{(\alpha)} , \quad i^{-|\alpha|} (\mathrm{ad}\, M)^{\alpha} A = A^{(\alpha)} , \quad [\alpha] \leq k \ .$$

The only change in the proof of Theorem 6.1 is that we must set up an analogous of Lemma 6.2 involving the operators $e^{iM_j t}$ instead of $e^{iD_j t}$, which will involve a simple Lemma analogous to Lemma 5.2, to show that conjugation with $e^{iM_j t}$ translates the symbol in the ξ_j-direction. Accordingly the integral of (6.11) must be replaced by

(8.4) $$F_j(A) = \int_0^{\infty} e^{-t} e^{iM_j t} A\, e^{-iM_j t} dt \ ,$$

the symbol of which is calculated as

(8.5) $$f_j = \int_0^{\infty} a(x, \xi - te_j)\, e^{-t} dt \ .$$

While we leave the remainder of the proof to the reader it may be of interest to note that the two kinds of operators $E_j \in L(\mathfrak{A}_0)$ and $F_l \in L(\mathfrak{A}_0)$ above commute (with E_j defined by (6.11)). Indeed one observes that

(8.6) $$e^{iD_j t} e^{iM_l \tau} e^{-iD_j t} = e^{it\tau} e^{iM_l \tau} \ .$$

Or,

(8.7) $$e^{iD_j t} e^{iM_l \tau} = \mu_{t\tau} e^{iM_l \tau} e^{iD_j t} , \quad \mu_{t\tau} = e^{it\tau} ,$$

which gives

$$e^{-iD_j t} e^{iM_l \tau} A e^{-iM_l \tau} e^{iD_j t} = e^{iM_l \tau} e^{-iD_j t} A e^{iD_j t} e^{-iM_l \tau} , \tag{8.8}$$

because the two scalar factors generated on either side of A cancel each other. That indeed shows that $E_j F_l = F_l E_j$.

This fact immediatedly results in another theorem, below.

Theorem 8.2. Let the space of all $a \in C(\mathbf{M})$ with $a_{(\alpha)} \in C(M)$, $a^{(\alpha)} \in C(\mathbf{M})$ for all $[\alpha] \le k$ be denoted by $C^{\{k\}}(\mathbf{M})$. Then, for $k < \infty$ and $a \in C^{\{k\}}$ there exists a collection $A^{(\alpha)}{}_{(\beta)} = A_{(\beta)}{}^{(\alpha)} \in \mathfrak{A}_o$, with $A_{(0)}{}^{(0)} = A$ such that

$$\sigma_{A^{(\alpha)}{}_{(\beta)}} = a^{(\alpha)}{}_{(\beta)}, \quad A^{(\alpha)}{}_{(\beta)} = i^{|\beta|-|\alpha|} (\mathrm{ad}\ D)^{\alpha} (\mathrm{ad}\ M)^{\beta} A, \quad [\alpha],[\beta] \le k. \tag{8.9}$$

Again we leave the proof to the reader as an exercise. Also we do not care to carry out a version of Theorem 8.2 for $k=\infty$.

9. Derivatives of a linear operator in $L(S,S')$.

In this section we come back to the discussion started in section 6 where we investigated the existence of commutators $(\mathrm{ad}\ D)^{\alpha}A$, for an operator $A \in \mathfrak{A}_0$. We found the existence of such commutators as operators in $\mathfrak{A}_0$ quite basically linked to the existence of the corresponding derivative $a_{(\alpha)}$ of the symbol $a = \sigma_A$. In fact if $(\mathrm{ad}\ D)^{\alpha}A \in \mathfrak{A}_0$ for all α then it was seen that A must be in $\mathfrak{A}_{\infty}$ and that the symbol a of A must be in $C^{\infty}(\mathbb{M})$, and we obtained a Leibnitz formula valid for commutation of A with a Fourier multiplier $b(D)$. Also in section 8 we were observing similar links between the commutators $(\mathrm{ad}\ M)^{\alpha}A$ and the ξ-derivatives $a^{(\alpha)}$ of the symbol.

In the following we shall deal more generally with the operators

$$(9.1)\qquad \begin{aligned} A^{(\alpha)}_{(\beta)} &= i^{|\beta|-|\alpha|}(\mathrm{ad}\ D)^{(\beta)}(\mathrm{ad}\ M)^{(\alpha)}A \\ &= i^{|\beta|-|\alpha|}(\mathrm{ad}\ M)^{(\alpha)}(\mathrm{ad}\ D)^{(\beta)}A \end{aligned}$$

where α,β are arbitrary multi-indices. Let $A \in L(S,S')$, and let us assume the same formal attitude as in sections 4,5,6 by interpreting the operators D_j and M_j as maps either $S \to S$ or $S' \to S'$ depending on whether they are applied before or after application of A. This will make the successive commutators formally well defined but it is to be kept in mind that D_j and M_j really are two different operators depending on their application before of after A.

We observe that the operations $\mathrm{ad}\ D_j$ and $\mathrm{ad}\ M_l$ commute, even though the operators D_j and M_l do not necessarily commute: We get the Heisenberg commutator relation

$$(9.2)\qquad [D_j,M_l] = -i\delta_{jl} \quad ,$$

but we also have the formal relation

$$(9.3)\qquad [\mathrm{ad}\ P,\ \mathrm{ad}\ Q] = \mathrm{ad}\ [P,Q]$$

valid for arbitrary operators P and Q, so that

$$(9.4)\qquad [\mathrm{ad}\ D_j,\ \mathrm{ad}\ M_l] = \mathrm{ad}\ \delta_{jl} = 0,$$

since $\mathrm{ad}\ I = 0$ (with the identity operator I). Accordingly we do not have to pay attention to the order of the various commutations in (9.1).

<u>Definition 9.1.</u> The operators $A^{(\alpha)}_{(\beta)}$ of (9.1) will be called the <u>(formal) derivatives</u> of the operator $A \in L(S,S')$. If for $A \in L(\mathfrak{H}_s,\mathfrak{H}_t)$ or $A \in O(r)$ the derivative $A^{(\alpha)}_{(\beta)}$ is in $L(\mathfrak{H}_s,\mathfrak{H}_t)$ [or $O(r)$] again. We will express this by saying that $A^{(\alpha)}_{(\beta)}$ <u>exists</u> (in $L(\mathfrak{H}_s,\mathfrak{H}_t)$) [or in $O(r)$, respectively] .

Notice that the formal derivative obeys Leibnitz formula again - only that we have to pay attention to the order in the products:

(9.5) $$(AB)^{(\alpha)}_{(\beta)} = \sum_{\gamma\le\alpha,\,\delta\le\beta} \binom{\alpha}{\gamma}\binom{\beta}{\delta} A^{(\gamma)}_{(\delta)} B^{(\alpha-\gamma)}_{(\beta-\delta)} .$$

Theorem 9.1. The collection of all $A \in \mathfrak{A}_0$ with $A_{(\alpha)} \in \mathfrak{A}_0$ for all α forms a subalgebra $\mathcal{L}_\infty$ of $\mathfrak{A}_0$, which is contained in $\mathfrak{A}_\infty$ as well. We have

(9.6) $$\sigma_{A_{(\alpha)}} = a_{(\alpha)} \quad \text{with} \quad a = \sigma_A ,$$

with the x-derivative $a_{(\alpha)}$ as defined in section 6.

Proof. The fact that $\mathcal{L}_\infty$ is an algebra is an immediate consequence of Leibnitz formula (9.5). Also (9.6) is a consequence of Theorem 9.2, below. Then it follows from Theorem 7.1 that $\mathcal{L}_\infty \subset \mathfrak{A}_\infty$.

Theorem 9.2. Let $A \in L(\mathfrak{H}_0)$ have all its derivatives $A^{(\alpha)}_{(\beta)}$ existing in $L(\mathfrak{H}_0)$. Then the function

(9.7) $$A(s,t) = e^{iDs}e^{-iMt}Ae^{+iMt}e^{-iDs} = e^{-iMt}e^{iDs}Ae^{-iDs}e^{+iMt}$$

is in $C^\infty(\mathbb{R}^n \times \mathbb{R}^n, L(\mathfrak{H}_0))$ [that is all partial derivatives of A(s,t) exist in norm convergence of $L(\mathfrak{H}_0)$] . Moreover, we have

(9.8) $$i^{|\alpha+\beta|}D_s^\alpha D_t^\beta A(s,t) = A^{(\beta)}_{(\alpha)}(s,t) = e^{iDs}e^{-iMt}A^{(\beta)}_{(\alpha)}e^{-iDs}e^{iMt} .$$

Vice versa, if the function A(s,t) of (9.7) is in $C^\infty(\mathbb{R}^n \times \mathbb{R}^n, L(\mathfrak{H}_0))$, then all derivatives $A^{(\alpha)}_{(\beta)}$ of A exist in $L(\mathfrak{H}_0)$.

Proof. First it is confirmed that the two expressions in (9.7) indeed define the same function: one finds that

(9.9) $$e^{iDs}e^{-iMt} = e^{-ist}e^{-iMt}e^{iDs}$$

but the factors e^{-ist} and e^{ist} on both sides of A cancel each other. Next one confirms that the successive difference quotients for the 2n variables $s_1,\dots,s_n$, $t_1,\dots,t_n$ converge in norm if and only if the corresponding commutator exists as an operator in $L(\mathfrak{H}_0)$, q.e.d.

Let the collection of all $A \in L(\mathfrak{H}_0)$ satisfying the assumptions of Theorem 9.2 be denoted by $\mathcal{L}_\infty$. Clearly $\mathcal{L}_\infty$ is a topological algebra with the Frechet topology induced by the norms

(9.10) $$\|A\|_{0,\alpha,\beta} = \|A^{(\alpha)}_{(\beta)}\|_0 .$$

Theorem 9.3. Let $A \in O(\infty)$ be such that the derivatives $A^{(\alpha)}$, with $[\alpha] \leq 1$ (c.f. section 6) are in $O(\infty)$. Then there exists a function $a \in C(\mathbb{R}^n \times \mathbb{R}^n)$ such that $a_{(\alpha)} \in C(\mathbb{R}^n \times \mathbb{R}^n)$ for all α and that $a(x,\eta) = O((1+|\eta|)^r)$ on compact x-sets for some r, and such that

$$(9.11) \qquad Au(x) = \int_{\mathbb{R}^n} a(x,\xi)e^{ix\xi}u^\wedge(\xi)d\!\!\!/\xi \quad , \quad u \in \mathcal{S}_\infty .$$

Proof. An operator $A \in O(\infty)$ is in some $O(r)$, for $r \in \mathbb{R}$. Let us ask the question about a possible construction of a with above properties if its existence is known. Formally we get

$$(9.12) \qquad \begin{aligned} AM_j u(x) &= \int a(x,\xi)e^{ix\xi}(x_j u)^\wedge(\xi)d\!\!\!/\xi \\ &= - \int a(x,\xi)e^{ix\xi}D_{\xi_j} u^\wedge(\xi)d\!\!\!/\xi \\ &= M_j Au(x) - i\int a_{\xi_j}(x,\xi)e^{ix\xi}u^\wedge(\xi)d\!\!\!/\xi \end{aligned}$$

which amounts to

$$(9.13) \qquad (A - i[M_j,A])u = \int (1+\partial/\partial\xi_j)a(x,\xi)e^{ix\xi}u^\wedge(\xi)d\!\!\!/\xi$$

and it is clear that the left hand side operator equals $A+A^{(e_j)} = B_j$ Let us introduce the operator B resulting from n such transformations of A, for $j=1,\dots,n$. Clearly we get

$$(9.14) \qquad B = \sum_{[\alpha]\leq 1} c_\alpha A^{(\alpha)}$$

with certain constants c_α. Also we get

$$(9.15) \qquad Bu(x) = \int b(x,\xi)e^{ix\xi}u^\wedge(\xi)d\!\!\!/\xi$$

with the function

$$(9.16) \qquad b(x,\xi) = \prod_{j=1}^{n} (1 + \partial/\xi_j)a(x,\xi) \quad .$$

We repeat that the above is intended formally, it only will serve to obtain a formula for the function $a(x,\xi)$ which then must be used to verify the result. In particular we will not have to involve derivatives of a for ξ, although formally these were used above.

Let us define the function $u = u_{z,\eta}(x)$ as inverse Fourier transform of $e^{-iz\xi}g(\xi,\eta)$ with respect to the variable ξ, for fixed z and η, where

$$(9.17) \qquad g(\xi,\eta) = \prod_{j=1}^{n} e^{\xi_j-\eta_j} H(\eta_j-\xi_j) \quad ,$$

with the Heaviside function $H(\tau) = 0,\ \tau \leq 0,\ H(\tau) = 1,\ \tau > 0$. If $u = u_{z,\eta}$ is substituted into (9.15) we get

$$(9.18) \qquad Bu_{z,\eta}(x) = \int d\!\!\!/\xi\, b(x,\xi)e^{i\xi(x-z)}g(\xi,\eta) \ .$$

Now it must be observed that the function g for fixed η is in every $L^2(\mathbb{R}^n, d\mu_s) = L^2_s$, $s \in \mathbb{R}$, and uniformly so on compact η-sets. Since the exponential function $e^{-i\xi z}$ is bounded and because we have $L^2_s = \mathcal{G}^\wedge_s$ (c.f. III, sections 1 and 2) it follows that $u = u_{z,\eta}$ above is in $\mathcal{G}_\infty$ for fixed z and η. Moreover all z-derivatives of u evidently are in $C(\mathbb{R}^n \times \mathbb{R}^n, \mathcal{G}_\infty)$, with respect to the 2n variables z and η. In particular a differentiation for z will only produce a factor ξ for the Fourier transform which is readily absorbed by the exponential decay of the function g. Furthermore the Sobolev norms of u and $D^\alpha_z u$ may be estimated:

$$(9.19) \qquad \| D^\alpha_z u_{z,\eta} \|_r = \| \xi^\alpha g(\cdot,\eta) \|_{L^2_r} \leq c_r (1 + |\eta|)^{s(r)}$$

with real-valued s(r) which can be explicitly determined by looking at the integral

$$(9.20) \qquad \left\{ \prod_{j=1}^n e^{-2\eta_j} \int_{\xi \leq \eta} \xi^{2\alpha} \prod_{j=1}^n e^{2\xi_j} (1 + \xi^2)^r d\xi \right\}^{1/2}$$

It therefore is concluded that $w_{z,\eta} = Bu_{z,\eta} \in C(\mathbb{R}^n \times \mathbb{R}^n, \mathcal{G}_s)$ for some $s \in \mathbb{R}$, because $B \in \mathcal{O}(\infty)$, by (9.14) and the assumptions of the theorem, so that B must be in some $\mathcal{O}(t)$. Similarly for $D^\alpha_z w_{z,\eta} = B(D^\alpha_z u_{z,\eta})$. Now by Sobolev's Lemma (III, Theorem 1.2) this implies that $D^\alpha_z w_{z,\eta} \in C(\mathbb{R}^n \times \mathbb{R}^n, CO^\infty(\mathbb{R}^n))$. Moreover we get

$$(9.21) \qquad \| D^\alpha_z w_{z,\eta} \|_{C^k(\mathbb{R}^n)} \leq c \, \| D^\alpha_z u_{z,\eta} \|_s = O((1+|\eta|)^{t(s)}) ,$$

as follows from a combination of (9.19) with the fact that B has an order, and with the estimate of Theorem 1.2 of Chapter III. Also we have the complex-valued function

$$(9.22) \qquad c(x,z,\eta) = w_{z,\eta}(x) \quad , \quad x,z,\eta \in \mathbb{R}^n$$

well defined and continuous in all 3 arguments, and

$$(9.23) \qquad D^\alpha_x D^\beta_z c(x,z,\eta) = O((1 + |\eta|)^t)$$

uniformly in x and z, with t depending on α and β.

Notice that this function c of (9.22) is well defined even if b in (9.16) has no meaning, assuming only the requirements of Theorem 9.3. On the other hand since the function c is continuous it is meaningful to set $x = z$ in formula (9.18). We notice that

$$(9.24) \qquad \int_{-\infty}^{\tau} e^{\rho-\tau} (1 + \partial/\partial\rho) p(\rho) d\rho = p(\tau)$$

under reasonable assumptions on the function $p(\tau)$. Accordingly formula (9.18) with $x = z$ formally yields

$$(9.25) \qquad \begin{aligned} Bu_{x,\eta}(x) &= \int d\!\!{}^-\xi \; b(x,\xi) g(\xi,\eta) \\ &= \int_{\xi_j \leq \eta_j} \prod_{j=1}^n (1+\partial/\partial\xi_j) a(x,\xi) \prod_{j=1}^n e^{\xi_j - \eta_j} d\!\!{}^-\xi = (2\pi)^{-n/2} a(x,\xi) . \end{aligned}$$

This clarifies our choice of the function a: We must chose

$$a(x,\eta) = (2\pi)^{n/2}\, Bu_{x,\eta}(x) = (2\pi)^{n/2} c(x,x,\eta)\ . \tag{9.26}$$

We already have seen that this choice is possible under the assumptions of the theorem, and that it yields a function $a(x,\eta)$ with all the properties stated.

Next it is observed that the linear expression

$$Qv(x) = \int d\!\!{}^-\eta\ a(x,\eta)\, e^{ix\eta}\, v^\wedge(\eta) \tag{9.27}$$

is meaningful for all $v \in S$. In fact it is evident that this is a continuous map $S \to C(\mathbb{R}^n)$, and it is not hard to confirm that Q maps from S to T, for example, although we will not offer details since this is not required here.

For our theorem it remains to be shown that Q coincides with A on S or on some suitable dense subspace of S. First we calculate explicitly the function $u_{z,\eta}(x)$ from its Fourier transform:

$$\begin{aligned} u_{z,\eta}(x) &= \int e^{i\xi(x-z)} g(\xi,\eta) d\!\!{}^-\eta \\ &= (2\pi)^{-n/2} \prod_{j=1}^{n} \left\{ e^{-\eta_j} \int_{-\infty}^{\eta_j} e^{\xi_j(1+i(x_j-z_j))} d\xi_j \right\} \\ &= (2\pi)^{-n/2}\, e^{i\eta(x-z)} \prod_{j=1}^{n} (1+i(x_j-z_j))^{-1}\ . \end{aligned} \tag{9.28}$$

First we look at the integral

$$\begin{aligned} I(x,z) &= \int d\eta\ c(x,z,\eta) e^{i\eta x} v^\wedge(\eta) = \int d\eta B u_{z,\eta}(x) e^{ix\eta} v^\wedge(\eta) \\ &= \int d\!\!{}^-\eta\ e^{ix\eta} v^\wedge(\eta) B(e^{i\eta(x-z)}/k_z(x)), \quad k_z(x) = \prod_{j=1}^{n} (1+i(x_j-z_j)). \end{aligned} \tag{9.29}$$

where the operator x under the integral sign acts on the variable x. Here it seems useful to come back to the concept of kernel products as introduced in I,9. This, however is understood with the difference that now we consider the ordinary kernel product, not the Fourier kernel product as in section 7, (c.f. I,9).

$$B = k(y-x) \triangledown A, \qquad k(z) = k_0(z) = \prod_1^n (1+iz_j) \tag{9.30}$$

with A,B as in (9.14) .

Then it becomes evident, using the laws of I,9 that

$$Qv(z) = (\delta_z, (\psi_z \triangledown A) 1_z), \quad \psi_z(x,y) = v(x+y-z) k(y-x), \quad \delta_z(x) = \delta(x-z), \tag{9.31}$$

with $1_z = 1/k_z$. Now it is only necessary to write down the corresponding distribution equation. Observe that $\delta_z \in \mathcal{G}_{-n} \subset \mathcal{G}_{-\infty}$, and that $1_z \in \mathcal{G}_\infty$ (Calculate the derivatives). Accordingly the expression at right of (9.31) is meaningful, and we get (with the distribution kernel of A denoted by q_A):

(9.32) $$Qv(z) = \int q_A(x,y)\delta_z(x)1_z(y)\psi_z(x,y)\,dx\,dy ,$$

with a distribution integral at right. Observe that

(9.33) $$\delta_z(x)1_z(y)\psi_z(x,y) = v(y)\delta_z(x) ,$$

so that we get

(9.34) $$Qv(z) = (\delta_z, Av) = Av(z) , \text{ q.e.d.}$$

Remark: For a more precise discussion one should replace δ_z above by a family $\delta_{\varepsilon,z}$ of C_0^∞-functions converging to δ_z in $\mathfrak{H}_{-n}$ and then observe that for $v \in S$ we get

(9.35) $$\int \delta_{\varepsilon,z}(x)(1_z(y)\psi_z(x,y) - v(y))q_A(x,y)dx\,dy \to 0, \quad \varepsilon \to 0 .$$

However, it is readily seen that for any $A \in \mathcal{O}(\infty)$ we can write

(9.36) $$A = C(1 - \Delta)^p ,$$

where p is a (sufficiently large) integer, while the distribution kernel of C is a bounded and continuous function of x and y. This may be used in (9.35) to write the left hand expression as

(9.37) $$\int \delta_{\varepsilon,z}(x)(1-\Delta_y)^p \{(1_z(y)\psi_z(x,y) - v(y))\} c(x,y)dx\,dy$$

which now indeed has a Riemann integral instead of a distribution integral and is readily seen to tend to zero as $\varepsilon \to 0$.

10. About infinitely differentiable operators of $L(\mathcal{G}_0)$.

In the following let $\mathcal{D}_\infty$ denote the algebra of $A \in L(\mathcal{G}_0)$ with $A^{(\alpha)}{}_{(\beta)} \in L(\mathcal{G}_0)$ for all α, β. The discussion, below, is formal, but can easily be made rigorous.

Theorem 10.1. An operator $A \in L(\mathcal{G}_0)$ is in $\mathcal{D}_\infty$ (or, is infinitely differentiable) if and only if there exists a $C^\infty(\mathbb{R}^{2n})$-function $a(x,\xi)$, called the MD-symbolof A which is bounded with all its derivatives $a^{(\alpha)}{}_{(\beta)}$ over $\mathbb{R}^{2n}$, and such that

$$(10.1) \qquad A^{(\alpha)}{}_{(\beta)}u(x) = \int a^{(\alpha)}{}_{(\beta)}(x,\xi)\, e^{ix\xi} u^\wedge(\xi) d\!\!\!/\xi \,, \qquad u \in \mathcal{G}_\infty.$$

Proof. We first establish (10.1) for $\alpha = \beta = 0$. To that extent we proceed somewhat similar to the proof of theorem 9.3. First let us ask how we can find a if it exists. Let us define

$$(10.2) \qquad b(x,\xi) = \prod_{j=1}^{n} ((2 + \partial/\partial x_j)^2 (2 + \partial/\partial \xi_j)^2) a(x,\xi)$$

and correspondingly the operator

$$(10.3) \qquad B = \prod_{j=1}^{n} ((1 + \text{ad } D_j)^2 (1 - i \text{ ad } M_j)^2) A \quad .$$

If a with the properties stated exists then we get

$$(10.4) \qquad Bu(x) = \int b(x,\xi) e^{ix\xi} u^\wedge(\xi) d\!\!\!/\xi \quad , \quad u \in \mathcal{G}_\infty \,,$$

as readily seen, following a verification proceedure similar to that in section 4. Moreover, we have

$$(10.5) \qquad a(y,\eta) = \int dx\, d\!\!\!/\xi\, b(x,\xi) e^{ix\xi} (e^{-ix\xi} g_2(x,y) g_2(\xi,\eta))$$

where g_2 must be chosen as the Greens function of the differential operator occurring in (10.2).

First we will calculate g_2 in details. Then we must show that the right hand side can be expressed by B only, using a suitable kernel product. This will give a formula for $a(x,\xi)$ which then should be verified in details and should be used to derive the properties stated.

Regarding construction of g we observe that

$$(10.6) \qquad \int_{-\infty}^{\tau} e^{2(\rho-\tau)} (\rho-\tau)(2 + \partial/\partial\rho) p(\rho)\, d\rho = p(\tau)$$

for bounded continuous $p(\rho)$, for example. This means that we must chose $g_2(x,y) = 0$ if $x_j \geq y_j$ for at least one j, and

$$(10.7) \qquad g_2(x,y) = \prod_{j=1}^{n} (x_j - y_j) e^{2(x_j - y_j)} \,, \quad x_j < y_j,\ j=1,\dots,n.$$

Next we will write

$$\phi_{y,\eta}(x,\xi) \quad = \quad e^{-ix\xi} g_1(x,y) g_1(\xi,\eta) \tag{10.8}$$

with

$$g_1(x,y) \quad = \quad \prod_{j=1}^{n} (x_j - y_j) e^{(x_j - y_j)} \quad ,\text{or} \quad = 0 \ , \ \text{resp.} \tag{10.9}$$

This function $\phi_{y,\eta}$ clearly is in $L^1(\mathbb{R}^{2n})$, keeping y,η fixed. Thus we may calculate its Fourier transform by a Riemann Integral:

Lemma 10.2. The Fourier transform

$$\psi_{y,\eta}(s,t) \quad = \int đx\, đ\xi \ e^{-i(tx+s\xi)} \phi_{y,\eta}(x,\xi) \tag{10.10}$$

is a function of the 2n variables y+s and ξ+t only; it may be written as

$$\psi_{y,\eta}(s,t) \quad = \quad \mu(y+s, \xi+t) \ , \tag{10.11}$$

where the function μ is in $L^1(\mathbb{R}^{2n}) \cap C^\infty(\mathbb{R}^{2n})$.

Proof. A simple calculation shows that

$$\psi_{y,\eta}(s,t) \quad = \quad e^{ist} \int đx\, đ\xi \ e^{-i(x+s)(\xi+t)} g_1(x,y) g_1(\xi,\eta) \ . \tag{10.12}$$

But $g_1(x,y)$ is a function of the difference x-y only, by definition (10.9). Therefore we conclude that

$$\psi_{y,\eta}(s,t) \quad = \quad e^{ist} \ \psi_{y+s,\eta+t}(0,0) \quad , \tag{10.13}$$

by a simple substitution of integration variable. On other words, we have confirmed (5.10) with

$$\mu(s,t) \quad = \quad \psi_{s,t}(0,0) \ , \tag{10.14}$$

if only we can show that this function is $C^\infty \cap L^1$, as required. On the other hand it also is clear that the function (10.14) is a product of n factors, each of them depending only on one pair of variables s_j, t_j, for fixed j. It thus is sufficient to show these factors to be in $L^1(\mathbb{R}^2) \cap C^\infty(\mathbb{R}^2)$. All n of them are of the same form, represented by

$$\int_{-\infty}^{s} d\rho\,(\rho-s) e^{\rho-s} \int_{-\infty}^{t} dr (r-t) e^{r-t} e^{ir\rho} \quad = \quad \int_{-\infty}^{0} \rho e^{\rho} d\rho \int_{-\infty}^{0} r e^{r} dr \ e^{i(r+t)(\rho+s)} \ . \tag{10.15}$$

From here it is readily seen that these factors are $C^\infty(\mathbb{R}^2)$. We still calculate the right hand side of (10.15) to equal

$$\begin{aligned} & e^{ist} \int_{-\infty}^{0} \rho d\rho \ e^{\rho(1+it)} \int_{-\infty}^{0} r dr \ e^{r(1+i(\rho+s))} \\ &= \quad e^{ist} \int_{-\infty}^{0} \rho d\rho \ e^{\rho(1+it)} / (1+i(\rho+s))^2 \ . \end{aligned} \tag{10.16}$$

Now a two-fold partial integration yields (with e^{ist} omitted)

$$(10.17) \qquad = -(1+is)^{-2}(1+it)^{-2} + \int_{-\infty}^{0} d\rho \; e^{\rho(1+it)}(1+it)^{-2} d^2/d\rho^2 \; \{\rho(1+i(\rho+s))^{-2}\}$$

It is evident that this expression is $O((1+s^2)^{-1}(1+t^2)^{-1})$, and therefore $L^1(\mathbb{R}^2)$, q.e.d.

Returning to the proof of the Theorem we now find that the right choice of $a(x,\xi)$ must be given by

$$(10.18) \qquad a(y,\eta) = \int đs \; đt \; e^{ist}\mu(y+s,\xi+t)(\overline{f}_{y,t}, Bf_{\eta,s})$$

with B as in (5.4) and $f_{y,t}(x) = g(x,y)e^{ixt}$, where g is as in section 4, that is $g(x,y) = 0$ for $x_j > y_j$ and $g(x,y) = \prod_1^n e^{x_j - y_j}$ elsewhere. It is clear that

$$(10.19) \qquad \| f_{y,t} \|_0 = 1 .$$

Therefore we get a uniformly bounded:

$$(10.20) \qquad |a(x,\xi)| \le c \; \|B\|_0 , \qquad c = \int |\mu(s,t)| đs \; đt .$$

Clearly also $a \in C(\mathbb{R}^{2n})$. In fact, alittle calculation shows that $a \in C^\infty(\mathbb{R}^{2n})$.

Now, the uniqueness of such function a is quite evident from relation (10.1) (assuming that it is verified.). We may repeat the same procedure with the derivatives $B^{(\alpha)}_{(\beta)}$ which clearly exist under the assumptions of the Theorem (and may be expressed as finite linear combinations of the $A^{(\alpha)}_{(\beta)}$.). In this manner we obtain functions $a^{(\alpha)}_{(\beta)}(x,\xi)$ which all are bounded but still must be shown to be the derivatives of a defined by (10.18).

References

[1] H.O. Cordes, Banach algebras, singular integral operators and partial differential equations; Lecture Notes, Lund (1971)

[2] --- , A global parametrix for pseudo-differential operators over $\mathbb{R}^n$, with Applications. Bonner Mathematische Schriften (No. 90, SFB 72 (1976)).

[3] --- and E. Herman, Gelfand theory of pseudo-differential operators. Amer.J.Math. 90 (1968) 681-717.

[4] --- and R.C. Mc. Owen, The C^*-algebra of a singular elliptic problem on a non-compact Riemannian manifold, Math. Zeitschrift 153, 101-116 (1977).

[5] --- --- , Remarks on singular elliptic theory for complete Riemannian manifolds; Pacific J. Math. 77 (1977).

[6] P. Dirac, Principles of Quantum Mechanics; Oxford 1949.

[7] J. Dixmier, Les C^*-algèbres et leurs representations; Gauthier Villars, Paris (1964).

[8] R.E. Edwards, Functional analysis, theory and applications; Holt Rinehart and Winston, New York 1965.

[9] L. Garding, Dirichlet's problem for linear elliptic partial differential operators. Math.Scand. 1,55-72 (1953).

[10] E. Herman, The symbol of the algebra of singular integral operators; Journ.Math.Mech. 15 (1966) 147-156.

[11] D. Hilbert, Grundzüge einer allgemeinen Theorie der Linearen Integralgleichungen; Chelsea Publ.Comp. New York, N.Y. 1953.

[12] F. Riesz, Über Lineare Funktionalgleichungen; Acta Math. 41 (1918) 71-98.

[13] C.E. Rickart, General theory of Banach algebras; van Nostrand, Princeton 1960.

[14] M. Taylor, Gelfand theory of pseudo-differential operators and hypoelliptic operators. Trans.Amer.Math.Soc. 153, 495-510 (1971)

Chapter V: Elliptic Boundary Problems.

0. Introduction.

In the present chapter we shall turn to the study of elliptic boundary problems. Note that in chapter IV we were solving "singular elliptic problems", involving a differential equation $Lu = f$ over $\mathbb{R}^n$, for functions $u, f \in L^2(\mathbb{R}^n)$, typically. For an md-elliptic problem then it is found that the problem is normally solvable, with L acting from $\mathcal{H}_N$ to $\mathcal{H}_0$, for example.

A similar theory may be obtained for more general subdomains $\Omega \subset \mathbb{R}^n$, if the coefficients of L are sufficiently singular near the boundary $\Gamma = \partial\Omega$ of Ω. In fact even for an infinitely differentiable manifold Ω without boundary one may introduce a Laplace comparison algebra (or even a comparison algebra with respect to another second order differential operator H on Ω satisfying certain conditions) (c.f. [14], [19], [20], [26]).

In contrast to these results requiring no boundary conditions we presently will turn to the other case, where we have a differential equation over a proper subdomain $\Omega \subset \mathbb{R}^n$ with coefficients continuous (and even C^∞) up to the boundary. Then, to arrive at a normally solvable problem, one must impose additional restrictions on u at or near the boundary- so-called boundary conditions.

Our approach is two-fold. First, we shall consider the Dirichlet problem for a general domain $\Omega \subset \mathbb{R}^n$ and a strongly elliptic differential operator (c.f. section 2). Here we use a conventional method of apriori estimate, by first proving Gårding's inequality from the results of chapter IV.

Second we consider a general elliptic boundary problem, involving boundary conditions of Lopatinskij-Shapiro type (c.f. [24], [1], [3], [7], [8], [17]). For simplicity in that case we consider only a half space $\Omega = \mathbb{R}^{n+1}_+$ (of n+1 dimensions).

This second approach attempts the use of the same technique as in chapter III and IV, involving structure of Banach algebras. A Laplace comparison algebra $\mathfrak{A}$ for the half space is generated from certain L^2-bounded operators generated from the Dirichlet and Neumann problem of $1-\Delta$ over $\mathbb{R}^{n+1}_+ = \Omega$, together with multiplications by bounded continuous functions.

There are interesting differences to the case of $\mathbb{R}^n$ discussed in chapter IV: Commutators are not necessarily compact, if $n \geq 1$, but the commutator ideal $\mathcal{E}$ of the Laplace comparison algebra $\mathfrak{A}$ itself is a C^*-algebra without unit, allowing the definition of a compact operator valued symbol, over the space $\mathbb{B}^n \times \mathbb{B}^n - \mathbb{R}^n \times \mathbb{R}^n = \mathbb{M}_2$, with $\mathbb{B}^n$ as in IV, 1. In other words, we get

$$(0.1) \qquad \mathfrak{A}/\mathcal{E} \cong C(\mathbb{M}_1)\ , \quad \mathcal{E}/K \cong C(\mathbb{M}_2, K(\mathfrak{f})),$$

with $\mathbb{M}_2$ as above and a compact space $\mathbb{M}_1$ to be described later on (Theorem 10.3); also, with $K = K(\mathcal{H})$, $\mathcal{H} = L^2(\Omega)$ and $\mathfrak{f} = L^2((0,\infty))$. As to be expected the

operators of $\mathcal{E}$ are nontrivial - that is non-compact - only over the boundary $\Gamma = \partial\Omega$, in a sense to be specified.
Moreover, the commutator ideal $\mathcal{E}$ is unitarily equivalent to the topological tensor product $K(\mathcal{H}) \hat{\otimes} \mathfrak{A}_0^n$, with a C^*-subalgebra $\mathfrak{A}_0^n$ of the algebra $\mathfrak{A}_0$ of chapter IV having symbol space $\mathbb{M}_1$ above.

This again allows a description of the Fredholm operators of the Laplace comparison algebra $\mathfrak{A}$. Also the Laplace comparison algebra is a postliminary algebra in the sense of Dixmier [21], and (0.1) gives the structure decompositions of the successive quotients of the ideal chain $\mathfrak{A} \supset \mathcal{E} \supset K$, as discussed there.

The information on the structure of $\mathfrak{A}$ then again will be used for criteria regarding normal solvability of the general elliptic boundary problem of the form (p) (in section 3; c.f. Theorem 3.1). We shall seek a comparison with a problem (p_0) involving the constant coefficient differential operator $D_0^N + \rho^N(1-\vec{\Delta})^{N/2}$, for even N, and large $\rho > 0$, with the normal and tangential operators D_0 and $\vec{\Delta}$ (c.f. section 12), but the same variable coefficient boundary conditions as (p).

The investigation of $\mathfrak{A}$ requires rather extensive preparations. We discuss the basic facts on $\mathfrak{A}$ in section 4. The key to $\mathfrak{A}$, for general dimension, is the Laplace comparison algebra $\mathfrak{P}$ of the half line (n=0), and an extension of $\mathfrak{P}$, called $\mathfrak{G}$, closely related with the two "convolution products" on the half line. The even Wiener-Hopf convolutions are discussed in section 4. They are put into relation with a commutative subalgebra $\mathfrak{W}$ of $L(\mathcal{H})$, $\mathcal{H} = L^2(\mathbb{R}_+)$, diagonalized by the Fourier cosine transform. The Mellin convolutions are diagonalized by the Mellin transform. They correspond to the convolution product of the locally compact Abelian group $\mathbb{R}_+$, with multiplication as group operation. Again we derive a commutative subalgebra $\mathfrak{M} \subset L(\mathcal{H})$, diagonalized by the Mellin transform. The algebra $\mathfrak{G}$ is generated by $\mathfrak{W}$, $\mathfrak{M}$ and by the algebra $\mathfrak{Z}$ of multiplications by functions in $C([0,\infty])$. It is proven that $\mathfrak{G}$ has compact commutator and contains $K(\mathcal{H})$, and that $\mathfrak{G}/K(\mathcal{H})$ is isometrically isomorphic to the algebra of continuous functions over a hexagon, with each of the generating algebras $\mathfrak{W}$, $\mathfrak{M}$, $\mathfrak{Z}$ corresponding to a pair of sides of the hexagon.

Then it is possible to analyze the two subalgebras $\mathfrak{P}$ and $\mathfrak{P}^{\#}$, generated by the "convolutions" of $\mathfrak{P}$ only. These algebras will have to be used for $\mathfrak{A}$.

In section 8 and 9 we include a short discussion of topological tensor products of C^*-algebras, with emphasis on algebras of the form

$$K(\mathcal{H}) \hat{\otimes} \mathfrak{T} \subset L(\mathcal{G}) \quad , \quad \mathcal{G} = \mathcal{H} \hat{\otimes} \mathcal{K} \quad , \tag{0.2}$$

with two infinite dimensional separable Hilbert spaces $\mathcal{H}$ and $\mathcal{K}$, and with a C^*-algebra $\mathfrak{T}$ having compact commutators, and containing $K(\mathcal{K})$. (C.f. also [6], [8] and [14]). As mentioned above the commutator ideal $\mathcal{E}$ of $\mathfrak{A}$ is unitarily equivalent to an algebra of that kind.

The theory in sections 8 and 9 also is interesting because it allows to extend the theory of Chapter IV to systems of equations (even of certain $\infty\times\infty$-systems)(c.f. 9, problems 1) and 2)).

With these preparations we continue the study of $\mathfrak{a}$ in sections 10 and 11. In section 12 we discuss theory of the problem (p_0) as mentioned and then carry out the comparison of the two problems (p) and (p_0) which will prove Theorem 3.1.

Our result will imply the apriori estimates by Browder [7] and Agmon-Douglis-Nirenberg [1] for L^2,Vice versa, normal solvability of (p) under the conditions of Theorem 3.1 does not follow from those estimates, vice versa. Meanwhile our result was reproven by Erkip [22] even for the case of more general Sobolev spaces, using a more conventional approach together with certain interpolation estimates.

We notice that our argument could also be considerably simplified if only Theorem 3.1 is desired, because most of the detailed structure discussion for then will be superfluous.

I am grateful to Phillip Colella and Albert Erkip for stimulating discussions on the subject.

1. On Gårding's inequality. Elliptic and strongly elliptic ψdo-s.

A ψdo $L\in\Psi D_s$ will be called elliptic (of order s) at $x^0\in\mathbb{R}^n$ if its s-symbol quotient does not vanish for any $(x^0,\xi)\in\mathbf{M}$. If $L=\sum_{|\alpha|\le N} a_\alpha(M)D^\alpha$ is a differential operator it follows at once that

$$(1.1)\qquad \tau_L^N(x^0,\xi) = \sum_{|\alpha|=N} a_\alpha(x^0)(\xi/|\xi|)^\alpha \ ,$$

because for $x^0\in\mathbb{R}^n$ we get $(x^0,\xi)\in\mathbf{M}$ if and only if $\xi\in\partial\mathbf{B}^n$ so that

$$(1.2)\qquad \xi^\alpha(1+\xi^2)^{-N/2}=0, \text{ as } |\alpha|<N, \text{ and } = (\xi/|\xi|)^\alpha, \text{ as } |\alpha|=N,$$

with $\xi/|\xi|$ denoting the unit vector in the direction of ξ.
Thus for differential operators of order N the condition of ellipticity at a point $x^0\in\mathbb{R}^n$ simply means that the differential operator with constant coefficients

$$(1.3)\qquad L_{x^0} = \sum_{|\alpha|\le N} a_\alpha(x^0)D^\alpha$$

is elliptic (as previously defined in IV, 2 pbm. 5).

Similarly we define $L\in\Psi D_s$ to be strongly elliptic at $x^0\in\mathbb{R}^n$ if its s-symbol quotient has positive real part for every $(x^0,\xi)\in\mathbf{M}$. With the above (1.2) we see that for a differential operator this amounts to the condition

$$(1.4)\qquad \mathrm{Re} \sum_{|\alpha|=N} a_\alpha(x^0)\ \xi^\alpha > 0$$

for all real $|\xi| = 1$.

An operator $K\in\Psi D_s$ will be called elliptic (or uniformly elliptic) (without reference to a specific point x^0) if its s-symbol quotient does never vanish on the entire principal symbol space $\mathbf{M}_p$. Here the principal symbol space $\mathbf{M}_p$ is defined as earlier in section 6: $\mathbf{M}_p = \{(x,\xi)\in\mathbf{M} : |\xi| = \infty\}$. Similarly we call $K\in\Psi D_s$ strongly elliptic if $\mathrm{Re}\ \tau_K^s > 0$ on all of $\mathbf{M}_p$.

<u>Theorem 1.1.</u> (Gårding's inequality). Let $L\in\Psi D_s$, $s>0$, be strongly elliptic, and let its s-symbol quotient be in $C^{[\infty]}(\mathbf{M})$ (c.f. IV, 6). Then

$$(1.5)\qquad \mathrm{Re}(Lu,u)_t \ge c(u,u)_{t+s/2} - \gamma(u,u)_t + (u,Fu)_t\ ,\ u\in\mathcal{G}_{t+s}\ ,$$

for every positive c and γ satisfying

$$(1.6)\qquad c<\mathrm{Re}\ \tau_L^s(x,\xi) \text{ on } \mathbf{M}_p, \gamma > -(1+\xi^2)^{-s/2}(\mathrm{Re}\ \tau_L^s(x,\xi)-c) \text{ on } \mathbf{M}_s\cup\partial\mathbf{M}_p$$

with a suitable $\mathcal{O}(-\infty)$-dyad $F = F(c,\gamma)$.

<u>Proof.</u> Introduce the operator $A = \Lambda^{s/2}L\Lambda^{s/2} \in \mathfrak{O}_\infty$ with symbol $\sigma_A = \tau_L^s \in C^{[\infty]}(\mathbf{M})$, and conclude that the operator

(1.7) $\quad Q = A + A^{<t>} + 2\gamma\,\Lambda^{s} \;\varepsilon\; \mathfrak{A}_{\infty}$

has its symbol in $C^{[\infty]}(\mathbb{M})$ and >2c on all of $\mathbb{M}$ where c and γ satisfy (1.6). In particular it is possible to chose γ according to (1.6) since Re τ_L^s > c on the boundary of $\mathbb{M}_s$ where the factor $(1+\xi^2)^{-s/2}$ tends to $+\infty$, so that the function of (x,ξ) at right of (1.6) is continuous over $\mathbb{M}_s$ and tends to $-\infty$ at the boundary of $\mathbb{M}_s$. [which implies that it must have a finite upper bound].

Using Theorem 7.1 we now construct a t-self-adjoint $A\varepsilon\mathfrak{A}_{\infty}$ such that

(1.8) $\quad \sigma_P(x,\xi) > 0 \quad , \quad (\sigma_P(x,\xi))^2 = \sigma_Q(x,\xi)$ on $\mathbb{M}$.

This is possible in particular since the positive square root of σ_Q clearly is in $C^{[\infty]}(\mathbb{M})$. It follows that

(1.9) $\quad Q = P^2 + C \quad , \quad C \;\varepsilon\; \mathcal{L}_0$

because Q and P^2 have the same symbol in each algebra $\mathfrak{A}_t$, so that $C=Q-P^2 \;\varepsilon\; K_t$ for all t, or that $C \;\varepsilon\; \mathcal{L}_0$.

Thus we may write

(1.10) $\quad (u,Qu)_t = \|Pu\|_t^2 + (u,Cu)_t \quad , \quad u \;\varepsilon\; \mathfrak{H}_t$.

But the t-self-adjoint operator P has its symbol bounded below by $\sqrt{c}+\varepsilon$ for sufficiently small ε>0. It follows that there is no continuous spectrum of P below $\sqrt{2c}+\varepsilon$. In fact there can be at most finitely many point eigenvalues of finite multiplicity below $\sqrt{2c}+\varepsilon$. Otherwise there would be either an infinite dimensional point eigenvalue or an accumulation point of eigenvalues in $(-\infty,\sqrt{2c}+\varepsilon)$. At such point λ_0 $P-\lambda_0$ cannot be Fredholm, while its symbol would be different from zero at all of $\mathbb{M}$, which gives a contradiction.

Thus we conclude that the spectral family $E(\mu)$ of P consist of projections of finite rank for $\mu \le \sqrt{2c}$. In fact, we get $E(\sqrt{2c}+\varepsilon/2)\varepsilon\;\mathcal{F}_{\infty}$, because all eigenvectors to eigenvalues $\le \sqrt{2c}+\varepsilon/2$ are in $\mathfrak{H}_{\infty}$ (c.f. IV, Problem 4.4). Thus we have

(1.11) $\quad \|Pu\|_t^2 \ge (\sqrt{2c}+\varepsilon/2)^2\, \|u\|_t^2 - \|F_1u\|_t^2 \;,\; F_1 \;\varepsilon\; \mathcal{F}_{\infty} \quad ,$

with

(1.12) $\quad F_1 = \int_{0-}^{\sqrt{c}+\varepsilon/2} (c'-\mu^2)^{1/2} dE(\mu) \quad , \quad c' = (\sqrt{2c}+\varepsilon/2)^2$.

(the integral is a finite sum of operators in $\mathcal{F}_{\infty}$).

On the other hand we may apply IV, lemma 3.2 for construction of an $O(-\infty)$-dyad F_2 with

(1.13) $\quad \|F_2-C\|_t \le \varepsilon/4$.

It follows that

(1.14) $$(u,Qu)_t \geq 2c \|u\|_t^2 + (u,F_2u)_t - (u,F_1^{<t>}F_1u)_t \quad .$$

Finally substitute Q from (10.7) for

(1.15) $$2\mathrm{Re}(u,Au)_t \geq 2c \|u\|_t - 2\gamma \|u\|_{t-s/2} + 2(u,F_3u)_t$$

with an $\mathcal{O}(-\infty)$-dyad $2F_3 = F_2 - F_1^{<t>}F_1$. In (1.15) substitute $u = \Lambda^{-s/2}v$ and set $\Lambda^{-s/2}F_3\Lambda^{-s/2} = F$, then replace v by u again, for the desired estimate (1.5), q.e.d..

Our inequality (1.5) is a sharp version of the original Gårding inequality [22], which became known as one of the first tools to attack N-th order elliptic boundary problems, for N>2. There are other sharp Gårding inequalities, obtained by using standard ψdo-calculus (c.f. [5], for example).

Problems: 1) Discuss a proof of the Sobolev inequality for $L^2(\mathbb{R}^n)$: If $K\varepsilon\ \Psi D_s$ has its symbol quotient τ_L^s in $C^{[\infty]}(\mathbb{M})$ and is md-elliptic then

(1.16) $$\gamma^2 \|u\|_{s+t}^2 \leq \|Ku\|_t^2 + \|Fu\|_t^2 \quad , \quad u\ \varepsilon\ \mathcal{G}_{s+t} \ ,$$

for any $t\varepsilon\mathbb{R}$ and any γ satisfying

(1.17) $$\gamma < \tau_K^s(x,\xi) \text{ on all of } \mathbb{M},$$

with a suitable $\mathcal{O}(-\infty)$-dyad $F = F(t,\gamma)$.

2. On the Dirichlet problem for strongly elliptic equations.

In this section we will consider the simplest kind of elliptic boundary problem - the so-called Dirichlet problem. Let $\Omega \subset \mathbb{R}^n$ be a domain - an open connected set, which is not necessarily bounded. At the moment we make no assumptions on smoothness of the boundary Γ, but assume it a nulset of $\mathbb{R}^n$, under Lebesgue measure.

Let us generally define, for $s \geq 0$,

$$(2.1) \qquad \mathfrak{H}_s(\Omega) = \{u \in \mathcal{D}'(\Omega) : \text{there exists } v \in \mathfrak{H}_s(\mathbb{R}^n) \text{ with } u = v|\Omega\} \quad .$$

This definition is equivalent to the conventional one if the boundary Γ of Ω is a simple infinitely smooth surface. Usually one only defines the Sobolev spaces of integral order $s = k$,as the classes of functions with derivatives up to order k existing in weak or strong L^2-sense(c.f. 27)(equivalent under a cone condition).

On the other hand, let $\overset{O}{\mathfrak{H}}_s(\Omega)$ denote the closure of $C_0^\infty(\Omega)$ in $\mathfrak{H}_s(\mathbb{R}^n)$. More precisely, the functions of $C_0^\infty(\Omega)$ may be extended to $\mathbb{R}^n$, by setting them zero outside Ω. In this manner $C_0^\infty(\Omega)$ is identified with a subspace of $C_0^\infty(\mathbb{R}^n) \subset \mathfrak{H}_s(\mathbb{R}^n)$, which admits a closure $\overset{O}{\mathfrak{H}}_s(\Omega)$. Since we assume $s \geq 0$ it consists of functions in $L^2(\mathbb{R}^n)$. They vanish outside $\Omega \cup \Gamma$, and they are fully determined by their values in Ω, since Γ is a nulset. Therefore $\overset{O}{\mathfrak{H}}_s(\Omega)$ may be regarded as a subset of $\mathfrak{H}_s(\Omega)$, where we restrict the functions to Ω again - the natural extension v to $\mathfrak{H}_s(\mathbb{R}^n)$ (minimizing $\|v\|_s$) then will be zero outside of Ω.

In section 1 we have discussed the property of being strongly elliptic, for a ψdo $L \in \Psi\mathcal{D}_{2s}$ $s>0$. Actually, let us assume for simplicity that L is defined over $\mathbb{R}^n$ (i.e. $L \in \Psi\mathcal{D}_{2s}$, as stated), although it would be easy to work with a differential operator L defined only in a suitable neighbourhood of $\Omega \cup \Gamma$.

We then have Gårding's inequality, in the somewhat specialized form

$$(2.2) \qquad c(u,u)_s \leq \operatorname{Re}\,(Lu,u)_0 + \gamma(u,u)_0 + (u,Fu)_0 \;,\; u \in C_0^\infty(\Omega) \,,$$

with c,γ and F as in (1.5).

Let us now assume that $L: C_0^\infty(\Omega) \to \mathfrak{H}_0(\Omega)$, $L^*: C_0^\infty(\Omega) \to \mathfrak{H}_0(\Omega)$, (correct if L is a differential operator in $\Psi\mathcal{D}_{2s}$). Then we also may assume, without loss of generality that

$$(2.3) \qquad F = \sum \phi_j\rangle\langle\psi_j \quad , \quad \phi_j,\psi_j \in \mathfrak{H}_\infty(\Omega) = \bigcap_s \mathfrak{H}_s(\Omega) \quad .$$

Observe that $K=L+L^*$ will define an unbounded hermitian operator of the Hilbert space $\mathfrak{H}_0(\Omega) = L^2(\Omega)$, with domain $C_0^\infty(\Omega)$, and that Gårding's inequality (1.2) implies

$$(2.4) \qquad c(u,u)_s \leq (Ku,u)_0 + \gamma(u,u)_0 + (u,Fu)_0 \leq ((K+\gamma')u,u)_0 \,,$$

for $u \in C_0^\infty(\Omega)$, with a suitable γ'.

As another assumption which is always satisfied for a differential operator

$L\in\psi\mathcal{D}_{2s}$, at least if s is an integer, let us suppose that

(2.5) $\quad |(Lu,v)_0| \le \|u\|_s \|v\|_s \quad , \quad u,v \in C_0^\infty(\Omega)$.

Then it follows that

(2.6) $\quad c(u,u)_s \le ((K+\gamma')u,u)_0 \le c'(u,u)_s \quad , \quad u \in C_0^\infty(\Omega)$.

Now we apply a bit of functional analysis, in the Hilbert space $\mathfrak{H}_0(\Omega)$, with inner product $(u,v) = (u,v)_0$. The hermitian operator K with domain $C_0^\infty(\Omega)$ admits self-adjoint extensions $\mathring{K}$, since it is semi-bounded below, by (2.4). Distinguished among all extensions is the so-called Friedrichs extension (c.f. [30]). It may be obtained as the restriction of the Hilbert-space-adjoint $\tilde{K}$ of K to the intersection dom $\mathring{K}$ = (dom $\tilde{K}$) $\cap \mathfrak{H}_s^0(\Omega)$, since by (2.6) $\mathfrak{H}_s^0(\Omega)$ is the space obtained by completing dom K = $C_0^\infty(\Omega)$ with the metric

(2.7) $\quad \{((\mathring{K}+\gamma')u,u)\}^{1/2} = \|u\|_K$.

Moreover the well defined positive square root of the unbounded positive definite self-adjoint operator $(\mathring{K}+\gamma')$ admits a bounded inverse, denoted by Λ_K, and we have Λ_K: $\mathfrak{H}_0(\Omega) \to \mathfrak{H}_s^0(\Omega)$, a surjective isomorphism. Let

(2.8) $\quad M = L - L^*$,

then (2.5) implies that

(2.9)
$$|(Mu,v)| = |(Lu,v)-\overline{(Lv,u)}| \le c' \|u\|_s \|v\|_s \le c'' \|\Lambda_K^{-1}u\|_{L^2(\Omega)} \|\Lambda_K^{-1}v\|_{L^2(\Omega)},$$
$$u,v \in C_0^\infty(\Omega) .$$

Here we replace u and v by $\Lambda_K u$ and $\Lambda_K v$, respectively, for

(2.10) $\quad |(\Lambda_K M \Lambda_K u,v)| \le c'' \|u\| \|v\| \quad , \quad u,v \in \Lambda_K C_0^\infty(\Omega)$

where the $L^2(\Omega)$-norm is denoted by $\|\cdot\|$, for shortness.
The space $\Lambda_K^{-1} C_0^\infty(\Omega)$ is dense in $\mathfrak{H}_0(\Omega)$, by standard theory (c.f. [30]). In fact, it is known that Λ_K^{-1} in the domain $C_0^\infty(\Omega)$ is essentially self-adjoint.

But (1.10) implies that the operator $\Lambda_K M \Lambda_K$, so far defined only in $\Lambda_K^{-1} C_0^\infty(\Omega)$, extends continuously to a bounded operator of $\mathfrak{H}_0(\Omega)$, denoted by B. Note that $2L = (L+L^*)+(L-L^*) = K+M$, for $u\in C_0^\infty(\Omega)$. This implies $2\Lambda_K(L+\gamma'')\Lambda_K u=(1+B)u$, $u \in \Lambda_K^{-1} C_0^\infty(\Omega)$, $\gamma'' = \gamma'/2$, hence (c.f. AI,6 for "$\subset$")

(2.11) $\quad 2(L+\gamma'')\subset \Lambda_K^{-1}(1+B)\Lambda_K^{-1} = 2(\mathring{L}+\gamma'')$.

Also, as a trivial consequence of the definition it follows that B is skew symmetric, that is,

(2.12) $\quad (Bu,u) + (u,Bu) = (\Lambda_K u,\{(L-L^*) + (L^*-L)\}\Lambda_K u) = 0$.

First this follows for $\Lambda_K u \in C_0^\infty(\Omega)$, then for all $u \in \mathfrak{H}_0(\Omega)$, by continuity.

But for a bounded skew symmetric operator B we have $(1+B): \mathfrak{H}_0(\Omega) \to \mathfrak{H}_0(\Omega)$, a surjective isomorphism, because the spectrum of B is entirely on the imaginary axis, and -1 is in the resolvent set. Accordingly the operator $\mathring{L}+\gamma''$ of (2.11) is closed, as a product of 3 closed unbounded Fredholm operators, for example. (C.f. AI, Theorem 6.3. The operators of (2.11) are invertible, hence Fredholm.) Also that operator is surjective and 1-1, evidently, and it is an extension of $L+\gamma$, by (2.11). Furthermore its domain is a linear subspace of $\mathfrak{H}_s^0(\Omega)$, evidently.

Note that we have $(\mathring{L}+\gamma'')u=f$, $u \in \operatorname{dom} \mathring{L}$, $f \in \mathfrak{H}_0(\Omega)$ if and only if

(2.13) (i) $u \in \mathfrak{H}_s^0(\Omega)$, (ii) $(u,(L^*+\gamma'')v) = (f,v)$, for all $v \in C_0^\infty(\Omega)$.

Indeed, $(L+\gamma'')u=f$ clearly implies (2.13), by definition of $\mathring{L}$ (i.e., by (2.11)). Vice versa, (2.13) implies that

$$(2.14) \qquad (\Lambda_K^{-1}u, (1-B)\Lambda_K^{-1}v) = 2(f,v) \quad , \quad v \in C_0^\infty(\Omega) \quad ,$$

using that

$$(2.15) \qquad 2(L^*+\gamma'')v = \Lambda_K^{-1}(1-B)\Lambda_K^{-1}v \quad \text{for } v \in C_0^\infty(\Omega) \quad ,$$

as easily derived. But (2.14) gives

$$(2.16) \qquad ((1+B)\Lambda_K^{-1}u, \Lambda_K^{-1}v) = (2f,v) \quad , \quad v \in C_0^\infty(\Omega) \quad ,$$

or, $(L+\gamma'')u=f$, since Λ_K^{-1} in $C_0^\infty(\Omega)$ is essentially self-adjoint, as we know.

Next we note that under our assumptions (2.13) is equivalent to the statement (i) $u \in \mathfrak{H}_s^0(\Omega)$; (ii) u is a distribution solution of the equation $(L+\gamma'')u=f$, for given $f \in \mathfrak{H}_0(\Omega)$, in the domain Ω.

The problem to find a solution of (i) and (ii), above, for a given $f \in L^2(\Omega) = \mathfrak{H}_0(\Omega)$ is called the <u>Dirichlet problem for $L+\gamma''$ (in $L^2(\Omega)$)</u>.

We have proven (a generalization of) the result, below.

<u>Theorem 2.1.</u> Let L be a strongly elliptic differential operator in $\Psi\mathcal{D}_{2s}$, with an integer $s>0$. There exists a real constant γ'' such that the Dirichlet problem is <u>well posed</u> for every open subdomain Ω of $\mathbb{R}^n$, with boundary of measure zero, but not necessarily a bounded domain, for the operator $L+\gamma''$.

Here the constant γ'' may be chosen as any γ'' satisfying

$$(2.17) \qquad \gamma + \|F\|_{L^2(\Omega)} < \gamma''$$

where γ and F are as in Gårding's inequality.

We note the two corollaries, below.

<u>Corollary 2.2.</u> The Dirichlet problem for $L+z$ and Ω, as in Theorem 2.1, is normally solvable (c.f. AI, 1) (although not necessarily well posed) if only $z \in \mathbb{C}$ and $\gamma=\operatorname{Re} z$ is a constant as in Garding's inequality, i.e., as in (2.2), for any suitable $c>0$.

In particular γ may be chosen 0 if the symbol quotient has positive real part also on the secondary symbol space (c.f. IV, 0, or IV, (6.4)).

Corollary 2.3. For a bounded domain Ω with boundary of measure 0 the Dirichlet problem for L+z as in theorem 2.1 is normally solvable whatever the constant $z\in\mathbb{C}$ may be chosen.

It is clear that a strongly elliptic symbol over a bounded domain may be modified arbitrarily outside an open neighbourhood of a bounded domain, without changing the operator $\mathring{L}$. Therefore Corollary 2.3 is a consequence of Corollary 2.2, since γ only depends on the secondary symbol space behaviour of τ_L^s.

Regarding the first corollary, we must observe that the first inequality (2.6) no longer remains true for γ instead of γ', except if we get restricted to a subspace of $C_0^\infty(\Omega)$, of finite codimension.
But the method of constructing the Friedrichs extension remains unchanged. We now find that the operator $K+\gamma$ still is self-adjoint, but no longer positive definite. On the other hand, that operator's spectrum on the negative axis consists of finitely many point eigenvalues of finite multiplicity. Thus, although the positive square root no longer is defined, one may instead take the positive square root of the operator's restriction to the positive spectrum, and define Λ_K as the self-adjoint operator inverting that square root on its image, and define it zero on the nulspace. Similarly Λ_K^{-1} will stand for that positive square root of the restriction itself, so that $\Lambda_K \Lambda_K^{-1} = \Lambda_K^{-1}\Lambda_K$ is not identity, but the orthogonal projection P onto the space corresponding to the positive part of the spectrum.

The above argument may now be repeated, but always modulo a finite dimensional exception space. One will get

$$(2.18) \qquad 2\Lambda_K(L+\gamma)\Lambda_K = (P+B)u \quad , \quad u \in \Lambda_K^{-1} C_0^\infty(\Omega) \quad ,$$

where now that space is not dense in $\mathfrak{H}_0(\Omega)$, but dense in im P, etc.
We leave the details of the argument to the reader.

The Dirichlet problem is considered a boundary problem. In fact, we shall want to look at it as of a special case of the general elliptic boundary problem, to be investigated, at least for $\Omega = \mathbb{R}_+^n$, a half space, in the following sections. Now, while certainly the condition that $u\in\mathfrak{H}_s^0(\Omega)$ of (2.13) may be perceived as a condition on $u\in\mathfrak{H}_s(\Omega)$ near the boundary of Ω, one should ask about a more detailed analysis of the type of condition imposed on u at the boundary. The lemma, below, will answer such questions. For simplicity of the argument we only consider the case $\Omega = \mathbb{R}_+^n = \{x=(x_1,\dots,x_n) \in \mathbb{R}^n : x_n > 0\}$.

Lemma 2.4. Let s=k be an integer ≥ 0. Then the following conditions are equivalent.

(i) $u \in \mathfrak{H}_s^0(\Omega)$; (ii) the function $v(x) = u(x)$, as $x_n \geq 0$, $v(x) = 0$, as $x_n < 0$, is in $\mathfrak{H}_s(\mathbb{R}^n)$; (iii) $u \in \mathfrak{H}_s(\Omega)$ and $D_n^j u = 0$ on $\Gamma = \partial\Omega = \{x: x_n = 0\}$, as $j=0,1,\dots,s-1$.

Regarding (iii), we perhaps should note that $D_n^j u \in \mathfrak{H}_{s-j}(\Omega) \subset \mathfrak{H}_1(\Omega)$ admits an extension to $\mathbb{R}^n$, which is in $\mathfrak{H}_1(\mathbb{R}^n)$, and admits a restriction to Γ which is independent of the extension, since the restrictions to the hyperplanes parallel to Γ depend continuously on the hyperplane, as discussed in III, lemma 1.5. Therefore the conditions $D_n^j u = 0$ are meaningful, for $j=0,\dots,s-1$.

To prove the lemma it is noticed first that (i) => (ii) is evident, by definition of $\mathfrak{H}_s^0$. Then (i) => (ii) => (iii) is a consequence of III, lemma 1.5. Next let us show that (iii) => (ii). For $u \in \mathfrak{H}_s$ we have an extension $v \in \mathfrak{H}_s(\mathbb{R}^n)$, and then (iii) amounts to $D_n^j v = 0$ at $x_n = 0$. For an induction proof let first s=k=1. Then it is sufficient to show that the function $w \in L^2(\mathbb{R}^n)$ defined by w=u=v, as $x_n \geq 0$, $w = 0$, as $x_n < 0$, has all its first order distribution derivatives in $L^2(\mathbb{R}^n)$. That again is accomplished if we show that $D_l w = D_l u$ as $x_n \geq 0$, $D_l w = 0$ as $x_n < 0$, $l=1,\dots,n$. That is, more precisely, with Lebesgue integrals,

$$(2.19) \quad \int_{x_n \geq 0} u D_l \phi dx = - \int_{x_n \geq 0} D_l u \, \phi \, dy \quad , \phi \in \mathcal{D}(\mathbb{R}^n) \quad .$$

But we claim that we have Green's formula satisfied :

$$(2.20) \quad \int_{x_n \geq 0} z D_l \phi \, dx = - \int_{x_n \geq 0} D_l z \, \phi \, dx + \delta_{l1} \int_{x_n = 0} z\phi \, d\tilde{x} \ ,$$

as $z \in \mathfrak{H}_1(\mathbb{R}^n)$, $\phi \in \mathcal{D}(\mathbb{R}^n)$, where $d\tilde{x} = dx_1 \dots dx_{n-1}$ and III, lemma 1.5 insures the existence of the restriction $z|\Gamma \in L^2(\Gamma)$. To verify (2.20) let $\sigma_\varepsilon(x_n) = 0$ or $= x_n/\varepsilon$, or $=1$, as $x_n \leq 0$, or $0 \leq x_n \leq \varepsilon$, or $x_n \geq \varepsilon$, respectively. Since z and $D_l z$ are in $L^2(\mathbb{R}^n)$ one derives that

$$(2.21) \quad \int_{\mathbb{R}^n} z \, D_l(\sigma_\varepsilon \phi) \, dx = - \int_{\mathbb{R}^n} D_l z \, \sigma_\varepsilon \phi \, dx \ , \ \phi \in \mathcal{D} \ , \ \varepsilon > 0 \quad ,$$

using that σ_ε may be approximated by a sequence of C^∞-functions, σ_ε^r , such that $\sigma_\varepsilon^r \phi \to \sigma_\varepsilon \phi$, $(\sigma_\varepsilon^r \phi)_{x_l} \to (\sigma_\varepsilon \phi)_{x_l}$, both in $L^2(\mathbb{R}^n)$, as easily verified. Then (2.20) follows by letting ε tend to 0, using III, lemma 1.5. Now it is clear that (2.20) implies (2.19) whenever u=0 at $x_n = 0$. Thus (iii) => (ii), for s = 1. The general case now follows by induction.

Finally let us prove that (ii) => (i). If the function v of (ii) is in $\mathfrak{H}_s$, then one shows first that the sequence of regularized functions $v_j(x) = (\phi_j * v)(\tilde{x}, x_n - 2j)$, with $\phi_j(x) = j^n \phi(jx)$, where $\phi \in C_0^\infty(\mathbb{R}^n)$, satisfies $\phi(x) \geq 0$, $\int \phi(x) dx = 1$, $\phi(x) = 0$ as $|x| \geq 1$, satisfies $v_j \in C^\infty(\mathbb{R}^n)$, supp $v_j \subset \{x : x_n \geq 1/j\}$ and that $v_j \to v$ in $\mathfrak{H}_s(\mathbb{R}^n)$. Then let $v_{jl}(x) = \chi(x/l) v_j$, with some $\chi \in C_0^\infty(\mathbb{R}^n)$, $0 \leq \chi \leq 1$, $\chi = 1$, near 0, and show that also $v_{jl} \to v_j$ in $\mathfrak{H}_s$. Thus a suitable sequence v_{jl_j} will converge to v in $\mathcal{H}_s(\mathbb{R}^n)$, and we have $v_{jl_j} \in C_0^\infty(\Omega)$, so that $v \in \mathcal{H}_s(\Omega)$. For the convergence $v_j \to v$ one uses III, (1.1) and that

(2.22) $\quad v_j^\wedge(\xi) = e^{-i\xi/j}\phi^\wedge(\xi/j)v^\wedge(\xi)$,

as easily derived. The convergence $v_{jl} \to v_j$, on the other hand follows by proving that $v_{jl}^{(\alpha)} \to v_j^{(\alpha)}$ in $L^2(\mathbb{R}^n)$, for all $|\alpha| \leq s$, using Leibnitz' formula. Q.e.d.

Finally we make some brief comments regarding the necessity of the condition 'strongly elliptic' for normal solvability, and the statements of theorem 2.1 and its corollaries.

From IV, 2, problem 10 we have precise information about the Fredholm spectrum of an elliptic operator, which tells us that in the special case of $\Omega = \mathbb{R}^n$ the Fredholm spectrum is fully determined as the set of values of the symbol of L on the secondary symbol space. As the other extreme consider the bounded domain, for which Corollary 2.3 implies that the Dirichlet problem is normally solvable for every strongly elliptic $L\varepsilon\psi\mathcal{D}_{2s}$, and every $\gamma\varepsilon\mathbb{C}$. The same is true for a general elliptic operator, not necessarily strongly elliptic, as shall not be discussed in details.

If the domain Ω becomes "thin" for large x it will be entirely possible that a given $L\varepsilon\psi\mathcal{D}_{2s}$ still has a Dirichlet problem with empty Fredholm spectrum. On the other hand one may consider the case of a straight circular cylinder $C_\rho = \{x_1^2+x_2^2 < \rho^2,\ x_3 > 0\} \subset \mathbb{R}^3$, and the Laplace operator $\Delta = \partial_{x_1}^2 + \partial_{x_2}^2 + \partial_{x_3}^2$ (c.f. Problem 2, below). Clearly $-\Delta$ is strongly elliptic, but $-\Delta-\gamma$ is not Fredholm (regarding the Dirichlet problem in the Cylinder) for every $\gamma \geq c_\rho$ with some $c_\rho > 0$ depending on the diameter ρ of C_ρ.

Problems: 1) Solve the Dirichlet problem for the Laplace operator Δ for the following two- or three-dimensional domains: a) a rectangle $a<x_1<b$, $c<x_2<d$; b) a circle $x_1^2+x_2^2<r^2$, $r>0$; c) a ball $x_1^2+x_2^2+x_3^2 < r^2$. (Hint: Use separation of variables plus the appropriate functions of mathematical Physics, in case of c)).

2) Discuss the Dirichlet problem for the Helmholtz-operator $\Delta+z$ and a straight circular cylinder as above in 3 dimensions, for the various $z\varepsilon\mathbb{C}$. Use again separation of variables to exactly specify for what $z\varepsilon\mathbb{C}$ the problem is normally solvable.

3. The general elliptic boundary problem on a half space.

In the following sections we will discuss the so-called elliptic boundary problem for a half space $\Omega = \mathbb{R}^{n+1}_{+}$ and its boundary $\Gamma = \mathbb{R}^{n+1}_{+}$ in n+1-dimensional space $\mathbb{R}^{n+1}$. This will involve an md-elliptic differential operator together with boundary conditions of a much more general type than the Dirichlet conditions imposed in section 2. The problem will be of the form

(p) $\quad u \varepsilon \mathfrak{H}_N(\Omega)$, $\langle a\rangle u = f$ on $\Omega \cup \Gamma$, $\langle b^j\rangle u = \phi_j$ on Γ, $j=1,\ldots,r$,

with the L^2-Sobolev space $\mathfrak{H}_N(\Omega)$ of (2.1), and with a differential expression $\langle a\rangle$ of the (even) order $N = 2r$. Also there are r boundary conditions imposed, involving the r differential expressions $\langle b^j\rangle$ of order $N_j<N$, defined in some neighbourhood of Γ.

We assume that $f\varepsilon L^2(\mathbb{R}^{n+1}_{+})$ and $\phi_j \varepsilon \mathfrak{H}_{N-N_j-1/2}(\Gamma)$ are given functions, as well as the coefficients of the expressions $\langle a\rangle$ and $\langle b^j\rangle$. The problem is, to find u satisfying (p), or to prove its existence. After the experiences with elliptic operators of $\mathbb{R}^n$ one will expect a result of the Fredholm type. In other words, one will expect a theorem asserting normal solvability of (p), assuming that certain conditions are satisfied. Such a theorem is stated below (theorem 3.1).

Before we engage in its discussion we must ask the question about the meaning of a boundary condition "$\langle b^j\rangle u = \phi_j$ on Γ", for a function in $\mathfrak{H}_N(\Omega)$. One finds trivially, from III, lemma 1.3 and from (2.1) above, that $u^{(\alpha)} \varepsilon \mathfrak{H}_{N-|\alpha|}(\Omega)$, as $|\alpha| \le N_j < N$. We will assume the coefficiemts of b^j to be C^∞ in some neighbourhood of Γ, and then may consider them extended to $\mathbb{R}^{n+1}$ again. In fact it will be in agreement with our later assumptions to require that the coefficients are $CB^\infty(\mathbb{R}^{n+1})$ (p.83). Accordingly one finds that $u\varepsilon\mathfrak{H}_N(\Omega)$ implies $\langle b^j\rangle u \varepsilon \mathfrak{H}_{N-N_j}(\Omega) \subset \mathfrak{H}_1(\Omega)$. Then for a function in $\mathfrak{H}_1(\mathbb{R}^{n+1})$ the restriction to any hyperplane will be meaningful as a function in L^2 of that hyperplane and will vary continuously with the hyperplane, as it is translated, by III, lemma 1.5. The function $\langle b^j\rangle u \;\varepsilon \mathfrak{H}_1(\Omega)$ admits an extension w to $\mathbb{R}^{n+1}$ which is in $\mathfrak{H}_1(\mathbb{R}^{n+1})$, hence has a meaningful restriction $w|\Gamma \;\varepsilon\; L^2(\Gamma)$. For any two such extensions w_1, w_2 we have $w_1-w_2 = v$ equal to zero in Ω. Hence $v|\tilde{\Gamma} = 0$ for any hyperplane $\tilde{\Gamma}$ parallel to Γ and above Γ. Since the restriction various continuously with the hyperplane we find that also $w_1|\Gamma = w_2|\Gamma$. This makes the restriction $\langle b^j\rangle u|\Gamma$ meaningful as a function in $L^2(\Gamma)$. Later on (12, pbm. 6) we shall see that even $\langle b^j\rangle u \;\varepsilon\; \mathfrak{H}_{N-N_j-1/2}(\Gamma)$, a fact which often is referred to as the trace theorem, hence our assumption that $\phi_j \varepsilon \mathfrak{H}_{N-N_j-1/2}(\Gamma)$.

Perhaps we should mention, in this connection, that the Dirichlet problem of section 2 must be associated to the boundary conditions

(3.1) $u = \nabla_\nu u = (\nabla_\nu)^2 u = \dots (\nabla_\nu)^{r-1} u = 0 \quad \text{on } \Gamma$,

as easily verified, with the first order differential operator ∇_ν of differentiating in the normal direction to Γ.

We make the following general assumptions on the differential expressions involved in (p). (For convenience one assumes that

$$\mathbb{R}^{n+1} = \{(x_0,\vec{x}) : x_0 \in \mathbb{R},\ \vec{x}\in\mathbb{R}^n\} = \mathbb{R}\times\mathbb{R}^n ,$$

(3.2)

$$\Omega = \mathbb{R}^{n+1}_+ = \{x : x_0 > 0\},\ \Gamma = \partial\Omega = \{x : x_0 = 0\}$$

and uses the notation $\xi = (\xi_0,\vec{\xi})$ $(D=(D_0,\vec{D})$, $M = (M_0,\vec{M})$, etc.) Then we assume that

$$\langle a\rangle = a(M_1,D) \ ,\ \langle b^j\rangle = b^j(M_1,D) \ ,$$

(3.3) $$a(x,\xi) = \sum_{|\alpha|\le N} a_\alpha(x) \ ,\ b^j(x,\xi) = b^j(\vec{x},\xi) = \sum_{|\alpha|+k\le N_j} b^j_{k,\alpha}(\vec{x})\,\xi_0^k\vec{\xi}^\alpha \ ,$$

$$a_\alpha^{(\beta)} \in CS(\mathbb{R}^{n+1}),\ b^{j(\beta)}_{k,\alpha} \in CS(\mathbb{R}^n) \text{ for all } \alpha,\beta,j,k \ ,$$

with $CS(\mathbb{R}^d)$ as in IV, (1.14). We note that (3.3) implies that $a_\alpha^{(\beta)} \in CO(\mathbb{R}^{n+1})$, $b^{j(\beta)}_{k,\alpha} \in CO(\mathbb{R}^n)$, as $\beta\neq 0$, as will be left to the reader to verify.

<u>Theorem 3.1.</u> Under the assumptions (3.1) problem (p) will be normally solvable if $\langle a\rangle$ and $\langle b^j\rangle$ are <u>md-elliptic</u>, (a sharpening of the well known condition of ellipticity (or Lopatinskij-Shapiro)), as follows:

$\langle a\rangle$ is called md-elliptic (of order N) if

(3.4) $0<c\le|a(x,\xi)|(1+\xi^2)^{-N/2}$, for x , $\xi\in\mathbb{R}^{n+1}$, $|x| + |\xi| \ge \delta_0$,

with c independent of x,ξ, and $\delta_0>0$ sufficiently large.

For md-elliptic $\langle a\rangle$, the system $\langle b^j\rangle$, $j=1,\dots,r = N/2$ of boundary expressions is called md-elliptic (of orders (N_j)), if for all $x = (0,\vec{x}) \in \Gamma$ and all $\xi = (\eta,\vec{\xi}) \in \mathbb{R}^{n+1}$, with $|\vec{x}| + |\vec{\xi}| \ge \delta_0$ (sufficiently large) the polynomials

$$p(\eta) = p_{\vec{x},\vec{\xi}}(\eta) = (1+\vec{\xi}^2)^{-N/2} a(x,\eta\sqrt{1+\vec{\xi}^2},\vec{\xi})$$

(3.5)

$$q_j(\eta) = q_{j,\vec{x},\vec{\xi}}(\eta) = (1+\vec{\xi}^2)^{-N_j/2} b_j(\vec{x},\eta\sqrt{1+\vec{\xi}^2},\vec{\xi}),\ j=1,\dots,r \ ,$$

satisfy the following conditions.

(i) $p(\eta)$ (which cannot have real roots due to (3.2)) has precisely $r=N/2$ roots, counting multiplicities, in the complex upper half plane $\operatorname{Im}\eta>0$, denoted by $\eta_1^+,\dots,\eta_r^+$.

(ii) Let $p^+(\eta) = \prod_{l=1}^{r} (\eta-\eta_l^+)$, then the r polynomials $q_j(\eta)$, $j=1,\dots,r$, are linearly independent modulo $p^+(\eta)$, and uniformly so, in x,ξ. That is, if $r_j(\eta) = \sum_{l=0}^{r-1} r_{jl}\eta^l$ denote the remainders of q_j, modulo p^+, then we have

$0<c\leq|\det((r_{jl}))|$, with c independent of $\vec{x},\vec{\xi}$.

The proof will require rather extensive preparations (sections 4 through 9). We then will use a 'comparison' with a simpler problem (p_0), discussed in section 12, with the same boundary conditions, but a different (constant coefficients) equation $\langle a_0 \rangle u = f$.
Problem (p_0) is easily reduced to a system of equations over $\mathbb{R}^n$, with coefficients in one of the algebras $\mathfrak{A}_s$ of Chapter IV. A similar technique as in chapter IV provides necessary and sufficient criteria for normal solvability. This uses more or less standard techniques, involving the tangential Fourier transform. On the other hand, our comparison between (p) and (p_0) reduces Theorem 3.1 to the discussion of a certain bounded operator Λ over $\mathfrak{H} = L^2(\mathbb{R}_+^{n+1})$. One finds that (p) is normally solvable if A is Fredholm. Moreover A belongs to a C^*-algebra investigated in section 10, in particular in view of its Fredholm theory.

The algebra $\mathfrak{A}$ is formed by means of operators relating to a pair of special boundary problems, the so-called Dirichlet and Neumann problem of the Laplace operator only, except for multiplications by continuous functions. It again will be called the Laplace comparison algebra of $L^2(\mathbb{R}_+^{n+1})$. To construct a Fredholm inverse of A will require the investigation of two symbols, σ and τ, where τ is matrix-valued. The inversion of τ requires a somewhat lengthy calculation (section 14).

In sections 4 through 9 we define the basic concepts, and recall some results required from [6] and [12]. In section 13 we study the action of an n+1-dimensional singular integral operator on the symbol τ.

The assumptions of Theorem 3.1 will restrict enough to imply the generalized Sobolev estimates of Agmon-Douglis-Nirenberg [1], and F. Browder [7] (for L^2). However, since the domain and its boundary are non-compact, these do not imply finiteness of the nulspace or even normal solvability of the problem.

4. The Laplace comparison algebra $\mathfrak{A}$ of $L^2(\mathbb{R}^{n+1}_+)$.

In the following we will use the notation $\mathfrak{H} = L^2(\mathbb{R}^{n+1}_+)$, and we also will rename several other Hilbert spaces, as follows:

$\mathfrak{f} = L^2(\mathbb{R}_+)$, $\mathfrak{K} = L^2(\mathbb{R}^n)$, $\tilde{\mathfrak{K}} = L^2(\mathbb{R}^{n+1})$.

From classical theory it is well known that the Dirichlet problem and the Neumann problem in the half space Ω are well posed for the differential operator $1-\Delta$, with the n+1-dimensional Laplace operator Δ. In the form (p) these are the two boundary problems

(d) $u \in \mathfrak{H}_2(\Omega)$, $u-\Delta u = f$, in $\Omega \cup \Gamma$, $u = 0$ on Γ ,

and

(n) $u \in \mathfrak{H}_2(\Omega)$, $u-\Delta u = f$, in $\Omega \cup \Gamma$, $D_0 u = 0$ on Γ .

In fact, (d) and (n) induce two unbounded closed self-adjoint positive operators Δ_d, Δ_n, of $\mathfrak{H}$, respectively, with

$$\text{dom}\,\Delta_d = \{u \in \mathfrak{H}_2(\Omega):\ u=0 \text{ on } \Gamma\} = H_2(\Omega) \cap \overset{0}{H}_1(\Omega),\quad \Delta_d u = \Delta u,\ u \in \text{dom}\,\Delta_d ,$$

(4.1)

$$\text{dom}\,\Delta_n = \{u \in \mathfrak{H}_2(\Omega):\ D_0 u=0 \text{ on } \Gamma\},\quad \Delta_n u = \Delta u ,\ u \in \text{dom}\,\Delta_n .$$

The positive inverse square roots

(4.2) $\quad \Lambda_d = (1-\Delta_d)^{-1/2}$, $\quad \Lambda_n = (1-\Delta_n)^{-1/2}$,

exist as bounded operators of $\mathfrak{H}$. Moreover, all the operators

(4.3) $\quad S_d = D_0 \Lambda_d$, $S_n = D_0 \Lambda_n$, $S_{d,j} = D_j \Lambda_d$, $S_{n,j} = D_j \Lambda_n$, $j=1,\dots,n$,

are well defined bounded operators of $\mathfrak{H}$.

In analogy to the discussion of III and IV we will generate a C^*-subalgebra of $L(\mathfrak{H})$ from the operators (4.2) and (4.3) and the multiplications by functions of $CS(\mathbb{R}^{n+1})$, restricted to Ω. That algebra we call the Laplace comparison algebra of $\mathfrak{A}$.

The wellposedness of (d) and (n), and all other above statements, regarding the operators (4.2) and (4.3) may be derived very simply and directly by a method using even and odd reflections. Let the two isometries E_o, E_e: $\mathfrak{H} \to \tilde{\mathfrak{K}}$ be defined by

$$(E_o u)(y,\vec{x}) = 2^{-1/2} u(y,\vec{x}),\ y \geq 0,\quad = -2^{-1/2} u(-y,\vec{x}) ,\ y<0 ,$$

(4.4)

$$(E_e u)(y,\vec{x}) = 2^{-1/2} u(y,\vec{x}),\ y \geq 0,\quad = 2^{-1/2} u(-y,\vec{x}),\ y<0 .$$

The adjoints E_o^*, E_e^*: $\tilde{\mathfrak{K}} \to \mathfrak{H}$ are partial isometries, explicitly given by

(4.5) $\quad E_o^* v = 2^{1/2} v_o\Big|_{\mathbb{R}^{n+1}_+}$, $\quad E_e^* v = 2^{1/2} v_e\Big|_{\mathbb{R}^{n+1}_+}$, where

$$v_o(y,\vec{x}) = 1/2(v(y,\vec{x}) - v(-y,\vec{x}))$$

(4.6)

$$v_e(y,\vec{x}) = 1/2(v(y,\vec{x}) + v(-y,\vec{x})) \quad , \quad v \in \breve{\mathcal{R}} \quad .$$

The isometries E_o, E_e satisfy $E_o^* E_o = E_e^* E_e = 1$. It is readily seen that

(4.7) $$E_e E_e^* v = v_e \quad , \quad E_o E_o^* v = v_o \quad .$$

Or, in other words, $E_e E_e^*$ and $E_o E_o^*$ are the projections onto the spaces of even and odd functions, respectively, which form an orthogonal direct decomposition of the Hilbert space $\breve{\mathcal{R}}$. Here a function $u=u(y,\vec{x})$ over $\mathbb{R}^{n+1}$ is called even or odd, respectively, if $u(-y,\vec{x}) = u(y,\vec{x})$, or $u(-y,\vec{x}) = -u(y,\vec{x})$, for all $(y,\vec{x}) \in \mathbb{R}^{n+1}$. The functions v_e and v_o of (4.6) are always even and odd, respectively. Every function $v: \mathbb{R}^{n+1} \to \mathbb{C}$ can be written as $v=v_e+v_o$, with v_e and v_o as in (4.6), and a pair of one even and one odd function is always orthogonal.

Consider the Laplace operator Δ as an unbounded operator of the Hilbert space $\breve{\mathcal{R}}$, with domain dom $\Delta = \mathcal{H}_2(\mathbb{R}^{n+1})$. It is clear from the definition of $\mathcal{H}_2$ (c.f. III, (2.4)) that im $(1-\Delta)=\breve{\mathcal{R}}$, and that $1-\Delta = F^{-1}(1+|M|^2)F$, with the unbounded operator $|M|^2u(x) = |x|^2u(x)$ in dom $|M|^2 = \{u\in\breve{\mathcal{R}}: (1+|M|^2)u\in\breve{\mathcal{R}}\}$. The operator $|M|^2$ is trivially a self-adjoint operator of $\breve{\mathcal{R}}$. Its spectral family is given explicitly by $\{\chi_\mu(M^2) = p_\mu(M): \mu\in\mathbb{R}\}$, with the characteristic function χ_μ of the half line $(-\infty,\mu]$. Since F is a unitary operator of $\breve{\mathcal{R}}$, by Parsevals relation, we conclude that $-\Delta$, with the domain $\mathcal{H}_2(\mathbb{R}^{n+1})\subset\breve{\mathcal{R}}$, is a self-adjoint operator of $\breve{\mathcal{R}}$, and has the spectral family $p_\mu(D)$.

But the Fourier transform leaves the spaces of even or odd functions invariant each. Therefore the projections of the spectral family of $-\Delta$ also leaves the spaces invariant. In other words, the self-adjoint operator $-\Delta$ is reduced by the orthogonal decomposition

(4.8) $$\breve{\mathcal{R}} = \breve{\mathcal{R}}_e \oplus \breve{\mathcal{R}}_o \quad , \quad \breve{\mathcal{R}}_e = \text{im } E_e \quad , \quad \breve{\mathcal{R}}_o = \text{im } E_o \quad .$$

In other words, $-\Delta$ induces a self-adjoint operator $-\Delta_e$ and $-\Delta_o$ in $\breve{\mathcal{R}}_e$ and $\breve{\mathcal{R}}_o$, with spectral families being the restrictions of the $p_\mu(D)$ to K_e and K_o, respectively.

However E_e acts as a unitary operator $\mathcal{H} \to \breve{\mathcal{R}}_e$, and $E_e^*|\breve{\mathcal{R}}_e$ is its inverse. We claim that the operators Δ_n and Δ_d of (4.1) are given by

(4.9) $$\Delta_n = E_e^*\Delta E_e \quad , \quad \Delta_d = E_o^*\Delta E_o \quad .$$

Indeed, first of all we can write $E_o^*\Delta E_o = (E_o^*|\breve{\mathcal{R}}_o)\Delta_o E_o$, and similar for $E_e^*\Delta E_e$. Therefore dom $E_o^*\Delta E_o$ = dom $\Delta_o E_o = \{u: v=E_o u \in \mathcal{H}_2(\mathbb{R}^{n+1}) \cap \breve{\mathcal{R}}_o\}$. We know from III, Lemma 1.5 that the $L^2(\mathbb{R}^n)$-valued functions $v(y,\cdot)$, $y\in\mathbb{R}$ and $D_O v(y,\cdot)$, $y\in\mathbb{R}$ are continuous for $v \in \mathcal{H}_2(\mathbb{R}^{n+1})$. Since v must be odd, by definition, it follows that $v(0,.) = 0$. Accordingly u must satisfy the Dirichlet condition, and we conclude that dom $E_o^* \Delta E_o \subset$ dom Δ_d. Vice versa, a function $u \in$ dom Δ_d admits an extension w to $\mathcal{H}_2(\mathbb{R}^{n+1})$ vanishing at y=0. Then we check that the function $z(y,\vec{x}) = u(y,\vec{x})+w(-y,\vec{x})$, $y \geq 0$ is in $\mathcal{H}_2(\mathbb{R}^{n+1})$ and satisfies $z = D_0 z = 0$ as y=0, which implies $z \in \overset{0}{\mathcal{H}}_2(\Omega)$, by lemma 2.4. Moreover, the function z, extended zero for y<0, is in $\mathcal{H}_2(\mathbb{R}^{n+1})$. Call that extension z again, and observe that $w(y,\vec{x}) - z(-y,\vec{x})$ is in $\mathcal{H}_2(\mathbb{R}^{n+1})$ again. Also we have

$$(4.10) \qquad w(y,\vec{x}) - z(-y,\vec{x}) = u(y,\vec{x}), \text{ as } y \geq 0, \quad = -u(-y,\vec{x}), \text{ as } y \leq 0 \ .$$

In other words,

$$(4.11) \qquad w(y,\vec{x}) - z(-y,\vec{x}) = 2(E_o u)(y,\vec{x}) \in \mathcal{H}_2(\mathbb{R}^{n+1}) \ .$$

Therefore indeed we get the two domains equal and (4.9) follows. (For the first relation we may proceed similarly.)

Now it follows trivially from (4.9) that Δ_d and Δ_n are self-adjoint, since it follows that they are unitary conjugates of Δ_o and Δ_e, respectively. Moreover one will get the spectral families related similarly. Recalling that

$$(4.12) \qquad \Lambda = (1-\Delta)^{-1/2} \ , \quad S_j = D_j \Lambda \ , \quad j=0,1,\dots,n \ ,$$

are bounded operators of $\mathcal{H}$ we get

$$(4.13) \qquad \Lambda_d = E_o^* \Lambda E_o \ , \quad \Lambda_n = E_e^* \Lambda E_e \ ,$$

$$S_d = D_0 \Lambda_d = E_e^* S_0 E_o \ , \quad S_n = D_0 \Lambda_n = E_o^* S_0 E_e \ ,$$

$$S_{d,j} = D_j \Lambda_d = E_o^* S_j E_o \ , \quad S_{n,j} = D_j \Lambda_n = E_e^* S_j E_e \ ,$$

$$j = 1,\dots,n \ .$$

(In (4.13) we have used that $E_o^* D_0 u = D_0 E_e^* u$, but $E_o^* D_j u = D_j E_o^* u$, for appropriate functions u, and similar for E_e^*).

<u>Proposition 3.1.</u> In $\mathbb{R}^d$, d=1,2,..., we have, for $u \in S$,

$$(4.14) \qquad \begin{aligned} (1+|D|^2)^{-t/2} f(z) &= \int G_{d,t}(z-z') f(z') dz' \ , \quad t > 0 \ , \\ D_j (1+|D|^2)^{-1/2} f(z) &= \lim_{\varepsilon \to 0, \varepsilon > 0} \int k_{d,j}(z-z') u(z') dz' \ , \quad j=1,\dots,d \ , \end{aligned}$$

where the integral kernels $G_{d,t}$ and $k_{d,j}$ are $C^\infty(\mathbb{R}^d - \{0\})$. Moreover, these kernels are in S^0_{as} of II, definition 5.2 (v). (They coincide with a function in S each, for $|x|>1$, and have a rational point at 0. Also they allow an asymptotic expansion mod D at 0 with lowest degree of homogeneity equal to -d in case of $k_{d,j}$ and to

$-d+t$ in case of $G_{d,t}$.)

Proposition 3.1 is a direct consequence of II, theorem 5.2, applied to the function $(1+|x|^2)^{-t/2}$ and $x_j(1+|x|^2)^{-1/2}$. One obtains asymptotic expansions mod $|x|$ at infinity for these functions from the Laurent expansion of $(1+r^2)^{-t/2}$ in the variable r at infinity. Then one may take the Fourier transform term by term, applying II, lemma 5.1. Also one must use formula I, (4.23).

Remark: The kernels in (3.1) may be explicitly obtained in terms of modified Hankel functions (c.f. I, 4, problem 3). That in particular results in an improvement of the statement of proposition 3.1, stating that the kernels decay exponentially with all their derivatives, using the Hankel asymptotic expansion for Bessel functions.

From (4.13) and (4.14) one derives integral representations for the operators Λ_d, Λ_n, S_d, S_n, $S_{d,j}$, $S_{d,n}$, using (4.5), (4.6), (4.7). These are conveniently written as

(4.15)
$$\Lambda_d = \Lambda_- - \Lambda_+ \ , \quad \Lambda_n = \Lambda_- + \Lambda_+ \ , \quad S_d = S_- - S_+ \ ,$$
$$S_n = S_- + S_+ \ , \quad S_{d,j} = S_{j,-} - S_{j,+} \ , \quad S_{n,j} = S_{j,-} + S_{j,+} \ ,$$

with

(4.16) $$(\Lambda_\pm u)(y,x) = \int_{\mathbb{R}^{n+1}_+} G_{n+1,1}(y\pm y', \vec{x}-\vec{x}')u(y',\vec{x}')\,d\vec{x}'dy'$$

and similar formulas for $S_\pm$, $S_{j,\pm}$, involving the kernels $k_{n+1,j}$.

For n=0 we will denote $\Lambda_\pm = Q_\pm$, $S_\pm = P_\pm$. We note the explicit representations

(4.17)
$$(P_\pm u)(y) = i/\pi \int_{\mathbb{R}_+} K_1(y\pm y')\,\mathrm{sgn}(y\pm y')u(y')dy' \ ,$$
$$(Q\pm u)(y) = 1/\pi \int_{\mathbb{R}_+} K_0(y\pm y')u(y')dy) \ ,$$

with modified Hankel functions ([24]).

The C^*-algebras generated by (taking operator norm closure in $L(\mathfrak{H})$ of the finitely generated algebra of the operators) (4.13), (or of the operators (4.13) together with the multiplications $\lambda(M)$, $s_j(M)$, j=0,1,...,n) will be denoted by $\mathfrak{A}^\#$ and $\mathfrak{A}$, respectively. We shall refer to $\mathfrak{A}$ as of the Laplace comparison algebra of $L^2(\mathbb{R}^{n+1}_+)$.

In sections 10 and 11 we investigate some questions about the structure of $\mathfrak{A}$ and $\mathfrak{A}^\#$, which may be of interest although it only partly seems relevant in pursuit of the proof of theorem 3.1. In particular the commutator ideal $\mathfrak{E}$ of $\mathfrak{A}$ no longer is the compact ideal $K(\mathfrak{H})$, except as n=0. Hence we get an ideal chain $\mathfrak{A} \supset \mathfrak{E} \supset K(\mathfrak{H})$, where it will be found that both quotients $\mathfrak{A}/\mathfrak{E}$ and $\mathfrak{E}/K(\mathfrak{H})$ are isometrically isometric to spaces of continuous functions over spaces to be determined. The first quotient gives complex-valued functions again. But the second quotient will yield functions taking values in $K(\mathfrak{h})$ for the Hilbert space mentioned initially.

First we will investigate a similar one-dimensional algebra, called $\mathfrak{F}$, important for the n-dimensional study (section 7). For this we shall need two generalized convolution products, the Wiener-Hopf convolutions (section 5) and the Mellin convolutions (section 6). The one-dimensional algebra above (i.e. for n=0) will be denoted by $\mathfrak{P}$. It is a subalgebra of $\mathfrak{F}$ and will play a special part in the analysis of the structure of $\mathfrak{A}$ for general n.

Also, to investigate the commutator ideal $\mathfrak{E}$ of $\mathfrak{A}$ we shall need an extension to systems of equations in the algebra of III and IV. In fact one needs compact-operator-valued symbols, in this context, which will be investigated in sections 8 and 9. This, of course is of independent interest in view of the investigation of systems of md-elliptic differential equations.

5. Wiener-Hopf convolutions.

A Wiener-Hopf convolution operator is defined to be an integral operator of the form

$$(5.1)\qquad H_\phi u = (2\pi)^{-1/2} \int_0^\infty \phi(x-y)u(y)\,dy \quad , \quad u \in C_0^\infty(\mathbb{R}_+) ,$$

with a function $\phi(s)$, generally assumed to be in $L^1(\mathbb{R})$.
In some cases we shall also allow functions $\phi \notin L^1(\mathbb{R})$, and then will speak of singular Wiener-Hopf convolutions.

Note that Wiener-Hopf convolution operators, with $\phi \in L^1$, are bounded operators of $\mathfrak{H} = L^2(\mathbb{R}_+)$. For, if the operator $E_+ : \mathfrak{H} \to \tilde{\mathfrak{H}} = L^2(\mathbb{R})$ is defined by

$$(5.2)\qquad E_+u(x) = u(x) \quad , \quad x \in \mathbb{R}_+ \quad , \quad = 0 \quad , \quad x \in \mathbb{R} - \mathbb{R}_+ \quad ,$$

then we get

$$(5.3)\qquad E_+^* u = u|\mathbb{R}_+ \quad , \quad u \in \tilde{\mathfrak{H}} \quad ,$$

hence

$$(5.4)\qquad H_\phi = E_+^* K_\phi E_+ \quad ,$$

with the proper convolution operator K_ϕ, as in I, (3.6).
Therefore H_ϕ is bounded, and, in fact, we must have

$$(5.5)\qquad \|H_\phi\|_{L^2} \leq \|\phi^\wedge\|_{L^\infty} \quad ,$$

by I, (3.26) or I, (4.23) and by Parsevals relation.

We shall speak of an even (L^1-) Wiener-Hopf convolution H_ϕ if the function ϕ is even (L^1). For an even Wiener-Hopf convolution we also consider the operator

$$(5.6)\qquad H_\phi^\# = H_\phi + H_\phi^+ = E_e^* K_\phi E_e$$

with the operator E_e defined as in the preceeding section, and with

$$(5.7)\qquad H_\phi^+ u(x) = (2\pi)^{-1/2} \int_0^\infty \phi(x+y)u(y)\,dy \quad .$$

Note that H_ϕ^+ is meaningful, if ϕ is defined only over $\mathbb{R}_+$.

Lemma 5.1. For any $\phi \in L^1(\mathbb{R})$ we have H_ϕ^+ in $K(\mathfrak{H})$.

Proof. First notice that H_ϕ^+ is an integral operator with bounded Schmidt-norm , that is

$$(5.8)\qquad \iint_{0\leq x,y\leq\infty} |\phi(x+y)|^2 dx\,dy = \int_0^\infty s|\phi(s)|^2 ds < \infty \quad ,$$

if and only if $s^{1/2}\phi(s) \in L^2(\mathbb{R}_+)$. But the space of all such ϕ is dense in $L^1(\mathbb{R}_+)$, since it contains the dense subspace $C_0^\infty([0,\infty))$, (c.f. I, lemma 3.5.). Moreover, we get

$$(5.9)\qquad \|H_\phi^+\|_{L^2} \leq (2\pi)^{-1/2} \|\phi\|_{L^1(\mathbb{R}_+)} \quad ,$$

by I, lemma 1.1. Since $K(\mathcal{f})$ is norm-closed in $L(\mathcal{f})$, this implies the statement.

Lemma 5.2. The operators $H^{\#}_{\phi}$, for even $\phi \epsilon L^1(\mathbb{R})$, form a commutative sub-algebra of $L(\mathcal{f})$, which is adjoint invariant. Specifically

$$H^{\#}_{\phi} H^{\#}_{\psi} = H^{\#}_{\phi * \psi} \quad ,$$

(5.10) $$cH^{\#}_{\phi} + c'H^{\#}_{\psi} = H^{\#}_{c\phi + c'\psi}$$

$$H^{\#*}_{\phi} = H^{\#}_{\phi^*} \quad , \quad \phi^*(x) = \bar{\phi}(x) \quad .$$

That is, (5.6) induces a $*$-isomorphism

(5.11) $$L^1_e(\mathbb{R}) \to L(\mathcal{f})$$

from the algebra $L^1_e(\mathbb{R})$ of all even L^1-functions over $\mathbb{R}$, with the convolution product I, (3.1) to the algebra $L(\mathcal{f})$ of bounded operators over $\mathcal{f}$.

Proof. The boundedness of the operators $H^{\#}_{\phi}$ follows from (5.5) and (5.9), and the algebra property is a matter of (4.4). The adjoint invariance follows by direct computation. Q.E.D.

As a consequence of lemma 5.1 and the fact that $K(\mathcal{f})$ is a Riesz ideal of $L(\mathcal{f})$ it will be sufficient, in the investigation of Fredholm properties of even L^1-Wiener-Hopf convolutions, to investigate the commutative algebra of lemma 5.2.

The commutative C^*-subalgebra of $L(\mathcal{f})$ obtained by adjoining the identity and closing (under operator norm) of the algebra of lemma 2.2 above will be called the Wiener-Hopf algebra, and shall be denoted by $\mathfrak{M}$, in the following.

Theorem 5.3. The Wiener-Hopf algebra $\mathfrak{M}$ has structure space $\mathbf{M}(\mathfrak{M})$ (homeomorphic to) the closed intervall $[0,\infty]$ (i.e., the one-point-compactification of $[0,\infty)$). Moreover, if we define the operator F_c by

(5.12) $$F_c = E^*_e F E_e$$

with the Fourier transform F of $\mathbb{R}$, then F_c is a unitary operator of $\mathcal{f}$, explicitly given by

(5.13) $$F_c u(x) = (2/\pi)^{1/2} \int_0^\infty u(\xi) \cos x\xi d\xi = F^*_c u(x) \quad , \quad u \epsilon C^\infty_0([0,\infty)) \quad ,$$

and we have

(5.14) $$H^{\#}_{\phi} = F^*_c (\phi^{\sim}) F_c \quad , \quad \phi \ \epsilon\ L^1_e(\mathbb{R}) ,$$

with $\phi^{\sim} = \phi^{\wedge}|\mathbb{R}_+$.

Proof. First we confirm that F_c, defined by (5.12) indeed is a unitary operator, as a consequence of Parsevals realtion I, (4.18) and of (4.4), and of the fact that the Fourier transform takes even functions into even functions. Moreover, get

(5.15) $$F_c H^{\#}_{\phi} F^*_c = E^*_e F E_e E^*_e K_{\phi} E^*_e \bar{F} E_e = E^*_e F K_{\phi} \bar{F} E_e = E^*_e \phi^{\wedge} E_e$$

and confirm that the right hand side amounts to multiplication by $\phi^{\sim}$, as defined above. Clearly the assignment $H_\phi^\# \to \phi^\sim$ defines an isometric $*$-isomorphism, which may be extended onto all of $\mathfrak{M}$, to supply an isometric $*$-isomorphism between $\mathfrak{M}$ and $C([0,\infty])$. By our discussion of AII, 4 this must be the structure isomorphism. Formula (5.13) may be derived from (5.12) by straightforward computation.

Remark. Please observe that F_c of (5.13) is a well known integral transform - the Fourier cosine transform.

Corollary 5.4. We have equality in (5.5), for even ϕ. That is,

$$(5.16) \qquad \|H_\phi\|_{L^2} = \|\phi^\sim\|_{L^\infty} , \quad \phi \in L^1_e(\mathbb{R}) .$$

Proof. Relation (5.16) is a consequence of the above theorem, if H_ϕ is replaced by $H_\phi^\# = H_\phi + H_\phi^+$. Also we know that $H^+ \in K(\mathfrak{f})$. But we have $\inf_{C\in K(\mathfrak{f})} \|\phi^\sim(M)+C\|_{L^2} = \|\phi^\sim\|_{L^\infty}$, where $\phi^\sim+C$ denotes the linear operator $u \to \phi^\sim u+Cu$ of $\mathfrak{f}$. Applying the Fourier cosine transform we therefore get

$$\|H_\phi\|_{L^2} = \|H_\phi - H_\phi^+\|_{L^2} \geq \inf \|H_\phi^\# + C\|_{L^2} = \|\phi^\sim\|_{L^\infty},$$

which implies (5.16), together with (5.5). Q.e.d.

Remark. The assertion of corollary 5.4 also is correct for general operators of the form $c+H_\phi$, $c\in\mathbb{C}$, $\phi\in L^1(\mathbb{R})$, as we note without proof.

6. Mellin convolutions.

We define a second type of linear operator T_ψ on $\mathfrak{H} = L^2(\mathbb{R}_+)$ by setting

$$(6.1) \quad T_\psi u(x) = 2^{-1}\pi^{-1/2} \int_0^\infty \psi(x/y)u(y)dy \quad , \quad u \in C_0^\infty(\mathbb{R}_+) \quad .$$

These operators will be called Mellin convolutions. It is clear that they are the convolution operators of the convolution product

$$(6.2) \quad (u \circledast v) = 2^{-1}\pi^{-1/2} \int_0^\infty u(x/y)v(y)dy/y$$

belonging to the locally compact abelian group $\mathbb{R}_+$, with the group operation defined by multiplication (of positive real numbers).

The product $u \circledast v$ may be defined for $u,v \in L^1(\mathbb{R}_+, x^{-1/2}dx)$, due to

$$(6.3) \quad \begin{aligned} &\int_0^\infty x^{-1/2}dx \left|\int_0^\infty u(x/y)v(y)dy/y\right| \\ &\leq \int_0^\infty |v(y)|dy/y \int_0^\infty |u(x/y)|x^{-1/2}dx \\ &= \left(\int_0^\infty |v(y)|y^{-1/2}dy\right)\left(\int_0^\infty |u(t)|t^{-1/2}dt\right) \end{aligned}$$

with the substitution $x=yt$, $dx=ydt$ of the variable of the inner integral.

The collection $\mathfrak{M}' = L^1(\mathbb{R}_+, x^{-1/2}dx)$ forms an algebra (a Banach algebra) under this product, which we shall call the Mellin algebra. Note that the Mellin algebra is not the group algebra of the group $\mathbb{R}_+$, since $x^{-1/2}dx$ is not an invariant measure. This algebra will be handier for our purpose, however.

The notation Mellin convolution and Mellin algebra is motivated by the fact, that the Fourier-Plancherel transform of the product (6.2) coincides with a well investigated integral transform - the Mellin transform M and its inverse $M^{-1} = M^*$, defined by

$$(6.4) \quad \begin{aligned} Mu(t) &= 2^{-1}\pi^{-1/2} \int_0^\infty u(y)y^{-(1+it)/2}dy \quad , \quad u \in C_0^\infty(\mathbb{R}_+) \ , \ t \in \mathbb{R} \ , \\ M^* v(x) &= 2^{-1}\pi^{-1/2} \int_0^\infty v(t)x^{-(1-it)/2}dt \ , \ v \in C_0^\infty(\mathbb{R}) \ , \ x \in \mathbb{R}_+ \ . \end{aligned}$$

This fact may be verified by introducing the substitution operator $U: \mathfrak{H} \to \tilde{\mathfrak{H}}$ defined by

$$(6.5) \quad Uu(t) = 2^{1/2}e^t u(e^{-2t}) \ , \ u \in \mathfrak{H} \ , \ t \in \mathbb{R} \quad .$$

It is a matter of a variable substitution, to confirm that U constitutes an isometric isomorphism from $\mathfrak{H}$ onto $\tilde{\mathfrak{H}}$, which is inverted by $U^*: \tilde{\mathfrak{H}} \to \mathfrak{H}$, explicitly given by

$$(6.6) \quad (U^*v)(x) = (2x)^{-1/2}v(1/2 \log x) \quad , \quad v \in \tilde{\mathfrak{H}} \ , \quad x \in \mathbb{R}_+ \quad .$$

We verify that

$$(6.7) \quad U(u \circledast v) = (Uu) * (Uv) \quad , \quad u,v \in C_0^\infty(\mathbb{R}_+) \quad ,$$

with the ordinary convolution product (for n=1) of I, (3.1), at right. Let $M = FU$, with the Fourier transform I, (3.9) then one calculates explicitly that M allows the representation (6.4) above (and M^* the representation (6.4)), and (6.7) and I, (3.26) imply that

(6.8) $$M(u \circledast v)(t) = (Mu)(t).(Mv)(t) \quad , \quad u,v \in \mathfrak{M}' \quad , \quad t \in \mathbb{R} \ .$$

In particular we also get

(6.9) $$\int_0^\infty |u(x)|x^{-1/2}dx = 2^{1/2} \int_0^\infty |(Uu)(t)|dt \quad ,$$

so that U also maps the space $\mathfrak{M}'$ - that is, the Mellin algebra - isometrically onto the ordinary convolution algebra $L^1(\mathbb{R})$, up to a factor $2^{1/2}$. This becomes a proper isometry, if we introduce the norm

(6.10) $$2^{-1}\pi^{-1/2} \int_0^\infty |u(x)|^{-1/2}dx$$

in $\mathfrak{M}'$, which also is the proper norm for the Banach algebra, satisfying AII, (1.1).

In particular, for our Mellin convolutions T_ψ of (6.1) we get

(6.11) $$MT_\psi M^* = \psi^\# \quad , \quad \psi^\# = M\psi \quad ,$$

with the right hand side denoting the multiplication operator ψ (M) over $\tilde{\mathfrak{f}}$. Clearly (6.11) implies

(6.12) $$T_\psi = M^* \psi^\# M \ . \qquad \psi \in \mathfrak{M}' \quad .$$

Since M and M^* are isometries, we conclude that T_ψ, $\psi \in \mathfrak{M}'$, is a bounded operator of $\mathfrak{f}$, and that

(6.13) $$\|T_\psi\| = \|\psi^\#\|_{L^\infty} \quad , \quad \psi \in \mathfrak{M}' \quad .$$

In the following we will distinguish the special Wiener-Hopf convolution

(6.14) $$M_- = H_\mu \quad , \quad \mu(s) = (2/\pi)^{-1/2}e^{-|s|} \quad ,$$

and the Mellin convolutions

(6.15) $$K_\pm = (4\pi)^{1/2}T_{\psi_\pm} \quad , \quad \psi_\pm(x) = (1 \pm x)^{-1} \quad .$$

However, while ψ_+ is in $\mathfrak{M}'$, the norm (6.10) being finite for $u=\psi_+$, this is not correct for the function ψ_-, with its singularity at x=1. The convolution integral (6.1), for ψ_-, will therefore have to be regarded as a singular integral,

(6.16) $$T_{\psi_-} u(x) = 2^{-1}\pi^{-1/2} \lim_{\varepsilon \to 0} \int_0^\infty \chi((x-y)/\varepsilon)u(y)dy/(x-y)$$

with the characteristic function χ of the set $|x| \geq 1$.

It follows that

(6.17) $$UT_{\psi_\pm}U^* = 2^{-1/2}K_{\tau_\pm} \ , \ \tau_+(t) = 1/\cosh t, \ \tau_-(t) = 1/\sinh t \quad ,$$

with the hyperbolic sine and cosine. Here again the operator K_{τ_-} must be interpreted as a singular integral operator. Both operators $K_{\tau_\pm}$ are defined by I, (3.8).

The distribution (represented by the principal value of) τ_- is in S^O_{ad} of II, 5. Therefore we still get formulas (6.11) and (6.12) for ψ_- as well as for ψ_+. The Fourier transforms of $\tau_\pm$ may be computed by a complex curve integral, and we obtain the formulas

$$\text{(6.18)}\qquad \begin{aligned} MK_-M^* &= -\, i\pi \tanh (t\pi/2) \\ MK_+M^* &= -\, \pi\, (\cosh (t\pi/2))^{-1} \,. \end{aligned}$$

It follows that the singular Mellin convolution operator K_- also is $L^2(\mathbb{R}_+)$-bounded, and commutes with (all operators of) the algebra $\mathfrak{M}'$. Let $\mathfrak{M}$ denote the closed sub-algebra of $L(\mathfrak{f})$, generated by the two operators K_+ and K_-. By (6.18) and the Stone Weierstrass theorem this is a commutative C^*-subalgebra of $L(\mathfrak{f})$ and we have the theorem, below.

Theorem 6.1. The algebra $\mathfrak{M}$ has its maximal ideal space $\mathbb{M}(\mathfrak{M})$ (homeomorphic to) the closed interval $[-\infty,+\infty]$, that is, the intervall $(-\infty,+\infty)$, compactified by adding two points $+\infty$ and $-\infty$, with the corresponding open sets. The structure isomorphism is effected by

$$\text{(3.19)}\qquad T_\psi \rightarrow \psi^{\#} \,,$$

for

$$\text{(3.20)}\qquad \psi = \psi_\pm \,,\; \psi_+^{\#} = (\pi/4)^{1/2}(\cosh (t\pi/2))^{-1} \,,\; \psi_-^{\#} = i(\pi/4)^{1/2}\tanh (t\pi/2) \,.$$

In particular $\mathfrak{M}$ contains the identity and all of $\mathfrak{M}'$, and $\psi^{\#}$ coincides with the function defined in (6.11) for $\psi\in\mathfrak{M}'$.

The proof of theorem 6.1 is straightforward after the above remarks and the proof of theorem 5.3.

Remark. While the definition of T_ψ, M,M^* and formula (6.7) were given only for C_0^∞-functions, their extension to the larger spaces required, by continuous extension, is evident in view of the close connection to the Fourier transform and ordinary convolution product, effected by the isometry U.

7. The algebra $\mathcal{Q}$ and its structure.

In addition to the two commutative algebras $\mathfrak{M}$ and $\mathfrak{N}$ of sections 5 and 6 we now introduce the algebra

$$(7.1) \qquad \mathfrak{Z} = \{a(M) : a \in C([0,\infty])\}$$

of bounded multiplication operators of $L(\mathfrak{h})$, where $[0,\infty]$ again means the one-point compactification of $[0,\infty)$, and where the continuous functions a over $[0,\infty]$ are the continuous functions over $[0,\infty)$, having a limit at ∞. For $a\in C([0,\infty])$ we write $a(M)$ for the multiplication operator $a(M)u(x) = a(x)u(x)$, $0\le x<\infty$. Occasionally we will shortly write $a(M)u = au$.

Lemma 7.1. All commutators between elements H_ψ, T_ψ, a of the three algebras $\mathfrak{M}$, $\mathfrak{N}$ and $\mathfrak{Z}$ are compact operators of $\mathfrak{h}$.

Proof. We first note that each of the algebras $\mathfrak{M}$, $\mathfrak{N}$ and $\mathfrak{Z}$ may be generated by just one of its elements (and I). In particular, such single generators are $H_{e^{-|x|}}$, K_-, and $(1+x^2)^{-1}$, respectively.

Proposition 7.2. Define

$$(7.2) \qquad \mu(x) = \frac{1}{2} e^{-|x|}, \phi(x) = 1/x, \kappa(x) = (1+x^2)^{-1}, \quad x \in \mathbb{R},$$

then K_μ, K_ϕ and the multiplication κ have commutators in $K(\mathfrak{h})$. (K_μ, K_ϕ as in I, (3.6), with a principal value integral for K_ϕ.)

Proof. Let again $\chi,\omega\in C^\infty(\mathbb{R})$, $0\le\chi$, $\omega\le1$, $\omega=0$ near 0, $\chi=0$ near ∞, $\omega+\chi=1$ on $\mathbb{R}$, and write

$$(7.3) \qquad \phi = \phi_+ + \phi_\infty,\ \phi_0 = \chi\phi, \quad \phi_\infty = \omega\phi .$$

Then it simply is a consequence of IV, theorem 1.1., that the commutator $[K_{\phi_0},\kappa]$ is compact. Also the three convolution operators K_{ϕ_0}, K_{ϕ_∞} and K_μ commute strictly, and for the commutator $[K_\mu,\kappa]$ we again have IV, theorem 1.1. Hence we are left with $[K_{\phi_\infty},\kappa]$, and note to this end, that the two operators $K_{\phi_\infty}\kappa$ and κK_{ϕ_∞} both have bounded Schmidt norm, hence are compact. Indeed, we get

$$(7.4) \qquad (K_{\phi_\infty}\kappa u)(x) = \int_{-\infty}^{+\infty} \phi_\infty(x-y)\kappa(y)u(y)dy ,$$

so that for the Schmidt norm of $K_{\phi_\infty}\kappa$ we get

$$(7.5) \qquad \begin{aligned} \iint_{-\infty}^{+\infty} |\phi_\infty(x-y)\kappa(y)|^2 dx\,dy &\le (2\pi)^{1/2} \|\phi_\infty^2 * \kappa^2\|_{L^1} \\ &\le \|\phi_\infty^2\|_{L^1} \|\kappa^2\|_{L^1} < \infty . \end{aligned}$$

Similarly for κK_{ϕ_∞}, hence also the commutator is compact, q.e.d.

To complete the proof of lemma 7.1, note that we may write

$$2K_- = E_o^* K_\phi E_e + E_e^* K_\phi E_o \quad ,$$

(7.6) $$M_- = E_e^* K_\mu E_e - D = E_o^* K_\mu E_o + D$$

$$\kappa(M) = E_e^* \kappa(M) E_e = E_o^* \kappa(M) E_o \quad ,$$

with the operator

(7.7) $$Du(x) = 1/2 \int_0^\infty e^{-x-y} u(y)dy$$

of finite rank, and where the same notation has been used for multiplication by the function κ in the two spaces $\mathfrak{f}$ and $\tilde{\mathfrak{f}}$. For definition and rules of calculation of the operators E_e and E_o leading to the derivation of (7.6) we refer to section 4 (for n=0).

Recall that $E_o^* E_o = 1$ on odd functions, and that $E_e^* E_e = 1$ on even functions, and conclude that

$$(E_o^* K_\phi E_e) M_- = E_o^* K_\phi E_e (E_e^* K_\mu E_e - D) = E_o^* K_{\phi * \mu} E_e + C$$

$$M_-(E_o^* K_\phi E_e) = (E_o^* K_\mu E_o + D)\, E_o^* K_\phi E_o = E_o^* K_{\phi * \mu} E_e + C' \quad ,$$

with compact operators C, C'. Hence the commutator $[E_o^* K\, E_e, M_-]$ is compact. Similarly one may proceed with all other commutators of the operators in (7.6), and so one obtains compactness of

$$[K_-, M_-] \quad , \quad [K_-, \kappa] \quad , \quad [M_-, \kappa] \quad ,$$

which proves lemma 7.1.

Let us now introduce the C^*-algebra $\mathfrak{q}$ generated by all three commutative algebras $\mathfrak{M}$, $\mathfrak{W}$ and $\mathfrak{Z}$, (and the compact operators of $K(\mathfrak{f})$). Clearly $\mathfrak{q}$ can be obtained from the finite set of generators (mod $K(\mathfrak{f})$)

(7.8) $$K_- \quad , \quad H_{e^{-|x|}} \quad , \quad \kappa(M) \quad .$$

By Lemma 7.1 it is clear that $\mathfrak{q}$ is a C^*-subalgebra of $L(\mathfrak{f})$, with unit, containing $K(\mathfrak{f})$, and with commutators contained in $K(\mathfrak{f})$. In the following we will obtain its <u>symbol space</u> - that is, the Gelfand space of the quotient $\mathfrak{q}/K(\mathfrak{f})$.

Recall that the Gelfand spaces of $\mathfrak{W}$, $\mathfrak{M}$ and $\mathfrak{Z}$ are given by

(7.9) $$\mathbf{M}(\mathfrak{W}) = [0,\infty] \quad , \quad \mathbf{M}(\mathfrak{M}) = [-\infty,+\infty] \quad , \quad \mathbf{M}(\mathfrak{Z}) = [0,\infty] \quad .$$

We shall adopt the convention, from now on, to denote the real variable in these three intervals by ξ, t and x, respectively, so that we have

(7.10) $$\mathbf{M}(\mathfrak{W}) = \{0 \le \xi \le \infty\} \quad , \quad \mathbf{M}(\mathfrak{M}) = \{-\infty \le t \le \infty\} \quad , \quad \mathbf{M}(\mathfrak{Z}) = \{0 \le x \le \infty\} \quad .$$

After the experiences made in chapter IV we will expect the symbol space $\mathbf{M}(\mathfrak{q})$ to be a compact subset of the cartesian product

(7.11) $$[0,\infty] \times [-\infty,+\infty] \times [0,\infty] = \{(x,t,\xi) : 0 \le x,\ \xi \le \infty,\ -\infty \le t \le \infty\} = \mathcal{Q} \quad .$$

This expectation is confirmed by the result stated below.

Theorem 7.3. The symbol space $\mathbf{M}(\mathfrak{A})$ of the C^*-algebra $\mathfrak{A}$ is (homeomorphic to) the subset of the cube $\mathcal{Q}$ of (7.11), described in fig. 7.1., below, - that is, in details, the set as follows.

$$(7.12)\quad \begin{aligned} (x,t,\xi) &= (0,t,\infty),\ -\infty\le t\le\infty\ ,\ \text{or}\ = (\infty,t,0)\ ,\ -\infty\le t\le\infty\ , \\ \text{or}\ &= (x,-\infty,\infty)\ ,\ 0\le x\le\infty\ ,\ \text{or}\ = (x,+\infty,\infty)\ ,\ 0\le x\le\infty\ , \\ \text{or}\ &= (\infty,-\infty,\xi)\ ,\ 0\le\xi\le\infty\ ,\ \text{or}\ = (\infty,+\infty,\xi)\ ,\ 0\le\xi\le\infty\ . \end{aligned}$$

The homeomorphism $\sigma: \mathfrak{A} \to C(\mathbf{M}(\mathfrak{A}))$ is explicitly given by

$$(7.13)\quad \begin{gathered} \sigma_{H_\psi}(x,t,\xi) = \psi^{\sim}(\xi),\ \sigma_{T_\psi}(x,t,\xi) = \psi^{\#}(t),\ \sigma_a(x,t,\xi) = a(x) \\ \sigma_C(x,t,\xi) = 0 \end{gathered}$$

for

$$(7.14)\quad H_\psi \in \mathfrak{W}\ ,\ T_\psi \in \mathfrak{M}\ ,\ a \in \mathfrak{z}\ ,\ C \in K(\mathfrak{h})\ ,$$

where $\psi^{\sim}$ and $\psi^{\#}$ are defined as in theorem 5.3 and theorem 6.1, respectively.

We prepare the proof with some Lemmata. First one obtains the following by direct extension of Herman's Lemma.

Lemma 7.4. The symbol space $\mathbf{M}(\mathfrak{A})$ is (homeomorphic to) a compact subset of the cube $\mathcal{Q}$ of (4.11), in such a way that (7.13) is satisfied.

The details are left to the reader. We also extend the technique of IV, 2-3 for the following.

Lemma 7.5. Let $A\in\mathfrak{z}$, $B\in\mathfrak{W}$, $C\in\mathfrak{M}$. Then the product ABC is compact if and only if the symbol space, under the homeomorphism of Lemma 7.4, maps into a subset of

$$(7.15)\quad \{(x,t,\xi) : \sigma_A(x)\sigma_B(t)\sigma_C(\xi) = 0\} \quad .$$

Again we leave the details to the reader.

Lemma 7.6. The two products

$$(7.16)\quad \kappa M_- \quad \text{and} \quad x\kappa K_+$$

are compact.

Proof. Using (7.6) we confirm that κM_- is the sum of a dyad and the operator $E_e^* \kappa K_\mu E_e$, which is compact, by III, lemma 8.1. Also, one may check by straight computation that $x\kappa K_+$ has finite Schmidt norm, hence is a compact integral operator.

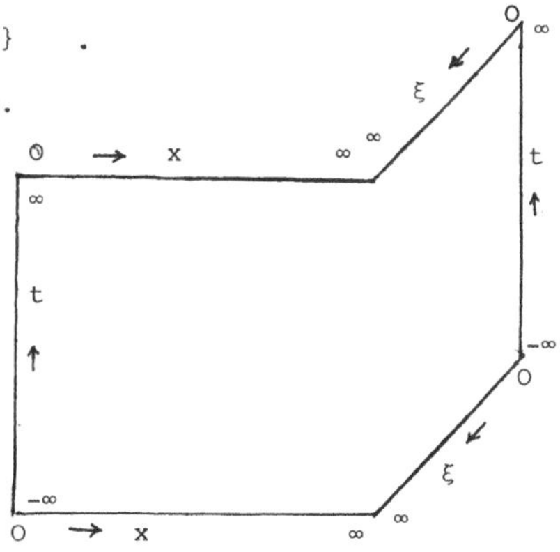

Figure 7.1. (The symbol space $\mathbf{M}(\mathfrak{A})$, as a compact subset of the cube $\mathcal{Q}$ of (4.11)).

Lemma 7.7. Let $\psi\in L^1(\mathbb{R})$ be defined as inverse Fourier transform of the function

$$(7.17)\qquad \psi^\wedge(\xi) = (1+2\xi^2)/(1+\xi^2)^2 \quad , \ \xi\in\mathbb{R}\ ,$$

and let

$$(7.18)\qquad \alpha(x) = x^{-1/2}e^{-(\log x)^2} \quad ,$$

$$\beta(x) = e^{-1/x^2}\ ,\quad x\in\mathbb{R}_+ \quad .$$

Then the operator

$$(7.19)\qquad P = \beta T_\alpha(1-H_\psi)\beta$$

is compact.

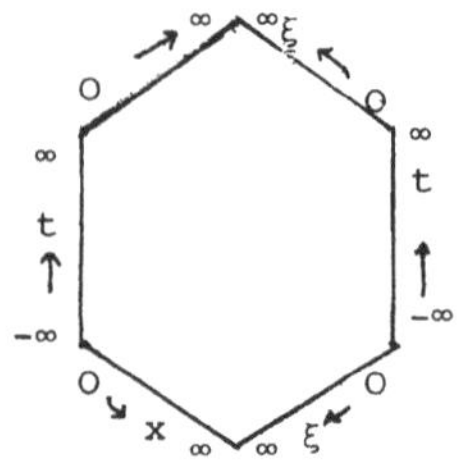

Figure 7.2. (The same symbol space $\mathbf{M}(\mathfrak{A})$ straightened into the boundary of a regular hexagon, henceforth called the characteristic hexagon of the algebra $\mathfrak{A}$).

Proof. First we confirm that P is an integral operator with kernel

$$(7.20)\qquad p(x,y) = (4\pi)^{-1/2}\beta(x)\beta(y)\left[(1/y)\alpha(x/y)-(2\pi)^{-1/2}\int_0^\infty(1/t)\alpha(x/t)\psi(y-t)dt\right] \quad .$$

Define a function γ by

$$(7.21)\qquad \gamma(t) = \alpha(1/t)/t \text{ for } t>0, \quad = 0 \text{ for } t\le 0 \quad ,$$

and notice that $\gamma\in\mathcal{S}$ (c.f. I,3), because its derivatives are sums of terms each involving the factor $e^{-(\log t)^2}$, vanishing faster at 0 and ∞ than any power of t or log t. The term within the brackets of (7.20) becomes equal to

$$(7.22)\qquad \gamma_{1/x}(y)/x - 1/x\ (\gamma_{1/x} * \psi)(y)$$

with the convolution product "$*$" of I, 3, and the function $\gamma_{1/x}$ defined by $\gamma_{1/x}(y) = \gamma(y/x)$ It follows that

$$(7.23)\qquad (\gamma_{1/x})^\wedge(\xi) = x\gamma^\wedge(\xi x) \quad .$$

Therefore relation I, (3.26) and the Fourier inversion formula I, (3.20) may be applied to (7.22) to imply

$$\begin{aligned} &= (\gamma^\wedge(x\eta) - \gamma^\wedge(x\eta)\psi^\wedge(\eta))^\vee(y) \\ (7.24)\qquad &= (2\pi)^{-1/2}\int_{-\infty}^{\infty} e^{iy\eta}\gamma^\wedge(x\eta)(1-(1+2\eta^2)/(1+\eta^2)^2)d\eta \\ &= (2\pi)^{-1/2}\int_{-\infty}^{\infty} e^{iy\eta}\gamma^\wedge(x\eta)\eta^4/(1+\eta^2)^2 d\eta \quad . \end{aligned}$$

A two-fold partial integration yields

$$(7.25)\qquad p(x,y) = c_0\beta(x)\beta(y)y^{-2}\int_{-\infty}^{\infty} e^{iy\eta}\partial^2/\partial\eta^2(\gamma^\wedge(x\gamma)\eta^4/(1+\eta^2)^2)d\eta \quad .$$

Expressing the derivative at right by the chain rule and Leibnitz' formula we get the integral at the right hand side bounded by the sum of three terms

$$(7.26)\qquad x^k\int_{-\infty}^{+\infty}|\tau_k(x\eta)|\ |\eta|^{2+k}d\eta \quad , \quad k=0,1,2,$$

where τ_k denote functions in $\mathcal{S}$. Finally, the expressions (7.26) are equal to

$c_k x^{-3}$, with constants c_k, as follows from a substitution of integration variable. As a consequence we get

$$(7.27)\qquad |p(x,y)| \le c_4(\beta(x)/x^3)(\beta(y)/y^2),\ x,y \in \mathbb{R}_+ \quad ,$$

which shows that the Schmidt norm of the operator P, given by

$$(7.28)\qquad \Big| \iint_{0\le x,y\le\infty} |p(x,y)|^2 dx\, dy\Big|^{1/2} \quad ,$$

is finite, so that P is compact. Q.e.d.

Remark. As an immediate consequence of the above, it follows now that the symbol space $\mathbf{M}(\mathfrak{A})$ must be a compact subset of the space (7.12), mentioned in theorem 7.3. For, we compute

$$(7.29)\qquad \begin{aligned} \sigma_{K_+}(t) &= \pi(\cosh(t\pi/2))^{-1} && (\text{in } \mathfrak{M}) \quad , \\ \sigma_{M_-}(\xi) &= (1+\xi^2)^{-1} && (\text{in } \mathfrak{W}) \quad , \\ \sigma_{\beta(x)} &= e^{-1/x^2} && (\text{in } \mathfrak{Z}) \quad , \\ \sigma_{(1-H_\psi)}(\xi) &= \xi^4/(1+\xi^2)^2 && (\text{in } \mathfrak{W}) \quad , \end{aligned}$$

using (6.18), for K_+, a direct calculation of the Fourier integral for M_-, and the definition (7.17) for $1-H_\psi$. Also, using formulas (6.4) and (6.11) on the function of (7.18) get

$$(7.30)\qquad \begin{aligned} \sigma_{T_\alpha}(t) &= (4\pi)^{-1/2} \int_0^\infty \alpha(y) y^{-(1+it)/2} dy = \\ &= (4\pi)^{-1/2} \int_0^\infty dy/y\ e^{-(\log y)^2}\ e^{-it/2 \log y} \\ &= (4\pi)^{-1/2} \int_{-\infty}^\infty d\tau\ e^{-\tau^2 - it\tau/2} = 1/2\ e^{-t^2/16} \quad . \end{aligned}$$

The above information may be used to show that the symbol space $\mathbf{M}(\mathfrak{A})$ does not contain points of the cube Q outside the set of Fig.7.1. Indeed, we have κM_- compact, hence $[(1+x^2)(1+\xi^2)]^{-1}$ vanishes on $\mathbf{M}(\mathfrak{A})$. That is, we get either $x=\infty$ or $\xi=\infty$ on $\mathbf{M}(\mathfrak{A})$, so that it must be contained in the set marked in fig.7.3 below.

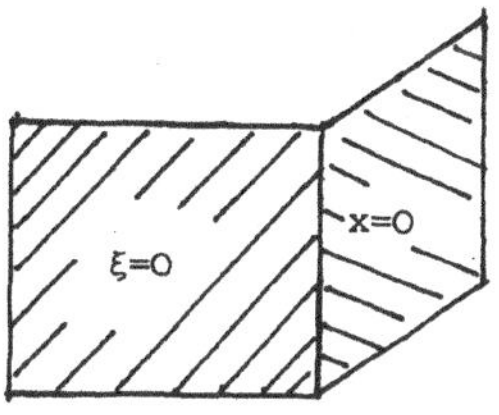

Fig. 7.3

Next, the compactness of $x\kappa K_+$ implies $x[(1+x^2)(\cosh \pi t/2)]^{-1} = 0$ on $\mathbf{M}(\mathfrak{A})$, so that we get either $x=0$, or $t=\pm\infty$ or $x=\infty$, which restricts us to the set marked in fig.7.4, below.

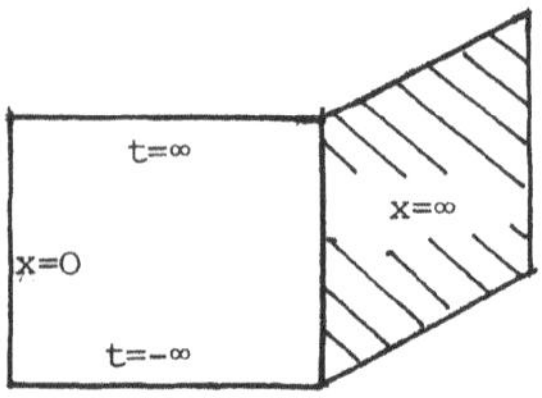

Fig. 7.4

Finally, compactness of the operator P of (7.19) yields

$$e^{-1/x^2}\, e^{-t^2/16}\xi^4/(1+\xi^2)^2 = 0 \quad \text{on} \quad \mathbf{M}(\mathfrak{O}) \quad ,$$

so that on $x=\infty$ we are restricted to either $t=\pm\infty$ or $\xi=0$, which brings us to the countour of Fig. 7.1.

Thus, applying lemma 7.6 on the three products (7.16) and (7.17), which are compact, by lemma 7.6 and lemma 7.7, we confirm that indeed no other point of the cube Q than listed in (7.12) may be contained in the symbol space $\mathbf{M}(\mathfrak{O})$.

Accordingly, for the proof of theorem 7.3 we now only must show that all points of the space (7.12) also are in $\mathbf{M}(\mathfrak{O})$.

In the following, let $\mathfrak{R}$ denote the C^*-algebra generated by only the elements of $\mathfrak{M}$ and $\mathfrak{Z}$, and of $K(\mathfrak{f})$, but not of $\mathfrak{W}$. With the isometry U defined in (6.5) and (6.6) define

(7.31) $\quad \mathfrak{T} = U\mathfrak{R}U^* = \{A: A\varepsilon L(\tilde{\mathfrak{f}}),\ A = U\,B\,U^*\ ,\ B\ \varepsilon\ \mathfrak{R}\}$.

By our above discussion $\mathfrak{R}$ may be generated by the two operators K_-, κ of lemma 4.1, and the compact operators only. But we get

(7.32) $\quad U\,\kappa\,U^* u(t) = (1+e^{4t})^{-1} u(t) \quad , \quad t\ \varepsilon\ \mathbb{R} \quad ,$

and also note (6.15) and (6.17), which show that

(7.33) $\quad UK_-U^* = c\,K_{\tau_-} \quad , \quad \tau_-(t) = 1/\sinh t$

with a nonvanishing constant c and the convolution operator I, (3.6), involving a singular integral. The Fourier transform of τ_- already was computed in section 6 to equal the function $\tanh(t\pi/2)$, up to a multiplicative nonvanishing constant. (C.f. (6.18)). It follows that the C^*-subalgebra $\mathfrak{T}$ of $L(\tilde{\mathfrak{f}})$ has the generators

	$C([-\infty,+\infty])$	(multiplicative)
(7.34)	$\tau_-(D)$	(convolutions)
	$C(\tilde{\mathfrak{f}})$	(compact operators) .

It is clear that (7.34) generate a subalgebra of $\mathfrak{A}_0$, of IV, 1-2, in the case of n=1. The techniques applied there may be used for the theorem, below. Details again are left to the reader.

Theorem 7.9. The C^*-operator algebra $\mathfrak{R}$ with symbol has symbol space $\mathbf{M}(\mathfrak{R})$ (homeomorphic to) the boundary

(7.35) $\quad [0,\infty] \times [-\infty,+\infty] - (0,\infty) \times (-\infty,+\infty)$

of the rectangle $0 \le x \le \infty$, $-\infty \le t \le \infty$, and the symbol of $a \varepsilon \mathfrak{Z}$, $T_\psi \varepsilon \mathfrak{W}$, $C \varepsilon K(\mathfrak{H})$ is given by

(7.36) $\quad \sigma_C = 0 \quad , \quad \sigma_a(x,t) = a(x) \quad , \quad \sigma_{H_\psi}(x,t) = \psi^{\#}(t) \quad .$

Corollary 7.10. The associate dual map $\mathbf{M}(\mathfrak{q}) \to \mathbf{M}(\mathfrak{R})$ of the injection $\mathfrak{R} \to \mathfrak{q}$ coincides with the projection

(7.37) $\quad (x,t,\xi) \to (x,t) \quad , \quad (x,t,\xi) \;\varepsilon\; \mathbf{M}(\mathfrak{q}) \quad .$

The proof of the corollary is a consequence of the properties of the associate dual map (c.f. AII, (5.5)), combined with the formulas (7.36) and (7.13).

Proof of theorem 7.3. After our above remark it is immediate that the four segments $(0,t,\infty)$, $(\infty,t,0)$, $-\infty<t<\infty$, and $(x,-\infty,\infty)$, (x,∞,∞), $0<x<\infty$ must be in $\mathbf{M}(\mathfrak{q})$, since the associate dual map of corollary 7.9 is surjective, by AII, prop. 5.8 (since a C^*-algebra is always completely regular). Since $\mathbf{M}(\mathfrak{q})$ is compact, the same may be stated for the closed segments. Next, we also get the projection $(x,t,\xi) \to \xi$ as an associate dual map of the canonical injection $\mathfrak{W} \to \mathfrak{q}$ (Note that theorem 5.3 implies that no Wiener-Hopf convolution is compact, except the trivial one.). Accordingly, for $0<\xi<\infty$ we always must have at least one of the two points $(\infty,\pm\infty,\xi)$ in $\mathbf{M}(\mathfrak{q})$. As a consequence of the next lemma, below, (lemma 7.11), the space $\mathbf{M}(\mathfrak{q})$ either must consist of the set (7.12) precisely, or we must obtain it by removing precisely one of the two segments (∞,∞,ξ), $0<\xi<\infty$ or $(\infty,-\infty,\xi)$, $0<\xi<\infty$. Finally, to exclude this last possibility, we will display an operator of the algebra $\mathfrak{q}$ with Fredholm index different from zero (lemma 7.12). If $\mathbf{M}(\mathfrak{q})$ would be missing one of the two ξ-segments mentioned, then it would be homeomorphic to a straight line segment. There would exist a homotopy

(7.38) $\quad \sigma : \mathbf{M}(\mathfrak{q}) \times [0,1] \to \mathbb{C}^* = \mathbb{C} - \{0\} \;, \quad \sigma = \sigma_{A_0} \text{ at } t=0 \;, \quad = 1 \text{ at } t=1 \quad .$

Since the symbol map is onto $C(\mathbf{M}(\mathfrak{q}))$ one then may pick a family A_τ, $0 \le \tau \le 1$ satisfying $A_\tau \varepsilon \mathfrak{q}$, $\sigma_{A_\tau} = \sigma|\{t=\tau\} = \sigma_\tau$, and we will get the Fredholm index of A_τ equal to a

constant, by AI, Theorem 3.7, and AI, theorem 4.9, and by the fact that for C^*-algebras, the structure homomorphism is an isometry, which yields

(7.39) $\quad \inf_{C \varepsilon K(\mathfrak{H})} \|A_\tau - A_{\tau'} + C\|_{L^2} = \|\sigma_\tau - \sigma_{\tau'}\|_{L^\infty} \to 0 \quad , \quad \tau - \tau' \to 0 \; .$

Since the Fredholm index of I is equal to zero, but that of A_0 is different from zero, this is a contradiction, and theorem 7.3 is established.

Lemma 7.11. For any $\tau>0$ define a unitary operator $W_\tau : \mathfrak{H} \to \mathfrak{H}$ by

(7.40) $\quad u(x) \to \tau^{1/2} u(\tau x) = v(x) \quad .$

Then the isomorphism $L(\mathfrak{H}) \to L(\mathfrak{H})$, defined by

(7.41) $A \to W_\tau^* A W_\tau$

leaves the four algebras $\mathfrak{z}$, $\mathfrak{m}$, $\mathfrak{M}$ and $\mathfrak{g}$ invariant, and the associate dual map of the induced automorphism $\mathfrak{g} \to \mathfrak{g}$ is represented by

$$(x,t,\xi) \to (\tau x, t, \xi/\tau) \quad .$$

<u>Lemma 7.12.</u> The operator

$$(7.42)\qquad \begin{aligned} A_0 &= a_0(M) + a_1(M)K_- \quad , \quad a_0(x) = \log x\ (\log x + 2i)^{-1} \\ a_1(x) &= 2/\pi\,(\log x + 2i)^{-1} \ , \end{aligned}$$

with K_- as in (6.15), has Fredholm index 1.

<u>Proof of lemma 7.11.</u> For maximal ideals $\mathfrak{m}_{x^0} = \{a: a(x^0) = 0\}$ of $\mathfrak{z}$ and

$\mathfrak{m}_{t^0} = \{T_\psi: \psi^{\#}(t^0) = 0\}$ of $\mathfrak{M}$, and $\mathfrak{m}_{\xi^0} = \{H_\psi: \psi^{\sim}(\xi^0) = 0\}$ of $\mathfrak{m}$ confirm that

(7.43) $W_\tau^* \mathfrak{m}_{x^0} W_\tau = \mathfrak{m}_{\tau x^0}$, $W_\tau^* \mathfrak{m}_{t^0} W_\tau = \mathfrak{m}_{t^0}$, $W_\tau^* \mathfrak{m}_{\xi^0} W_\tau = \mathfrak{m}_{\xi^0/\tau}$. Also recall A II,

prop. 5.4, which completes the proof.

<u>Proof of lemma 7.12.</u> Note that with the Mellin transform M and its adjoint ((6.4) and (6.5)) we get

$$(7.44)\qquad M[\log x/(2i+\log x)]M^* = F[x/(x+i)]F^* = d/dx[d/dx+1]^{-1}$$

and

$$(7.45)\qquad M[2/(2i+\log x)]M^* = F[(x+i)^{-1}]F^* = -i(d/dx+1)^{-1} \quad .$$

Note that the differential operator $d/dx+1$ possesses a bounded inverse in $\tilde{\mathfrak{f}} = L^2(\mathbb{R})$, explicitly given by

$$(7.46)\qquad (d/dx+1)^{-1}u(x) = e^{-x}\int_{-\infty}^{x} e^{y}u(y)dy \quad , \quad u \in \tilde{\mathfrak{f}} \quad ,$$

as easily derived by differentiation and noting that the right term at right is in $\tilde{\mathfrak{f}}$. Using (6.18) above we get

$$(7.47)\qquad \begin{aligned} MA_0M^* &= (1+d/dx)^{-1}[(d/dx - \tanh(x\pi/2)]) \\ MA_0^*M^* &= [d/dx - \tanh(x\pi/2)](-1+d/dx)^{-1} \end{aligned}$$

Let $a(x) = 2/\pi \log\cosh(x\pi/2)$, then $MA_0M^*u = 0$ for any distribution u implies

$$u(x) = c\,e^{a(x)} \quad , \quad \text{so that u is not in } L^2(\mathbb{R}) = \tilde{\mathfrak{f}} \text{ unless } c=0.$$

On the other hand $MA_0^*Mv = 0$ holds if $w = (-1+d/dx)^{-1}v$ is of the form $w = c\,e^{-a}$.

Note that $a \to +\infty$, as $x \to \pm\infty$. Hence

$$(7.49)\qquad v = dw/dx - w = c\,(\tanh(x\pi/2)-1)e^{-a} \in L^2(\mathbb{R}) = \tilde{\mathfrak{f}} \quad .$$

This implies that

(7.50) $\dim \ker A_0 = 0$, $\operatorname{codim} \operatorname{im} A_0 = 1$, $\operatorname{ind} A_0 = -1$,

q.e.d.

<u>Remarks.</u> 1) The above example is due to E. Herman (c.f. [18], p. 1672). 2) It is not known to the author whether or not the algebra $\mathfrak{A}$ is invariant under the substitution

(7.51) $u \to v(x) = u(1/x)/x$,

which represents a unitary operator of $\mathfrak{H}$ and leaves the two algebras $\mathfrak{Z}$ and $\mathfrak{M}$ invariant, but clearly not the algebra $\mathfrak{M}$. If it were, then the above index consideration would become superfluous, since in the symbol space of $\mathfrak{M}$ the corresponding homeomorphism is the reflection $t \to -t$.

<u>Problems.</u> 1) Let the function $\gamma: \mathbb{R} \to \mathbb{R}$ be defined by

(7.52) $\gamma(x) = \int_0^\infty \sin t \, dt/(x^2+t^2)$.

Prove that the operators $K^0_\pm : \mathfrak{H} \to \mathfrak{H}$, defined by

(7.53) $(K^0_\pm u)(y) = \int_0^\infty u(y') \, e^{-|y \pm y'|} dy'/(y \pm y')$,

with a Cauchy principal value integral in case of K^0_-, may be written in the form

(7.54) $K^0_\pm = -\pi(1-2/\pi^2 H_\gamma)K_\pm + C_\pm$,

with $C_\pm \in K(\mathfrak{H})$ and with H_γ, $K_\pm$ as in (5.1) and (6.15).

2) Show that $K^0_\pm$ (of problem 1) are operators in the algebra $\mathfrak{A}$, and that their symbol in $\mathfrak{A}$ is given by

(7.55)
$$\sigma_{K^0_-} = -2i \arctan \xi \cdot \tanh(t\pi/2) ,$$
$$\sigma_{K^0_-} = 2 \arctan \xi / \cosh(t\pi/2) .$$

3) Show that for an odd $\psi \in C_0^\infty(\mathbb{R})$ we get $H_\psi = -K_- H_w + C \in \mathfrak{A}$, with $C \in K(\mathfrak{H})$, and where

$\omega(x) = -1/\pi \int_{-\infty}^{+\infty} \psi(y)dy/(x-y)$, with a principal value integral, is even and in $L^1(\mathbb{R})$.

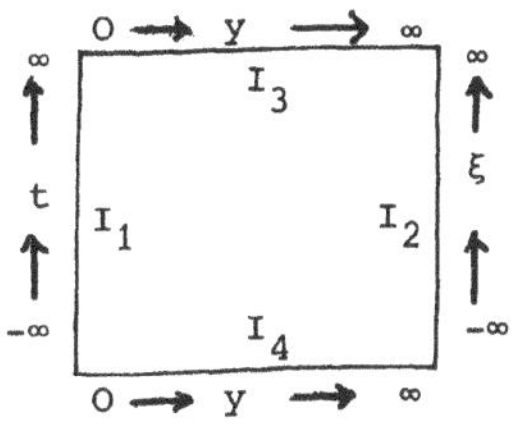

Fig. 7.5

4) Show that the algebra $\mathfrak{A}$ contains all Wiener-Hopf convolutions of the form (5.1) with $\phi \in L^1(\mathbb{R})$, not necessarily even. Moreover show that, with the Fourier transform $\phi^\wedge \in CO(\mathbb{R})$ of $\phi \in L^1(\mathbb{R})$, we get $\sigma_{H_\phi} = \phi^\wedge(\xi)$, $= \phi^\wedge(0)$, $= \phi^\wedge(-\xi)$, respectively, on the right three segments of the

hexagon $\mathbb{M}(\mathcal{O\!\!\!\!\!\!\!/})$ in Fig 7.2, proceeding downward, while $\sigma_{H_\phi} = 0$ otherwise on $\mathbb{M}(\mathcal{O\!\!\!\!\!\!\!/})$.

5) Show that the Laplace comparison algebra of the half-line, called $\mathcal{R}$ in section 4, is a subalgebra of $\mathcal{O\!\!\!\!\!\!\!/}$ above.

6) Show that the symbol space $\mathbb{M}(\mathcal{R})$ of the algebra $\mathcal{R}$ is given by the rectangle of Fig. 7.5, with the interpretation that

(7.56)
$$\sigma_{a(M)} = a(y) \text{ on } I_3 \cup I_4, \text{ continuous and constant otherwise ;}$$
$$\sigma_{Q_+} = 0 \ ; \ \sigma_{Q_-} = (1+\xi^2)^{-1/2} \text{ on } I_2, \ = 0 \text{ on } I_1 \cup I_3 \cup I_4 \ ;$$
$$\sigma_{P_-} = \tanh(t\pi/2) \text{ on } I_1, \ = \xi(1+\xi^2)^{-1/2} \text{ on } I_2, \text{ constant continuous}$$
$$\text{otherwise ; } \sigma_{P_+} = 1/\cosh(t\pi/2) \text{ on } I_1, \ = 0 \text{ on } I_2 \cup I_3 \cup I_4 \ .$$

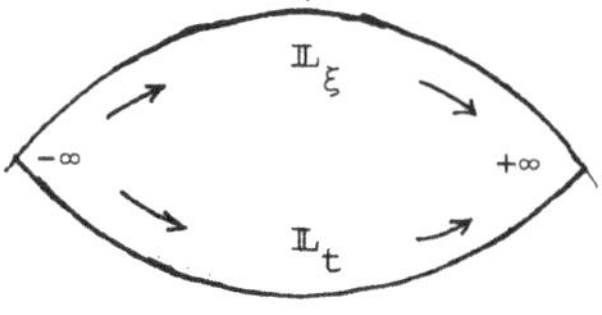

Fig. 7.6

7) Let $\mathcal{R}^\# \subset \mathcal{R}$ be the C^*-algebra with unit generated by $K(\mathfrak{f})$ and the "convolutions" $Q_\pm$, $P_\pm$ of $\mathcal{R}$ only. Show that the symbol space $\mathbb{M}(\mathcal{R}^\#)$ is given by the lens in Fig. 7.6, where

(7.57)
$$\sigma_{Q_+} = 0; \ \sigma_{Q_-} = (1+\xi^2)^{-1/2} \text{ on } \mathbb{L}_\xi, \ = 0 \text{ on } \mathbb{L}_t; \ \sigma_{P_+} = 1/\cosh(t\pi/2)$$
$$\text{on } \mathbb{L}_t, \ = 0 \text{ on } \mathbb{L}_\xi \ ; \ \sigma_{P_-} = \tanh(t\pi/2) \text{ on } \mathbb{L}_t,$$
$$= \xi(1+\xi^2)^{-1/2} \text{ on } \mathbb{L}_\xi \ .$$

8) Show that $\mathcal{R}^\#$ is generated by norm closing the finitely generated algebra with 1 containing $P_\pm$ and $Q_\pm$, so that the generators of $K(\mathfrak{f})$ are redundant. (Likewise for $\mathcal{R}$ and $\mathcal{O\!\!\!\!\!\!\!/}$). Hint: Recall the techniques of III, Lemma 10.1 and IV, Theorem 1.1 and Lemma 3.3.

8. C^*-algebras with matrix-valued or compact-operator-valued symbols

In the following we consider C^*-algebras of the form

(8.1) $$K(\mathfrak{f}) \hat{\otimes} \mathfrak{L} \subset L(\mathfrak{H})$$

in case where the Hilbert space $\mathfrak{H}$ is a direct product - i.e. a topological tensor product $\mathfrak{H} = \mathfrak{f} \hat{\otimes} \mathfrak{K}$ of two Hilbert spaces $\mathfrak{f}$ and $\mathfrak{K}$, where $\mathfrak{L}$ is a C^*-algebra of $L(\mathfrak{K})$ with complex valued symbol. That is, $\mathfrak{L}$ contains $K(\mathfrak{K})$ and the identity operator and has compact commutators so that the quotient $\mathfrak{L}/K(\mathfrak{K})$ again is commutative, hence $*$-isometrically isomorphic to $C(\mathbb{M})$ with some compact Hausdorff space $\mathbb{M} = \mathbb{M}(\mathfrak{L})$. For $A \in \mathfrak{L}$ the continuous function associated to its coset modulo $K(\mathfrak{K})$ is denoted by $\sigma_A = \sigma_A(m)$, $m \in \mathbb{M}$ again.

In [5] it was shown that algebras of type (8.1) are of general interest, because necessary and sufficient criteria can be obtained for an operator of the form $1 + B$, $B \in K(\mathfrak{f}) \hat{\otimes} \mathfrak{L}$, to be Fredholm, in terms of a symbol which happened to be a compact operator valued function over $\mathbb{M}$. This may be applied to the algebras $\mathfrak{A}_s = \mathfrak{L}$ of chapter IV, for example, to obtain criteria for systems of pseudo-differential equations, even for systems of infinitely many equations in infinitely many unknowns (c.f 9, problems 1, 2, 3).

There is another reason, however, to discuss these facts here.

It will be shown later on (Theorem 10.4) that the commutator ideal $\mathfrak{E}$ of our Laplace comparison algebra $\mathfrak{A}$ of the half space is isometrically isomorphic to an algebra of type (8.1), the isomorphism being established by conjugation with a unitary operator of $\mathfrak{H} = L^2(\mathbb{R}^{n+1}_+) = \mathfrak{f} \hat{\otimes} \mathfrak{K}$, with $\mathfrak{f}$ and $\mathfrak{K}$ as in section 4. Accordingly we usually will silently assume that $\mathfrak{f}$ and $\mathfrak{K}$ are explicitely given as in section 4. Also we will apply the results only to the case where $\mathfrak{L}$ is the C^*-subalgebra of $\mathfrak{A}_0$ in IV,1 generated by the operators (10.12), below, with $\mathbb{M}$ given by (10.13), for $d = n$.

A norm $\|f\|$ defined on the algebra tensor product $X \otimes \Xi = \mathfrak{X}$ of two Banach spaces X and Ξ is called a cross norm, if we have

(8.2) $$\|u \otimes \phi\| = \|u\| \cdot \|\phi\|, \quad u \in X, \ \phi \in \Xi.$$

If $\mathfrak{f}$ and $\mathfrak{K}$ denote two Hilbert spaces, then on the algebraic tensor product

(8.3) $$\mathfrak{H}' = \mathfrak{f} \otimes \mathfrak{K},$$

defined as the collection

(8.4) $$\mathfrak{H}' = \{\sum_{j=1}^{N} \phi_j \otimes u_j : \phi_j \in \mathfrak{f}, \ u_j \in \mathfrak{K}\}$$

of all finite formal sums of tensor products $\phi \otimes u$, with the obvious vector operations one may introduce an inner product

(8.5) $$(f,g) = \sum_{j,k=1}^{N,M} (\phi_j, \psi_k)(u_j, v_k),$$

for

$$(8.6)\qquad f = \sum_{j=1}^{N} \phi_j \otimes u_j \ , \ g = \sum_{k=1}^{M} \psi_k \otimes v_k \ \varepsilon\ \mathcal{G}',$$

with the inner products

$$(8.7)\qquad (\phi_j,\psi_k) \text{ and } (u_j,v_k)$$

of $\mathcal{J}$ and $\mathcal{R}$, respectively.

In order to confirm that (8.5) defines an inner product, we must show that it is a definite sesqui-linear form. That is, we must confirm that

$$(8.8)\qquad (f,f) = \sum_{j,k=1}^{N} (\phi_j,\phi_k)(u_j,u_k) > 0 \quad ,$$

unless f = 0. Note that it may be assumed that both of the collections $\phi_1,\phi_2,\ldots,\phi_N$ and $u_1,u_2,\ldots,u_N$, are linearly independent, unless f = 0 (in which case we may write $f = 0 \otimes 0$.). Indeed, if one of the sets ϕ_j or u_j above are linearly dependent, - assume u_j, for am instant, - then the u_j may be expressed linearly by fewer than N linearly independent vectors which results in another representation of the type (8.6), for the same f, with fewer, but independent u_j. Now, if in the new representation the ϕ_j are dependent, we will express these linearly by a system of still fewer vectors, which results in another representation of f again with still fewer terms in the sum. If $f \neq 0$ this recursive process must break off some times and supply a representation (8.6) for f, where both systems ϕ_j and u_j are linearly independent, respectively. Then the two matrices

$$(8.9)\qquad (((\phi_j,\phi_k)))_{j,k=1,\ldots,N}(((u_j,u_k)))_{j,k=1,\ldots,N}$$

both are positive definite. Using Schur's lemma (c.f. Polya-Szego [29],), we conclude that also the matrix

$$(8.10)\qquad ((c_{jk}))\ ,\ c_{jk} = (u_j,u_k)(\phi_j,\phi_k) \quad ,$$

is positive definite. Therefore we get

$$(8.11)\qquad (f,f) = \sum_{j,k=1}^{N} c_{jk} > 0 \quad ,$$

which indeed proves that the inner product (8.5) is positive definite.

In our case where

$$(8.12)\qquad \mathcal{J} = L^2(\mathbb{R}_+)\ ,\ \mathcal{R} = L^2(\mathbb{R}^n) \ ,$$

one finds that f and g of (8.6) take the form

$$(8.13)\qquad f(y,x) = \sum_{j=1}^{N} \phi_j(y)u_j(x)\ ,\ g(y,x) = \sum_{k=1}^{M} \psi_k(y)v_k(x) \quad ,y\in\mathbb{R}_+ \ ,\ x\in\mathbb{R}^n \ ,$$

in which case the inner product (8.5) simply takes the form

$$(8.14)\qquad (f,g) = \int_{\mathbb{R}_+^{n+1}} \bar{f}(y,x)g(y,x)\ dx\ dy$$

We note that the inner product (8.5) induces a norm on the algebraic tensor product $\mathfrak{H} \otimes \mathfrak{K}$, defined by

$$\|f\| = \{(f,f)\}^{1/2} \quad . \tag{8.15}$$

This norm is a cross-norm, as defined above, as follows immediately. The linear vector space $\mathfrak{G}'$ will become a pre-Hilbert space, with this norm (3.15) and may be completed to form a Hilbert space $\mathfrak{G}$, denoted as the topological tensor product

$$\mathfrak{G} = \mathfrak{H} \hat{\otimes} \mathfrak{K} \tag{8.16}$$

(or the direct product) of the Hilbert spaces $\mathfrak{H}$ and $\mathfrak{K}$.

Our discussion, below will follow closely the pattern laid out in [6], by M. Breuer and the author.

Lemma 8.1. If $\mathfrak{H}$ and $\mathfrak{K}$ are separable, then $\mathfrak{G}$ is separable.

Proof. Since $\mathfrak{G}'$ is dense in $\mathfrak{G}$ by definition, it will only have to be shown that there exists a countable and dense subset of $\mathfrak{G}'$. If $\{\omega_j\}$ and $\{w_j\}$ form dense and countable subsets of $\mathfrak{H}$ and $\mathfrak{K}$ respectively, then the collection

$$\{\sum_{l=1}^{R} \omega_{j_l} \otimes w_{k_l}\} \tag{8.17}$$

of all finite sums of tensor products $\omega_j \otimes w_k$ is countable, and also is dense, as becomes obvious from the inequality

$$\|f\| \le N \operatorname*{Max}_{j=1}^{N} \|\phi_j\| \operatorname*{Max}_{j=1}^{N} \|u_j\| \quad , \tag{8.18}$$

for f of the form (8.6). Q.e.d.

Lemma 8.2. We have

$$L^2(\mathbb{R}_+) \hat{\otimes} L^2(\mathbb{R}^n) = L^2(\mathbb{R}_+^{n+1}) \quad . \tag{8.19}$$

Proof. This is an immediate consequence of the Riesz-Fischer theorem.

For a linear operator $S \in L(\mathfrak{H})$ and a linear operator $T \in L(\mathfrak{K})$ we define a tensor product $S \otimes T \in L(\mathfrak{G})$ by setting

$$(S \otimes T)(\phi \otimes u) = (S\phi) \otimes (Tu) . \tag{8.20}$$

By linear extension this defines a linear operator over $\mathfrak{G}'$. We then must show that

$$\|(S \otimes T)f\| \le \|S\| \cdot \|T\| \cdot \|f\| , \ f \in \mathfrak{G}', \tag{8.21}$$

which will enable us to extend the operator $S \otimes T$, thus far only defined on $\mathfrak{G}'$, to all of the space $\mathfrak{G}$, by continuous extension. We also conclude from (8.21) that we have

$$\|S \otimes T\| \le \|S\| \cdot \|T\| \quad . \tag{8.22}$$

In fact, we must have equality in (8.22), as follows immediately, if we apply $S \otimes T$

to a sequence $\{\phi_l \otimes u_l\}$, such that

$$\|S\| = \lim_{l\to\infty} \|S\phi_l\|, \quad \|T\| = \lim_{l\to\infty} \|Tu_l\| \quad , \tag{8.23}$$

while

$$\|\phi_l\| = \|u_l\| = 1, \; l = 1,2,\ldots \quad . \tag{8.24}$$

For the proof of (3.21) we write

$$S \otimes T = (S \otimes 1)(1 \otimes T), \tag{8.25}$$

with the identity operator in $\mathfrak{f}$ or $\mathfrak{k}$ denoted by "1", and then note that it is sufficient to prove that (8.21) holds in the case of S = 1. This becomes evident, if we notice that among the various representations of $f \in \mathfrak{g}'$ there are some (of the form (8.6)) with the vectors ϕ_j, j = 1,...,N, forming an orthonormal system:

$$(\phi_j, \phi_l) = \delta_{jl} \quad . \tag{8.26}$$

Indeed, for any given such representation the ϕ_j may be expressed linearly by any orthonormal base of the space spanned by them, which will give such special representation.

Now, for a representation of f in the form (8.6) with (8.26) we will get

$$\|f\|^2 = \sum_{j=1}^{N} \|u_j\|^2, \quad \|(1 \otimes T)f\|^2 = \sum_{j=1}^{N} \|Tu_j\|^2 \quad . \tag{8.27}$$

This implies (8.21), for the operator $1 \otimes T$.

By taking formal finite sums of tensor products $S \otimes T$ of the above form, we arrive at an isometric imbedding of the tensor product

$$L(\mathfrak{f}) \otimes L(\mathfrak{k}) = \{\sum_{j=1}^{N} S_j \otimes T_j : S_j \in L(\mathfrak{f}), T_j \in L(\mathfrak{k})\} \tag{8.28}$$

into the algebra $L(\mathfrak{g})$, where the norm of an operator in $L(\mathfrak{f}) \otimes L(\mathfrak{k}) \subset L(\mathfrak{g})$ is defined to be the operator norm of $L(\mathfrak{g})$. Please observe that we have proven that the operator norm of $L(\mathfrak{g})$ induces a cross-norm in the algebraic tensor product (8.28) of $L(\mathfrak{f})$ and $L(\mathfrak{k})$. We thus may complete $L(\mathfrak{f}) \otimes L(\mathfrak{k})$, that is, take the norm closure in $L(\mathfrak{g})$, to obtain a closed subalgebra of $L(\mathfrak{g})$ denoted by

$$L(\mathfrak{f}) \hat{\otimes} L(\mathfrak{k}) \tag{8.29}$$

and called the <u>topological tensor product</u>, or <u>direct product</u> of $L(\mathfrak{f})$ and $L(\mathfrak{k})$.

We note without proof that $L(\mathfrak{f}) \hat{\otimes} L(\mathfrak{k})$ is a proper subalgebra of $L(\mathfrak{g})$, whenever $\dim \mathfrak{f} = \infty$, $\dim \mathfrak{k} = \infty$.

Similarly we may consider the tensor products $\mathfrak{T} \otimes \mathfrak{Z}$ of subalgebras $\mathfrak{T} \subset L(\mathfrak{f})$ and $\mathfrak{Z} \subset (\mathfrak{k})$ as subalgebras of $L(\mathfrak{g})$ and again may define the <u>direct product</u> $\mathfrak{T} \hat{\otimes} \mathfrak{Z}$ of $\mathfrak{T}$ and $\mathfrak{Z}$ as the closure of the tensor product $\mathfrak{T} \otimes \mathfrak{Z}$ under the operator norm of $L(\mathfrak{g})$ which again defines a cross norm on $\mathfrak{T} \hat{\otimes} \mathfrak{Z}$.

It is clear that $L(\mathfrak{f}) \hat{\otimes} L(\mathfrak{k})$, as well as $\mathfrak{T} \hat{\otimes} \mathfrak{Z}$ define a Banach algebra each; in fact they each define C^*-operators algebras of $L(\mathfrak{g})$, if we still require that

$\mathfrak{p}$ and $\mathfrak{q}$ both contain their Hilbert space adjoints ,as easily confirmed.

Of special interest, in this respect, are the three algebras

(8.30) $$L(\mathfrak{h}) \hat{\otimes} K(\mathfrak{k}),\ K(\mathfrak{h}) \hat{\otimes} L(\mathfrak{k}),\ K(\mathfrak{h}) \hat{\otimes} K(\mathfrak{k}).$$

All three C^*-subalgebras (3.30) of $L(\mathfrak{g})$ are two-sided closed ideals of the C^*-algebra $L(\mathfrak{h}) \hat{\otimes} L(\mathfrak{k})$, as evident from the fact that the corresponding algebraic tensor products all are two-sided ideals in the algebraic tensor product $L(\mathfrak{h}) \otimes L(\mathfrak{k})$. If precisely one of the two spaces $\mathfrak{h}$ and $\mathfrak{k}$ is finite dimensional - for instance the space $\mathfrak{k}$ - then $L(\mathfrak{k}) = K(\mathfrak{k})$, then the last two of the algebras (8.30) coincide with the compact ideal $K(\mathfrak{g})$, while the first one equals $L(\mathfrak{h}) \otimes L(\mathfrak{k}) = L(\mathfrak{h}) \hat{\otimes} L(\mathfrak{k}) = L(\mathfrak{g})$, as follows from simple reflections.

If both spaces $\mathfrak{h}$ and $\mathfrak{k}$ are finite dimensional, then all the algebras (8.29) and (8.30) and the corresponding algebraic tensor products and the two algebras $L(\mathfrak{g})$ and $K(\mathfrak{g})$ all coincide. In this case, all the above reduces to matrix calculus.

If both spaces $\mathfrak{h}$ and $\mathfrak{k}$ are infinite dimensional, then the algebras (8.29) and (8.30) are distinct. For example the operators $1 \otimes E$ and $F \otimes 1$ with $0 \neq E \in E(\mathfrak{k})$, $0 \neq F$ $E(\mathfrak{h})$, are in $L(\mathfrak{h}) \hat{\otimes} K(\mathfrak{k})$ and $K(\mathfrak{h}) \hat{\otimes} L(\mathfrak{k})$, resp., but are not compact, hence not in $K(\mathfrak{h}) \hat{\otimes} K(\mathfrak{k})$, by lemma 8.3. The algebras $L(\mathfrak{h}) \hat{\otimes} K(\mathfrak{k})$ and $K(\mathfrak{h}) \hat{\otimes} L(\mathfrak{k})$ are proper closed two-sided ideals of $L(\mathfrak{h}) \hat{\otimes} L(\mathfrak{k})$, which properly contain the compact ideal. These algebras are R-algebras in the sense defined by the author [9]. They are defined by means of a "Rellich criterion", with the family of closed subspaces with finite co-dimension replaced by another family of closed subspaces of (called an R-set.). Also it seems that they play a not unimportant role in theory of generalized Fredholm operators (c.f. [9], [10] and [11], also, AI, 4).

<u>Lemma 8.3.</u> We have

(8.31) $$K(\mathfrak{h}) \hat{\otimes} K(\mathfrak{k}) = K(\mathfrak{h} \hat{\otimes} \mathfrak{k}).$$

<u>Proof.</u> For a Hilbert space the compact ideal equals the norm closure of the ideal E of all linear operators of finite rank (of all dyads, that is.). Now, it is evident that we have

(8.32) $$E(\mathfrak{h}) \otimes E(\mathfrak{k}) = \tilde{E}(\mathfrak{h} \otimes \mathfrak{k}) = \tilde{E}(\mathfrak{g}')$$

where $\tilde{E}(\mathfrak{h} \otimes \mathfrak{k})$ denotes the collection of all dyads

(8.33) $$F = \sum_{j=1}^{N} f_j \rangle\langle g_j \quad , f_j, g_j \in \mathfrak{h} \otimes \mathfrak{k} = \mathfrak{g}' \quad .$$

Also we note that (the algebra) $\tilde{E}(\mathfrak{g}')$ is dense in $E(\mathfrak{g})$. For if $Z \in E(\mathfrak{g})$ has a representation of the form

(8.34) $$Z = \sum_{j=1}^{N} f_j \rangle\langle g_j \quad , f_j, g_j \in \mathfrak{g},$$

then the f_j and g_j may be approximated by elements of $\mathfrak{g}'$ to any degree of accuracy, which leads to corresponding approximations of Z by dyads of $\tilde{E}(\mathfrak{g}')$ in operator norm of $L(\mathcal{G})$. Accordingly, if we take closures in relation (8.32), we obtain (8.31), q.e.d.

Let us define the C^*-subalgebra $\mathcal{L}^\infty$ of $L(\mathcal{G})$ by setting

(8.35) $$\mathcal{L}^\infty = K(\mathcal{f}) \mathbin{\hat{\otimes}} \mathcal{L} \quad .$$

Lemma 8.4. The algebra

(8.36) $$\mathcal{L}^{\infty'} = E(\mathcal{f}) \otimes \mathcal{L}$$

is a dense sub-algebra of $\mathcal{L}^\infty$, under operator norm.

The proof of this lemma is a consequence of the fact that in any Hilbert space the compact ideal is the norm closure of the algebra of dyads, so that we conclude that $E(\mathcal{f}) \otimes \mathcal{L}$ is dense in $K(\mathcal{f}) \otimes \mathcal{L}$. Then the statement follows by taking closure, q.e.d.

Next, we note that it is natural to introduce a symbol

(8.37) $$\sigma_A : \mathbb{M} \to E(\mathcal{f})$$

defined as a function from the space $\mathbb{M}$ to the algebra $E(\mathcal{f}) \subset K(\mathcal{f}) \subset L(\mathcal{f})$, for every operator

(8.38) $$A = \sum_{j=1}^{N} E_j \otimes A_j \in \mathcal{L}^{\infty'}$$

by setting

(8.39) $$\sigma_A(m) = \sum_{j=1}^{N} \sigma_{A_j}(m) E_j \quad .$$

This definition is independent of the special representation (8.39), and therefore consistent.

It is our intention now to prove that the homomophism

(8.40) $$\dot{\sigma} : \mathcal{L}^{\infty'} \to C(\mathbb{M}, E(\mathcal{f}))$$

defined by (8.37), or, explicitely, by(8.39), is continuous, as a map from (a subset of) $\mathcal{L}^\infty$ to $C(\mathbb{M}, K(\mathcal{f}))$, under the norm topologies of these algebras. This will be done in the next following section. Then the homomorphism $\dot{\sigma}$ of (8.40) may be extended to all of $\mathcal{L}^\infty$, to form a continuous homomorphism

(8.41) $$\sigma : \mathcal{L}^\infty \to C(\mathbb{M}, K(\mathcal{f})) \quad .$$

For an operator $A \in \mathcal{L}^\infty$ the function of $C(\mathbb{M}, K(\mathcal{f}))$ associated to A by the homomorphism σ will be denoted by σ_A again, and will be called the symbol of A.

The symbol of $A \in \mathcal{L}^\infty$ is a continuous function $\sigma_A(m)$ defined on the space $\mathbb{M}$ taking its values in the algebra $K(\mathcal{f})$ of compact operators over $\mathcal{f}$.

Again we shall have to ask the question then,whether the symbol σ_A is related to the Fredholm properties of the operator A, and we shall except (and prove) results similar to those of chapter IV.

Problems. 1) For a norm continuous function $A: \mathbb{B}^n \to L(\mathcal{f})$, with $\mathbb{B}^n$ as in IV, 1, and

$\mathfrak{f} = L^2(\mathbb{R}_+)$, $\mathfrak{K} = L^2(\mathbb{R}^n)$, we define a linear operator B: $\mathfrak{f} \otimes \mathfrak{K} \to \mathfrak{g} = \mathfrak{f} \hat{\otimes} \mathfrak{K}$ by setting , for almost all x,

(8.42) $$(Bu)(y,x) = (A(x)u(\cdot,x))(y) \quad .$$

Show that B extends continuously to an operator in $L(\mathfrak{g})$, denoted by A(M), and that

(8.43) $$\|A(M)\| = \sup \{ \|A(x)\| : x \in \mathbb{B}^n\} \quad .$$

2) Prove that the isometric imbedding $C(\mathbb{B}^n, L(\mathfrak{f})) \to L(\mathfrak{g})$, given by $A(x) \to A(M)$, of problem 1 defines a map

(8.44) $$C(\mathbb{B}^n, K(\mathfrak{f})) \to K(\mathfrak{f}) \hat{\otimes} \mathfrak{Q},$$

with the algebra $\mathfrak{Q} = \{a(M) : a \in C(\mathbb{B}^n)\} \subset L(\mathfrak{K})$.

3) For a C^*-subalgebra $\mathfrak{T} \subset L(\mathfrak{f})$, containing $\{K(\mathfrak{f}) + 1\}$ and with compact commutators show that $C(\mathbb{B}^n, \mathfrak{T}) \to \mathfrak{W} \subset L(\mathfrak{g})$, under the isometry (8.42), with a C^*-algebra $\mathfrak{W}$ containing $\mathfrak{T} \hat{\otimes} \mathfrak{Q}$.

4) Show that, in problem 3, we have $\mathfrak{W} = \mathfrak{T} \hat{\otimes} \mathfrak{Q}$, and that

(8.45) $$(\mathfrak{T} \hat{\otimes} \mathfrak{Q}) / (K(\mathfrak{f}) \hat{\otimes} \mathfrak{Q}) \cong C(\mathbb{M}_{\mathfrak{T}} \times \mathbb{B}^n),$$

with the Gel'fand space $\mathbb{M}_{\mathfrak{T}}$ of $\mathfrak{T}/K(\mathfrak{f})$, Hint: Use Stone-Weierstrass to show that the commutative C^*-algebra $\mathfrak{W}/(K(\mathfrak{f}) \hat{\otimes} \mathfrak{Q})$ has Gel'fand space $\mathbb{M}_{\mathfrak{T}} \times \mathbb{B}^n$. Then note that $C(\mathbb{M}_{\mathfrak{T}} \times \mathbb{B}^n) = C(\mathbb{M}_{\mathfrak{T}}) \hat{\otimes} C(\mathbb{B}^n)$, again using Stone-Weierstrass. Finally, if the coset $A^{\vee}(\vec{M})$ of $A(\vec{M})$ can be approximated by elments of $(\mathfrak{T} \hat{\otimes} \mathfrak{Q}) / (K(\mathfrak{f}) \hat{\otimes} \mathfrak{Q})$ then the same must be true for $A(\vec{M}) \in \mathfrak{W}$ (by elements of $\mathfrak{T} \hat{\otimes} \mathfrak{Q}$).

9. Fredholm and structure theory of C^*-algebras with compact (operator valued) symbol.

We return to the discussion of the preceeding section and note at first, that an operator in $\mathcal{L}^{\infty'} = E(\mathfrak{f}) \otimes \mathcal{L}$ has its symbol identically zero on $\mathbb{M}$ if and only if it is compact. Indeed, we may resort to a representation (8.38) again, in which all the N dyads E_j are linearly independent, and then find that σ_A, defined by (8.39) vanishes identically if and only if all N functions $\sigma_{A_j}(m)$ vanish identically, which happens if and only if $A_j \in K(\mathfrak{k})$, $j = 1,\dots,N$. This proves that $A = \sum_{j=1} E_j \otimes C_j$ must be compact.

Let us first consider the sub-algebra of $\mathcal{L}^{\infty'}$ given by

(9.1) $$\mathcal{L}^{\mathfrak{r}} = \{\sum_j E_j \otimes A_j : \operatorname{im} E_j \subset \mathfrak{r},\ \operatorname{im} E_j^* \subset \mathfrak{r},\ j = 1,\dots,N\}.$$

where $\mathfrak{r}$ denotes any closed subspace of $\mathfrak{f}$ of finite dimension R.

Clearly the operators of $\mathcal{L}^{\mathfrak{r}}$ are reduced by the direct decomposition

(9.3) $$\mathfrak{H} = (\mathfrak{r} \otimes \mathfrak{k}) \oplus (\mathfrak{r}^{\perp} \hat{\otimes} \mathfrak{k}) ,$$

and are zero on the orthogonal complement $(\mathfrak{r} \otimes \mathfrak{k})^{\perp} = \mathfrak{r}^{\perp} \hat{\otimes} \mathfrak{k}$ of $\mathfrak{r} \otimes \mathfrak{k}$. Thus $\mathcal{L}^{\mathfrak{r}}$ is isometrically isomorphic to the algebra $\mathcal{L}_0^{\mathfrak{r}}$ of the restrictions of its operators to $\mathfrak{H}^{\mathfrak{r}} = \mathfrak{r} \otimes \mathfrak{k}$. Also $\mathcal{L}_0^{\mathfrak{r}} \subset L(\mathfrak{r}) = L(\mathfrak{r}) \otimes L(\mathfrak{k})$ in effect is simply the algebra of all R × R-matrices with entries in $\mathcal{L}$,

(9.4) $$B = ((B_{jk}))_{j,k=1,\dots,R}, \quad B_{jk} \in \mathcal{L} ,$$

relative to any given orthonormal base of $\mathfrak{r}$. Thus we get the isometric isomorphism

(9.5) $$\mathcal{L}^{\mathfrak{r}}/K(\mathfrak{H}^{\mathfrak{r}}) \cong C(\mathbb{M}, L(\mathfrak{r}))$$

which is just a matter of matrix calculus: We get

(9.6) $$K(\mathfrak{H}^{\mathfrak{r}}) = \{((C_{jk}))_{j,k=1,\dots,R}: C_{jk} \in K(\mathfrak{f}), j,k=1,\dots,R\} .$$

Hence

(9.7) $$\{A + K(\mathfrak{H}^{\mathfrak{r}})\} = \{((A_{jk} + K(\mathfrak{k})))\} = \{((\sigma_{A_{jk}}))\}$$

establishes a $*$-isomorphism between the two sides of (9.5), which must be an isometry, by AII, theorem 7.17.

Now let $P = P_{\mathfrak{r}}$ denote the orthogonal projection of $\mathfrak{H}$ with image $\mathfrak{H}^{\mathfrak{r}}$ and note that for $A \in \mathcal{L}^{\mathfrak{r}}$ we get $A = PAP$, hence

(9.8) $$\|A + PCP\| = \|P(A + C)P\| \le \|A + C\| ,$$

for every $C \in K(\mathfrak{H})$. Taking infima we get

(9.9) $$\inf\{\|A + C\| : C \in K(\mathfrak{H})\} = \inf\{\|A + PCP\| : C \in K(\mathfrak{H})\} ,$$

since the infimum at right is taken over the smaller set of compact operators of the form PCP, hence cannot be smaller than the left hand side infimum.

But the collection

(9.10) $\{PCP: C \in K(\mathcal{G})\}$

is contained in $\mathcal{L}^{\mathfrak{J}}$ and corresponds to $K(\mathcal{G}^{\mathfrak{J}})$ under the isometry $\mathcal{L}^{\mathfrak{J}} \leftrightarrow \mathcal{L}_0^{\mathfrak{J}}$. Therefore (9.5) and (9.9) imply an isometric isomorphism

(9.11) $$\mathcal{L}^{\mathfrak{J}}/K(\mathcal{G}) \cong C(\mathbf{M}, L(\mathfrak{J})) \quad .$$

Or, in details: For every $A \in \mathcal{L}^{\mathfrak{J}}$ we have

(9.12) $$\inf \{ \|A + C\| : C \in K(\mathcal{G}) \} = \sup \{ \|\sigma_A(m)\| : m \in \mathbf{M}\}$$

However, it may be noticed that every operator $A \in \mathcal{L}^{\infty'}$ is in some $\mathcal{L}^{\mathfrak{J}}$ with suitably chosen finite dimensional space $\mathfrak{J}$. Accordingly, we have relation (9.12) satisfied throughout all of the algebra $\mathcal{L}^{\infty'}$ and therefore have proven the lemma, below.

<u>Lemma 4.1.</u> The symbol homomorphism $\dot{\sigma}$ of (8.40), as defined by (8.38) and (8.39) is a contractive (hence continuous) map from the algebra $\mathcal{L}^{\infty'} \subset \mathcal{L}^{\infty}$ to the algebra $C(\mathbf{M}, E(\mathfrak{f})) \subset C(\mathbf{M}, K(\mathfrak{f}))$, under their respective natural norms, which induces an isometry

(9.13) $$\dot{\kappa}: \mathcal{L}'/K(\mathcal{G}) \to C(\mathbf{M}, K(\mathfrak{f})).$$

Moreover, the image of the homomorphism $\dot{\sigma}$ contains every continuous function

(9.14) $$\omega: \mathbf{M} \to E^{\mathfrak{J}}$$

where $E^{\mathfrak{J}}$ denotes the algebra of all finite rank operators $E: \mathfrak{f} \to \mathfrak{f}$ with im E and im E^* contained in the given finite dimensional space $\mathfrak{J}$.

We now indeed may proceed by extending the homomorphism $\dot{\sigma}$ continuously onto all of the algebra $K(\mathfrak{f}) \hat{\otimes} \mathcal{L} = \mathcal{L}^{\infty}$, to obtain

(9.15) $$\sigma: \mathcal{L}^{\infty} \to C(\mathbf{M}, K(\mathfrak{f})),$$

with the property that σ induces an isometric isomorphism

(9.16) $$\mathcal{L}^{\infty}/K(\mathcal{G}) \longleftrightarrow C(\mathbf{M}, K(\mathfrak{f})) \quad ,$$

which is surjective, and a $*$-isomorphism.

Next we observe that the C^*-algebra $\mathcal{L}^{\infty}$ does not possess an identity, if $\dim \mathfrak{f} = \infty$. Indeed, since $A \in \mathcal{L}^{\infty}$ can be approximated arbitrarily by a finite sum

(9.17) $$A = \sum E_j \otimes A_j \quad , \qquad E_j \in E(\mathfrak{f}),$$

which vanishes in the space

(9.18) $$\mathfrak{q} \hat{\otimes} \mathfrak{k} \quad , \mathfrak{q} = \bigcap \ker E_j \quad , \quad \dim \mathfrak{q} = \infty \quad ,$$

we would get the contradiction

(9.19) $$\|f\| = \|(1 \otimes 1)f\| = \|((1 \otimes 1) - A)f\| \leq \varepsilon \|f\|,$$

for all $f \in \mathfrak{q} \hat{\otimes} \mathfrak{k}$, where ε could be chosen 1/2, for example. Hence the identity operator $1 = (1 \otimes 1)$ indeed cannot be in $\mathcal{L}^{\infty}$, whenever $\mathfrak{f}$ has finite dimension.

In the following, let us assume that $\mathfrak{f}$ is separable and infinite dimensional.

Then the Banach-algebra $\mathcal{L}_1^\infty = \mathcal{L}^\infty + \{1\}$, as discussed in AII, 1 will also be a C^*-algebra, and a subalgebra of $L(\mathfrak{H})$. All its elements are of the form $c + A$, with $c \in \mathbb{C}$, and $A \in \mathcal{L}^\infty$. Instead of introducing the norm AII, (1.3) we will prefer the operator norm in $\mathcal{L}_1^\infty$, which must be equivalent to that norm: Suppose $\| c_1 + A_1 \| \to 0$, $c_1 \in \mathbb{C}$, $A_1 \in \mathcal{L}^{\infty'}$, then, if not $c_1 \to 0$ we get either $c_1 \to \infty$ or $c_1 \to c_0 \neq 0$ (by passing to a subsequence). In either case we get $\|1 + A_1/c_1\| \to 0$, resulting in a contradiction, since every A_1 has a nontrivial nulspace. Hence we also get $\|A_1\| \to 0$ which proves the desired equivalence (c.f. also AII, prop. 8.2).

In the C^*-algebra

$$\mathcal{L}_1^\infty = 1 + \mathcal{L}^\infty = \{A+c: A \in \mathcal{L}^\infty, c \in \mathbb{C}\} \tag{9.20}$$

we may introduce the symbol homomorphism

$$\sigma_B(m) = c + \sigma_A(m) \quad , \quad m \in \mathbf{M} \quad , B = c + A \in \mathcal{L}_1^\infty \quad , \tag{9.21}$$

which is uniquely defined in $\mathcal{L}_1^\infty$, and extends the symbol homomorphism of $\mathcal{L}^\infty$, since the decomposition $B = A + c$ is unique for $B \in \mathcal{L}_1^\infty$ and $A \in \mathcal{L}^\infty$.

Note that the values of the symbol for the C^*-algebra $\mathcal{L}_1^\infty$ are in the algebra

$$K_1(\mathfrak{f}) = \{1\} + K(\mathfrak{f}) = \{C + c : C \in K(\mathfrak{f}), c \in \mathbb{C}\} \quad . \tag{9.22}$$

Again we note that $K_1(\mathfrak{f})$, above, is a closed sub-algebra of $L(\mathfrak{f})$, containing its adjoints, by an argument like that above for $\mathcal{L}_1^\infty$, (under the operator norm of $L(\mathfrak{f})$, of course.). The symbol does not assume all continuous functions

$$\sigma : \mathbf{M} \to K_1(\mathfrak{f}) \quad , \tag{9.23}$$

but precisely those of the form $c + C(m)$, with constant $c \in \mathbb{C}$ and arbitrary continuous $C: \mathbf{M} \to K(\mathfrak{f})$.

Clearly we get $\sigma_B = 0$ if and only if $B = c + A$, with $c = 0$ and $\sigma_A(m) = 0$, so that $\sigma_B = 0$ if and only if $B \in K(\mathfrak{f})$. This shows that the extended symbol still induces a $*$-isomorphism

$$\mathcal{L}_1^\infty / K(\mathfrak{H}) \to C_1(\mathbf{M}, K_1(\mathfrak{f})) \tag{9.24}$$

with the expression, at right, denoting the algebra of all continuous functions $c + C(m)$, $m \in \mathbf{M}$ with constant $c \in \mathbb{C}$. Accordingly, the isomorphism (9.24) still is an isometry, again by AII, theorem 7.17. Again the algebra $\mathcal{L}_1^\infty$ is Fredholm closed, and an operator $A \in \mathcal{L}_1^\infty$ is Fredholm if and only if its coset modulo $K(\mathfrak{H})$ is invertible. We have proven the result, below.

<u>Theorem 9.2.</u> If $\mathcal{L}^\infty = K(\mathfrak{f}) \hat{\otimes} \mathcal{L} \subset L(\mathfrak{H})$, where $\mathfrak{H} = \mathfrak{f} \hat{\otimes} \mathfrak{k}$, and $\mathcal{L} \subset L(\mathfrak{k})$ is a C^*-algebra with compact commutator, containing $K_1(\mathfrak{f}) = 1 + K(\mathfrak{f})$, then the homomorphism $\sigma: \mathcal{L}_1^\infty \to C_1(\mathbf{M}, K_1(\mathfrak{f}))$ of (9.21) has the following properties.

(i) An operator $c + A$, with $A \in \mathcal{L}^\infty$, $c \in \mathbb{C}$, is Fredholm if and only if all operators $c + \sigma_A(m)$, for $m \in \mathbf{M}$ admit a (bounded) inverse in $L(\mathfrak{f})$, which then is also of the form $c^{-1} + C(m)$, with $C(m) \in K(\mathfrak{f})$;

(ii) we have

(9.25) $$\ker \sigma = K(\mathcal{G}), \ \mathrm{im}\ \sigma = C_1(\mathbb{M}, K(\mathcal{J}));$$

(iii) we have

(9.26) $$\inf \{\|c + A + C\| : C \in K(\mathcal{G})\} = \sup \{\|\sigma_A(m) + c\| : m \in \mathbb{M}\} \quad ,$$

for all $A \in \mathcal{L}^\infty$, and $c \in \mathbb{C}$.

Problems. 1) Let $\mathcal{G}_\infty$ and $\mathcal{O}\!\mathcal{l}_\infty$ be as in IV, 3. Show that the linear operators of the algebra $\mathcal{L}^N_\infty = L(\mathbb{C}^N) \otimes \mathcal{O}\!\mathcal{l}_\infty$ which is a C^*-subalgebra of $L(\mathbb{C}^N \otimes \mathcal{G}_\infty)$ can be provided with an $N \times N$-matrix-valued symbol by setting

(9.27) $$\sigma_A = ((\sigma_{A_{jk}})) \text{ for } A = ((A_{jk}))_{j,k=1,\dots,N} \in \mathcal{L}^N_\infty.$$

Show that (9.27) provides a homomorphism $\mathcal{L}^N_\infty \to C(\mathbb{M}, L(\mathbb{C}^N))$, with $\mathbb{M}$ as in IV, (1.23), having kernel $\mathcal{L}_0$ (of IV, (3.10)).

2) With the matrix-valued symbol on $\mathcal{L}^N_\infty$ of problem 1, show that $A \in \mathcal{L}^N_\infty$ admits a special Green inverse in $L(\mathbb{C}^N) \otimes \mathcal{O}(0) = \mathcal{O}^N(0)$ if and only if $\sigma_A(x,\xi)$ is a nonsingular $N \times N$-matrix for all $(x,\xi) \in \mathbb{M}$. Here a special Green inverse is defined to be an operator $B = ((B_{jk})) \in \mathcal{O}^N(\infty) = L(\mathbb{C}^N) \otimes \mathcal{O}(\infty)$ such that

(9.28) $$AB = 1 - P_{N^*} \quad , \ BA = 1 - P_N \quad ,$$

with projections P_N, P_{N^*} onto the spaces $N = \ker A$, $N^* = \ker A^*$ being $N \times N$-matrices of $\mathcal{O}(-\infty)$-dyads. Again the $\mathcal{O}(\infty)$-adjoint A^* of A is given by $A^* = ((A^*_{kj}))$, with the (∞)-adjoints A^*_{jk} of A_{jk} as in III, 5.

3) Let $\Psi\mathcal{D}^N = \bigcup_{s\in\mathbb{R}} \Psi\mathcal{D}^N_s \subset \mathcal{O}^N(\infty)$ with $\Psi\mathcal{D}^N_s = \{A^{-s} : A \in \mathcal{L}^N_\infty\}$.

Show that $A \in \Psi\mathcal{D}^N_s$ admits a (special) Green inverse in $\mathcal{O}^N(-r)$ if and only if it is md-elliptic (of order r). Here we denote $\mathcal{O}^N(s) = L(\mathbb{C}^N) \otimes \mathcal{O}(s)$ and call $A \in \Psi\mathcal{D}^N_s$ md-elliptic (of order r) if $\sigma_{A\Lambda^r}(x,\xi)$ is nonsingular for all $(x,\xi) \in \mathbb{M}$.

10. The Laplace comparison of $\mathfrak{H} = L^2(\mathbb{R}^{n+1}_+)$, continued.

In the following sections we use the term symbol space of $\mathfrak{X}$, and the notation $\mathbf{M}(\mathfrak{X})$ for the Gel'fand space of the quotient $\mathfrak{X}/\mathfrak{Y}$ of the C*-algebra $\mathfrak{X}$ modulo its commutator ideal $\mathfrak{Y}$.

As a first step we examine the symbol space structure of $\mathfrak{A}^\#$, the C*-algebra generated by the operators (4.13). Let $\vec{F}$ be the Fourier transform in the last n variables, i.e.,

$$(10.1)\qquad (\vec{F}\, u)(y,\vec{x}) = (2\pi)^{-n/2}\int_{\mathbb{R}^n} e^{-i\vec{x}\vec{\xi}}u(y,\vec{\xi})\, d\vec{\xi} \quad .$$

$\vec{F}$ defines a unitary operator of $\mathfrak{H}$ and $\mathfrak{K}$ each, and, in that sense, commutes with E_e and E_o : $E_e\vec{F}u = \vec{F}E_e u$, $E\,\vec{F}u = \vec{F}E\, u$, $u \in \mathfrak{H}$. Introduce $\tau(\vec{x}) = (1+\vec{x}^2)^{1/2}$ and then the unitary substitution operator T: $\mathfrak{H} \to \mathfrak{H}$ (or $\mathfrak{K} \to \mathfrak{K}$)

$$(10.2)\qquad (Tu)(y,\vec{x}) = \tau^{-1/2}(\vec{x})u(y/\tau(\vec{x}),\vec{x})$$

which commutes with E_e, E_o in a similar sense. Define $U = T\vec{F}$. Then

$$(10.3)\qquad U \Lambda U^* = (1+D_0^2)^{-1/2} \otimes \tau^{-2}(\vec{M}) \quad ,$$

with the operator $(1+ D_0)^{-1/2}$ acting on $L^2(\mathbb{R})$, and with respect to the tensor decomposition $\mathfrak{K} = L^2(\mathbb{R}) \hat{\otimes} \mathfrak{k}$. Indeed, we have $\vec{F} = G^*F$ with the 1-dimensional Fourier transform G: $\mathfrak{K} \to \mathfrak{K}$, acting on y.
Also, $GT = T^*G$, $G^*T = T^*G^*$, by a calculation. It follows that

$$U \Lambda U^* = TG^*(\tau^2(\vec{M}) + M_0^2)^{-1/2}GT^* = G^* T^*(\tau^2(\vec{M}) + M_0^2)^{-1/2}TG$$
$$= \tau^{-1}(\vec{M})G^*(1+M_0^2)^{-1/2}G,$$

which proves (10.3) since the right hand is only another way of writing the right hand side of (10.3).

Applying E_e or E_o we get

$$(10.5)\qquad \begin{aligned} &U\Lambda_d U^* = Q_d \otimes \tau^{-1}(\vec{M}), \quad && U\Lambda_n U^* = Q_n \otimes \tau^{-1}(\vec{M})\\ &US_n U^* = P_n \otimes 1 \quad , && US_d U^* = P_d \otimes 1,\\ &US_{n,j}U^* = Q_d \otimes M_j\tau^{-1}(\vec{M}) \quad , && US_{d,j}U^* = Q_n \otimes M_j\tau^{-1}(\vec{M}) \quad , \end{aligned}$$

with P_n, P_d, Q_n, $Q_d = P_- \pm P_+$, $Q_- \pm Q_+$, respectively.

Proposition 10.1. The algebra $U\,\mathfrak{A}^\#\,U^*$ coincides with the subalgebra of

$$(10.6)\qquad \mathfrak{P}^\#\hat{\otimes}\,\mathfrak{Z}_n \quad , \quad \mathfrak{Z}_n = \{a(\vec{M}) \in L(\mathfrak{k}) : a \in C(\mathbb{B}^n)\} \quad ,$$

generated by the operators (10.5), where $\mathbb{B}^n$ denotes the smallest compactification of $\mathbb{R}^n$ onto which s_j, $j = 1,\ldots,n$ and λ all extend to continuous functions. Moreover, if $\mathfrak{E}^\#$ is the commutator ideal of $\mathfrak{A}^\#$, then

$$(10.7)\qquad U\;\mathfrak{E}^\#\,U^* = K(\mathfrak{f})\;\hat{\otimes}\;\mathfrak{Z}_n \quad .$$

Proof. The first part was discussed above. Also, it is trivial that '$\subset$' holds in

(10.7). We find that $U \mathcal{L}^{\#} U^*$ contains $C \otimes \lambda(\vec{M})$, $C \otimes s_j(\vec{M})$, for all $C \in K(\mathfrak{f})$ (c.f. 11, problem 1). Let $C = P$ be a 1-dimensional projection. Then the above operators generate the algebra of all $P \otimes a(\vec{M})$, $a \in C(\mathbb{B}^n)$, which is contained in $U \mathcal{L}^{\#} U^*$. Then Lemma 8.4 may be used for the second statement.

Proposition 10.2. Let $\mathbb{L} = \mathbb{L}_t \cup \mathbb{L}_\xi = \mathbf{M}(\mathcal{P}^{\#})$, with the closed "t-segment" $\mathbb{L}_t$, and the open "ξ-segment" $\mathbb{L}_\xi$, according to fig. 7.6, of $\mathbb{L}$. The C^*-algebra $\mathfrak{A}^{\#}/\mathcal{L}^{\#}$ is isometrically isomorphic to a subalgebra of $C(\mathbb{L} \times \mathbb{B}^n)$. In fact, to the subalgebra generated by $\sigma_A(l) \cdot b(\vec{x})$, $l \in \mathbb{L}$, $x \in \mathbb{B}^n$, where $A \otimes b(\vec{M})$ is any one of the tensor products in (10.5). Moreover, the symbol space $\mathbf{M}(\mathfrak{A}^{\#})$ is homeomorphic to the space obtained from $\mathbb{L} \times \mathbb{B}^n$ by collapsing each set $l \times \mathbb{B}^n$ into a point, for all $l \in \mathbb{L}_t$. Also the symbol of the generator $G = U^*((A \otimes b(\vec{M}))U$ of the collection (4.13) then is equal to the function induced by $\sigma_A(l) \cdot b(\vec{x})$ on $\mathbf{M}(\mathfrak{A}^{\#})$ as described by above homeomorphism.

Proof. By 8, problem 4, $(\mathcal{P}^{\#} \hat{\otimes} \mathfrak{Z}_n)/(K(\mathfrak{f}) \hat{\otimes} \mathfrak{Z}_n) \cong C(\mathbb{L} \times \mathbb{B}^n)$; thus by prop. 10.1, $\mathfrak{A}^{\#}/\mathcal{L}^{\#}$ is isometrically isomorphic to a subalgebra of $C(\mathbb{L} \times \mathbb{B}^n)$. The reminder of the proposition then is a matter of explicit calculations, involving Stone-Weierstrass, and the dual of the above injection.

Remark: It is immediate that (i) $\mathbb{B}^n$ is homeomorphic to the n-ball $\{x \in \mathbb{R}^n : |x| \le 1\}$, with $\{|x| = 1\}$ representing the infinite points of $\mathbb{B}^n$. Also that $\mathbf{M}^{\#} = \mathbf{M}(\mathfrak{A}^{\#})$ is homeomorphic to an n + 1-ball.

$$\mathbf{M}_1 = \{(\xi_0, \vec{\xi}) : \xi_0 \in \mathbb{R},\ \vec{\xi} \in \mathbb{R}^n,\ \xi_0^2 + |\vec{\xi}|^2 \le 1\} \quad , \tag{10.8}$$

with a 1-dimensional segment $\mathbf{M}_2^{\#} = \{-1 \le \mu \le 1\}$ attached as a handle, with its endpoints $\mu = \pm 1$ identified with the north and south pole $\xi_0 = \pm 1$ of the ball, respectively. (fig. 10.1),

Let $\mathfrak{A}$ be the C^*-algebra generated by $\mathfrak{A}^{\#}$ above, and the multiplication operators

$$\lambda(M),\ s_j(M) : \mathfrak{H} \to \mathfrak{H} \quad , \ j = 0, \ldots, n \tag{10.9}$$

with the functions λ, s_j in n+1-dimensions, like in section 1.

The operators (10.9) generate a commutative C^*-algebra isometrically isomorphic to $C(\mathbb{H}^{n+1})$, with the compactification

$$\mathbb{H}^{n+1} = \{x = (y, \vec{x}) \in \mathbb{B}^{n+1} : y \ge 0\} \tag{10.10}$$

of $\mathbb{R}_+^{n+1}$. Applying Lemma we conclude that $\mathbf{M}(\mathfrak{A})$ is homeomorphic to a subspace of the product $\mathbb{H}^{n+1} \times \mathbf{M}^{\#}$.

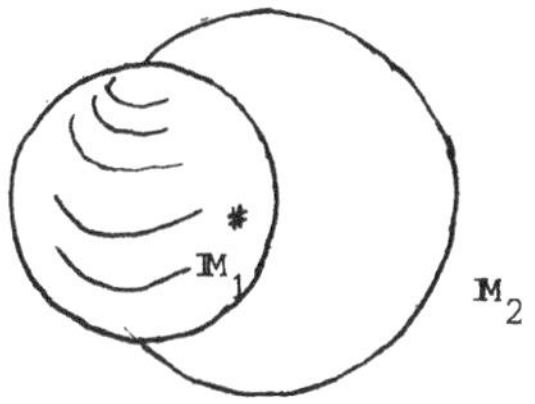

Fig. 10.1

Theorem 10.3. $\mathbf{M}(\mathfrak{A})$ consists of the following union of subsets of $\mathbb{H}^{n+1} \times \mathbf{M}^{\#}$:

(10.11) $\mathbb{M}(\mathfrak{A}) = ((\mathbb{H}^{n+1}-\overline{\mathbb{R}^{n+1}_+})\times \mathbb{M}^{\#}_1) \cup (\overline{\mathbb{R}^{n+1}_+}\times \partial\mathbb{M}^{\#}_1) \cup (\partial\mathbb{R}^{n+1}_+ \times \mathbb{M}^{\#}_2) \cup (\partial\partial\mathbb{R}^{n+1}_+ \times \mathbb{M}^{\#}_2)$,

with $\partial\mathbb{M}^{\#}_1$, $\partial\mathbb{R}^{n+1}_+$, $\overline{\mathbb{R}^{n+1}_+}$, $\partial\partial\mathbb{R}^{n+1}_+$ denoting the boundary of the n + 1-ball (10.8), the boundary of $\mathbb{R}^{n+1}_+$ in $\mathbb{R}^{n+1}$, the closure of $\mathbb{R}^{n+1}_+$, and the boundary of $\partial\mathbb{R}^{n+1}_+$ in $\mathbb{H}^{n+1}$, respectively.

The proof is postponed to the end of section 10. To also get an insight into the structure of the commutator ideal $\mathfrak{E}$ of $\mathfrak{A}$ we introduce the C^*-algebra $\mathfrak{A}^d_0$, acting on $L^2(\mathbb{R}^d)$, and generated by

(10.12) $\lambda(D)$, $\lambda(M)$, $s_j(D)$, $s_j(M)$, $j = 1,\dots,d$.

It is clear at once that $\mathfrak{A}^d_0$ is a subalgebra of $\mathfrak{A}_0$ as in IV, 1, for d = n. As noted in IV, 1, problems, $\mathfrak{A}^d_0$ has commutator ideal K, and we get

(10.13) $\mathbb{M}(\mathfrak{A}^d_0) = (\partial\mathbb{B}^d \times \mathbb{B}^d) \cup (\mathbb{B}^d \times \partial\mathbb{B}^d) = \mathbb{B}^d \times \mathbb{B}^d - \mathbb{R}^d \times \mathbb{R}^d \subset \mathbb{B}^d \times \mathbb{B}^d$,

and

(10.14) $\sigma_{a(M)} = a(x)$, $\sigma_{a(D)} = a(\xi)$, $(x,\xi) \in \mathbb{M}(\mathfrak{A}^d_0)$.

Theorem 10.4. Let $W = \vec{F}^*T\vec{F}$ (as in (10.1), (10.2); we have $K(\mathfrak{J}) \subset \mathfrak{E}$, and

(10.15) $W\mathfrak{E}W^* = K(\mathfrak{J}) \hat{\otimes} \mathfrak{A}^n_0$, and $\mathfrak{E}/K(\mathfrak{H}) \cong C(\mathbb{M}(\mathfrak{A}^n_0), K(\mathfrak{J}))$,

with the class $C(\mathbb{M},K)$ of continuous functions from $\mathbb{M}$ to K.

We prove Theorem 10.4 in a series of Lemmata.

Lemma 10.5. Let $a \in C(\mathbb{H}^{n+1})$, $A \in \mathfrak{A}^{\#}$, then $[a(M),A] = a(M)A - Aa(M) \in K(\mathfrak{J})$.

Proof. It suffices to show this for a generator of $\mathfrak{A}^{\#}$. Let $b: \mathbb{R}^{n+1} \to \mathbb{C}$ be the even extension of a. Clearly $E_\kappa a(M) = b(M)E_\kappa$ for $\kappa = e,o$. The generators (4.13) are all of the form $E^*_\kappa B\, E_\gamma$, $B = \Lambda$, S_j, κ, $\lambda = e$, we get $[a(M),E^*_\kappa BE_\gamma] = E^*_\kappa[b(M),B]E_\gamma =$ $= E^*_\kappa CE_\gamma$, $C \in K(\mathfrak{H})$, by III, Lemma 9.4 , which proves the Lemma.

The following is an easy consequence of the Lemma.

Corollary 10.6. The set of all operators of the form

(10.16) $\sum_{i=1}^{N} a_i(M)E_i + C$, $a_i \in C(\mathbb{H}^{n+1})$, $E_i \in \mathfrak{E}^{\#}$, $C \in K(\mathfrak{H})$,

is dense in $\mathfrak{E}$.

The following two Lemmata essentially complete the proof of Theorem 10.4. We defer their proofs until the next section.

Lemma 10.7. For $a \in C(\mathbb{H}^{n+1})$ let $\tilde{a} \in C(\mathbb{R}^{n+1}_+) \cap L^\infty(\mathbb{R}^{n+1}_+)$ be defined by $\tilde{a}(y,\dot{x}) = a(0,\dot{x})$, $x = (y,\dot{x}) \in \mathbb{R}^{n+1}_+$. Then, for all $E \in \mathfrak{E}^{\#}$, $\tilde{a}(M)E - a(M)E \in K(\mathfrak{H})$.

By Corollary 10.6 and Lemma 10.7, $\mathfrak{E}$ is the closed linear span mod $K(\mathfrak{H})$ of $\{\tilde{a}(M)E : \tilde{a} \in C(\mathbb{B}^n), E \in \mathfrak{E}^{\#}\}$ which equals the closed linear span, mod $K(\mathfrak{H})$

(10.17) $\{(1 \otimes a_0(\vec{M})) . U^*C \otimes b_0(\vec{M})U: a_0,b_0 \in C(\mathbb{B}^n), C \in K(\mathfrak{J})\}$, by Lemma 10.5 and Proposition 10.1. The following Lemma says that the generators listed in (10.17) are equal, mod $K(\mathfrak{g})$ to $U^*(C \otimes (a_1(\vec{D})b_0(\vec{M}))U$, with $a_1(x) = a_0(-x)$.

Lemma 10.8. Let $a_0 \in C(\mathbb{B}^n)$, $C \in K(\mathfrak{f})$. Then,

(10.18) $$(T(1 \otimes a_0(\vec{D}))T^*)(C \otimes 1) - C \otimes a_0(\vec{D}) \in K(\mathfrak{g}).$$

Assuming Lemma 10.8, for the moment, we find that $W\mathfrak{E}W^*$ is the closed linear span, mod $K(\mathfrak{g})$, of

(10.19) $$WU^*(C \otimes a_1(\vec{D})b_0(\vec{M})UW^* = F^*(C \otimes a_1(\vec{D})b_0(\vec{M}))F = C \otimes a_0(\vec{M})b_0(\vec{D}) \quad .$$

But this precisely amounts to the first relation (10.15). Also we noted above that $\mathfrak{A}_0 = \mathfrak{A}_0^n$ contains $K(\mathfrak{R})$. Thus $W\mathfrak{E}W^*$ contains $K(\mathfrak{g}) = K(\mathfrak{f}) \hat{\otimes} K(\mathfrak{R})$, using Lemma 8.3, and hence also $\mathfrak{E} \supset K(\mathfrak{g})$. Furthermore, the second part of (10.15) then is a consequence of the investigation of section 9, in particular formula (9.16). This proves Theorem 10.4.

Proof of Theorem 10.3. First, $\mathbb{M} = \mathbb{M}(\mathfrak{A})$ does not contain the points $(x,\xi) \in \mathbb{H}^{n+1} \times \mathbb{M}^{\#}$ with $x \in \mathbb{R}_+^{n+1}$, $\xi \in \mathring{\mathbb{M}}_1^{\#} = \mathbb{M}_1^{\#} - \partial\mathbb{M}_1^{\#}$, since, for $\psi \in C_0(\mathbb{R}_+^{n+1})$, we get $\psi(M)\Lambda_-\psi(M) \in K(\mathfrak{g})$, as easily confirmed. Since $K(\mathfrak{g}) \subset \mathfrak{E}$, we must get the symbol $\psi(x)\sigma_{\Lambda_-}\psi(x) = 0$ at all points of $\mathbb{M}$. But σ_{Λ_-} is > 0 in $\mathring{\mathbb{M}}_1^{\#}$, by (10.5).

For any of the above points ψ can be chosen such that the above product does not vanish. Hence the point cannot belong to $\mathbb{M}$.

Next let $\varepsilon > 0$, and $\mathbb{H}_\varepsilon^{n+1} = \{x \in \mathbb{H}^{n+1} : y \geq \varepsilon\tau(\vec{x})\}$, $\mathbb{R}_\varepsilon^{n+1} = \mathbb{R}^{n+1} \cap \mathbb{H}_\varepsilon^{n+1}$, $\tau = (1+\vec{x}^2)^{1/2}$, $\mathbb{M}_\varepsilon = \mathbb{M} \cap (\mathbb{H}_\varepsilon^{n+1} \times \mathbb{M}^{\#})$. Then we can show that no (x,ξ) with $y > 0$ $\xi \in \mathbb{M}_2^{\#}$ is in $\mathbb{M}$, as follows. By a calculation we get

(10.20) $$\phi(M)U^*(P_+ \otimes 1)U\phi(M) = \phi(M)U^*(\omega(M_0)P_+\omega(M_0) \otimes 1)U\phi(M)$$

whenever $\phi \in C_0(\mathbb{R}_\varepsilon^{n+1})$ and $\omega \in C_0^\infty((0,\infty])$ equals 1 for $y \geq \varepsilon$. But the integral operator $\omega(M_0)P_+\omega(M_0)$ is in $K(\mathfrak{f})$, because its symbol in $\mathfrak{R}$ vanishes. Hence (10.20) is in $\mathfrak{E}$ which implies the statement, by an argument as above. So we get $\mathbb{M}_\varepsilon \subset \mathbb{M}'_\varepsilon$, with

(10.21) $$\mathbb{M}'_\varepsilon = \{y \geq \varepsilon\tau, |x| = \infty, \xi \in \mathbb{M}_1^{\#}\} \cup \{y \geq \varepsilon\tau, |x| < \infty, \xi \in \partial\mathbb{M}_1^{\#}\} .$$

We now show that $\mathbb{M}_\varepsilon = \mathbb{M}'_\varepsilon$. Let

(10.22) $$\mathfrak{A}_\varepsilon = \{A \in \mathfrak{A} : \text{supp}\ \sigma_A \subset \mathbb{M}_\varepsilon\} \quad ,$$
$$\mathfrak{A}_{0,\varepsilon} = \{A \in \mathfrak{A}_0^{n+1} : \text{supp}\ \sigma_A \subset (\mathbb{H}^{n+1} \times \mathbb{B}^{n+1}) \cap \mathbb{M}(\mathfrak{A}_0^{n+1}) = \mathbb{M}'\} \quad .$$

Then $\mathfrak{A}_\varepsilon$ and $\mathfrak{A}_{0,\varepsilon}$ are both C^*-algebras, and $\mathfrak{A}_\varepsilon/\mathfrak{E}$ and $\mathfrak{A}_{0,\varepsilon}/K(\mathfrak{R}_{n+1})$ are both isomorphic to C^*-subalgebras of $C_\varepsilon(\mathbb{H}_\varepsilon^{n+1} \times \mathbb{M}_1^{\#})$, the algebra of continuous functions on $\mathbb{H}_\varepsilon^{n+1} \times \mathbb{M}_1^{\#}$ vanishing at $y = \varepsilon$, because the two balls $\mathbb{B}^{n+1}$ and $\mathbb{M}_1^{\#}$ are homeomorphic.

Let us observe that the above homeomorphism between the balls may be chosen such that the symbols in $\mathfrak{A}$ and $\mathfrak{A}_0^{n+1}$ agree, respectively. Recall, in that respect, that $\mathbb{M}_1^{\#}$ was constructed from the product $[-\infty,+\infty] \times \mathbb{B}^n$ by collapsing each of the two sets $\pm\infty \times \mathbb{B}^n$ into a point. The generators (4.13) of $\mathfrak{A}^\#$ occur in pairs. In fact, these generators are derived from the "convolution generators" of $\mathfrak{A}_0^{n+1}$ by applying E_e and E_o. Comparing the functions

$$(10.23) \qquad \sigma_{\Lambda_d}|\mathbb{M}_1^\# = \sigma_{\Lambda_n}|\mathbb{M}_1^\# \;,\quad \sigma_{S_d}|\mathbb{M}_1^\# = \sigma_{S_n}|\mathbb{M}_1^\#, \sigma_{S_{j,d}}|\mathbb{M}_1^\# = \sigma_{S_{j,n}}|\mathbb{M}_1^\# \;,$$

(where the symbols are taken in $\mathfrak{A}^\#$) with

$$(10.24) \qquad \sigma_\Lambda,\ \sigma_{S_0} \quad ,\quad \sigma_{S_j} \quad ,$$

(with symbols of the commutative C^*-algebra $\mathfrak{A}^\#$ generated by these operators) then we find that the functions obtained agree on the interior of their balls of definition, after the transformation of variables

$$(10.25) \qquad \xi = (\xi_0,\vec{\xi}) \to (\xi_0(1+\vec{\xi}^2)^{1/2},\vec{\xi}) = (\xi_0',\vec{\xi}) = \xi'$$

has benn carried out. Note that this is the transformation also underlying the unitary map T, above. Since the functions (10.23) and (10.24) are generators of $C(\mathbb{M}_1^\#)$ and $C(\mathbb{B}^{n+1})$, by Stone-Weierstrass, it follows that the homeomorphism (10.25) of the interior extends into a homeomorphism $\mathbb{M}_1^\# \longleftrightarrow \mathbb{B}^{n+1}$. as the dual map of the corresponding isomorphism of the function algebras. Henceforth we consider $\mathbb{M}_1^\#$ and $\mathbb{B}^{n+1}$ identified, by this homeomorphism. Then it follows that

$$(10.26) \qquad \sigma_\Lambda = \sigma_{\Lambda_d} = \sigma_{\Lambda_n} \;,\ \sigma_S = \sigma_{S_d} = \sigma_{S_n} \;,\ \sigma_{S_j} = \sigma_{S_{j,d}} = \sigma_{S_{j,n}} \;,\ \text{as } y > 0 \quad .$$

In particular (10.26) holds for $\mathbb{M}_\varepsilon \subset \mathbb{M}_\varepsilon'$. One obtains a map $\pi\colon C(\mathbb{M}_\varepsilon') \to C(\mathbb{M}_\varepsilon)$, defined by restriction $a \to a|\mathbb{M}_\varepsilon$ which is a $*$-homomorphism and such that (10.26) holds. The map π may be interpreted as a continuous $*$-homomorphism

$$(10.27) \qquad \pi : \mathfrak{A}_{0,\varepsilon}/K(\mathfrak{H}) \to \mathfrak{A}_\varepsilon/\mathfrak{E} \quad .$$

If we can show π to be an injection, then its dual must be surjective, which means that $\mathbb{M}_\varepsilon = \mathbb{M}_\varepsilon'$, as stated.

But the same map (10.27) may be obtained as follows. Let $\phi \in C(\mathbb{H}^{n+1})$ be zero near $y = 0$ and $= 1$ near $\mathbb{H}^{n+1}$. Regard ϕ extended zero into $y \leq 0$, to obtain an extension into $\mathbb{B}^{n+1}$. The operators $\phi(M)\Lambda\phi(M)$, $\phi(M)S_j\phi(M)$, $\phi(M)\Lambda_\kappa\phi(M)$, $\phi(M)S_\kappa\phi(M)$, $\phi(M)S_{\kappa,j}\phi(M)$, $\kappa = d, n$; $j = 0,\dots,n$, may be regarded as operators either $\mathfrak{H} \to \mathfrak{H}$ or $\tilde{\mathfrak{K}} \to \tilde{\mathfrak{K}}$. In that sense a calculation confirms that

$$(10.28) \qquad \phi(M)\Lambda\phi(M) \equiv \phi(M)\Lambda_n\phi(M) \equiv \phi(M)\Lambda_d\phi(M) \quad (\text{mod } K(\mathfrak{H})),$$

and similar congruences for the S_j etc. Note that the operators

$$(10.29) \qquad \phi(M)\Lambda\phi(M) \;,\ \phi(M)S_j\phi(M),\ j = 0,\dots,n \quad ,$$

generate a subalgebra of $\mathfrak{A}_0^{n+1}$, mod $K(\tilde{\mathfrak{K}})$, and a subalgebra of $\mathfrak{A}$, mod $\mathfrak{E}$, which contain $\mathfrak{A}_{0,\varepsilon}$ and $\mathfrak{A}_\varepsilon$, respectively, where the cosets of finitely generated elements correspond to each other by the map π. Note that, with the orthogonal projection $P_\delta\colon \mathfrak{H} \to L^2(\mathbb{R}_\delta^{n+1})$,

$$\begin{aligned}(10.30)\quad & \inf_{E\in\mathcal{E}} \|A+E\| = \inf_{E\in\mathcal{E}} \|B_\delta(A+E) + (1-P_\delta)E\| \geq \inf_{E\in\mathcal{E}} \|A+P_\delta E\| \\ & \geq \inf_{C\in K(\mathfrak{g})} \|A+C\| \geq \inf_{C\in K(\mathfrak{K})} \|A+C\|\end{aligned}$$

for sufficiently small δ, depending on ϕ, for any operator A finitely generated from (10.29). This shows that indeed π is injective, so that indeed $\mathbf{M}_\varepsilon = \mathbf{M}'_\varepsilon$. Letting $\varepsilon \to 0$, and taking closure in $\mathbf{M}$ we find that $\mathbf{M}$ contains all of the first and second set of the union (10.11).

Finally let us consider the last two sets of that union. In that respect let us look for an operator with symbol having support in the set $\{(x,\xi):\xi \in \mathbf{M}_2^{\#}\}$ which is not in $\mathcal{E}$. In that respect consider the algebra $\mathfrak{A}' = U\mathfrak{A}U^*$ which has $U\mathcal{E}U^* = K(\mathfrak{f}) \hat{\otimes} \mathfrak{A}_0^n$ as a subalgebra, by (10.15). From (10.5) it is clear that $\mathfrak{A}'$ contains $P_\pm \otimes 1$, and thus also the C^*-algebra generated by the operators $P_+ \otimes 1$ and $P_+.P_- \otimes 1$. These two operators have symbol 0 on $\mathbf{M}_1^{\#}$ -in the algebra $\mathfrak{A}^{\#}$ -, but separate interior points of the interval $\mathbf{M}_2^{\#}$. It follows the existence of $B_f \otimes 1 \in \mathfrak{A}'$ with $\sigma_{B_f} = f$ on $\mathbf{M}_2^{\#}$, $= 0$ on $\mathbf{M}_1^{\#}$ where f may be any continuous function over $\mathbf{M}_2^{\#}$ vanishing at the boundary. Likewise, $\mathfrak{A}'$ contains $Ua(\vec{M})b(M_0)U^*$ whenever $a \in C_0^\infty(\mathbb{R}^n)$, $b \in C_0^\infty([0,\infty))$, since $ab \in C_0(\mathbb{R}_+^{n+1}) \subset \mathfrak{A}$. But,
$A_f = Ua(\vec{M})b(M_0)U^*(B_f \otimes 1) = a(-\vec{D})b(M_0/\tau(\vec{M}))(B_f \otimes 1)$ is never in $K(\mathfrak{f}) \hat{\otimes} \mathfrak{A}_0^n$, unless either $f = 0$ or $a = 0$ or $b(0) = 0$. Otherwise for every $\phi \in C_0(\mathbb{R}^n)$ the operator $(\phi, A_f\phi) = C_f \in \mathcal{L}(\mathfrak{f})$ defined by $(u, C_f v) = (u \otimes \phi, A_f(v \otimes \phi))$ would be compact. Indeed, the latter is not true, because a calculation shows that $C_f = B_f \cdot q$ with $q \in C([0,\infty))$ defined by $q(y) = (\phi, a(-\vec{D})b(y/\tau(\vec{M}))\phi)$ (with inner product in $\mathcal{R}$). Taking symbols in $\mathcal{R}$ it is clear that C_f is compact if and only if either $f = 0$ or $q(0) = 0$. But if $f \neq 0$, and $a \neq 0$ and $b(0) \neq 0$ then ϕ always may be chosen such that $q(0) = (\phi, a(-\vec{D})\phi)b(0) \neq 0$.

Suppose that some (x,ξ) with $y = 0$, $|\vec{x}| < \infty$ and $\xi \in \mathring{\mathbf{M}}_2^{\#}$ is not in $\mathbf{M}$. Then $a \neq 0$, $f \neq 0$ may be chosen with supports in a sufficiently small neighbourhood of x and ξ such that the product still vanishes on $\mathbf{M}$ (because is closed). It follows that $U^*A_fU \in \mathcal{E}$, a contradiction if only $b(0) \neq 0$ is chosen. Hence $\mathbf{M}$ contains all such points. Since $\mathbf{M}$ is closed it therefore also contains the last two sets of the union (10.11). This completes the proof of Theorem 10.3.

11. Proof of the left-over auxiliary results.

Proof of Lemma 10.7. We may restrict our attention to functions of the form $a(x) = P(x)/(1+x^2)^{k/2}$, with a polynomial P of degree k, since $C(\mathbb{H}^{n+1})$ is the closed linear span of such functions. Then $a(x) - a(0,\vec{x}) = s_0(x)q(x)$, with a bounded function q, by a calculation. Also by Corollary 10.6 we may chose $E = P_{\phi,\psi}$. Then, using Proposition 10.1, we find that it is sufficient to show compactness of $s_0(M)U^*(P_{\phi,\psi}\otimes 1)U$ for any operator $P_{\phi,\psi}u = \phi\cdot(\psi,u)$, ϕ, ψ, $u \in \mathcal{J}$. Or, equivalently, we must show that, for ϕ, ψ of the form $\phi = 1$ in $[0,p]$, $= 0$ in (p,∞),

$$(11.1)\qquad (P_{\phi,\psi}\otimes 1)U\, s_0^2(M)U^*(P_{\phi,\psi}\otimes 1) = H \in K(\mathcal{G}).$$

We may write $s_0^2(y,\vec{x}) = (\mu y)^2(1+\mu^2\vec{x}^2)^{-1}$, $\mu = (1+y^2)^{-1/2}$, so that

$$(\vec{F}s_0^2(M)\vec{F}^*u)(x) = ((\mu y)^2(1+\mu^2\vec{D}^2)^{-1}u)(x) =$$

$$(11.2)\qquad = (2\pi)^{-n/2}y^2\mu^{2-n}\int G_{n,2}((\vec{x}-\vec{x}')/\mu)u(y,\vec{x}')d\vec{x}'.$$

Then, by a calculation, $Hu = (2\pi)^{-n/2}(P_{\psi,\psi}\otimes Z)u$ with the integral operator

$$(11.3)\qquad Zu(x) = \int_{\mathbb{R}^n} d\vec{x}'\int_0^\infty ds\,\phi(s\tau)\,\phi(s\tau')\,s^2\mu^{2-n}(s)G_{n,2}((\vec{x}-\vec{x}')/\mu(s))\,(\tau\tau')^{1/2}u(\vec{x}') \quad .$$

Again it suffices to show that $Z \in K(\mathcal{R})$. The interchange in integration leading to (11.3) can be justified, since ϕ has compact support, and from the properties of $G_{n,2}$, as listed in section 4. (Note that in (11.3) we introduced $\tau = \tau(\vec{x})$, $\tau' = \tau(\vec{x}')$.)

Substituting ϕ in details we get Z to have the kernel

$$z(\vec{x},\vec{x}') = (\tau\tau')^{1/2}\int_0^{p\gamma} s^2(1+s^2)^{n/2-1}G_{n,2}((\vec{x}-\vec{x}')(1+s^2)^{1/2})ds$$

$$= (\tau\tau')^{1/2}\int_0^{p\gamma} (p\gamma-s)G(s,\vec{x}-\vec{x}')ds \quad ,$$

with $\gamma = \gamma(\vec{x},\vec{x}') = \operatorname{Min}\{\tau^{-1},\tau'^{-1}\}$, and with

$$(11.5)\qquad G(s,\vec{\xi}) = \partial/\partial s\,\{s^2(1+s^2)^{n/2-1}G_{n,2}(\vec{\xi}(1+s^2)^{1/2})$$

by a partial integration. Using the estimates (4.11) and the fact that $(\tau\tau')^{1/2} \le 1/\gamma$ one finds that, for every $\varepsilon > 0$, $z = z_{1,\varepsilon} + z_{2,\varepsilon}$, with $|z_{1,\varepsilon}(\vec{x},\vec{x}')| \le \lambda(\vec{x})f_\varepsilon(\vec{x}-\vec{x}')$, $z_{2,\varepsilon}(\vec{x},\vec{x}') \le g_\varepsilon(\vec{x}-\vec{x}')$, $\|g_\varepsilon\|_{L^1} < \varepsilon$, with an $f_\varepsilon \in L^2(\mathbb{R}^n)\cap L^1(\mathbb{R}^n)$ and $\lambda(\vec{x}) = (1+\vec{x}^2)^{-1/2}$. If χ_q denotes the characteristic function of the ball $\{|\vec{x}| < q\}$, then it follows that $\chi_q(\vec{x})z_{1,\varepsilon}(\vec{x},\vec{x}') = O(\chi_q(\vec{x})f_\varepsilon(\vec{x}-\vec{x}')) \in L^2(\mathbb{R}^{2n})$, so that the operator $\chi_q(\vec{M})Z_{1,\varepsilon}$ is a Hilbert Schmidt operator and therefore compact. On the other hand we get $(1-\chi_q(\vec{M}))Z \to 0$ in norm convergence of $\mathcal{R}$, by Schur's criterion. This implies compactness of Z.

Proof of Lemma 10.8. It suffices to consider the assertion for the case where a_0 is either $\lambda(\vec{x})$ or $s_j(\vec{x})$, $j = 1,\dots,n$. Also we may assume $C = P_{\phi,\psi}$ again, as in the preceeding proof, and with the same choice of ϕ, even with $p = 1$. With these simplification we are reduced to showing that $K = K_1 + K_2$ is compact, with

$$(11.6)\quad (K_1u)(x) = \int (\tau'/\tau)^{1/2}\{\phi(y\tau'/\tau)-\phi(y)\}k(\vec{x}-\vec{x}')\psi(t)u(t,\vec{x}')dtd\vec{x}' \ ,$$

$$(K_2u)(x) = \int ((\tau'/\tau)^{1/2}-1)\phi(y)k(\vec{x}-\vec{x}')\psi(t)u(t,\vec{x}')dtd\vec{x}' \ ,$$

where k is the convolution kernel of $a_0(\vec{D})$, as in (4.11).

Observe that $K_2 = P_{\phi,\psi} \otimes V$, with an integral operator $V: \mathbb{R} \to \mathbb{R}$ having kernel $((\tau'/\tau)^{1/2}-1)k(\vec{x}-\vec{x}')$. The compactness of V may be proven with the method used for Z in the preceeding proof.

We turn to K_1. Let $\phi_\eta(v) = \phi(\eta v)$ and write, for $\eta > 1$,

$$(11.7)\qquad K_1 = \phi_\eta(M_0)K_1 + (1-\phi_{1/\eta}(M_0))K_1 + (\phi_{1/\eta}(M_0) - \phi_\eta(M_0))K_1.$$

We shall show that the first two operators at right of (11.7) are Hilbert Schmidt, and that the last one tends to zero in norm, as $\eta \to 1$. The kernel of K_1 is non-zero only if $y\tau'/\tau > 1$ as $y \le 1$ (case (a)), or, $y\tau'/\tau \le 1$, as $y > 1$ (case (b)). In case (a) we get $1 + \vec{x}^2 \le y^2(1 + \vec{x}'^2)$, $y < 1$, or $0 \le (1 - y^2)(1 + \vec{x}'^2) \le \vec{x}'^2 - \vec{x}^2 \le 2|\vec{x}'|\,|\vec{x}-\vec{x}'|$, so

$$(11.8)\qquad |\vec{x} - \vec{x}'|^{-1} \le 2|\vec{x}'|((1-y^2)(1+\vec{x}'^2))^{-1} \le c\,\lambda(\vec{x}')(1-y^2)^{-1} \ .$$

Similarly in case (b) we get

$$(11.9)\qquad |\vec{x}-\vec{x}'|^{-1} \le c(1-y^{-2})^{-1}\lambda(\vec{x}) \ .$$

Let $\eta < 1$, then $\phi_\eta(M_0)K_1$ has kernel zero for case (a), otherwise,

$$(11.10)\qquad \kappa_1(y,t,\vec{x},\vec{x}') = \phi_\eta(y)k(\vec{x}-\vec{x}')\psi(t)(\tau'/\tau)^{1/2} \ .$$

For arbitrary $y,t,\vec{x},\vec{x}'$ we get, using $k(z) = O(|z|^{-r}\lambda^r(z))$, $r \ge n$, (c.f. (4.11)),

$$(11.11)\qquad \kappa_1(y,t,\vec{x},\vec{x}') = O(\phi_\eta(y)(1-y^2)^{-r}\psi(t)\lambda^{1/2}(\vec{x})\lambda^{r-1/2}(\vec{x}')\lambda^r(\vec{x}-\vec{x}'))$$

for all $r \ge n$. This implies that $\kappa_1 \in L^2(\mathbb{R}^{2n+2})$, i.e., is a Schmidt kernel. Accordingly $\phi_\eta(M_0)K_1 \in K(\mathfrak{H})$.

Similarly for $L_\eta = (1-\phi_{1/\eta}(M_0))K_1$ the kernel is $\ne 0$ only in case (b), and we get its kernel estimated by

$$(11.12)\qquad \kappa_2(y,t,\vec{x},\vec{x}') = O((1-\phi_{1/\eta}(y))(1-y^{-2})^{-r}\psi(t)\lambda^r(\vec{x}-\vec{x}')\lambda^{r-1/2}(\vec{x})\lambda^{1/2}(\vec{x}'))$$

which again implies L_η to be a Schmidt-operator, hence compact.

Finally, regarding the third term in (11.7) -with kernel κ_3 - we note that $\kappa_3 \ne 0$ only as $1/\eta < y < \eta$, which is a small interval as $\eta \to 1$. For such y we may use either (11.8) or (11.9) for the estimate $|\vec{x}-\vec{x}'|^{-1} \le c|1-y|^{-1}$. Then we may estimate, with (4.11) again,

$$\kappa_3 = O(\psi(t)(\phi_{1/\eta}(y)-\phi_\eta(y))|\vec{x}-\vec{x}'|^{-n+\varepsilon}|y-1|^{-\varepsilon}e^{-\varepsilon|\vec{x}-\vec{x}'|}) \ ,$$

where also Peetre's inequality was used, and where $0 < \varepsilon < 1$.
It then is a consequence of Schur's Lemma that this operator tends to zero in operator norm, as $\eta \to 1$. This proves Lemma 10.8.

<u>Problems:</u> 1) Prove that $U \mathfrak{C}^\# U^*$ contain $C \otimes \lambda(\vec{M})$ and $C \otimes s_j(\vec{M})$, for all $C \in K(\mathfrak{H})$

and λ, s_j, j = = 1,...,n.

<u>Hint</u>: Use techniques as in III, Lemma 10.1 and IV, Theorem 11. Lemma 3.3.

12. General boundary conditions, for a constant coefficient equation.

Let $\rho > 0$, and consider the boundary problem

$$(p_0)\qquad \begin{array}{l} u \in \mathcal{G}_N(\Omega) \quad , \langle a_0\rangle u = D_0^N u + \rho^N (1-\vec{\Delta})^{N/2} u = f \quad \text{on } \Omega \cup \Gamma \quad , \\ \langle b^j\rangle u = 0 \text{ on } \Gamma,\ j = 1,\dots,r \quad , \end{array}$$

where $f \in \mathcal{G}$, Ω and Γ denote $\mathbb{R}^{n+1}_+$ and its boundary, respectively, and where $\langle b^j\rangle$ denote differential expressions of order $N_j < N = 2r$:

$$(12.1)\qquad \langle b^j\rangle = \sum_{k=0}^{N_j} \sum_{|\alpha| \le N_j - k} b^j_{k,\alpha}(\vec{x}) \vec{D}^\alpha D_0^k \quad ,$$

with coefficients in $C^\infty(\Gamma)$ satisfying $b^{j\,(\beta)}_{k,\alpha} \in C(\mathbb{B}^n)$, for all β.

Application of the tangential Fourier tranform will convert all tangential derivatives into multiplications, but also to multiplications with $b^j_{k,\alpha}(\vec{x})$ into convolutions. Also application of T (as in (10.2)) will take D_0 into $\lambda^{-1}(\vec{M})D_0$. Therefore, if we introduce the function $\tilde{u} = \lambda^{-N}(\vec{M})Uu$, problem (p_0) proves equivalent to

$$(12.2)\qquad \begin{array}{l} \tilde{u} \in \mathcal{J}_N \hat{\otimes} \mathcal{R} \quad , (D_0^N + \rho^N) = g \quad \text{on } \Omega \cup \Gamma \quad , \\ \displaystyle\sum_{k=0}^{N_j} \tilde{B}_{jk}\tilde{u}_k = 0 \quad j = 1,\dots,r \quad , \end{array}$$

with $\mathcal{J}_N = \mathcal{G}_N(\mathbb{R}_+)$, $g = Uf$, $\tilde{u}_k = D_0^k \tilde{u}|\Gamma$, and with the linear operators $\tilde{B}_{jk} \in L(\mathcal{R})$ defined by

$$(12.3)\qquad \tilde{B}_{jk} = \lambda^{-N+N_j+1/2}(\vec{M})\, B_{jk}\, \lambda^{N-N_j-1/2}(\vec{M})$$

with

$$(12.4)\qquad B_{jk} = \sum_{|\alpha| \le N_j - k} b^j_{k,\alpha}(-\vec{D})\, s^\alpha(\vec{M})\, \lambda^{N_j-k-|\alpha|}(\vec{M}) \in \mathfrak{A}^n_0 \quad ,$$

where $s(\vec{\xi}) = \vec{\xi}\lambda(\vec{\xi})$. In fact, also the $\tilde{B}_{jk}$ are in $\mathfrak{A}^n_0$, due to III, Lemma 9.1, for $s = N-N_j-1/2$, and $\sigma^n_{B_{jk}} = \sigma^n_{\tilde{B}_{jk}}$, since the difference $B_{jk}-\tilde{B}_{jk}$ is compact.

Now, for a given $g \in \mathcal{G}$ the ordinary differential equation $(D_0^N + \rho^N)v = g$ admits precisely a family of solutions $v \in \mathcal{J}_N \hat{\otimes} \mathcal{R}$, with r arbitrary functions $c_l \in \mathcal{R}$,explicitely given in the form

$$(12.5)\qquad v = \sum_{l=1}^{r} c_l(\vec{\xi})\, e^{i\kappa_l y} + \int_0^\infty G(y-y')\, g(y',\vec{\xi})\, dy' \quad ,$$

where

$$(12.6)\qquad G(t) = \sum_{l=1}^{r} i\gamma_l\, e^{i\kappa_l |t|} \quad , \gamma_l^{-1} = 2\kappa_l \prod_{\nu \ne l} (\kappa_l^2 - \kappa_\nu^2) \quad ,$$

and where κ_l, in any order, denote the r distinct roots of $\kappa^N + \rho^N = 0$ with positive imaginary part. Moreover, since the "Greens function" G must have (piecewise) continuous derivatives up to order N-1, it follows that (12.5) may be differentiated under the integral sign, up to N-1 times.

Accordingly we may assume $\tilde{u}$ of the form (12.5) and substitute into the second equation (12.2), to determine the functions c_l. We get

$$(12.7)\quad \begin{aligned} \tilde{u}_k &= \sum_{l=1}^{r} \kappa_l^k \{c_l + (-1)^k i\gamma_l g_l\} \quad , k = 0,1,\dots,N_j \quad , \\ g_l(\vec{\xi}) &= \int_0^\infty e^{i\kappa_l y'} g(y',\vec{\xi})\,dy' \quad , \end{aligned}$$

so that the second equation (12.2) takes the form

$$(12.8)\quad \sum_{l=1}^{r} \tilde{Q}_{jl} c_l = \sum_{l=1}^{r} \tilde{P}_{jl} g_l \quad , j = 1,\dots,r \quad ,$$

with

$$\begin{aligned} \tilde{Q}_{jl} &= \lambda^{-t_j}(\vec{M}) Q_{jl} \lambda^{t_j}(\vec{M}) \quad , \tilde{P}_{jl} = \lambda^{-t_j}(\vec{M}) P_{jl} \lambda^{t_j}(\vec{M}) \quad , \\ t_j &= N-N_j-1/2 \ , \ Q_{jl} = b^l(-\vec{D},(\kappa_l/\lambda)(\vec{M}),\vec{M})\lambda^{N_j}(\vec{M}) \quad , \\ P_{jl} &= -i\gamma_l b^j(-\vec{D},(-\kappa_l/\lambda)(\vec{M}),\vec{M})\lambda^{N_j}(\vec{M}) \\ b^j(\vec{x},\xi) &= b^j(\vec{x},\eta,\vec{\xi}) = \sum_{|\alpha|+k\le N_j} b^j_{k,\alpha}(\vec{x})\eta^k \xi^\alpha \quad . \end{aligned}$$

Clearly (12.8) is a (singular integral) equation within the C^*-algebra $\mathfrak{A}_0^n$ - or rather, within the tensor product $L(\mathbb{C}^r) \otimes \mathfrak{A}_0^n$, an operator algebra over $\mathbb{C}^r \otimes \mathcal{R}$. There exists a Fredholm inverse (and even a special Green inverse, as described in section 9, problem 3) $((\tilde{Z}_{jl}))_{j,l=1,\dots,r}$ of the matrix $((\tilde{Q}_{jl}))$ if and only if the matrix of symbols

$$(12.10)\qquad ((b^j(\vec{x},\kappa_l/\lambda(\vec{\xi}),\vec{\xi})\lambda^{N_j}(\vec{\xi})))_{j,l=1,\dots,r}$$

is non-singular for every $(\vec{x},\vec{\xi}) \in \mathbb{M}_0^n$ - that is for $\vec{x}$, $\vec{\xi} \in \mathbb{B}^n$ with $|\vec{x}| + |\vec{\xi}| = \infty$. The explicit calculation of symbols is easily done using (10.14). For a somewhat informal surveying first discussion, using a special Green inverse $((Z_{jl}))$ one will (normally) solve the boundary problem (p_0) by solving (12.8) for the c_l (modulo finite rank) and substituting the formula obtained into (12.5). It is likely then from the derivation that the boundary problem (p_0) will be normally solvable <u>if and only if</u> the determinant of the matrix (12.10) does not vanish for every $\vec{x},\vec{\xi} \in \mathbb{B}^n$ with $|\vec{x}| + |\vec{\xi}| = \infty$. In that case a Fredholm inverse T: $\mathcal{G} \to \mathcal{G}_N(\Omega)$ is given explicitely in the form

$$(12.11)\quad \begin{aligned} \tilde{u}(y,\vec{\xi}) &= \int_0^\infty G(y-y')g(y',\vec{\xi})\,dy' + \sum_{k,l=1}^{r} \tilde{R}_{kl}\, e^{i\kappa_k y} g_l(\xi) \quad , \\ \tilde{R}_{kl} &= \sum_{j=1}^{r} \tilde{Z}_{kj}\tilde{P}_{jl} \in \mathfrak{A}_0^n \quad . \end{aligned}$$

In order to formulate a theorem we introduce the polynomials

$$(12.12)\qquad q_j(\eta) = q_{j,\vec{x},\vec{\xi}}(\eta) = b^j(\vec{x},\eta/\lambda(\vec{\xi}),\vec{\xi})\lambda^{N_j}(\vec{\xi}),\ j = 1,\dots,r \quad ,$$

of the complex variable η, and observe that q_j is of degree N_j and has coefficients depending continuously on $\vec{x}$ and $\vec{\xi}$ over the compact space $\mathbb{B}^n \times \mathbb{B}^n$ with boundary $\partial(\mathbb{B}^n \times \mathbb{B}^n) = \mathbb{M}_0^n$, after a continuous extension. Then the matrix (12.10) may be written as $((q_j(\kappa_l)))$, and similarly $((P_{jl}))$ has symbols $((-i\gamma_l q_l(-\kappa_l)))$.

Lemma 12.1. Let the r polynomials (12.12) be linearly independent for every $|\vec{x}| + |\vec{\xi}| = \infty$ (i.e. for $(\vec{x},\vec{\xi}) \in M_0^n$). Then we have

$$\det((q_j(\kappa_l))) \neq 0, \qquad (x,\xi) \in M_0^n \qquad (12.13)$$

provided that the constant ρ in the differential equation of (p_0) is chosen sufficiently large.

Proof. It suffices to consider a single point of M_0^n. If for each such point (12.13) holds for $\rho \geq \rho_0$ then by compactness of M_0^n and continuity of coefficients a uniform ρ_0, valid for all of M_0^n, may be found. At a given fixed point we may replace the polynomials by others, applying "row operations", in such a way that all of them have highest coefficient 1, and any two of them have different degrees. Clearly, $|\kappa_1| = \rho \to \infty$, as $\rho \to \infty$, so that for large ρ the determinant (12.13) is approximated by

$$\rho^{N_1+\ldots+N_r} \det((\exp(2\pi i N_j \cdot l/N)))\exp(i\pi(N_1+\ldots+N_r)/N) \quad . \qquad (12.14)$$

The determinant in (12.14) is the van der Monde of the r distinct unit roots $e^{2\pi i N_j/N}$, $j = 1,\ldots,r$, and therefore is $\neq 0$, q.e.d.

Theorem 2.2. If and only if (12.12) are linearly independent, for all $|\vec{x}| + |\vec{\xi}| = \infty$, and if ρ in (p_0) is sufficiently large then there exists a Fredholm inverse $T: \mathfrak{H} \to \vartheta \subset \mathfrak{H}_N(\Omega)$ of the linear operator $L_0: \vartheta \to \mathfrak{H}$ associated to the problem (p_0), where

$$\vartheta = \{u \in \mathfrak{H}_N(\Omega): \langle b^j \rangle u = 0 \text{ on } \Gamma,\ j = 1,\ldots,r\} \quad ,$$
$$L_0 u = \langle a_0 \rangle u \quad , u \in \vartheta \quad , \qquad (12.15)$$

such that $L_0 T = 1 - F$, $TL_0 = 1 - F'$, $F \in E(\mathfrak{H})$, $F' \in E(\vartheta)$, with $E(\mathfrak{X}) \subset K(\mathfrak{X})$ denoting the ideal of continuous operators of finite rank, in the space $\mathfrak{X}$. (Note that ϑ is a closed subspace of $\mathfrak{H}_N(\Omega)$, and thus a Hilbert space.)

Moreover we get $D^\alpha T \in \mathcal{O}\!\ell$ for all $|\alpha| \leq N$, if T is properly chosen. In fact such a choice is explicitely given by

$$T = T_0 + K + C \quad , UT_0U^* = G \otimes \lambda^N(\vec{M}) \quad ,$$
$$UKU^* = \sum_{k,l=1}^{r} (e_k\rangle\langle e_l) \otimes (\lambda^N(\vec{M})\tilde{R}_{kl}),\ UCU^* = \sum_{j=1}^{\nu} \psi_j\rangle\langle \phi_j \quad , \qquad (12.16)$$

with the operators

$$(Gv)(y) = \int_0^\infty G(y-y')v(y')dy' \quad , v \in \mathfrak{f} \quad , \tilde{R}_{kl} \in \mathcal{O}\!\ell_0^n \quad ,$$
$$(\phi\rangle\langle\psi)w = \phi \cdot \int_X \psi w dz \quad , \phi, \psi \ , \quad w \in L^2(X),\ X \subset \mathbb{R}^m \quad ,$$
$$e_j(y) = e^{i\kappa_j y} \quad , y \in \mathbb{R}_+,\ \phi_j(y,\vec{\xi}) = \sum_{k=1}^{r} e_k(y)\omega_{jk}(\vec{\xi}) \quad , \qquad (12.17)$$
$$\psi_j(y,\vec{\xi}) = \sum_{k=1}^{r} (ye_k(y)\chi_{jk}(\vec{\xi}) + e_k(y)\Theta_{jk}(\vec{\xi})),\ j = 1,\ldots,\nu \quad ,$$

(12.17) $\omega_{jk},\chi_{jk},\theta_{jk} \in \mathcal{K}_\infty$, $((\sigma^n_{\tilde{R}_{kl}}(\vec{\xi},-\vec{x}))) = ((q_{k,\vec{x},\vec{\xi}}(\kappa_1)))^{-1}((-i\gamma_1 q_{k,\vec{x},\vec{\xi}}(-\kappa_1)))$,

where $\bar{\phi}_1,\dots,\bar{\phi}_\nu$ form an orthogonal base in $\mathfrak{H}$ of the (finite dimensional) orthogonal complement of im $UL_O U^* \subset \mathfrak{H}$. We also used $\mathcal{K}_\infty = \bigcap_s L^2(\mathbb{R}^n,(1+x^2)^s dx)$.

Moreover we have a corresponding representation for $T^{k,\alpha} = D_O^k \vec{D}^\alpha T$:

$$T^{k,\alpha} = T_O^{k,\alpha} + K^{k,\alpha} + C^{k,\alpha} \;,\; UT_O^{k,\alpha}U^* = (D_O^k G) \otimes (\vec{M}^\alpha \lambda^{N-k}(\vec{M}))$$

(12.18) $$UK^{k,\alpha}U^* = \sum_{j,l=1}^{r} (\kappa_j^k e_j\rangle\langle e_l) \otimes (\vec{M}^\alpha \lambda^{N-k}(\vec{M})\tilde{R}_{jl}) \;,\; UC^{k,\alpha}U^* = \sum_{j=1}^{\nu} (D_O^k \vec{M}\lambda^{-k}(\vec{M})\psi_j)\rangle\langle \phi_j$$

$$, \quad |\alpha| + k \le N.$$

Proof. Suppose (12.17) can be verified, then (12.18) follows by differentiation, using that $D_O^k \vec{D}^\alpha U = UD_O^k \lambda^{-k}(\vec{M})\vec{M}^\alpha$. Now, using the properties of our special Green inverse $((\tilde{Z}_{kl}))$ of the operator matrix $((\tilde{Q}_{kl}))$ we first introduce the operators $\tilde{T}: \mathfrak{H} \to \mathfrak{f}_N \hat{\otimes} \mathcal{K}$ by

(12.19) $$g \to (G \otimes 1)g + \sum_{k,l=1}^{r} ((e_k\rangle \langle e_l) \otimes \tilde{R}_{kl})g = \tilde{T}g = \tilde{u} \quad .$$

The operator $\tilde{T}$ pertains to the transformed boundary problem (12.2). In particular:
(i) if u solves (12.2) then $u = \tilde{T}g + \sum_{l=1}^{r} d_l e_l$, with $(d) = (d_l) \in \mathbb{C}^r \otimes \mathcal{K}_\infty$ solving $\tilde{Q}(d) = 0$.

(ii) For general $g \in \mathfrak{H}$ we have $\tilde{u} = \tilde{T}g$ solving

(12.20)
$$\tilde{u} \in \mathfrak{f}_N \hat{\otimes} \mathcal{K} \;,\; (D_O^N + \rho^N)\tilde{u} = g \text{ on } \Omega \cup \Gamma \quad ,$$
$$\sum_{k=1}^{r} \tilde{B}_{jk}\tilde{u}_k = -(1-P_{\mathrm{im}\,\tilde{Q}})\tilde{P}(g) \quad , \; (g) = (g_l)$$

Therefore $\tilde{u} = \tilde{T}g$ solves (12.2) if and only if $(1-P_{\mathrm{im}\,\tilde{Q}})\tilde{P}(g) = 0$, which amounts to a finite set of conditions $(\phi_j,g) = 0$, $j = 1,\dots,\nu$, with ϕ_j exactly of the form as in (12.17). If these vectors are orthogonalized, they will preserve that form. From (i) and (ii) together it follwos that

(1) $\tilde{u} = \tilde{T}g$ solves (12.2) if and only if g is orthogonal to $\phi_1,\dots,\phi_\nu$ and (2) then all solutions of (12.2) are of the form $\tilde{u} = \tilde{T}g + \sum_{l=1}^{r} e_l \otimes d_l$ with $(d_l) \in \ker \tilde{Q} \cap \mathcal{K}_\infty$.

Let $\tilde{\vartheta} = \lambda^{-N}(\vec{M})U\vartheta$, then $\tilde{T}$ does not map $\mathfrak{H}$ into $\tilde{\vartheta}$, but the operator $\hat{T} = \tilde{T}(1-F)$, $F = \sum_{j=1}^{\nu} \phi_j\rangle\langle \phi_j$ will do this. From (1) and (2) it follwos that L_O is a Fredholm operator. Since 1-F projects onto the orthogonal complement of the ϕ_j we get $\tilde{L}_O\hat{T} = 1 - F$ (with properly defined $\tilde{L}_O$, as a transformation of L_O). Hence a Fredholm inverse exists and $\hat{T}$ is a right Fredholm inverse, therefore also must be a left Fredholm inverse. All remaining statements of the theorem now follow by transforming back onto the original form, applying U and $\lambda^N(\vec{M})$, etc. Q.e.d.

Corollary 12.3. In the decompositions (12.6) and (12.7) we have

(12.21) $\quad C^{k,\alpha} \in E(\mathfrak{H}),\ K^{k,\alpha} \in \mathfrak{L}' \subset \mathfrak{L},\ T_0^{k,\alpha} = \pi_+^+((D_0^k \vec{D}^\alpha (D_0^N + \rho^N(1-\vec{\Delta})^r)^{-1})$,

with

(12.22) $\quad \mathfrak{L}' = U^*(E(\mathfrak{f}) \otimes \mathfrak{A}_0^n)U$

and π_+^+ of (13.5). In particular $T_0^{k,\alpha}$ depend on ρ and N, but otherwise are entirely independent of the boundary conditions imposed in (p_0).

This is a trivial consequence of (12.18).

Problems. 1) Let the space $\tilde{\mathfrak{H}}_N^0$, for an integer $N \geq 1$, be defined by

(12.23) $\quad \tilde{\mathfrak{H}}_N^0 = \{u \in \mathfrak{H}_N(\mathbb{R}^{n+1}) : D_0^j u = 0 \text{ at } y = 0,\ j = 0,1,\dots,N-1\}$,

where III, Lemma 1.5 must be used to define the restrictions $D_0^j u|_{y=0}$. Show that $\tilde{\mathfrak{H}}_N^0$ is a closed subspace of $\mathfrak{H}_N = \mathfrak{H}_N(\mathbb{R}^{n+1})$.

2) The space $\mathfrak{H}_N$ of problem 1) also is a Hilbert space under the norm

(12.24) $\quad \|u\|^{\bullet}_N = \|(D_0^N + (1-\vec{\Delta})^{N/2})u\|_{L^2(\mathbb{R}^{n+1})}$

equivalent to $\|u\|_N$ (That is, $c\,\|u\|_N \leq \|u\|^{\bullet}_N \leq C\,\|u\|_N$ with $0 < c \leq C < \infty$ independent of u). Obtain the corresponding inner product $(u,v)^{\bullet}_N$.

3) For $u \in \mathfrak{H}_N$ there is a unique decomposition

(12.25) $\quad u = w + v,\ v \in \tilde{\mathfrak{H}}_N^0,\ w \mathrel{\dot{\perp}} \tilde{\mathfrak{H}}_N^0$,

where "$\dot{\perp}$" denotes orthogonality with respect to $(.,.)^{\bullet}_N$. Here w is uniquely characterized, for given u, as the function $w \in \mathfrak{H}_N$ minimizing the functional $V(u) = \|u\|^{\bullet\,2}_N$, within the set

(12.26) $\quad \{f \in \mathfrak{H}_N : D_0^j f = D_0^j u \text{ at } y = 0,\ j = 0,\dots,N-1\}$.

Moreover, we also have w uniquely characterized as the solution of

$$\text{(i)}\quad w \in \mathfrak{H}_N;\quad \text{(ii)}\quad (D_0^{2N} + (1-\vec{\Delta})^N)w = 0 \text{ as } y \neq 0;$$
$$\text{(iii)}\quad D_0^j w = \phi_j \text{ at } y = 0,\ j = 0,\dots,N-1 ,$$

with

(12.28) $\quad \phi_j = D_0^j u|_{y=0} \quad , j = 0,\dots,N-1$,

where the differential equation (ii) is to be interpreted in the distribution sense, in the open set $y \neq 0$.

4) Reduce the Dirichlet problem (12.27) by application of the transformation $U = T\vec{F}$, above, to the problem

$$\text{(12.29)}\quad \begin{array}{l}\text{(i)}\quad \tilde{w} \in \mathfrak{H}_N(\mathbb{R}) \hat{\otimes} k;\quad (D_0^{2N} + 1)\tilde{w} = 0 \quad ;\\ \text{(iii)}\quad D_0^j \tilde{w} = \psi_j \text{ at } y = 0,\ j = 0,\dots,N-1 \quad ,\end{array}$$

for $\tilde{w} = \lambda^{2N}(\vec{M})Uw,\quad \psi_j = \lambda^{2N-j}(\vec{M})U\phi_j$.

5) Solve (12.29) explicitly in the form

(12.30) $$\hat{w}(y,\vec{\xi}) = \sum \alpha_{1k}\psi_k(\vec{\xi})e^{i\kappa_1 y} \quad ,$$

with the N roots $\kappa_1,\ldots,\kappa_N$ of the equation $\kappa^{2N}+1 = 0$, having positive real part, and with complex constants α_{1k} .

Derive from (12.30) that

(12.31) $$c\,\|\lambda^{-N}(\vec{M})\tilde{w}\|_{\mathcal{G}_N(\mathbb{R})\hat{\otimes}\mathcal{R}} \leq \sum_{j=0}^{N-1} \|\lambda^{N-j}(\vec{M})U\phi_j\|_{\mathcal{R}} \leq C\,\|\lambda^{-N}(\vec{M})\tilde{w}\|_{\mathcal{G}_N(\mathbb{R})\otimes\mathcal{R}}$$

with constants $0 < c < C$ independent of u. Translate (12.31) into

(12.32) $$c\,\|w\|_{\mathcal{G}_N(\mathbb{R}^{n+1})} \leq \sum_{j=0}^{N-1} \|\phi_j\|_{\mathcal{G}^{N-j-1/2}(\mathbb{R}^n)} \leq C\,\|w\|_{\mathcal{G}_N(\mathbb{R}^{n+1})} \quad .$$

6) Using the preceeding problems prove the <u>Trace Theorem</u>: Let $N \geq 1$. The linear map

(12.33) $$u \to \tilde{u}^N(a) = (u(a,.),\ldots,D_0^j u(a,.),\ldots,D_0^{N-1}u(a,.))$$

for $a \in \mathbb{R}$, with the restriction map of III, Lemma 1.5, defines a continuous operator

(12.34) $$\mathcal{G}_N(\mathbb{R}^{n+1}) \to CB(\mathbb{R},\mathcal{G}_{N-1/2}(\mathbb{R}^n)\times \ldots \times\mathcal{G}_{N-j-1/2}(\mathbb{R}^n)\times \ \ldots \times\mathcal{G}_{1/2}(\mathbb{R}^n)) \quad .$$

Moreover, for every given fixed $a \in \mathbb{R}$ the induced map $\mathcal{G}_N(\mathbb{R}^{n+1}) \to \mathcal{G}_{N-1/2}(\mathbb{R}^n) \times \ldots$ $\ldots \times \mathcal{G}_{1/2}(\mathbb{R}^n)$ is surjective.

13. Relation between $\mathfrak{A}_0^{n+1}$ and $\mathfrak{A}$, and its effect on symbols.

Let us recall the isometries E_e, E_o from section 4 and define $E_{\pm}: \mathfrak{H} \to \mathfrak{K}$ by

$$(13.1)\quad \begin{aligned} &(E_+u)(y,\vec{x}) = u(y,\vec{x}) \quad \text{on } \Omega, \quad = 0 \text{ as } y < 0 \;, u \in \mathfrak{H} \;, \\ &(E_-u)(y,\vec{x}) = u(-y,\vec{x}), \text{ as } y < 0, \quad = 0 \text{ on } \Omega \;, u \in \mathfrak{H} \;, \end{aligned}$$

so that

$$(13.2)\quad E_e = 2^{-1/2}(E_+ + E_-),\; E_o = 2^{-1/2}(E_+ - E_-) \quad .$$

Clearly E_e and E_o correspond to even and odd extension from $\mathbb{R}_+^{n+1}$ into $\mathbb{R}^{n+1}$ while E_+ simply extends as zero. All maps are isometries but nor unitary maps. Thus the adjoints are only partial isometries, explicitely given as the restrictions

$$(13.3)\quad E_\gamma^* v = \sqrt{2} v_\gamma|\Omega,\; \gamma = e,o,\; E_\pm^* v = v(\pm y,\vec{x})|\Omega \quad , v \in \mathfrak{K} \;,$$

with the even and odd part

$$(13.4)\quad v_e(y,\vec{x}) = (v(y,\vec{x})+v(-y,\vec{x}))/2,\; v_o(y,\vec{x}) = (v(y,\vec{x})-v(-y,\vec{x}))/2 \quad .$$

We define the linear contraction maps (not homomorphisms) $\Pi_\pm^\pm: L(\mathfrak{K}) \to L(\mathfrak{H})$ (with $L(\mathfrak{X})$ a Banach space with operator norm) by

$$(13.5)\quad \Pi_\gamma^\kappa L = E_\kappa^* L E_\gamma,\; L \in L(\mathfrak{K}),\; \kappa,\quad \gamma = +,- \quad .$$

<u>Proposition 13.1.</u> $\Pi_\gamma^\kappa: \mathfrak{A}_0^{n+1} \to \mathfrak{A}$, $\Pi_\gamma^\kappa: K(\mathfrak{K}) \to K(\mathfrak{H})$, for $\kappa,\gamma = +,-$.

<u>Proof.</u> Since $E_\pm$ and $E_\pm^*$ are partial isometries it is clear that compactness is preserved by all the maps Π_γ^κ. Also it suffices to show that Π_γ^κ maps the generators $a(M)$ (with $a \in C(\mathbb{B}^{n+1})$) and Λ, $S_0,\dots,S_n$ of $\mathfrak{A}_0^{n+1}$ into $\mathfrak{A}$. For if $A, B \in \mathfrak{A}_0^{n+1}$, and $\Pi_\gamma^\kappa A$, $\Pi_\gamma^\kappa B \in \mathfrak{A}$, then we use the relation, which is easily verified,

$$(13.6)\quad E_+E_+^* + E_-E_-^* = 1$$

to show that

$$(13.7)\quad \Pi_+^+(AB) = (\Pi_+^+A)(\Pi_+^+B) + (\Pi_-^+A)(\Pi_+^-B) \in \mathfrak{A} \quad .$$

Similarly for all other operators $\Pi_\gamma^\kappa(AB)$, with a slightly different formula (13.7).

But for a multiplication $a(M)$ we get $\Pi_+^+a(M) = b_+(M)$, $\Pi_-^-a(M) = b_-(M)$, with $b_\pm(y,x) = a(\pm y,x)|\Omega$, while $\Pi_+^-a(M) = \Pi_-^+a(M) = 0$, so that indeed $\Pi_\gamma^\kappa a(M) \in \mathfrak{A}$. Moreover we observe that

$$(13.8)\quad \begin{aligned} &E_o^*\Lambda E_o = \Lambda_d \;,\; E_e^*\Lambda E_e = \Lambda_n,\; E_o^*S_jE_o = S_{j,d},\; E_e^*S_jE_e = S_{j,n},\; j = 1,\dots,n \;, \\ &E_e^*S_0E_o = S_{0,d},\; E_o^*S_0E_e = S_{0,n} \\ &E_o^*\Lambda E_e = E_e^*\Lambda E_o = E_o^*S_jE_e = E_e^*S_jE_o = E_e^*S_0E_e = E_o^*S_0E_o = 0,\; j = 1,\dots,n \;, \end{aligned}$$

using the fact that Λ, $S_1,\dots,S_n$ preserves the spaces of even and odd functions, while S_0 takes even into odd functions, and vice versa. But (3.2) may be solved for $E_\pm$:

$$(13.9)\qquad E_{\pm} = 2^{-1/2}(E_e \pm E_o) \qquad ,$$

which implies that $\Pi^{\kappa}_{\gamma}A$ is a linear combination of $E^*_{\mu}AE_{\nu}$, $\mu,\nu=e,o$. This completes the proof, since continuity of Π^{κ}_{γ} allows closing.

It is clear, from Proposition 13.1. that Π^{κ}_{γ} induce four continuous maps $\mathfrak{A}^{n+1}_{o}/K(\mathfrak{H}) \to \mathfrak{A}/K(\mathfrak{g})$. The first algebra is equal (isometrically isomorphic) to $C(\mathbb{M}^{n+1}_{o})$. On the other hand, $\mathfrak{A}/K(\mathfrak{g})$ induces algebras of linear operators on its closed two-sided ideal $\mathfrak{L}/K(\mathfrak{g}) = C(K(\mathfrak{f}),\mathbb{M}^{n}_{o})$, a Banach space, by left and right multiplication. Accordingly, for every $\gamma,\kappa = +,-$, and every symbol $a\varepsilon C(\mathbb{M}^{n+1}_{o})$ there exists an operator $A^{\kappa}_{\gamma}: C(K(\mathfrak{f}),\mathbb{M}^{n}_{o}) \to C(K(\mathfrak{f}),\mathbb{M}^{n}_{o})$ such that

$$(13.10)\qquad \begin{aligned} &A\varepsilon\,\mathfrak{A}^{n+1}_{o}\ ,\ E,F^{\kappa}_{\gamma}\ \varepsilon\ \mathfrak{L}\ ,\quad \sigma^{n+1}_{A} = a\ ,\ F^{\kappa}_{\gamma} = (\Pi^{\kappa}_{\gamma}A)E \\ &\text{implies } \tau_{F^{\kappa}_{\gamma}} = A^{\kappa}_{\gamma}\tau_{E} \qquad , \end{aligned}$$

with the symbol τ_E of $E\ \varepsilon\ \mathfrak{L}$. Similarly for the right multiplication of symbols of $E\ \varepsilon\ \mathfrak{L}$ with the coset of A modulo $K(\mathfrak{H})$.

Theorem 3.2. The operators A^{κ}_{γ} of (13.10) coincide with those defined by right multiplication and are explicitly given by

$$(13.11)\qquad A^{\kappa}_{\gamma}(\vec{x},\vec{\xi}) = \pi^{\kappa}_{\gamma}a(0,\vec{x},\sqrt{1+\xi^2}D_o,\vec{\xi}) \quad , \quad a\varepsilon C(\mathbb{M}^{n+1}_{o})\ \gamma,\kappa \quad , \quad = +,- \quad ,$$

where $\pi^{\kappa}_{\gamma}: L(L^2(\mathbb{R})) \to L(\mathfrak{f})$ denote the operators Π^{κ}_{γ} for n=0. That is, for $A\varepsilon\,\mathfrak{A}^{n+1}_{o}$, $E,F^{\kappa}_{\gamma},G^{\kappa}_{\gamma}\varepsilon\ \mathfrak{L}$, with $\sigma_A = a$, $F^{\kappa}_{\gamma} = (\Pi^{\kappa}_{\gamma}A)E$, $G^{\kappa}_{\gamma} = E(\Pi^{\kappa}_{\gamma}A)$, we have (for $(\vec{x},\vec{\xi})\ \varepsilon\ \mathbb{M}^{n}_{o}$)

$$(3.12)\qquad \tau_{F^{\kappa}_{\gamma}}(\vec{x},\vec{\xi}) = A^{\kappa}_{\gamma}(\vec{x},\vec{\xi})\tau_{E}(\vec{x},\vec{\xi})\ ,\ \tau_{G^{\kappa}_{\gamma}}(\vec{x},\vec{\xi}) = \tau_{E}(\vec{x},\vec{\xi})A^{\kappa}_{\gamma}(\vec{x},\vec{\xi}) \quad .$$

Proof. First, for the operators $E = U^*(C\otimes B)U$, $C\varepsilon K(\mathfrak{f})$, $B\varepsilon\,\mathfrak{A}^{n}_{o}$, and a multiplication $a(M)$, $a\varepsilon C(\mathbb{B}^{n+1})$, the assertion is a direct consequence of lemma 10.7 and lemma 10.8, ba a calculation. Also for E, as above, and one of the other generators Λ, S_j of $\mathfrak{A}^{n+1}_{o}$ one will use (13.8) and (10.5) to directly calculate the symbols τ_E, $\tau_{F^{\kappa}_{\gamma}}$, $\tau_{G^{\kappa}_{\gamma}}$, thus confirming (13.12). Finally, formula (13.6) is valid for dimension 1 just as well as for dimension n+1, which makes the π^{κ}_{γ} behave just like the Π^{κ}_{γ}. For example we get

$$(13.13)\qquad \pi^{+}_{+}(PQ) = \pi^{+}_{+}(P)\pi^{+}_{+}(Q) + \pi^{+}_{-}(P)\pi^{-}_{+}(Q)\ ,\ P,Q\ \varepsilon\ L(L^2(\mathbb{R})) \quad ,$$

similar as (13.7). This implies that (13.11) and (13.12) are valid for $AB\varepsilon\,\mathfrak{A}^{n+1}_{o}$, if they hold for A and $B\ \varepsilon\ \mathfrak{A}^{n+1}_{o}$. All maps introduced clearly are continuous, so that the proof is completed by taking closures. Also the corollary, below, is evident.

Corollary 13.3. The operators $A^{\kappa}_{\gamma} = \Pi^{\kappa}_{\gamma}A$, $A\ \varepsilon\ \mathfrak{A}^{n+1}_{o}$, take $\mathfrak{L}' + K(\mathfrak{g})$ to itself.

14. The operator A = <a>T and its normal solvability.

We now approach the boundary problem (p), as posed in section 3, by substituting $u = Tv$, $v \in \mathfrak{H}$, with the Fredholm inverse $T: \mathfrak{H} \to \mathfrak{D}$ constructed for the problem (p_0), in section 12. We want the boundary expressions $\langle b^j \rangle$ for (p) and (p_0) to be identical, and also $N=2r$. Also we assume $\phi_j = 0$, $j=1,\dots,r$, which is no restriction of generality : For general $\phi_j \in \mathfrak{H}_{N-N_j-1/2}$ it is possible to determine $w \in \mathfrak{H}_N(\Omega)$ satisfying the boundary conditions $\langle b^j \rangle w = \phi_j$ on Γ, but not necessarily the differential equation, using the trace Theorem (12, problem 6) and linear independence of the polynomials q_j. Then $v=u-w$ will satisfy (p), with f replaced by $f - \langle a \rangle w$, and with homogeneous boundary conditions.

We assume $\langle a \rangle$ and $\langle b^j \rangle$ to be md-elliptic, as formulated in theorem 3.1 . This will imply linear independence of the q_j, as required for construction of w above, and for Proposition 12.1.

The substitution $u=Tv$ will automatically satisfy $u \in \mathfrak{H}_N(\Omega)$, and the boundary conditions $\langle b^j \rangle u = 0$, since $Tv \in \mathfrak{D}$, by construction. Since $T: \mathfrak{H} \to \mathfrak{D} \subset \mathfrak{H}_N(\Omega)$ is Fredholm, its image will have finite codimension in $\mathfrak{D}$. Therefore we always can write $u=Tv$ if only finitely many conditions are satisfied. Moreover, the second equation of (p) takes the form

$$(14.1) \qquad Av = f \quad , \quad A = \sum_{|\beta| \le N} a_\beta(M)(D^\beta T) \in \mathfrak{A} \quad ,$$

using that $D^\beta T = T^{\beta_0, \vec{\beta}} \in \mathfrak{A}$, by Theorem 12.2, and that $a_\beta(M) \in \mathfrak{A}$, due to the assumptions (3.3) on our coefficients. (By the Stone-Weierstrass theorem (AII, thm. 6.2) $a \in CS(\mathbb{R}^{n+1})$ may be approximated uniformly over $\mathbb{R}^{n+1}$ by a polynomial in $\lambda(x)$, $s_j(x)$ contained among the generators of $\mathfrak{A}$). This makes the proposition, below, evident.

Proposition 14.1. Problem (p), with homogeneous boundary conditions, is normally solvable if and only if the operator A of (14.1) is Fredholm, as a map from $\mathfrak{H}$ to $\mathfrak{H}$.

Proposition 14.2. Let a be md-elliptic, and assume that $\langle b^j \rangle$ and ρ satisfy the assumptions of Lemma 12.1, so that T of Theorem 12.2 exists. Then there exists an inverse of A (in (14.1)) modulo $\mathfrak{L}$.

Proof. From Corollary 12.3 it follows that

$$(14.2) \qquad \begin{aligned} A &= \Pi_+^+ A_0 + K_0 + C_0 \ , \ K_0 = \sum_{|\alpha|+k \le N} a_{(k,\alpha)}(M) K^{k,\alpha} \in \mathfrak{L}' + K(\mathfrak{H}) \ , \\ C_0 &= \sum_{|\alpha|+k \le N} a_{(k,\alpha)}(M) C^{k,\alpha} \in K(\mathfrak{H}) \ , \end{aligned}$$

while

$$(14.3) \qquad A_0 = \sum_{|\alpha| \le N} a_\alpha^0(M) D^\alpha (D_0^N + \rho^N)(1+\vec{D}^2)^r)^{-1} \quad , \quad a_\alpha^0 = \sqrt{2}\, E_e a_\alpha \quad ,$$

Since C_O, $K_C \varepsilon \mathfrak{L}$, it clearly suffices to construct an inverse mod $\mathfrak{L}$ of $\Pi_+^+ A_O$ only. But $A_O \varepsilon \mathfrak{A}_O^{n+1}$ has symbol

(14.4) $\sigma_{A_O}^{n+1}(x,\xi) = \sum_{|\alpha|\leq N} a_\alpha^O(x)\xi^\alpha(\xi_O^N+\rho^N\lambda^{-N}(\vec{\xi}))^{-1}$, as $x,\xi\varepsilon\mathbf{R}^{n+1}$, $|x|+|\xi| = \infty$.

But $\sigma_A^{n+1} \neq 0$ on $\mathbf{M}_O^{n+1}$, since a is md-elliptic. Thus there exists a Fredholm inverse $B_O \varepsilon \mathfrak{A}_O^{n+1}$ such that $1-A_O B_O$, $1-B_O A_O \varepsilon E(\mathfrak{K})$. Using (13.7) it follows that

(14.5) $(\Pi_+^+ B_O)(\Pi_+^+ A_O) \equiv 1 - (\Pi_-^+ B_O)(\Pi_+^- A_O)$ (mod $K(\mathfrak{H})$)

However, a calculation shows that $\Pi_+^- A_O \varepsilon \mathfrak{L}'$, and also $\Pi_-^+ A_O \varepsilon \mathfrak{L}'$, so that the last term in (14.5) is in $\mathfrak{L}' \subset \mathfrak{L}$, and similar for the other order of multiplication. This proves the proposition.Moreover, we even may explicitly calculate the perturbation terms, as stated in the Corollary, below.

Corollary 14.3. With $B = \Pi_+^+ B_O$, as constructed, we have

(14.6) $BA = 1+E_1$, $AB = 1 + E_2$,

with E_1, E_2 explicitly given in the form

$$E_1 \equiv (\Pi_+^+ B_O)(K_O+C_O) - (\Pi_-^+ B_O)(\Pi_+^- A_O) \pmod{K(\mathfrak{H})} ,$$

(14.7)

$$E_2 \equiv (K_O+C_O)(\Pi_+^+ B_O) - (\Pi_-^+ A_O)(\Pi_+^- B_O) \pmod{K(\mathfrak{H})} ,$$

where we have explicitly, using lemma 10.7 and lemma 10.8 ,

$$U(\Pi_+^- A_O)U^* \quad \sum_{l=1}^{r} i\gamma_l(e_l><e_l) \otimes p_{-\vec{D},\vec{M}}(-\kappa_l) \pmod{K(\mathfrak{H})} ,$$

(14.8)

$$U(\Pi_-^+ A_O)U^* \quad \sum_{l=1}^{r} i\gamma_l(e_l><e_l) \otimes p_{-\vec{D},\vec{M}}(\kappa_l) \pmod{K(\mathfrak{H})} ,$$

with γ_l, e_l as in section 12, and $p_{-\vec{D},\vec{M}}(\eta)\varepsilon\mathfrak{A}_O^n$ defined by

(14.9) $$p_{-\vec{D},\vec{M}}(\eta) = \sum_{j=0}^{N} p_j(-\vec{D},\vec{M})\eta^j ,$$

where

(14.10) $$p(\eta) = p_{\vec{x},\vec{\xi}}(\eta) = \sum_{j=0}^{N} p^j(\vec{x},\vec{\xi})\eta^j$$

is the polynómial, defined in (3.3). It is clear that $e_l><e_l \varepsilon E(\mathfrak{f})$, so that we indeed get $\Pi_+^- A_O$, $\Pi_-^+ A_O \varepsilon \mathfrak{L}'$. We also have C_O and B_O explicitly as in (14.2), and application of Corollary 13.3 completes the proof.

Note that theorem 13.2 allows calculation of the symbols τ_{E_1}, τ_{E_2}, from the symbols of A_O, B_O in $\mathfrak{A}_O^{n+1}$: First we get $\tau_{(\Pi_+^+ B_O)C_O}=0$, since C_O is compact. Similarly

$\tau_{C_O(\Pi_+^+ B_O)} = 0$. Also

(14.11) $$\tau_{K_O}(\vec{x},\vec{\xi}) = \sum_{j,l=1}^{r} (e_j\rangle\langle e_l)\; p_{\vec{x},\vec{\xi}}(\kappa_j) r_{jl}(\vec{x},\vec{\xi}) \quad ,$$

where (c.f. (12.7))

(14.12) $$\begin{aligned} ((r_{jl}(\vec{x},\vec{\xi}))) &= ((\sigma^n_{\tilde{R}_{jl}}(\vec{\xi},-\vec{x}))) \\ &= ((q_{j,\vec{x},\vec{\xi}}(\kappa_l)))^{-1}((-i\gamma_l q_{j,\vec{x},\vec{\xi}}(-\kappa_l))) \quad . \end{aligned}$$

Also (14.8) allows calculation of

(14.13) $$\begin{aligned} \tau_{(\Pi_+^- A_O)}(\vec{x},\vec{\xi}) &= \sum_{l=1}^{r} i\gamma_l (e_l\rangle\langle e_l)\; p_{\vec{x},\vec{\xi}}(-\kappa_l) \quad , \\ \tau_{(\Pi_-^+ A_O)}(\vec{x},\vec{\xi}) &= \sum_{l=1}^{r} i\gamma_l (e_l\rangle\langle e_l)\; p_{\vec{x},\vec{\xi}}(\kappa_l) \quad . \end{aligned}$$

Next we may explicitly calculate the operators $A^\kappa_\gamma(\vec{x},\vec{\xi}) = B^\kappa_\gamma(\vec{x},\vec{\xi})$ of (13.11) for $a=b=\sigma^{n+1}_{B_O}$. We get

(14.14) $$b(0,\vec{x},\eta/\lambda(\vec{\xi}),\vec{\xi}) = (\eta^N+\rho^N)/p_{\vec{x},\vec{\xi}}(\eta) = h_{\vec{x},\vec{\xi}}(\eta) \quad ,$$

and then

(14.15) $$B^\kappa_\gamma(\vec{x},\vec{\xi}) = \pi^\kappa_\gamma h_{\vec{x},\vec{\xi}}(D_O) = \pi^\kappa_\gamma\{(D_O^N+\rho^N)^{-1} p_{\vec{x},\vec{\xi}}(D_O)\} \quad .$$

From now on we omit explicit notation of $\vec{x}$ and $\vec{\xi}$, in all formulas to follow, but keep in mind that expressions will depend on $\vec{x},\vec{\xi}$, and will be defined for $|\vec{x}|+|\vec{\xi}|=\infty$. Using (14.11), (14.12), (14.13), (14.15) we get, with $\langle B^{\kappa t}_\gamma e,v\rangle = \langle e,B^\kappa_\gamma v\rangle$, $\langle e,v\rangle = \int_0^\infty evdy$, $e,v \in \mathcal{J}$,

(14.16) $$\begin{aligned} \tau_{E_1} &= \sum_{l=1}^{r}\Big(\sum_{k=1}^{r} p(\kappa_k) r_{kl} B_+^+ e_k - i\gamma_l p(-\kappa_l) B_-^+ e_l\Big)\rangle\langle e_l = \sum_{l=1}^{r} v_l\rangle\langle e_l \\ \tau_{E_2} &= \sum_{l=1}^{r} e_l\rangle\langle\Big(\sum_{k=1}^{r} p(\kappa_l) r_{lk} B_+^{+t} e_k - i\gamma_l p(\kappa_l) B_+^{-t} e_l\Big) = \sum_{l=1}^{r} e_l\rangle\langle w_l \quad . \end{aligned}$$

We must check for invertibility of $1+\tau_{E_j}$, for $|\vec{x}|+|\vec{\xi}|=\infty$. In each case the corresponding equation is of the form

(14.17) $$v + \sum_{l=1}^{r} \zeta_l \langle\theta_l,v\rangle = w \quad , \quad v,w,\zeta_l,\theta_l \in \mathcal{J} \quad .$$

As a matter of linear algebra we note that (14.17) is always uniquely solvable if and only if

(14.18) $$\det((\delta_{jk} + \langle\theta_j,\zeta_k\rangle)) \neq 0 \quad .$$

Accordingly we must check whether the two matrices (14.19) below have the eigenvalue -1, or not.

(14.19)
$$\langle e_j, v_l\rangle = \sum_{k=1}^{r} p(\kappa_k) r_{kl} \langle e_j, B_+^+ e_k\rangle - i\gamma_l p(-\kappa_l)\langle e_j, B_-^+ e_l\rangle$$
$$\langle w_j, e_l\rangle = \sum_{k=1}^{r} p(\kappa_j) r_{jk} \langle e_k, B_+^+ e_l\rangle - i\gamma_j p(\kappa_k)\langle e_j, B_+^- e_l\rangle$$

By a calculation,

(14.20)
$$\langle e_j, B_+^+ e_k\rangle = 1/2\pi \int_{-\infty}^{+\infty} h(\eta)d\eta \int_0^{\infty} dxdy\ e^{i\eta(x-y)} e_j(x) e_k(y)$$
$$= 1/2\pi \int_{-\infty}^{+\infty} h(\eta)d\eta/((\eta+\kappa_j)(\eta-\kappa_k)) ,$$
$$\langle e_j, B_-^+ e_k\rangle = -1/2\pi \int_{-\infty}^{+\infty} h(\eta)d\eta/((\eta+\kappa_j)(\eta+\kappa_k)) ,$$
$$\langle e_j, B_+^- e_k\rangle = -1/2\pi \int_{-\infty}^{+\infty} h(\eta)d\eta/((\eta-\kappa_j)(\eta-\kappa_k)) .$$

Therefore with $r_{kl} = -i\tilde{r}_{kl}$,

(14.21)
$$\langle e_j, v_l\rangle = 1/2\pi i \int_{\mathbb{R}} h(\eta)d\eta/(\eta+\kappa_j)\left(\sum_{k=1}^{r} p(\kappa_k)\tilde{r}_{kl}/(\eta-\kappa_k) - \gamma_l p(-\kappa_l)/(\eta+\kappa_l)\right)$$
$$\langle w_j, e_l\rangle = 1/2\pi i \int_{\mathbb{R}} h(\eta)d\eta/(\eta-\kappa_l)\left(\sum_{k=1}^{r} p(\kappa_j)\tilde{r}_{jk}/(\eta+\kappa_k) - \gamma_j p(\kappa)/(\eta-\kappa_j)\right)$$

First we look at the second matrix. The integrand is of order $O(\eta^{-2})$, at $\eta=\infty$, thus the integral may be converted into $\int_{\Gamma_+} + \int_{\Gamma_+^0}$ with closed positively oriented countours Γ_+, Γ_+^0 containing the upper half plane roots of $p(\eta)$ and of $\eta^N+\rho^N$ in the interior, respectively (and all other roots in the outside). But $\int_{\Gamma_+^0} = -\delta_{jl}$, by a simple calculation of residues, using that the first term of $\langle w_j, e_l\rangle$ has its integrand regular inside Γ_+^0, as ρ is large enough, as from now on to be assumed. Hence

(14.22) $$\delta_{jl} + \langle w_j, e_l\rangle = 1/2\pi i \int_{\Gamma^+} h(\eta) J_j(\eta) d\eta/(\eta-\kappa_l) = \alpha_{jl}$$

with

(14.23) $$\sum_{l=1}^{r} (q_m/p)(\kappa_l) J_l(\eta) = \sum_{l=1}^{r} (q_m(-\kappa_l)\gamma_l/(\eta+\kappa_l) - \gamma_l q_m(\kappa_l)/(\eta-\kappa_l)) .$$

A calculation of residues shows that (14.23) is the partial fractions decomposition of the function $-q_m(\eta)/(\eta^N+\rho^N)$. Also we have seen before that the matrix $(((q_j/p)(\kappa_l)))$ is nonsingular. Therefore $1+\tau_{E_2}$ is Fredholm, by (14.22) and (14.23), if and only if

(14.24) $$\beta_{mk} = \int_{\Gamma_+} q_m(\eta)/p(\eta) d\eta/(\eta-\kappa_k)$$

gives a nonsingular matrix (for all $|\vec{x}|+|\vec{\xi}|=\infty$.) A simple calculation shows that this is true if and only if the q_j are linearly independent modulo p^+ (i.e., if the boundary conditions are md-elliptic).

Finally, for E_1, one observes that, for a countour encircling the roots of $p(\eta)$ in the lower half-plane, positively oriented and with $\pm\kappa_j$ and all other roots of p in the outside we get

$$(14.25)\qquad 1/2\pi i \int_{\Gamma_-} J_{jl}\,d\eta = -\langle e_j, v_l\rangle - \delta_{jl} = \mu_{jl}$$

with the integrand J_{jl} of the first relation (14.21), therefore we must investigate if (14.25) defines a nonsingular matrix. This follows because again $J_{jl}(\eta)$ is of order $O(\eta^{-2})$ at infinity, and by evaluation of residues inside a suitable Γ_-^o in which the first term of J_{jl} stays regular. Let $\phi(\eta)$ be a polynomial of degree < r, and observe the partial fractions decomposition

$$(14.26)\qquad p^-(\eta)\phi(\eta)(\eta^N+\rho^N)^{-1} = \sum_{l=1}^{r} \gamma_l((p^-\phi)(\kappa_l)/(\eta-\kappa_l)-(p^-\phi)(-\kappa_l)/(\eta+\kappa_l)) ,$$

with $p^-(\eta) = p(\eta)/p^+(\eta)$ the part of $p(\eta)$ belonging to the roots within Γ_-. Multiply μ_{jl} of (14.25) with $(\phi/p^+)(-\kappa_l)$ and sum over l, and observe that the last term in $\int_{\Gamma_-} J_{jl}dy$ supplies the term

$$(14.27)\qquad \begin{aligned} &-\int_{\Gamma_-} h(\eta)d\eta/(\eta+\kappa_j)\sum_{l=1}^{r}\gamma_l(\phi p^-)(-\kappa_l)/(\eta+\kappa_l) \\ &= -\int_{\Gamma_-} h(\eta)d\eta/(\eta+\kappa_j)\sum_{l=1}^{r}\gamma_l(\phi p^-)(\kappa_l)/(\eta-\kappa_l) \quad , \end{aligned}$$

using that

$$(14.28)\qquad \int_{\Gamma_-} h(\eta)d\eta/(\eta+\kappa_j)(\phi p^-)(\eta)(\eta^N+\rho^N)^{-1} = 0$$

Therefore

$$(14.29)\qquad \begin{aligned} &2\pi i\sum_{l=1}^{r}\mu_{jl}(\phi/p^+)(-\kappa_l) = \sum_{k=1}^{r}\theta_{jk}\Psi_k \ , \quad \theta_{jk} = \int_{\Gamma_-} h(\eta)((\eta+\kappa_j)(\eta-\kappa_k))^{-1}d\eta \\ &\Psi_k = p(\kappa_k)(\sum_{l=1}^{r}\tilde{r}_{kl}(\phi/p^+)(-\kappa_l)-\gamma_k(\phi/p^+)(\kappa_k)) \quad . \end{aligned}$$

Letting $\phi_m(\eta) = \eta^{m-1}$, m=1,...,r, and observing that $(((\phi_m/p^+)(-\kappa_l)))$ is a non-singular Van der Monde, and that $((\theta_{jk}))$ is nonsingular we focus on the correspon-ding $\Psi_k = \Psi_{km}$. We also have $((s^+_{jl})) = (((q_j/p)(\kappa_k)))$ nonsingular, and thus focus on

$$(14.30)\qquad \sum_{r=1}^{r} s^+_{jk}\Psi_{km} = \sum_{l=1}^{r}\gamma_l((q_j\phi_m/p^+)(-\kappa_l) - (q_j\phi_m/p^+)(\kappa_l)) \quad .$$

The right hand side of (14.30) is the sum of residues of the function $q_j(\eta)\eta^{m-1}/(p^+(\eta)(\eta^N+\rho^N))$, within some annulus $\rho-\varepsilon\le|\eta|\le\rho+\varepsilon$. That rational function has residue at infinity equal to zero. Therefore (14.30) takes the form (up to a non-vanishing constant)

$$(14.31)\qquad \int_{\Gamma_+} q_j(\eta)\,\eta^{m-1}/(p^+(\eta)(\eta^N+\rho^N))\,d\eta \qquad ,$$

Again (14.31) defines a nonsingular matrix if and only if the boundary conditions $<b_j>$ are md-elliptic, with respect to $<a>$.

This completes the proof of Theorem 3.1.

15. References.

[1] S. Agmon, A. Douglis and L. Nirenberg, Estimates near the boundary for solutions of elliptic partial differential equations satisfying general boundary conditions I; Communications Pure Appl. Math. 12 (1959) 623-727.

[2] N. Aronszajn, Boundary values of functions with a finite Dirichlet integral; Conference on partial differential equations,U.of Kansas,No.14 (1954),77 - 94 .

[3] N. Aronszajn, On coercive integro-differential quadratic forms, Conf. on partial differential equations, U. Kansas, No 14 (1954) 94-106.

[4] N. Aronszajn and K.T. Smith, Theory of Bessel potentials I, Ann. Inst. Fourier (Grenoble) 11 (1961) 385-475.

[5] R. Beals, Square roots of nonnegative systems and the sharp Gårding inequality; J. Diff. Equ. 24 (1977) 235-239.

[6] M. Breuer and H.O. Cordes, On Banach algebras with σ-symbol II, J. f. Math. Mech. 14 (1965) 299-314.

[7] F. Browder, Estimates and existence theorems for elliptic boundary problems; Proc. Nat. Acad. Sci. USA, 365-372 (1959).

[8] P. Colella and H.O. Cordes, The C^*-algebra of the elliptic boundary problem; to appear in Pac. Journ. Math. 1978; preprint NO 160 SFB 72, U. of Bonn.

[9] H.O. Cordes, On a class of C^*-algebras; Math. Ann. 170 (1967) 283-313.

[10] ----, On a generalized Fredholm theory, J. f. r. u. angew. Math. 227 (1967) 121-149.

[11] ----, Über eine nichtalgebraische Charakterisierung von $\mathfrak{J}$-Fredholm Operatoren; Math. Ann 163 (1966) 212-229.

[12] ----, Pseudo-differential operators on a half-line, Journal f. Math. Mech. 18 (1969) 893-908.

[13] H.O. Cordes, An algebra of singular integral operators with two symbol homomorphisms. Bull. Amer. Math. Soc. 75 (1969) 37-42.

[14] ----, Lecture notes on Banach algebra methods in partial differential equations, Lund 1970/71.

[15] ----, Partial differential equations, lecture notes, Berkeley 1977.

[16] ----, On compactness of commutators of multiplications and convolutions, and boundedness of pseudo-differential operators, J. of Functional Analysis 18 (1975) 115-131.

[17] ---- and A. Erkip, On the general elliptic boundary problem in a half-space, to appear in Rocky Mountain J. 1978.

[18] ---- and E. Herman, Singular Integrals on the half-line; Proc, Nat. Akad. Sci. 56 (1966) 1668-1673.

[19] ---- and R.C. McOwen, The C^*-algebra of a singular elliptic problem on a non-compact Riemannian manifold, Math. Z. 153 (1977) 101-116.

[20] ----, Remarks on singular elliptic theory for complete Riemannian manifolds, to appear in Pacific J. Math.

[21] J. Dixmier, Les C^*-algebres et leurs representations, Paris: Gauthier-Villars 1964.

[22] A. Erkip, On the elliptic boundary problem in a half space. Commun. Part. Diff. Equ. To appear.

[23] L. Gårding, Dirichlet's problem for linear elliptic partial differential operators. Math. Scand. 1 (1953) 55-72.

[24] Y.B. Lopatinskij, On a method of reducing boundary problems for a system of differential equations of elliptic type to regular equations; Ukrain. Math. Z. 5 (1953) 123-151.

[25] W. Magnus and F. Oberhettinger, Formeln und Saetze fuer die speziellen Funktionen der Mathematischen Physik, 2nd ed., Springer, Berlin-Goettingen-Heidelberg, 1948.

[26] R. McOwen, Thesis, Berkeley 1978, to appear.

[27] C.B. Morrey, Multiple integrals and the Calculus of variations; Springer, Berlin Göttingen Heidelberg 1966.

[28] J. Peetre, Another approach to elliptic boundary problems; Comm. Pure Appl. Math. 14 (1969) 711-731.

[29] G. Polya-G. Szegö, Aufgaben und Lehrsätze aus der Analysis; Vol. I, Springer, Berlin 1925.

[30] F. Riesz-B.v.Sz.Nagy, Functional Analysis; Ungar Publ. Co. New York 1955.

[31] M. Schechter, Various types of boundary conditions for elliptic equations; Comm. Pure Appl. Math. 13 (1960) 407-425.

[32] E.M. Stein, Singular integrals and differentiability properties of functions, Princeton Math. Ser. 30; Princeton NJ. 1970.

Appendix AI: Preparations on Linear Operators and Operator Algebras.

In the present appendix we are going to discuss some aspects of infinite dimensional linear algebra which are not contained in most of the standard books. In particular we will discuss Fredholm operators, the Rellich criteria for compact operators on Banach spaces, Fredholm closed algebras and Riesz ideals, and the Fredholm pairs, introduced by Kato [12].

The concept of Fredholm operator is defined and discussed in the first three sections. Section 2 considers Fredholm operators between general linear spaces without involvement of a topology. Only in section 3 we then introduce Banach spaces and extend the concept regarding its invariance with respect to small or compact perturbations. Although the Fredholm alternatives were observed rather early for integral operators and differential operators, there does not seem to exist a systematic abstract discussion of the concept as a property of linear operators until the 1950-s.

The Rellich criteria, discussed in section 4, seem to contradict on first glance with the fact that there exist Banach spaces in which the operators of finite rank are not dense in the ideal of compact operators. Their formulation for Banach spaces seems to be new, although not hard to prove.

The property of Fredholm closedness, discussed in section 5 is an important tool for our Laplace comparison algebra technique of chapter IV and V.

In sections 6 and 7 we shortly discuss unbounded operators of a Banach space, specifically unbounded Fredholm operators and Fredholm pairs, which seem to offer the shortest way of discussing invariance of the index and the Fredholm property for unbounded operators, apart from the Gohberg aperture topology.

The approach to the invariance of the Fredholm property and the index for unbounded operators seems different from other such discussions.It perhaps is easier than the use of the Gohberg-Krein Aperture topology in [3].In particular we introduce a new topology for the class of closed subspaces of a Hilbert space, called the projection topology . The discussion goes far beyond the level required in the earlier chapters.

In section 1 we give a discussion without proof of some of the historical examples of Fredholm operators,which served to formulate the concept.Actually the concept of Fredholm map,as developed in section 2 without topology,might be thought of as a part of a first lecture on linear algebra.Only the fact that it becomes trivial in finite dimensional spaces may have accounted for its late discovery.

1. Definition of Fredholm map; historical examples.

Let $A: \mathfrak{X} \to \mathfrak{Y}$ denote a linear map between the two complex vectorspaces $\mathfrak{X}$ and $\mathfrak{Y}$, and let us consider the linear equation

$$(1.1) \qquad Ax = f$$

where $f \in \mathfrak{Y}$ is assumed known, and the "solution" $x \in \mathfrak{X}$ is to be determined.

In the attempt to provide basic information about solutions of equation (1.1) the best, Mathematics can provide is an existence and uniqueness theorem, asserting the existence of a unique solution $x \in \mathfrak{X}$ for every $f \in \mathfrak{Y}$. In terms of the linear map A this simply amounts to existence of an inverse $A^{-1}: \mathfrak{Y} \to \mathfrak{X}$, that is, of a linear map $B = A^{-1}$ from $\mathfrak{Y}$ to $\mathfrak{X}$ such that $ABy = y$ and $BAx = x$ holds for all $x \in \mathfrak{X}$ and all $y \in \mathfrak{Y}$. Following a convention in theory of partial differential equations we then also will say that the linear problem (1.1) is "well posed".

Experience shows, however, that for many types of problems in theory of differential or integral equations the most general result asserts only that

$$(1.2) \qquad \alpha(A) = \dim \ker A < \infty \quad \text{and} \quad \beta(A) = \operatorname{codim} \operatorname{im} A = \dim(\mathfrak{Y}/\operatorname{im} A) < \infty.$$

That is, equation (1.1) is solvable only for all f satisfying finitely many given linear conditions, and the solution in general is not unique, but all solutions form a finite dimensional linear submanifold in $\mathfrak{X}$.

Apparently the problem is well posed if and only if both above dimensions (1.2) vanish. Therefore condition (1.2) may be called a "finite dimensionally degenerated well-posedness".

A linear map A satisfying the two conditions (1.2) will be called a Fredholm map, or, A will be said to have property F. The corresponding linear problem (1.1) then is called normally solvable.

The examples, below, are historical starting points of the theory.

Example 1. In the linear space $\mathfrak{X} = C([a,b])$ of all continuous complex-valued functions over the closed interval $[a,b]$ of the real line let us consider the linear map $A: \mathfrak{X} \to \mathfrak{X}$ defined by

$$(1.3) \qquad u(x) \to v(x) = u(x) + \int_a^b k(x,y)u(y)dy , \qquad u \in \mathfrak{X} .$$

This operator is well defined and Fredholm whenever the function $k: [a,b] \times [a,b] \to \mathbb{C}$ is continuous. The statement was proven by Fredholm [6], using a generalized kind of determinant defined for **integral operators.Moreover,Fredholm** established that for this type of operator the two dimensions (1.2) are always equal. The result is known as "the three Fredholm alternatives" and may be found in most older books on integral equations.

A modern proof first concludes the result for "degenerating kernels"

(1.4) $$k(x,y) = \sum_{j=1}^{N} a_j(x)b_j(y)$$

with 2N continuous functions $a_j(x)$, $b_j(y)$, where the integral equation degenerates into a finite system of N equations in N unknown constants, and then uses the Weierstrass approximation theorem and the Ascoli-Arzela theorem to extend to the general case. (See also the proof of Lemma 2.3, for more details.)

Example 2. Consider the boundary problem

(1.5) $$\Delta u = u_{xx} + u_{yy} = 0 \text{ in } \mathcal{D}$$
$$xu_x + yu_y + hu = \phi \quad \text{on } \Gamma$$

with $u_x = \partial u/\partial x$, etc. in the unit disc $\mathcal{D} = \{x^2 + y^2 \leq 1\}$ with boundary Γ, and with complex-valued functions h, ϕ defined on Γ.

A solution is assumed Hoelder continuous on $\mathcal{D}$ together with its partial derivatives up to the order $N \geq 2$ including. That is we have $|u(x+h,y+k) - u(x,y)| \leq c(|h|^\alpha + |k|^\alpha)$, for (x,y), $(x-h,y-k) \in \mathcal{D}$, with constants $\alpha, c > 0$, depending on u but not on the point, and correspondingly for the derivatives as listed. The space of these functions will be called $\mathfrak{M}_N(\mathcal{D})$ and we define $\mathfrak{M}_N(\Gamma)$ as the space of restrictions of the functions in $\mathfrak{M}_N(\mathcal{D})$ to Γ.

A solution of the Laplace equation $\Delta u = 0$ is determined by its boundary values, by the maximum principle. Moreover, using Poisson's integral formula, it can be shown that every $u \in \mathfrak{M}_N(\Gamma)$ possesses an extension $u(x,y)$ into $\mathfrak{M}_N(\mathcal{D})$ solving the Laplace equation. We define a linear map $Q: \mathfrak{M}_N(\Gamma) \to \mathfrak{M}_{N-1}(\Gamma)$ assigning to $\psi \in \mathfrak{M}_N(\Gamma)$ the function $\phi = [xu_x + yu_y + hu]_\Gamma$ with above extensions u, where we assume that $h \in \mathfrak{M}_{N-1}(\Gamma)$.

This linear map Q is Fredholm, and again, the dimensions (1.2) are equal.

The statement reflects a well known result of potential theory. A proof results from a "single layer integral" - that is, we assume the solution u of (1.5) in the form

(1.6) $$u(x,y) = \int_0^{2\pi} \log((x - \cos t)^2 + (y - \sin t)^2)\mu(t)\,dt$$

with a $\mu(t) \in \mathfrak{M}_{N-1}(\Gamma)$. The Laplace equation then is automatically satisfied in $\mathcal{D} - \Gamma$, and it is possible to compute the derivatives of u on $\mathcal{D}$ up to the order N as integral expressions of μ and its derivatives up to the order N-1. Thus it can be seen that $\mu \in \mathfrak{M}_{N-1}(\Gamma)$ implies $u \in \mathfrak{M}_N(\mathcal{D})$. Vice versa, differentiating (1.6) in the direction normal to Γ, one shows that $2\pi\mu$ differs from the exterior normal derivative of u on Γ only by a constant. Further, the boundary condition of (1.5) goes into an integral equation of the general form (1.3), but over the circle Γ, with (slightly) discontinuous kernel, and over the space $\mathfrak{M}_{N-1}(\Gamma)$, rather than $C([a,b])$. Investigation of the operator Q therefore can be reduced to the study of an integral equation and a proof of the statement results similar as for example 2.

If the function h is infinitely differentiable, the construction may be done for arbitrary $N \geq 2$. The linear map Q then may be considered from $C^{\infty}(\Gamma)$ to $C^{\infty}(\Gamma)$, and satisfies the Fredholm alternatives also under this new meaning.

Example 3. If Γ is as in example 2, let a,b denote complex-valued functions over Γ, and consider the Cauchy type singular integral operator $K: \mathfrak{M}_N(\Gamma) \to \mathfrak{M}_N(\Gamma)$ defined by

$$(1.7)\qquad u(t) \to v(t) = a(t)u(t) + b(t)/\pi \int_0^{2\pi} \cot (t-s)/2 \ u(s)ds$$

where a,b are in $\mathfrak{M}_N(\Gamma)$, $N \geq 0$, and where s and t measure the arc length on the circle Γ from $(x,y) = (1,0)$. The integral is a principal value $(= \lim_{\varepsilon \to 0} \int_{|s-t| \geq \varepsilon})$

This operator K is Fredholm if and only if $a^2 + b^2 \neq 0$ on all of Γ. As a significant difference to the previous examples we note that for K we do not have equality of the two dimensions (1.2) anymore, but instead get the formula

$$(1.8)\qquad \dim \ker K - \dim \mathfrak{M}_N(\Gamma)/ \operatorname{im} K) = \nu_- - \nu_+$$

where $\nu_{\pm} = (2\pi)^{-1} \int_0^{2\pi} d(\arg (a(t) \pm ib(t))$ denote the winding numbers of the two non-vanishing complex vector fields $a \pm ib$ over the circle Γ.

This topological relation is typical for the more general results expecting us. Essentially the above was first noted by F. Noether [17].

For any Fredholm map $A: \mathfrak{X} \to \mathfrak{Y}$ the difference

$$(1.9)\qquad \dim \ker A - \dim (\mathfrak{Y}/ \operatorname{im} A) = \operatorname{ind} A$$

is called the (Fredholm) index of A. The dimensions $\alpha(A)$ and $\beta(A)$ are called nullity and deficiency of A, respectively.

Example 4. Consider the boundary problem

$$(1.10)\qquad \begin{aligned} \Delta u &= 0 \text{ in } \mathcal{D} \\ au_x + bu_y &= \phi \text{ on } \Gamma , \end{aligned}$$

with $\mathcal{D}$ and Γ as in example 2. We assume $u \in \mathfrak{M}_N(\mathcal{D})$, $a,b,\ \phi \in \mathfrak{M}_{N-1}(\Gamma)$, and then define a linear map $R: \mathfrak{M}_N(\Gamma) \to \mathfrak{M}_{N-1}(\Gamma)$ by

(1.11) $$\psi \to [au_x + bu_y]_\Gamma$$

similar as in example 2, with the harmonic extension u of $\psi \in \mathfrak{M}_N(\Gamma)$ into $\mathcal{D}$.

With the same single layer integral as in example 2 one reduces the boundary problem to a singular integral equation of the form studied in example 3. As a consequence we find that R is Fredholm precisely if $a^2 + b^2 \neq 0$ on Γ and get the index formula

(1.12) $$\text{ind } R = \nu_- - \nu_+ + 2 ,$$

again with the winding numbers $\nu_\pm$ of $a \pm ib$ on Γ.

Also, the corresponding result follows for $C^\infty(\Gamma)$ instead of $\mathfrak{M}_N$ and $\mathfrak{M}_{N-1}$ again. We shall not offer a proof of the above facts, but mention that they are typical for the kind of theory discussed in all details in chapters III, IV and V.

Example 5. Perhaps the simplest type of Fredholm maps are the following "shift operators". In the space l^p of all sequences $x = (x_1, x_2, \ldots)$ with $\sum_{j=1}^\infty |x_j|^p < \infty$ (with $1 < p < \infty$) define the maps $T_N, N = 0, \pm 1, \pm 2, \ldots$ by the assignment

(1.13) $$x \to T_N x = (x_{N+1}, x_{N+2}, \ldots)$$

if $N \geq 0$, and by

(1.14) $$x \to T_N x = (0_1, 0_2, \ldots, 0_N, x_1, x_2, \ldots)$$

with $0_1 = 0_2 = \ldots = 0_N = 0$, if $N < 0$.

Note that all operators T_N are Fredholm maps and that $\text{ind } T_N = N$.

2. Algebraic theory of Fredholm operators.

It is clear that invertible linear maps have always property F, and have Fredholm index 0. Fredholm maps, on the other hand, may be characterized by the following weaker invertibility property.

Theorem 2.1. A linear map $A: \mathfrak{X} \to \mathfrak{Y}$ is Fredholm if and only if there exists a linear map $B: \mathfrak{Y} \to \mathfrak{X}$ such that $(1-AB): \mathfrak{Y} \to \mathfrak{Y}$ and $(1-BA): \mathfrak{X} \to \mathfrak{X}$ both are of finite rank, where "1" denotes the identity map in the corresponding space.

Recall that $F: \mathfrak{X} \to \mathfrak{Y}$ is said to be of finite rank if it has finite dimensional image. The collection of all such maps forms a linear vector space over $\mathbb{C}$, denoted by $E(\mathfrak{X},\mathfrak{Y})$. Let $L(\mathfrak{X},\mathfrak{Y})$ denote the space of all linear maps from $\mathfrak{X}$ to $\mathfrak{Y}$, then $A \in L(\mathfrak{Y},\mathfrak{Z})$, $B \in L(\mathfrak{M},\mathfrak{X})$, $F \in E(\mathfrak{X},\mathfrak{Y})$ implies $AF \in E(\mathfrak{X},\mathfrak{Y})$, $FB \in E(\mathfrak{M},\mathfrak{Y})$.

Let $L/E = L/E(\mathfrak{X},\mathfrak{Y})$ denote the quotient space of all co-sets $A^\vee = \{A + F: F \in E(\mathfrak{X},\mathfrak{Y})\}$, for $A \in L(\mathfrak{X},\mathfrak{Y})$. Clearly L/E is a linear vector-space over $\mathbb{C}$, and there is a product $A^\vee B^\vee \in L/E(\mathfrak{X},\mathfrak{Y})$ meaningfully defined for $A^\vee \in L/E(\mathfrak{Y},\mathfrak{Z})$ and $B^\vee \in L/E(\mathfrak{X},\mathfrak{Y})$, by setting $A^\vee B^\vee = (AB)^\vee$. Specifically $L/E(\mathfrak{X},\mathfrak{X}) = L/E(\mathfrak{X})$ is an algebra over $\mathbb{C}$ with this product. Note that $E(\mathfrak{X}) = E(\mathfrak{X},\mathfrak{X})$ is a two-sided ideal of the algebra $L(\mathfrak{X}) = L(\mathfrak{X},\mathfrak{X})$, and that $L/E(\mathfrak{X})$ above also may be interpreted as quotient of the algebra $L(\mathfrak{X})$ modulo its ideal $E(\mathfrak{X})$. The algebra $L/E(\mathfrak{X})$ has a unit 1 supplied by the co-set of the identity map $1: \mathfrak{X} \to \mathfrak{X}$. For $A^\vee \in L/E(\mathfrak{X},\mathfrak{Y})$ and $B^\vee \in L/E(\mathfrak{Y},\mathfrak{X})$ both products $A^\vee B^\vee$ and $B^\vee A^\vee$ are meaningful, and we will say that $A^\vee$ and $B^\vee$ are invertible, and mutually inverses if both products are the identity.

With these preparations we may restate Theorem 2.1 as follows.

Theorem 2.1'. A linear map $A \in L(\mathfrak{X},\mathfrak{Y})$ is Fredholm if and only if its co-set $A^\vee = A + E \in L/E(\mathfrak{X},\mathfrak{Y})$ is invertible.

Proof. Suppose first we have $B: \mathfrak{Y} \to \mathfrak{X}$ with $AB - 1 = F_1$, $BA - 1 = F_2$ then $x \in \ker A$ implies $x = F_2 x \in \operatorname{im} F_2$. Or, $\ker A \subset \operatorname{im} F_2$ is finite dimensional. Also for any $y \in \mathfrak{Y}$ we get $y + (F_1 y) = A(By) \in \operatorname{im} A$. That is, if $w_1, w_2, \ldots, w_N$ form a basis of $\operatorname{im} F_1$, then we get $y = \sum_{j=1}^N \alpha_j w_j + z$, $\alpha_j \in \mathbb{C}$, $z \in \operatorname{im} A$, for all $y \in \mathfrak{Y}$. Or, the vectors $w_1, \ldots, w_N$ span $\mathfrak{Y}$ mod $(\operatorname{im} A)$, i.e. $\dim (\mathfrak{Y}/\operatorname{im} A) \leq N < \infty$. Hence A is Fredholm.

Next we assume that A is Fredholm, and let

$$\mathfrak{X} = \ker A \oplus \mathfrak{Z}, \quad \mathfrak{Y} = \mathfrak{M} \oplus \operatorname{im} A, \tag{2.1}$$

that is in details

$$\begin{aligned} \mathfrak{X} &= \ker A + \mathfrak{Z}, \quad \ker A \cap \mathfrak{Z} = \{0\}, \\ \mathfrak{Y} &= \mathfrak{M} + \operatorname{im} A, \quad \mathfrak{M} \cap \operatorname{im} A = \{0\}. \end{aligned} \tag{2.1'}$$

with linear subspaces $\mathfrak{Z} \subset \mathfrak{X}$ and $\mathfrak{W} \subset \mathfrak{Y}$. The last decomposition is achieved by picking a basis $w_1,\ldots,w_N$ of $\mathfrak{Y}$ mod (im A) and denoting its span by $\mathfrak{W}$. For the first decomposition we pick a basis $v_1,\ldots,v_M$ of ker A and then a Hamel basis (c.f. [4]) $B = \{b\}$ of $\mathfrak{X}$ containing the v_j as elements. Then the span of $B' = B - \{v_1,\ldots,v_M\}$ is a space $\mathfrak{Z}$ as desired.

The above decompositions define projections $P: \mathfrak{X} \to \mathfrak{X}$ and $Q: \mathfrak{Y} \to \mathfrak{Y}$ assigning to $x \in \mathfrak{X}$ the (ker A)-component, and to $y \in \mathfrak{Y}$ the $\mathfrak{W}$-component, respectively, and P,Q are of finite rank. The linear map $A^O: \mathfrak{Z} \to \text{im } A$, defined as restriction of A to $\mathfrak{Z}$ is bijective and thus has an inverse $B^O: \text{im } A \to \mathfrak{Z}$. Define $B: \mathfrak{Y} \to \mathfrak{X}$ by $B = B^O(1 - Q)$ and notice that $AB = A^O B^O(1 - Q) = 1 - Q$ and $BA(1 - P) = B^O(1 - Q)A(1 - P) = 1 - P$, i.e., $BA = 1 - P$. Since $F_1 = -Q$ and $F_2 = -P$ are of finite rank, the theorem is established.

A linear map B satisfying the conditions of Theorem 2.1 shall be called a Fredholm inverse of A . Clearly a Fredholm inverse exists if and only if A satisfies property F and every Fredholm inverse has property F again. A and B then are mutually Fredholm inverses of each other.

The special type of Fredholm inverse B constructed for A in the proof of Theorem 2.1 may be described by the following diagram

$$\begin{array}{ccccccc}
\mathfrak{X} & = & \ker A & \oplus & \text{im } B & \oplus & \{0\} \\
A \downarrow \uparrow B & & 0 \,\big| & & B^O \uparrow \ \downarrow A^O & & \uparrow 0 \\
\mathfrak{Y} & = & \{0\} & \oplus & \text{im } A & \oplus & \ker B
\end{array} ,$$

the operators $A^O: \text{im } B \to \text{im } A$ and $B^O: \text{im } A \to \text{im } B$ being strict inverses of each other. Such Fredholm inverses will be referred to as special Fredholm inverses in the remainder of this section. The following then is evident.

Lemma 2.2. If A and B are special Fredholm inverses of each other then we have $\text{ind } A + \text{ind } B = 0$.

Lemma 2.3. A linear map of the form $A + F$ with invertible $A: \mathfrak{X} \to \mathfrak{Y}$ and $F \in \mathcal{E}(\mathfrak{X},\mathfrak{Y})$ is Fredholm and has index zero.

Proof. It is sufficient to consider operators of the form $(1 + F): \mathfrak{X} \to \mathfrak{X}$ only, because $A + F = A(1 + A^{-1}F)$ and the isomorphism A will not change property F nor index.

Lemma 2.4. For an $F \in \mathcal{E}(\mathfrak{X},\mathfrak{Y})$ we get

$$\text{codim ker } F = \dim \text{im } F = N < \infty \quad . \tag{2.2}$$

Indeed, if $w_j = F z_j$, $j=1,\ldots,N$, is a basis of im F then the z_j are linearly independent modulo ker F, and N+1 elements x_j linearly independent mod ker F would generate N+1 linearly independent elements Fx_j of ker F which cannot exist.

To continue the proof of lemma 2.3 note that by lemma 2.4 we get the unique representation

$$(2.3)\qquad x = \sum_{j=1}^{N} \xi_j z_j + z\ ,\quad \xi_j \in \mathbb{C}\ ,\quad z \in \ker F\ ,$$

for every $x \in \mathfrak{X}$, where $z_1,\ldots,z_N$ denotes a basis of $\mathfrak{X}$ mod ker F. The $\xi_j = \ell_j(x)$ determine linear functionals over $\mathfrak{X}$, and we have

$$(2.4)\qquad Fx = \sum_{j=1}^{N} \ell_j(x)\ Fz_j\ .$$

Thus the equation $(1+F)x = f$ reads

$$(2.5)\qquad x + \sum_{j=1}^{N} \ell_j(x)\ Fz_j = f\ .$$

Set $\xi_j = \ell_j(x)$ and apply the functional ℓ_k to (2.5) to see that (2.5) implies

$$(2.6)\qquad \xi_k + \sum_{j=1}^{N} \xi_j a_{kj} = \ell_k(f),\ k = 1,\ldots,N\ ,$$

with the matrix

$$(2.7)\qquad ((a_{kj})) = ((\ell_k(Fz_j))) = A$$

Equations (2.6) may be regarded as a system of N linear equations for the N unknown constants ξ_j.

For a given fixed $f \in \mathfrak{Y}$ and a solution $(\xi_1,\ldots,\xi_N)$ of (2.7) set

$$(2.8)\qquad x = f - \sum_{j=1}^{N} \xi_j\ F z_j$$

and note that (2.7) implies $\ell_k(x) = \ell_k(f) - \sum_{j=1}^{N} \xi_j \ell_k(Fz_j) = \xi_k$, or that (2.5) holds. Accordingly, for any given fixed f, the two equations (2.5) and (2.6) are equivalent. Especially for f=0 they have equally many linearly independent solutions, that is

$$(2.9)\qquad \dim\ker(1+F) = \dim\ker(1+A) = N - r$$

where r denotes the rank of the matrix A. Further (2.5) is solvable only for those f for which the N-tuple $(\ell_1(f),\ldots,\ell_N(f))$ belongs to the image of $(I + A)$, that is whenever f satisfies N-r linear conditions

$$(2.10)\qquad m_1(f) = m_2(f) = \ldots = m_s(f) = 0\ ,\ s = N - r\ ,$$

with linear functionals m_j over $\mathfrak{X}$. Since the functionals are independent, there must be s vectors $f_1,\dots,f_s$, such that the matrix $((m_j(f_k)))$ is non-singular, and it then follows that the f_j form a basis of $\mathfrak{X}$ modulo im $(1+F)$. Accordingly

(2.11) $\quad \operatorname{codim}(\operatorname{im}(1+F)) = s = N - r \quad .$

Relations (2.9) and (2.11) then imply the statement of lemma 2.3.

Lemma 2.5. If A and B are Fredholm and AB is defined, and if ind A ≥ 0, ind B ≥ 0, then we have AB also Fredholm and

$$\operatorname{ind}(AB) = \operatorname{ind} A - \operatorname{ind} B \ .$$

Proof. Let $A: \mathfrak{Y} \to \mathfrak{Z}$ and $B: \mathfrak{X} \to \mathfrak{Y}$ with nonnegative indices be given and let $\mathfrak{U}$ and $\mathfrak{W}$ denote two finite dimensional vector spaces with dimension ind A and ind B, respectively. Define the spaces $\tilde{\mathfrak{Y}} = \mathfrak{Y} \times \mathfrak{W}$ and $\tilde{\mathfrak{Z}} = \mathfrak{Z} \times \mathfrak{W} \times \mathfrak{U}$ of all ordered pairs and triples, respectively, and let

$$\mathfrak{X} = \mathfrak{Z}_B \oplus \ker B \ , \quad \mathfrak{Y} = \operatorname{im} B \oplus \mathfrak{M}_B \ ,$$
$$\mathfrak{Y} = \mathfrak{Z}_A \oplus \ker A \ , \quad \mathfrak{Z} = \operatorname{im} A \oplus \mathfrak{M}_A \ ,$$

denote the decompositions of type (2.1) for the two operators B and A, respectively. Define operators $\tilde{A} : \tilde{\mathfrak{Y}} \to \tilde{\mathfrak{Z}}$ and $\tilde{B} : \mathfrak{X} \to \tilde{\mathfrak{Y}}$ according to the following diagram

$$\begin{array}{ccccccc}
\mathfrak{X} & = & \mathfrak{Z}_B & \oplus & & \ker B & \\
\tilde{B}\downarrow & & \downarrow B^0 & & & \downarrow J & \\
\tilde{\mathfrak{Y}} & = & \operatorname{im} B & \oplus & \mathfrak{M}_B & \oplus & \mathfrak{W}
\end{array}$$

$$\begin{array}{ccccccccc}
\tilde{\mathfrak{Y}} & = & \mathfrak{Z}_A & \oplus & & \ker A & \oplus & \mathfrak{W} & \\
\tilde{A}\downarrow & & \downarrow A^0 & & & \downarrow K & & \searrow 1 & \\
\tilde{\mathfrak{Z}} & = & \operatorname{im} A & \oplus & \mathfrak{M}_A & \oplus & \mathfrak{U} & \oplus & \mathfrak{W}
\end{array}$$

Here J and K are isomorphisms between the spaces of equal dimension as marked.

It is clear that the operators $\tilde{A}$ and $\tilde{B}$ are invertible and thus have property F and index zero as isomorphisms between two spaces. Hence $\tilde{A}\,\tilde{B}$ is also invertible and has property F and index zero. Let further $A^\wedge$ and $B^\wedge$ defined by the same scheme but with J and K replaced by zero, then it is clear that $\operatorname{im}(\tilde{A} - A^\wedge) \subset \mathfrak{M}_A \oplus \mathfrak{U}$ and $\operatorname{im}(\tilde{B} - B^\wedge) \subset \mathfrak{M}_B \oplus \mathfrak{W}$, and hence that both differences have finite rank. It follows that $A^\wedge B^\wedge = \tilde{A}\,\tilde{B} + F$ with an F of finite rank and lemma 2.3 then implies that $A^\wedge B^\wedge$ is Fredholm and has index zero. Finally observe that $A^\wedge B^\wedge = A\,B\,W$ with $W : \mathfrak{Z} \to \tilde{\mathfrak{Z}}$ denoting the injection map. Since

$$\dim \tilde{\mathfrak{Z}}/\mathfrak{Z} = \dim \mathfrak{U} + \dim \mathfrak{W} = \operatorname{ind} A + \operatorname{ind} B$$

the statement of the lemma is immediate.

Theorem 2.6. If $A: \mathfrak{X} \to \mathfrak{Y}$ is Fredholm then A+F also is Fredholm, for every map $F: \mathfrak{X} \to \mathfrak{Y}$ of finite rank. Both operators A and A+F have the same index.

Proof. Any Fredholm inverse of A is also a Fredholm inverse of A + F, therefore A + F is Fredholm whenever A is. First assume ind A ≥ 0. Then also ind $(A+F) \geq 0$ follows, because if ind $(A+F) < 0$ and B is special Fredholm inverse of A+F, then ind B $= -$ ind $(A+F) > 0$, by lemma 2.2, and lemma 2.5 gives the contradiction

$$0 = \text{ind } AB = \text{ind } (1+F) = \text{ind } A + \text{ind } B > 0 .$$

Since both ind A and ind (A+F) are nonnegative we may apply lemma 2.5 and get

$$\text{ind } A = \text{ind } (A[B(A+F)]) = \text{ind } ([AB](A+F)) = \text{ind } (A+F) \quad .$$

Finally, if ind $A<0$ then let B and $B_1 = B + F_1$ denote special Fredholm inverses of A and A+F, respectively. Use lemma 2.2 to get ind B $= -$ ind $A > 0$ and hence conclude that ind B = ind B_1, hence again the statement of the theorem.

Corollary 2.7 If A and B are Fredholm inverses then ind A + ind B = 0.

Proof. Any Fredholm inverse B of A differs from a special Fredholm inverse only by an operator of finite rank. Thus the statement follows from lemma 2.2.

Theorem 2.8. If $A: \mathfrak{Y} \to \mathfrak{Z}$ and $B: \mathfrak{X} \to \mathfrak{Y}$ are Fredholm then $AB: \mathfrak{X} \to \mathfrak{Z}$ is Fredholm and

$$\text{ind } AB = \text{ind } A + \text{ind } B \quad .$$

Proof. Discriminate the 6 cases

(i) ind A, ind B ≥ 0 (ii) ind A, ind B ≤ 0

(iii) ind A ≥ 0, ind B ≤ 0 and ind AB ≥ 0

(iv) ind A ≥ 0 ind B ≤ 0 and ind AB ≤ 0

(v) ind A ≤ 0 ind B ≥ 0 and ind AB ≥ 0

(vi) ind A ≤ 0 ind B ≥ 0 and ind AB ≤ 0 .

Accordingly write

$$AB = A \cdot B \quad , \quad (AB)' = B' \cdot A' + F ,$$

$$A = (AB) \cdot B' + F \quad , \quad B' = (AB)' + F \quad ,$$

$$B = A' \cdot (AB)' + F \quad , \quad A' = B \cdot (AB)' + F ,$$

where L' denotes any Fredholm inverse of L and where the products at right always have factors with nonnegative index. Application of **lemma** 2.5 and theorem 2.6 therefore result in the statement.

3. Bounded Fredholm operators on Banach spaces.

All vector spaces important in Analysis carry a topology. Specifically all spaces occuring in chapters I through V are either Banach spaces or locally convex topological vector spaces. We take as our next objective the study of continuous linear maps between Banach spaces, regarding property F. Such linear maps also will be called "bounded operators". Later on we plan to extend the theory, towards unbounded closed operators (c.f. section 6). Some results also will extend to Frechet spaces (c.f. problem 4,2).

For a Banach space $\mathfrak{X}$ the norm of elements $x \in \mathfrak{X}$ will be denoted by $\|x\|$. A linear map $A: \mathfrak{X} \to \mathfrak{Y}$ is bounded - that is $\|Ax\| \leq c\,\|x\|$ for all $x \in \mathfrak{X}$ for some c if and only if A is continuous, and we then will denote

$$(3.1) \qquad \|A\| = \sup_{x \neq 0} \|Ax\| / \|x\| \quad .$$

Although linear maps between Banach spaces can be found which are not continuous, such examples are always pathological, and we adopt the general habit that $L(\mathfrak{X},\mathfrak{Y})$ and $E(\mathfrak{X},\mathfrak{Y})$ will denote the collections of all bounded linear maps (with finite rank, resp.) only, if we think of $\mathfrak{X}$ and $\mathfrak{Y}$ as Banach spaces.

Note that $\|A\|$ as in (3.1) provides a norm on the linear vector-space $L(\mathfrak{X},\mathfrak{Y})$ under which it is complete, thus becomes a Banach space.
For the composition product of section 2 we get

$$(3.2) \qquad \|AB\| \leq \|A\| \; \|B\| \; , \; A \in L(\mathfrak{Y},\mathfrak{Z}) \; , \; B \in L(\mathfrak{X},\mathfrak{Y}) \quad .$$

Specifically $L(\mathfrak{X}) = L(\mathfrak{X},\mathfrak{X})$ becomes a normed algebra - a Banach algebra - under this norm, since we also have $\|I\| = 1$ for the identity map $I \in L(\mathfrak{X})$. which clearly plays the role of a unit in $L(\mathfrak{X})$.

In contrast to the convention for vector spaces without topology in section 2 we now will write $\mathfrak{X} = \mathfrak{V} \oplus \mathfrak{W}$ or will speak of a corresponding direct decomposition, or call $\mathfrak{V}$ and $\mathfrak{W}$ complements of each other only if $\mathfrak{V}$ and $\mathfrak{W}$ are closed subspaces of $\mathfrak{X}$, in addition to the usual $\mathfrak{X} = \mathfrak{V} + \mathfrak{W}$, $\mathfrak{V} \cap \mathfrak{W} = \{0\}$. The cartesian product $\mathfrak{X} \times \mathfrak{Y}$ of two Banach spaces shall be considered a Banach space under its natural norm

$$(3.3) \qquad \|[x,y]\| = (\|x\|^2 + \|y\|^2)^{1/2} \quad \text{for } x \in \mathfrak{X} \text{ and } y \in \mathfrak{Y} \; .$$

For closed $\mathfrak{T} \subset \mathfrak{X}$ the quotient $\mathfrak{X}/\mathfrak{T}$ is a Banach space under the norm $\|x^\vee\| = \inf\{\|x\| : x \in x^\vee\}$, for $x^\vee \in \mathfrak{X}/\mathfrak{T}$.

A projection $P: \mathfrak{X} \to \mathfrak{Y}$ is defined as a bounded idempotent linear map (i.e. $P^2 = P$), and there is a 1-1-correspondence between projections and direct decompositions $\mathfrak{X} = \mathfrak{V} \oplus \mathfrak{W}$ established by assigning to P the direct decomposition $\mathfrak{X} = \operatorname{im} P \oplus \ker P$, and to $\mathfrak{X} = \mathfrak{V} \oplus \mathfrak{W}$ the projection P defined by $Px = x$ in $\mathfrak{V}$, and $Px = 0$ in $\mathfrak{W}$, (which defines a bounded operator, by the closed graph theorem).

We then will say that P annihilates $\mathfrak{W}$ = ker P and projects onto $\mathfrak{V}$ = im P.

For bounded Fredholm maps between Banach spaces defined by the purely algebraical condition (1.2) we now will show that all of section 2 still holds true if the additional condition is imposed that all occuring linear maps be bounded. Thereafter we will silently assume that only bounded operators occur, specifically that Fredholm inverses are bounded.

Theorem 3.1. a) A bounded operator between Banach spaces is Fredholm if and only if it has a bounded Fredholm inverse.

b) A Fredholm operator between Banach spaces has closed image.

c) If A: $\mathfrak{X} \to \mathfrak{Y}$ is a bounded Fredholm operator then for any complement $\mathfrak{Z}$ of ker A there exists a constant $c = c_{\mathfrak{Z}}$ such that $\|x\| \leq c\,\|Ax\|$ for all $x \in \mathfrak{Z}$.

Proof. The crucial point is that the two decompositions (2.1) can be chosen direct, in the sense specified above. If A is Fredholm, then the finite dimensional (closed) subspace ker A of the Banach space $\mathfrak{X}$ possesses a (closed) complement $\mathfrak{Z}$, which may be obtained as annihilator of N continuous linear functionals $\ell_1(x),\dots,\ell_N(x)$ over $\mathfrak{X}$ chosen to satisfy $\ell_j(x_k) = \delta_{jk}$, $j,k = 1,\dots,N$, with a basis $x_1,\dots,x_N$ of ker A.

With this meaning of $\mathfrak{Z}$ let now $\mathfrak{W}, A^O, B^O$ be as defined in the proof of theorem 2.1, and note that the operator B there can also be defined by setting $B = B^O$ on im A, = O on $\mathfrak{W}$. The graph $\{[z,Az]: z \in \mathfrak{Z}\}$ of A^O is a closed subspace of $\mathfrak{X} \times \mathfrak{Y}$, since $\mathfrak{Z}$ is closed and A is bounded and the graph of B is given by $\{[z,Az]+[0,w]: z \in \mathfrak{Z},\ w \in \mathfrak{W}\}$, which is an algebraic sum of a closed and a finite dimensional subspace of $\mathfrak{X} \times \mathfrak{Y}$ and therefore closed. By the closed graph theorem the operator B must be bounded and its restriction B^O is also bounded. Since B^O and A^O are inverses of each other, it follows that $\|x\| = \|B^O A^O x\| \leq \|B^O\| \cdot \|Ax\|$ for all $x \in \mathfrak{Z}$, which amounts to c) with $c_{\mathfrak{Z}} = \|B^O\|$.

It is an immediate consequence of c) now that im A must be closed, or that we have b). Thereafter, it is clear that also the second decomposition of (2.1') is direct. Accordingly the projections P and Q are bounded. It is clear then that B as defined above represents a bounded Fredholm inverse. This proves a) and the theorem is established.

For a Banach space $\mathfrak{X}$ let $\mathfrak{X}^*$ denote the space of all continuous linear functionals $\ell: \mathfrak{X} \to \mathbb{C}$. Instead of $\ell(x)$ for the value of ℓ at $x \in \mathfrak{X}$ we also shall write $\ell(x) = \langle x^*, x\rangle$, where x^* shall be another notation for the linear functional ℓ.

A bounded operator A: $\mathfrak{X} \to \mathfrak{Y}$ always has an adjoint A : $\mathfrak{Y}^* \to \mathfrak{X}^*$, defined by

$$\langle y^*, Ax\rangle = \langle A^* y^*, x\rangle \text{ for all } x \in \mathfrak{X} \text{ and all } y^* \in \mathfrak{Y}^* \quad . \tag{3.4}$$

We have $\|A\| = \|A^*\|$. The mapping $*: L(\mathfrak{X},\mathfrak{Y}) \to L(\mathfrak{Y}^*, \mathfrak{X}^*)$ defined by taking adjoints is an isometry. If the spaces $\mathfrak{X}$ and $\mathfrak{Y}$ are reflexive - that is if $\mathfrak{X}^{**}$ and $\mathfrak{Y}^{**}$ are isometrically isomorphic to $\mathfrak{X}$ and $\mathfrak{Y}$, resp. -, this mapping is inverted by itself.

In the general case, where $\mathfrak{X} \subset \mathfrak{X}^{**}$, $\mathfrak{Y} \subset \mathfrak{Y}^{**}$ via the canonical isometric imbeddings, we still have the restriction of $A^{**}: \mathfrak{X}^{**} \to \mathfrak{Y}^{**}$ to $\mathfrak{X}$ identical with A. Theorem 3.2, below, also should be compared with Corollary 7.7.

Theorem 3.2. A bounded operator $A \in L(\mathfrak{X}, \mathfrak{Y})$ is Fredholm only if its adjoint A^* is, and we have

(3.5) $\alpha(A) = \beta(A^*)$, $\beta(A) = \alpha(A^*)$, ind $A = -$ ind A^*. Specifically, if A and B are Fredholm inverses, then so are A^* and B^*, and vice versa.

Recall the definition of ortho-complement $\mathfrak{M}^{\perp}$ of a subset $\mathfrak{M} \subset \mathfrak{X}$ as the (closed) linear subspace $\mathfrak{M}^{\perp} \subset \mathfrak{X}^*$ containing all x^* vanishing on $\mathfrak{M}$.

Lemma 3.3. For any bounded linear map we have ker $A^* = (\text{im } A)^{\perp}$. If in addition im A is closed, then we also have $(\ker A)^{\perp} = \text{im } A^*$.

Proof. First, it is clear that $\|A^*z^*\| = 0$ simply means that $\langle A^*z^*, x\rangle = \langle z^*, Ax\rangle = 0$ for all $x \in \mathfrak{X}$, or that $z^* \in (\text{im } A)^{\perp}$. This proves the first relation of the lemma. If im A is closed, and $x^* \in (\ker A)^{\perp}$ is given, then a linear functional f over im A can be defined by setting

(3.6) $f(y) = \langle x^*, x\rangle$ for any x with $Ax = y$.

This is meaningful because $\langle x^*, z\rangle = 0$ for all $z \in \ker A$ implies that $\langle x^*, x\rangle$ is constant on the set of all x with the same image under A. This functional is bounded, because A induces a bounded operator $A^{\vee}: \mathfrak{X}/\ker A \to \text{im } A$ between Banach spaces, defined as $A^{\vee} = A\Pi^{-1}$, with the canonical projection $\Pi: \mathfrak{X} \to \mathfrak{X}/\ker A$, and $A^{\vee}$ has a bounded inverse $A^{\vee -1}$: im $A \to \mathfrak{X}/\ker A$, hence $f(y) \le \inf_{x=Ay} \|x^*\| \cdot \|x\|$ $= \|x^*\| \; \|A^{\vee -1}y\| \le \|A^{\vee -1}\| \; \|x^*\| \; \|y\|$. By the Hahn-Banach theorem f can be extended to all of $\mathfrak{Y} \supset \text{im } A$, to supply an element $y \in \mathfrak{Y}^*$ satisfying $\langle y^*, Ax\rangle = \langle x^*, x\rangle$ for all $x \in \mathfrak{X}$. In other words, $A^*y^* = x^*$, which amounts to $x^* \in \text{im } A^*$. Hence $(\ker A)^{\perp} \subset \text{im } A^*$ follows. Vice versa for $x^* \in \text{im } A^*$ we get $\langle x^*, x\rangle = \langle A^*y^*, x\rangle = \langle y^*, Ax\rangle = 0$ for all $x \in \ker A$, which proves the other inclusion, and Lemma 3.3 is established.

Lemma 3.4. For any closed subspace $\mathfrak{Z}$ of a Banach space $\mathfrak{X}$ we have the following natural isometric isomorphisms.

(3.7) $(\mathfrak{X}/\mathfrak{Z})^* = \mathfrak{Z}^{\perp}$, $\mathfrak{X}^*/\mathfrak{Z}^{\perp} = \mathfrak{Z}^*$.

Proof. (See also [13] 22.3) If $\Pi: \mathfrak{X} \to \mathfrak{X}/\mathfrak{Z}$ denotes the canonical projection then any $x^* \in \mathfrak{Z}^{\perp}$ vanishes on $\mathfrak{Z} = \ker \Pi$, hence $x^*\Pi^{-1}: \mathfrak{X}/\mathfrak{Z} \to \mathbb{C}$ is meaningful and in $(\mathfrak{X}/\mathfrak{Z})^*$. Also for $\ell^{\vee} \in (\mathfrak{X}/\mathfrak{Z})^*$ we may define $x^* = \ell\Pi$, as an element of $\mathfrak{Z}^{\perp} \subset \mathfrak{X}^*$. Both mappings invert each other and the relation

$$\sup_{\|x^{\vee}\| < 1} |\ell^{\vee}(x^{\vee})| = \sup_{\|x\| < 1} |\langle x^*, x\rangle|$$

shows that we have obtained the first isometry.

For $s^* \varepsilon\, \mathcal{Y}^*$ use Hahn-Banach to extend to $x_0^* \varepsilon\, \mathcal{X}^*$ with $\|s^*\| = \|x_0^*\|$. Any two bounded extensions of s^* to $\mathcal{X}$ can differ only by a $z^* \varepsilon\, \mathcal{Y}^\perp$ and vice versa $x_0^* + z^*$ also is an extension of s^* to $\mathcal{X}$, for every $z^* \varepsilon\, \mathcal{Y}^\perp$. This provides a 1-1-correspondence between the co-sets in $\mathcal{X}^*/\mathcal{Y}^\perp$ and the space $\mathcal{Y}^*$ adjoint to $\mathcal{Y}$ as a Banach space of its own: We get $\|x^{*\vee}\| \le \|x_0^*\| = \|s^*\|$, and vice versa that $\|s^*\| \le \|x^{*\vee}\|$, so again the above 1-1-correspondence is an isometric isomorphism, which proves the second relation.

Lemma 3.5. We have $\dim \mathcal{X} = \dim \mathcal{X}^*$ whenever one of the two dimensions is finite.

Proof. For a finite dimensional Banach space $\mathcal{X}$ with basis $x_1, \ldots, x_N$ the adjoint space $\mathcal{X}^*$ is isomorphic to the space $\mathbb{C}^N$ (of N-tuples of complex numbers) via the mapping $\ell(x) \leftrightarrow (\ell_1, \ldots, \ell_N)$ with $\ell_j = \ell(x_j)$ and

$\ell(x) = \sum_{j=1}^{N} \xi_j \ell_j$ for $x = \sum_{j=1}^{N} \xi_j x_j$. Clearly this gives the above equality. Vice versa, if $\mathcal{X}$ is infinite dimensional then for an arbitrary integer N an N-dimensional subspace may be picked the linear functionals of which can be extended to N linearly independent functionals in $\mathcal{X}^*$, by Hahn-Banach. Hence $\mathcal{X}^*$ cannot have finite dimension, and the lemma is proved.

Lemma 3.6. The adjoint A^* of a bounded operator is of finite rank if and only if A is of finite rank. Both ranks are equal.

Proof. If $A^*: \mathcal{Y}^* \to \mathcal{X}^*$ has finite rank then $\dim(\mathcal{Y}^*/\ker A^*) = \dim \operatorname{im} A^* = N$ by (2.2). By the first formula of lemma 3.3 we get $\ker A^* = (\operatorname{im} A)^\perp$ thus lemma 3.5 and lemma 3.4 yield $\dim \operatorname{im} A = \dim (\operatorname{im} A)^* = \dim(\mathcal{Y}^*/(\operatorname{im} A)^\perp) = \dim(\mathcal{Y}^*/\ker A^*) =$ $= \dim \operatorname{im} A^* = N$. That is $A = F \varepsilon\, E(\mathcal{X}, \mathcal{Y})$. Vice versa, if $F \varepsilon\, E(\mathcal{X}, \mathcal{Y})$, we get $\dim \operatorname{im} F = \dim(\mathcal{X}/\ker F) = \dim(\mathcal{X}/\ker F)^* = \dim(\ker F)^\perp$. Now the image of F is finite dimensional, hence closed, thus the second formula of lemma 3.3 implies $= \dim \operatorname{im} F^*$, and so we have F^* of finite rank. Since equality of ranks - that is equality of dim im F and dim im F^* has come out too the lemma is established.

Proof of theorem 3.2. If A and B are Fredholm inverses then so are A^* and B^* Fredholm inverses, and vice versa, in view of Lemma 3.6. This proves that A^* is Fredholm, whenever A is. Also we get $\alpha(A) = \dim \ker A = \dim (\ker A)^* =$ $= \dim(\mathcal{X}^*/(\ker A)^\perp) = \dim(\mathcal{X}^*/\operatorname{im} A^*) = \beta(A^*)$ and $\beta(A) = \dim(\mathcal{Y}/\operatorname{im} A) =$ $= \dim(\mathcal{Y}/\operatorname{im} A)^* = \dim(\operatorname{im} A)^\perp = \dim \ker A^* = \alpha(A^*)$. Theorem 3.2 is established.

Theorem 3.7. Property F and ind A are continuous in the norm topology, that is, under the norm (3.1) the collection $F(\mathcal{X}, \mathcal{Y})$ of all Fredholm operators in $L(\mathcal{X}, \mathcal{Y})$ is an open subset of $L(\mathcal{X}, \mathcal{Y})$, and the Fredholm index is constant on connected components of $F(\mathcal{X}, \mathcal{Y})$.

Proof. Let A' be a Fredholm inverse of $A \varepsilon F(\mathcal{X}, \mathcal{Y})$ and let $B \varepsilon L(\mathcal{X}, \mathcal{Y})$ be arbitrary, then $A+B = A(1+A'B)+F$ with some $F\epsilon E(\mathfrak{X},\mathfrak{Y})$. Clearly $\|A'\| \cdot \|B\| <1$ implies that $1+A'B$ is invertible, the inverse being given by the Neumann series $(1+A'B)^{-1} = \sum_0^\infty (-A'B)^n$. Accordingly the operator $A(1+A'B)+F$ is Fredholm, by theorem 2.6 and theorem 2.8, and has the same index as A, q.e.d. .

It may be observed that the linear subspace $E(\mathfrak{X},\mathfrak{Y})$ of all operators with finite rank of the space $L(\mathfrak{X},\mathfrak{Y})$ is not closed, unless $\mathfrak{X}$ or $\mathfrak{Y}$ have finite dimension. In many respects it is more natural to consider its closure instead, which will be denoted by $C_0(\mathfrak{X},\mathfrak{Y})$. The quotient-space $L/C_0(\mathfrak{X},\mathfrak{Y})$ of all co-sets $A^\vee = \{A+C : C\epsilon C_0(\mathfrak{X},\mathfrak{Y})\}$ is a Banach-space again, under the norm $\|\check{A}\| = \inf \|A+C\|$. We notice that, in view of property F, the space $E(\mathfrak{X},\mathfrak{Y})$ may be replaced by the closed space $C_0(\mathfrak{X},\mathfrak{Y})$.

Corollary. For any Fredholm inverse A' of A the sphere $\{B: \|B-A\| < \|A'\|^{-1}\}$ belongs to F.

Theorem 3.8. An operator $A\epsilon L(\mathfrak{X},\mathfrak{Y})$ is Fredholm if and only if there exists an operator $B\epsilon L(\mathfrak{Y},\mathfrak{X})$ such that $1 - AB$ and $1 - BA$ are in C_0.

Proof. It is sufficient to show that $1 - AB$, $1 - BA \;\epsilon\; C_0$ imply that A (and B) are Fredholm. Now, let $AB = 1+C$, $BA = 1+C'$ with $C,C'\epsilon C_0$. Write $C = E+F$ and $C' = E'+F'$ with $\|E\|$, $\|E'\| < 1$ and $F,F'\epsilon E$, as is possible by definition of C_0. Note that Theorem 3.7 implies property F for AB and for BA. So by theorem 2.1 we have Q and R such that $ABQ = 1+G$ and $RBA = 1+G'$ with $G,G'\epsilon E$. Then $RBABQ = BQ + G'BQ = RB + RBG$ shows that BQ and RB differ only by an operator of finite rank. Hence RB and BQ both are Fredholm inverses of A and therefore A has property F, q.e.d..

Theorem 3.9. If A is Fredholm and $C\epsilon C_0$, then A+C is Fredholm and has same index as A.

Proof. Write $C = E+F$ with $F\epsilon E$ and E so small that A+E is Fredholm, by theorem 3.7. Then $A+C = A+E+F$ is also Fredholm and all three operators have the same index, q.e.d.

With notations as in section 2, $E(\mathfrak{X},\mathfrak{Y})$ replaced by $C_0(\mathfrak{X},\mathfrak{Y})$ we can reformulate the above results as follows.

Theorem 3.10. a) $A\epsilon E(\mathfrak{X},\mathfrak{Y})$ is Fredholm if and only if its co-set $\check{A}$ mod $C_0(\mathfrak{X},\mathfrak{Y})$ possesses an inverse (in $L/C_0(\mathfrak{Y},\mathfrak{X})$)

b) The set $I(\mathfrak{X},\mathfrak{Y})$ of all invertible elements in $L/C_0(\mathfrak{X},\mathfrak{Y})$ is open,

c) The Fredholm index induces a map ind: $I(\mathfrak{X},\mathfrak{Y}) \to \mathbb{Z}$ which is continuous and leaves the algebraic structure invariant. Here $\mathbb{Z}$ denotes the additive group of integers with discrete topology.

4. Compact operators and Rellich Criteria.

Recall that a bounded operator $C: \mathfrak{X} \to \mathfrak{Y}$ is called compact (completely continuous) if for any bounded sequence $\{x_n\}$ a convergent subsequence of $\{Cx_n\}$ can be found. In the following we shall mostly rely on a different characterization of compact operator, which for Hilbert spaces is due to F. Rellich [18].

Criterion 4.1 (Rellich) A bounded operator $A: \mathfrak{X} \to \mathfrak{Y}$ is compact if and only if for any $\varepsilon > 0$ a closed subspace $\mathfrak{T} \subset \mathfrak{X}$ with finite codimension can be found such that the restriction of A to $\mathfrak{T}$ has operator norm not larger than ε.

Criterion 4.2 A bounded operator $A: \mathfrak{X} \to \mathfrak{Y}$ is compact if and only if for any $\varepsilon > 0$ a finite dimensional subspace $\mathfrak{Q} \subset \mathfrak{Y}$ may be found such that $\|\Pi_{\mathfrak{Q}} A\| \le \varepsilon$, where $\Pi_{\mathfrak{Q}} : \mathfrak{X} \to \mathfrak{X}/\mathfrak{Q}$ denotes the canonical projection.

Proof (of both criteria). a) Assume a sequence $\mathfrak{T}_m$ of closed subspaces with finite codimension given such that $\|A|\mathfrak{T}_m\| < 1/m$, $m=1,2,\dots$, and let $\Pi_m : \mathfrak{X} \to \mathfrak{X}/\mathfrak{T}_m$ denote the canonical projections, which clearly all have norm not larger than one. Then if a bounded sequence x_n (with $\|x_n\| \le 1$) is given, use Cantors diagonal scheme to obtain a subsequence y_l such that $\lim_{l\to\infty} \Pi_m y_l$ exists for every $m=1,2,\dots$. This is possible, because every Π_m has finite rank. For a given $\varepsilon > 0$, $\varepsilon <$ pick an m_0 with $m_0 \varepsilon > 10$ and then $N = N(\varepsilon)$ such that $\|\Pi_{m_0}(y_j - y_l)\| \le \varepsilon/(4\|A\|+1)$ for $j,l \ge N(\varepsilon)$. It follows existence of $t_{jl} \in \mathfrak{T}_{m_0}$ such that $\|y_j - y_l + t_{jl}\| \le \operatorname{Min}\{2\varepsilon/(4\|A\|+1), 3\}$ since also $\|y_j - y_l\| \le 2$. Clearly we get $\|t_{jl}\| \le 5$, hence $\|A(y_j - y_l)\| \le \|A\| \, \|y_j - y_l + t_{jl}\| + \|At_{jl}\| \le \varepsilon/2 + \|t_{jl}\|/m_0 \le \varepsilon$. Hence Ay_j converges and A is compact.

b) Now assume that we have

$$(4.1) \qquad \inf\{\|\Pi_{\mathfrak{Q}} A\| : \dim \mathfrak{Q} < \infty\} > \gamma > 0 \qquad ,$$

and let us assume we have n unit vectors $x_1, \dots, x_n$ with the property that $\|Ax_k - Ax_l\| \ge \gamma$ for $k > l$. Denote the span of $Ax_1, \dots, Ax_n$ by $\mathfrak{S}_n$, then we know $\|\Pi_{\mathfrak{S}_n} A\| > \gamma$ so that there exists a unit vector, called x_{n+1} such that $\|\Pi_{\mathfrak{S}_n} Ax_{n+1}\| \ge \gamma$. This implies $\|Ax_{n-1} - Ax_k\| \ge \gamma$, for $k=1,\dots,n$. Accordingly we have the system $x_1, \dots, x_n$ augmented by one vector x_{n+1} while maintaining the condition. By induction we may construct a sequence $x_1, x_2, \dots$ with $\|x_j\| = 1$ and $\|Ax_j - Ax_l\| \ge \gamma$ for $j > l$. It is clear now that the sequence $Ax_1, Ax_2, \dots$ cannot have a convergent subsequence. Hence A is not compact.

We have thus shown that $\rho_r(A) = 0$ implies "A compact", and that "A compact" implies $\rho_l(A) = 0$, where

$$\rho_r(A) = \inf\{\|A|\mathfrak{T}\| : \mathfrak{T} \subset \mathfrak{X}, \ \operatorname{codim} \mathfrak{T} < \infty\}$$

and

$$\rho_1(A) = \inf \{ \|\Pi_{\mathfrak{F}} A\| : \mathfrak{F} \subset \mathcal{Y} , \dim \mathfrak{F} < \infty\} .$$

It remains to be shown that $\rho_1(A) = 0$ implies $\rho_r(A) = 0$, which will be accomplished by the following series of Lemmata.

<u>Lemma 4.3.</u> Let $\mathcal{X}$ be a Banach space and let $\mathfrak{K}$ be a finite dimensional subspace of $\mathcal{X}$. Then for any $\varepsilon>0$ we can find a closed subspace $\mathcal{R}^* \subset \mathcal{X}^*$ of finite dimension such that $\|\Pi_{\mathfrak{K}} x\| \geq (1/2-\varepsilon) \|x\|$ for all x with $<\mathcal{R}^*,x> = 0$.

<u>Proof.</u> For any unit vector k_0 we may pick a unit vector $k_0^* \in \mathcal{X}^*$ with $<k_0^*,k_0> = 1$, by Hahn-Banach, and define the hyperplane

$\mathcal{R}_{k_0}' = \{x : <k_0^*,x> = 0\}$. Then we get

$$d(k_0,\mathcal{R}_{k_0}') = \inf \{ \|k_0+x\| : x\in\mathcal{R}_{k_0}'\} = 1 ,$$

due to $\|k_0+x\| \geq |<k_0^*,k_0+x>| = 1$.

If $\delta>0$ is given then we may pick N unit vectors $k_1,\dots,k_N$ in $\mathfrak{K}$ such that every point of the unit sphere $\mathfrak{K}_1$ in $\mathfrak{K}$ is within distance δ from one of these points, because $\mathfrak{K}_1$ is compact. Let $\mathcal{R}^*$ denote the span of the corresponding k_j^* and let $\mathcal{R}_\delta' = \bigcap \mathcal{R}_{k_j}' = \{x\in\mathcal{X} : <k_j^*,x> = 0, j=1,\dots,N\}$. For arbitrary $k\in\mathfrak{K}_1$ we have

$$\inf \{ \|k+x\| : x\in\mathcal{R}_\delta\} \geq \inf \{ \|k+x\| : x\in\mathcal{R}_{k_j}'\}$$

$$\geq \inf \{ \|k_j+x\| : x\in\mathcal{R}_{k_j}'\} - \|k-k_j\| \geq 1-\delta ,$$ with suitable chosen j. This gives

$$(4.2) \quad \inf \{ \|k+r\| / \|k\| : k\neq0, r\neq0, k\in\mathfrak{K}, r\in\mathcal{R}\} \geq 1-\delta .$$

Since the function a/(1-a) of the real argument a is decreasing, for a>0 we find that

$$\|k+r\| / \|r\| \geq \|k+r\| / (\|k+r\| + \|k\|)$$

$$= (\|k+r\| / \|k\|)(\|k+r\| / \|k\| +1)^{-1} \geq (1-\delta)/(2-\delta) ,$$

using (4.2). For $\varepsilon>0$ a $\delta>0$ can be found such that $(1-\delta)/(2-\delta) \geq 1/2-\varepsilon$. Hence the statement of the Lemma follows.

<u>Lemma 4.4.</u> We have

$$\inf \{ \|A|\mathfrak{T}\| : \mathfrak{T}\subset\mathcal{X} , \operatorname{codim}\mathfrak{T} < \infty\} \leq 2 \inf \{ \|\Pi_{\mathfrak{F}} A\| : \mathfrak{F}\subset\mathcal{Y} , \dim \mathfrak{F} < \infty\} .$$

<u>Proof.</u> Let us denote the infimum at right by r(A) and let $\varepsilon>0$ and a finite dimensional subspace $\mathfrak{F}\subset\mathcal{Y}$ be given. Then use Lemma 4.3 to construct $r_1^*,\dots,r_M^* \in \mathcal{Y}^*$ and $\mathcal{R} = \{y\in\mathcal{Y} : <r_j^*,y> = 0, j=1,\dots M\}$ such that $y\in\mathcal{R}$ implies $\|\Pi_{\mathfrak{F}} y\| \geq (1/2-\varepsilon) \|y\|$, and let

(4.3) $\quad \mathfrak{S} = \{ x \in \mathfrak{X} : \langle A^* r_j^*, x\rangle = 0 \ , \ j=1,\ldots,M\}$

Note that $\mathfrak{S}$ has finite codimension and that $\mathfrak{R} = A\mathfrak{S}$. Accordingly there exists a unit vector $x \in \mathfrak{S}$ with $\|Ax\| \geq r(A)-\varepsilon$ and we get

$$\|\Pi_{\mathfrak{Z}} A\| \geq \|\Pi_{\mathfrak{Z}} Ax\| \geq (1/2-\varepsilon)(r(A)-\varepsilon) \quad .$$

Since ε is arbitrary the statement of the lemma results.

Note that Lemma 4.4 and Criterion 4.2 imply the missing conclusions in the proof of Criterion 4.1 now, because $\rho_I(A)>0$ implies $\inf\{\|\Pi_{\mathfrak{Z}} A\| : \dim \mathfrak{Z} < \infty\} > 0$ and thus A is not compact by Criterion 4.2.

Theorem 4.5. A bounded operator $C: \mathfrak{X} \to \mathfrak{Y}$ is compact if and only if its adjoint is compact.

For the proof we consider two more Lemmata.

Lemma 4.6. Let $\mathfrak{S}$ denote a closed subspace of $\mathfrak{X}$ and let $I_{\mathfrak{S}} : \mathfrak{S} \to \mathfrak{X}$ denote the injection map. Then if $\mathfrak{S}^*$ is identified with $\mathfrak{X}^*/\mathfrak{S}^{\perp}$, via the isometry of Lemma 3.4, we have $(I_{\mathfrak{S}})^* = \Pi_{\mathfrak{S}^\perp}$ with $\Pi_{\mathfrak{S}^\perp} : \mathfrak{X}^* \to \mathfrak{X}^*/\mathfrak{S}^\perp$ denoting the canonical projection.

Proof. $\mathfrak{S}^* \cong \mathfrak{X}^*/\mathfrak{S}^\perp$ was achieved by assigning to a functional over $\mathfrak{S}$ the coset of all of its extensions to $\mathfrak{X}$, hence we have $\langle(\Pi_{\mathfrak{S}^\perp})x^*, s\rangle = \langle x^*, s\rangle$ for all $x^* \in \mathfrak{X}^*$ and all $s \in \mathfrak{S}$, which is just the defining relation of the adjoint $(I_{\mathfrak{S}})^*$, hence $(I_{\mathfrak{S}})^* = \Pi_{\mathfrak{S}^\perp}$, q.e.d..

Lemma 4.7. Any finite dimensional subspace $\mathfrak{S}^*$ of the adjoint space $\mathfrak{X}^*$ of a Banach space $\mathfrak{X}$ is orthogonal complement of a space $\mathfrak{Z} \subset \mathfrak{X}$ of finite codimension.

Proof. Let $x_1^*, \ldots, x_N^*$ be a basis of $\mathfrak{S}^*$, then we may pick $z_1, \ldots, z_N \in \mathfrak{X}$ such that the matrix $((\langle x_j^*, z_l\rangle)) = W$ is invertible. If $W^{-1} = ((\alpha_{jl}))$, then let $x_l = \sum_j \alpha_{jl} z_j$, and confirm that $\langle x_k^*, x_l\rangle = \delta_{kl}$. Let $\mathfrak{Z} = \{x \in \mathfrak{X} : \langle x_j, x\rangle = 0, \ j=1,\ldots,N\}$. Then every $x \in \mathfrak{X}$ can be written uniquely in the form

$$x = \sum_j \langle x_j, x\rangle \, x_j + t \ , \ t \in \mathfrak{Z} \quad .$$

Or, the x_j form of a basis of $\mathfrak{X}$ mod $\mathfrak{Z}$ or codim $\mathfrak{Z} = N < \infty$. For an arbitrary $x^* \in \mathfrak{X}^*$ vanishing on $\mathfrak{S}$ we get $\langle x^*, x\rangle = \sum_j \langle x_j^*, x\rangle\langle x^*, x_j'\rangle$. hat is we have $x^* = \sum_j \langle x^*, x_j\rangle \, x_j^* \in \mathfrak{S}^*$, which proves the lemma.

Proof of theorem 4.5. If C is compact, we get $\|CI_{\mathfrak{S}}\| \leq \varepsilon$, by Rellich's Criterion, with a suitable $\mathfrak{S} \subset \mathfrak{X}$ of finite codimension. Taking adjoints one has $\|(\Pi_{\mathfrak{S}^\perp})C^*\| \leq \varepsilon$, by Lemma 4.6, and $\mathfrak{S}^\perp$ has finite dimension, hence Criterion 4.2 implies that C^* is compact.

Vice versa, if C^* is compact, let $\mathfrak{S}^* \subset \mathfrak{X}^*$ denote a space of finite dimension with $\|(\Pi_{\mathfrak{S}^*})C^*\| \leq \varepsilon$, by Criterion 4.2, and note that $\mathfrak{S}^* = \mathfrak{Z}^\perp$, by lemma 4.7, where codim $\mathfrak{Z} < \infty$. Hence we have $\|CI_{\mathfrak{Z}}\| = \|(CI_{\mathfrak{Z}})^*\| \leq \varepsilon$ which proves that C is compact, by Criterion 4.1. Theorem 4.5 is proved.

In the following let $K(\mathfrak{X},\mathfrak{Y})$ denote the collection of all compact operators from $\mathfrak{X}$ to $\mathfrak{Y}$. Since a continuous linear map preserves both, boundedness and convergence of any sequence it is clear that any product between a bounded and a compact operator is compact again. Also it is an immediate consequence of the Rellich Criterion that any uniform limit of compact operators is compact again: $\|A - C_n\| \to 0$ and $C_n \in K$ implies $A \in K$. Specifically any operator with finite rank is compact and therefore also any operator in $C_0(\mathfrak{X},\mathfrak{Y})$, the norm closure of $E(\mathfrak{X},\mathfrak{Y})$.

For Hilbert spaces, as well as all other well known Banach spaces it is well known that $K(\mathfrak{X},\mathfrak{Y}) = C_0(\mathfrak{X},\mathfrak{Y})$, but in the general case this equality is an unproven conjecture of long standing. (C.f. [4])

Coming back to Fredholm operators, we note that $K(\mathfrak{X},\mathfrak{Y})$ may be substituted for $C_0(\mathfrak{X},\mathfrak{Y})$ in all results of section 3. Clearly the quotients $L/K(\mathfrak{X},\mathfrak{Y})$ may be formed again, and products of cosets are meaningful as before, and inverses are defined again.

Theorem 4.8. A bounded operator $A:\mathfrak{X}\to\mathfrak{Y}$ is Fredholm if and only if its coset $\check{A}$ mod K is invertible.

Proof. It is sufficient to show that existence of an operator $B\in L(\mathfrak{Y},\mathfrak{X})$ with $AB = 1+C$, $BA = 1+C'$ implies that A is Fredholm . Using the two Criteria, pick $\mathfrak{z}\subset\mathfrak{Y}$ of finite dimension and $\mathfrak{r}\subset\mathfrak{X}$ of finite codimension such that $\|\Pi_{\mathfrak{z}} C\| < 1$ and $\|C|\mathfrak{r}\| < 1$.
Note that both $\Pi_{\mathfrak{z}}$ and $I_{\mathfrak{r}}$ are Fredholm operators, that $I_{\mathfrak{r}}^* = \Pi_{\mathfrak{r}^\perp}$ and that $\Pi_{\mathfrak{z}}^* = I_{\mathfrak{z}^\perp}$, both via the isometries of lemma 3.4. It follows that $\|(1+C)I_{\mathfrak{r}}x\| \geq$ $\geq (1-\|CI_{\mathfrak{r}}\|)\,\|x\|$ which implies $\ker(1+C)I_{\mathfrak{r}} = \{0\}$ and closedness of im $(1+C)I_{\mathfrak{r}}$. The Fredholm operator $\Pi_{\mathfrak{z}}$ acts as topological isomorphism between any complement of (the finite dimensional space) im $(1+C)I_{\mathfrak{r}} \cap \ker \Pi_{\mathfrak{z}}$ in im $(1+C)I_{\mathfrak{r}}$ and im $\Pi_{\mathfrak{z}}(1+C)I_{\mathfrak{r}}$, hence this image is also closed. Accordingly lemma 3.3 implies $(\text{im } D)^\perp = \ker D^*$ for $D = \Pi_{\mathfrak{z}}(1+C)I_{\mathfrak{r}}$. Note that $D^* = (\Pi_{\mathfrak{r}^\perp})(1+C^*)(I_{\mathfrak{z}^\perp})$, and that again $\ker(1+C)^*(I_{\mathfrak{z}^\perp}) = 0$ for reason similar as above. Since $(\Pi_{\mathfrak{r}^\perp})$ is Fredholm (Theorem 3.2) its kernel is finite dimensional, and therefore $\dim\ker D^* < \infty$. It follows that $\dim(\text{im } D)^\perp = \dim(\mathfrak{Y}/\text{im } D)^* = \dim(\mathfrak{Y}/\text{im } D) < \infty$, using the lemmata of section 3. Hence D is Fredholm. Similarly $D' = \Pi_{\mathfrak{z}}(1+C')I_{\mathfrak{r}}$ must be Fredholm if $\mathfrak{r}$ and $\mathfrak{z}$ are suitable adjusted. Let P, Q, V, W be Fredholm inverses of D, D', $I_{\mathfrak{r}}$ and $\Pi_{\mathfrak{z}}$, respectively, then all of the operators $I_{\mathfrak{r}}P\Pi_{\mathfrak{z}}(1+C)I_{\mathfrak{r}}V$, $W\Pi_{\mathfrak{z}}(1+C')I_{\mathfrak{r}}Q\Pi_{\mathfrak{z}}$, $I_{\mathfrak{r}}V$ and $W\Pi_{\mathfrak{z}}$ are of the form $1+F_j$, $F_j\in E$. This shows that $I_{\mathfrak{r}}P\Pi_{\mathfrak{z}}B$ and $BI_{\mathfrak{r}}Q\Pi_{\mathfrak{z}}$ are left and right inverses of A mod E, and it follows that A is Fredholm, by the standard algebraic conclusion discussed in

details in the proof of theorem 3.8.

Theorem 4.9. If A is Fredholm, then A+C is Fredholm, for any compact operator C, and A and A+C have the same index.

Proof. Let A' be any Fredholm inverse of A, then write $A + C = A(1 + A'C) + F$, $F \in \mathcal{E}$. Since A'C is compact, we get 1+A'C Fredholm, because it clearly is invertible mod $\mathcal{K}$. Hence the right hand side above is Fredholm, by the results of section 2, and so A+C is Fredholm. By the same conclusion we may show that the continuous family $A_t = A + t C$, $0 \le t \le 1$, consists of Fredholm operators only. By Theorem 3.7 the index of A_t must be a constant, and hence A and A+C have the same index, q.e.d. .

Problems. 1) Show that for Hilbert spaces $\mathfrak{X}$ and $\mathfrak{Y}$ we have $K(\mathfrak{X},\mathfrak{Y}) = C_0(\mathfrak{X},\mathfrak{Y})$. 2) Let $\mathfrak{X}$ and $\mathfrak{Y}$ be Frechet spaces (F-spaces, c.f. Dunford-Schwartz [4]). Show that then also every continuous Fredholm operator admits a continuous Fredholm inverse (and that it has closed image). 3) Consider the three concepts (R), (B), and (H), as introduced in IV,3, for Frechet spaces. Discuss their possible equivalence for the Sobolev space $\mathfrak{H}_\infty$. 4) Discuss the relationship between the classes $\mathcal{L}_s$ of IV,3 and the compact ideal $K(\mathfrak{H}_\infty)$ of the space $\mathfrak{H}_\infty$, as introduced in IV,3.

5. Fredholm closed algebras and Riesz-ideals

Definition 5.1. We define an algebra $\mathfrak{A}$ of linear maps A: $\mathfrak{X} \to \mathfrak{X}$, containing the identity, to be Fredholm closed, if it contains all Fredholm inverses of its elements. In case of a topological vector space this again only refers to the continuous Fredholm inverses.

Definition 5.2. A (non-vanishing) 2-sided ideal $\mathfrak{J}$ of $L(\mathfrak{X})$, for a Banach space $\mathfrak{X}$, will be called a Riesz ideal, if $A \in F(\mathfrak{X})$, $J \in \mathfrak{J}$, implies $A+J \in F(\mathfrak{X})$.

We have learned about some Riesz ideals in the preceeding sections. Specifically E, C_0 and K are Riesz ideals. There are other Riesz ideals on certain Banach spaces. For example the class of strictly singular (bounded) operators $S: \mathfrak{X} \to \mathfrak{Y}$, for Banach spaces $\mathfrak{X}$ and $\mathfrak{Y}$, is defined by the property that $A|\mathfrak{Z}$ maps bi-uniquely from the closed subspace $\mathfrak{Z} \subset \mathfrak{X}$ to a closed subspace of $\mathfrak{Y}$ if and only if $\dim \mathfrak{Z} < \infty$ (c.f. Kato [11]). If $S(\mathfrak{X}, \mathfrak{Y})$ denotes this class of strictly singular operators, then the class $S(\mathfrak{X}) = S(\mathfrak{X}, \mathfrak{X})$ may be seen to be a closed Riesz ideal of $L(\mathfrak{X})$, distinct from $C(\mathfrak{X})$ or $C_0(\mathfrak{X})$ in certain spaces (c.f. Gamelin [7]).

Proposition 5.3. There exists only one proper closed ideal $\neq 0$ in separable infinite dimensional Hilbert space, the ideal of compact operators.

Proposition 5.4. Every non-vanishing two-sided ideal $\mathfrak{J}$ of the Banach algebra $L(\mathfrak{X})$, for a Banach space $\mathfrak{X}$, contains the ideal $E(\mathfrak{X})$ of all operators of finite rank.

As a consequence of prop. 5.3., which is due to Calkin [2], there exists only one closed Riesz ideal in separable Hilbert space - the ideal of compact operators. The same again is true in many well investigated Banach spaces. (C.f. Gamelin and Feldman [5], Gohberg-Markus [9]).

To prove prop. 5.4., let $A \in \mathfrak{J}$ be different from zero. The adjoint A^* then also does not vanish. Let $x^* \in \mathfrak{X}^*$ be chosen such that $z^* = A^* x^* \neq 0$ and let $B = z{>}{<}x^*$, with any $z \in \mathfrak{X}$, different from zero, and with Dirac's notation $z{>}{<}x^*$, defined as the bounded linear operator

$$(5.0) \qquad (z{>}{<}x^*)u = \langle x^*, u\rangle z \quad , \quad u \in \mathfrak{X} \quad .$$

It follows that $BA = z{>}{<}z^* = z{>}{<}A^* x$ is in the ideal $\mathfrak{J}$ and also is of rank one. Accordingly, since $D_1 B A D_2 = D_1 z{>}{<}D_2^* z^* \in \mathfrak{J}$, and will run through all operators of rank one, as D_1 and D_2 run through $L(\mathfrak{X})$, we conclude that $\mathfrak{J}$ contains all operators of rank one, hence all operators of finite rank, since every operator in E is a sum of finitely many operators of rank one, q.e.d..

Note that the assertion of proposition 5.4 also is correct, if $L(\mathfrak{X})$ is replaced by an arbitrary algebra $\mathfrak{A}$ of bounded linear operators of $\mathfrak{X}$ which contains at least one element of rank one and has the property that every element $x \neq 0$, $x \in \mathfrak{X}$, is cyclic - i.e. $\mathfrak{A}x = \mathfrak{X}$ for all $x \neq 0$.

Proof of prop. 5.3. For a Hilbert space there exist projections with norm 1 onto every closed subspace - for example the orthogonal projections. Using Criterion 4.1 construct a sub-space of finite co-dimension, denoted by $\mathcal{Y}$, for every compact operator C and $\varepsilon>0$ such that $\|C|\mathcal{Y}\| \leq \varepsilon$. If P is the orthogonal projection onto $\mathcal{Y}^{\perp}$, then we get

$$\|C - PC\| \leq \|C|\mathcal{Y}\| \cdot \|(1-P)\| < \varepsilon \quad .$$

It follows that every compact operator C is in C_0, so that $K = C_0$.

If an arbitrary non-vanishing ideal $\mathcal{J}$ of $L(\mathcal{G})$, for a Hilbert space $\mathcal{G}$, is given, then we get $E(\mathcal{G}) \subset \mathcal{J}$, by prop. 5.4. If $\mathcal{J}$ is closed, it therefore must contain $C_0 = K$. Suppose some non-compact operator $A \in L(\mathcal{G})$ is in $\mathcal{J}$. By Rellich's criterion, (crit. 4.1) there exists a $c>0$ such that $\|A|\mathcal{Y}\| > c$ is true for every closed subspace $\mathcal{Y}$ with finite co-dimension. We may construct a sequence x_j, $j=1,2,\ldots$, which is orthonormal and such that also $\langle x_j, A^*Ax_l\rangle = 0$, $j \neq l$, $\geq c^2$ for $j=l$. Indeed, if N such vectors have been constructed, let $\mathcal{Y}_N$ be the span of x_j, and A^*Ax_j, $j=1,\ldots,N$. Since $\mathcal{Z}_N = \mathcal{Y}_N^{\perp}$ has finite codimension, there exists a vector in $\mathcal{Z}_N$, called x_{N+1}, of norm 1, and with $\|Ax_{N+1}\| \geq c$. It follows that the vectors x_j, $j=1,\ldots,N+1$ also satisfy the assumptions, and by induction we get the desired sequence. The span $\mathcal{Y}$ of the orthogonal sequence x_j, $j=1,2,\ldots$, is an infinite dimensional subspace of $\mathcal{G}$, on which

$$\|Ax\|^2 = \sum_j |a_j|^2 \|Ax_j\|^2 \geq c^2 \sum_j |a_j|^2 = c^2 \|x\|^2, \; x = \sum_j a_j x_j \; .$$

Let finally $\mathcal{G}$ be separable, and let y_j, $j=1,2,\ldots$, be an orthonormal base of $\mathcal{G}$. Let $M : \mathcal{G} \to \mathcal{Y}$, an isometry, be defined by $y_j \to x_j$, $j=1,2,\ldots$, and then note that the operator M^*A^*AM is a bounded and invertible operator of $\mathcal{G}$, which is contained in $\mathcal{J}$. It follows that $\mathcal{J}$ coincides with $L(\mathcal{G})$. This proves proposition 5.3.

We note that any Riesz ideal may play the part of E, C_0 or K in the preceeding sections, as manifested by the result below.

Theorem 5.5. Let $\mathcal{J}$ be a Riesz ideal, then $A \in L(\mathcal{X})$ is Fredholm if and only if its coset modulo $\mathcal{J}$ is invertible in the algebra $L(\mathcal{X})/\mathcal{J}$.

Proof. Since every Riesz ideal contains the Riesz ideal E, by prop. 5.4, it suffices to show that the condition of the theorem is sufficient, in view of theorem 2.1, or theorem 3.1. a). Suppose there exists $B \in L(\mathcal{X})$, with

(5.1) $\quad AB = 1 + J \; , \; BA = 1 + J' \; , \; J,J' \in \mathcal{J} \quad .$

Then $1 + J$, $1 + J'$ are Fredholm, since $1 \in F(\mathcal{X})$, and by definition of the Riesz ideal. Hence there exists D,D' with

(5.2) $\quad ABD = (1 + J)D = 1 + E, \;\; D'AB = D'(1 + J') = 1 + E', \;\; E,E' \in E \quad .$

Accordingly BD and D'B are left and right Fredholm inverses of A, respectively. They must coincide mod E and constitute a Fredholm inverse each, hence A is Fredholm, and theorem 5.1 is established.

Lemma 5.6. Let $\mathcal{J}$ be a Riesz ideal, then $A \varepsilon F$, $J \varepsilon \mathcal{J}$ implies $A + J \varepsilon F$, and

(5.3) $$\text{ind } A = \text{ind } (A + J) \quad .$$

Proof. Clearly A+tJ is a continuous family of Fredholm operators. Since the index is a continuous function on $F(\mathfrak{X})$, it must be constant for all t.

Lemma 5.7. A Fredholm closed algebra $\mathfrak{A}$ contains the ideal $E(\mathfrak{X})$. If $\mathfrak{A}$ contains the Riesz ideal $\mathcal{J}$, then $\mathfrak{A}$ also contains every inverse mod $\mathcal{J}$ of one of its elements - that is, $B \varepsilon L(\mathfrak{X})$, $A \varepsilon \mathfrak{A}$, $AB - 1$, $BA - 1 \varepsilon \mathcal{J}$, implies $B \varepsilon \mathfrak{A}$.

Proof. Since 1 + E is a Fredholm inverse of 1, for every $E \varepsilon E$, we must have $1+E \varepsilon \mathfrak{A}$, that is $E \varepsilon \mathfrak{A}$, since by assumption $1 \varepsilon \mathfrak{A}$, for any Fredholm closed algebra.

Suppose AB = 1 + J, BA = 1 + J', for some $B \varepsilon L(\mathfrak{X})$. Let D be a Fredholm inverse of 1 + J. Then BD is a Fredholm inverse of $A \varepsilon \mathfrak{A}$, and hence is in $\mathfrak{A}$, since $\mathfrak{A}$ is Fredholm closed. Also $(BD)(1 + J) = B + E \varepsilon \mathfrak{A}$, since $BD \varepsilon \mathfrak{A}$ and $J \varepsilon \mathfrak{A}$, and $1 \varepsilon \mathfrak{A}$, q.e.d..

We note that a C^*-subalgebra $\mathfrak{A}$ of $L(\mathfrak{H})$, for a Hilbert space $\mathfrak{H}$, is necessarily Fredholm closed if it contains the ideal $K(\mathfrak{H})$ and the identity operator of $\mathfrak{H}$ (c.f. AII, corollary 7.16).

6. Unbounded linear operators on Banach spaces.

In this section we use the term "(unbounded) linear operator" (of a Banach space $\mathfrak{X}$) for any linear map A: $\mathfrak{D}(A) \to \mathfrak{X}$ from a dense linear subspace $\mathfrak{D}(A) \subset \mathfrak{X}$ to $\mathfrak{X}$. Then $\mathfrak{D}(A)$ is called the domain of A. Similarly we may speak of unbounded linear operators from $\mathfrak{X}$ to $\mathfrak{Y}$ as linear maps from (the domain) $\mathfrak{D}(A) \subset \mathfrak{X}$, dense in $\mathfrak{X}$, to $\mathfrak{Y}$. However, for convenience, we shall assume $\mathfrak{X} = \mathfrak{Y}$ here.

Since unbounded linear operators are not linear maps of $\mathfrak{X}$, their sum and product needs a special interpretation. For A with domain $\mathfrak{D}(A)$ and B with domain $\mathfrak{D}(B)$ (also denoted by dom(A) and dom(B), occasionally) we define

$$(6.1) \qquad \mathfrak{D}(A+B) = \mathfrak{D}(A) \cap \mathfrak{D}(B), \quad (A+B)u = Au + Bu \quad \text{for} \quad u \in \mathfrak{D}(A+B)$$

and

$$(6.2) \qquad \mathfrak{D}(AB) = \{ u \in \mathfrak{D}(B) : Bu \in \mathfrak{D}(A)\}, \quad (AB)u = A(Bu), \quad u \in \mathfrak{D}(AB).$$

Also we define $cA = (c1)A$ for $c \in \mathbb{C}$, and the identity 1 of $\mathfrak{X}$. The precise interpretation of the term "unbounded" should be "not necessarily bounded". That is, bounded operators are a special class of unbounded operators. An operator, in this respect, is called "bounded", if

$$(6.3) \qquad \|Au\| \le c\,\|u\| \quad , \qquad u \in \mathfrak{D}(A) \quad .$$

A bounded operator always possesses a unique continuous extension with domain $\mathfrak{X}$, satisfying (6.3) with the same constant c. Indeed, for any sequence u_m convergent to u we have

$$\|Au_m - Au_k\| \le c \;\|u_m - u_k\| \quad , \quad m,k \to \infty \quad .$$

Accordingly, Au_m is a Cauchy sequence, and converges to $v \in \mathfrak{X}$, which is uniquely determined by u, and thus defines a linear extension of A to $\mathfrak{X}$. Continuity and the estimate (6.3) for this extension then are evident.

The cartesian product $\mathfrak{X}^0 = \mathfrak{X} \times \mathfrak{X}$ of a Banach space with itself, that is, the collection of ordered pairs $[u,v]$, $u,v \in \mathfrak{X}$, is a Banach space again, with linear operations defined componentwise, and with norm defined by

$$(6.4) \qquad \|[u,v]\| = \{ \|u\|^2 + \|v\|^2\}^{1/2}.$$

For a linear operator A of $\mathfrak{X}$ the graph $\mathfrak{G}(A)$ is defined as the linear subspace of $\mathfrak{X}^0$ defined by

$$(6.5) \qquad \mathfrak{G}(A) = \{[u,Au] : u \in \mathfrak{D}(A)\} \quad .$$

Vice versa, if for any linear subspace $\mathfrak{Y}^0 \subset \mathfrak{X}^0$ we have the condition satisfied that $[u,v] \in \mathfrak{Y}^0$, u = 0 implies v = 0, the the assignment

$$(6.6) \qquad \mathfrak{D}(A) = \{u \in \mathfrak{X} : [u,v] \in \mathfrak{Y}^0, \text{ for some } v \in \mathfrak{X}\}, \; Au = v,$$

defines a linear operator A of $\mathfrak{X}$, provided that $\mathfrak{D}(A)$ above is dense.

For two linear operators A and B we say that B extends A if and only if

$\mathcal{G}(A) \subset \mathcal{G}(B)$, which we shall also indicate by writing $A \subset B$ (or $B \supset A$).

We shall say that a linear operator A is closed if its graph $\mathcal{G}(A)$ is a closed subspace of $\mathfrak{X}^0$. If the closure of the graph $\mathcal{G}(B)$ induces a linear operator, in the sense of (6.6), then B is said to be a pre-closed operator, and A, as determined by (6.6), will be called the closure of B. Clearly we get $B \subset A$.

For a pre-closed operator A we define the adjoint A^* by

$$\mathfrak{D}(A^*) = \{u^* \varepsilon \mathfrak{X}^* : \langle u^*, Az \rangle = \langle v^*, z \rangle,\ z \varepsilon \mathfrak{D}(A), \text{ with some } v^* \varepsilon \mathfrak{X}^*\} \quad ,$$

(6.7)

$$A^* u^* = v^* \ , \text{ for } u^* \varepsilon \ \mathfrak{D}(A^*) \quad .$$

In a reflexive Banach space, every pre-closed operator has an adjoint. Indeed, the assignment $u^* \to v^*$ above, defining A^*, is always unique, because $\langle v^*, v \rangle = 0$ for all $v \ \varepsilon \mathfrak{D}(A)$ implies $v^* = 0$, since $\mathfrak{D}(A)$ was assumed dense in $\mathfrak{X}$.

On the other hand, if $\mathfrak{X}^{**} = \mathfrak{X}$, then $\mathfrak{D}(A^*)$, as defined, also is dense, because then the space

$$\underset{\sim}{\mathfrak{z}}^0 = \{[-v^*, u^*], \text{ with } [u^*, v^*] \ \varepsilon \mathcal{G}(A^*)\}$$

is the precise orthogonal complement of the graph $\mathcal{G}(A)$ in the Banach space $\mathfrak{X}^0$. If A is pre-closed, the closure of $\mathcal{G}(A)$ - which equals the ortho-complement of $\mathfrak{z}^0$ above, may not contain elements of the form $[0, v]$, except if $v = 0$. If $\mathfrak{D}(A^*)$, as in (6.7), is not dense, however, there must exist $v \neq 0$, with $\langle u^*, v \rangle = 0$ for all $u^* \varepsilon \ \mathfrak{D}(A^*)$, which may be read as

$$\langle [-v^*, u^*],\ [0, v] \rangle = 0, \text{ for all } [-v^*, u^*] \ \varepsilon \ \mathfrak{z}^0 \quad ,$$

a contradiction, so that for reflexive Banach spaces the adjoint indeed has dense domain, and is an unbounded operator of $\mathfrak{X}$.

The above also implies that for reflexive Banach spaces the closure of a pre-closed operator A is given by A^{**}.

We note that there are examples of unbounded operators in non-reflexive spaces, for which the adjoint A^* does not exist, because the space $D(A^*)$ defined by (6.7) is not dense in $\mathfrak{X}^*$ (c.f. M.A. Kaeshoek [10], p. 15).

If a linear operator B of $\mathfrak{X}^*$ exists such that

(6.8) $\quad \langle v^*, Au \rangle = \langle Bv^*, u \rangle \quad , \quad \text{for all } u \ \varepsilon \ \mathfrak{D}(A),\ v^* \ \varepsilon \ \mathfrak{D}(B) \ ,$

then we shall say that A and B are in <u>adjoint relation</u>. In this case A must be pre-closed. For otherwise there exists $u_m \to 0$ with $Au_m \to w \neq 0$. Substituting u_m into (6.8) and passing to the limit, we get $\langle v^*, w \rangle = 0$ for all $v^* \ \varepsilon \ \mathfrak{D}(B)$, a dense space. By the Hahn-Banach theorem this is a contradiction.

An unbounded closed operator $A\colon\ \mathfrak{D}(A) \to \mathfrak{X}$ is called Fredholm, if it is Fredholm as a linear map from $\mathfrak{D}(A)$ to $\mathfrak{X}$. Here it is to be noted that for a closed operator A the domain $\mathfrak{D}(A)$ is a Banach space, under the <u>graph norm</u>

(6.9) $\quad \|u\|_A = (\|u\|^2 + \|Au\|^2)^{1/2} \quad , \quad u \in \mathfrak{D}(A) \quad ,$

which is nothing but the norm of the pair $[u,Au]$, in the space $\mathfrak{X}^{\circ}$. Since for a closed operator A the graph $\mathfrak{g}(A)$ is a closed subspace of $\mathfrak{X}^{\circ}$, $\mathfrak{D}(A)$ under (6.9) indeed is a Banach space.

Accordingly, in the above definition of unbounded Fredholm operator, $A: \mathfrak{D}(A) \to \mathfrak{X}$ is to be regarded as continuous operator between the two Banach spaces $\mathfrak{D}(A)$ and $\mathfrak{X}$. Continuity of this operator is a consequence of the (trivial) inequality

(6.10) $\quad \|Au\| \leq \|u\|_A \quad , \quad u \in \mathfrak{D}(A) \quad .$

It is evident, from the above, that an unbounded closed operator is Fredholm if and only if (1.2) is satisfied. The results below are immediate applications of the results in section 3, above.

Theorem 6.1. An unbounded closed operator is Fredholm if and only if there exists a bounded operator $B: \mathfrak{X} \to \mathfrak{D}(A) \subset \mathfrak{X}$, bounded with respect to both norms $\|\cdot\|_A$ and $\|\cdot\|$, in the image space, such that

(6.11) $\quad AB = 1 + F \quad , \quad BA = 1 + F', \quad F \in E(\mathfrak{X}), \quad F' \in E(\mathfrak{X}, \mathfrak{D}(A)) \ .$

Proof. According to our definition of the Fredholm property for unbounded closed operators, there exists a Fredholm inverse $B': \mathfrak{X} \to \mathfrak{D}(A)$ of the linear map $A: \mathfrak{D}(A) \to \mathfrak{X}$ if and only if the unbounded closed linear operator A is Fredholm. We will have

(6.12) $\quad AB' = 1 + G, \quad B'A = 1 + G', \quad G \in E(\mathfrak{X}), \quad G' \in E(\mathfrak{D}(A)) \ .$

Now, the operator G' can be written in the form

(6.13) $\quad G' = \sum_{j=1}^{N} (x_j\rangle\langle x_j^* + (x_j\rangle\langle y_j^*)A), \ x_j \in \mathfrak{D}(A), \ x_j, y_j \in \mathfrak{X}^*,$

because the Banach space $\mathfrak{D}(A)$ is isometrically isomorphic to the closed subspace

(6.14) $\quad \mathfrak{g}(A) = \{[u,Au] : u \in \mathfrak{D}(A)\}$

of the graph space $\mathfrak{X}^{\circ}$, so that the continuous linear functionals over $\mathfrak{D}(A)$ can be written in the form

(6.15) $\quad \langle x^*, x\rangle + \langle y^*, Ax\rangle \quad , \quad x^*, y^* \in \mathfrak{X}^* \quad .$

We can write (6.13) in the form

(6.16) $\quad G' = F' + G''A, \ F' = \sum_{j=1}^{N} x_j\rangle\langle x_j^* \ , \ G'' = \sum_{j=1}^{N} x_j\rangle\langle y_j^*, \ F', G'' \in E(\mathfrak{X}, \mathfrak{D}(A))$

Let $B = B' - G''$ and confirm that (6.11) is correct with F' as in (6.16), and with $F = G - AG'' \in E(\mathfrak{X})$. Vice versa, it is clear that $E(\mathfrak{D}(A))$ is a subset of $E(\mathfrak{X}, \mathfrak{D}(A))$, by virtue of the boundedness of the injection $\mathfrak{D}(A) \to \mathfrak{X}$. This completes the proof of theorem 6.1.

Let A be a closed operator. An (unbounded) operator B with domain $\mathcal{D}(B) \supset \mathcal{D}(A)$ will be called A-bounded, if

(6.17) $\quad \|Bx\| \le c(\|x\| + \|Ax\|) \quad , \quad x \in \mathcal{D}(A)$,

and B will be called A-compact, if the restriction of B to the Banach space $\mathcal{D}(A)$ (with <u>graph norm</u> $\|\cdot\|_A$) is a compact operator between the Banach spaces $\mathcal{D}(A)$ and $\mathfrak{X}$.

Clearly, A-bounded operators are simply those with restrictions to $\mathcal{D}(A)$ bounded from $\mathcal{D}(A)$ (with graph norm) to $\mathfrak{X}$. Vice versa, B is A-compact if and only if for $\varepsilon>0$ exists a subspace $\mathfrak{Y} \subset \mathcal{D}(A)$ with $\dim \mathcal{D}(A)/\mathfrak{Y} < \infty$, such that

(6.18) $\quad \|Bx\| \le \varepsilon(\|x\| + \|Ax\|) \, , \, x \in \mathfrak{Y}$.

The following result is evident, by the result of section 3, 4.

<u>Theorem 6.2.</u> Let A be an unbounded closed operator, let B and C be an A-bounded, and an A-compact operator, respectively. Then, if A is Fredholm, we have $A+\varepsilon B$ and $A+C$ Fredholm, and

(6.19) $\quad \text{ind } A = \text{ind } (A+\varepsilon B) = \text{ind } (A+C)$,

whenever ε is sufficiently small.

By (6.1) we only must show that $A+\varepsilon B$ and $A+C$ are closed, which is trivial, due to $u_n \to u$, $(A+\varepsilon B)u_n \to f$, $v_n \to v$, $(A+C)v_n \to g$ implying $Au_n \to h$, $Av_{n_k} \to w$, for a suitable subsequence v_{n_k}.

<u>Theorem 6.3.</u> The product of two unbounded Fredholm operators A and B, defined by (6.2), is an unbounded Fredholm operator, and we have

(6.20) $\quad \text{ind } AB = \text{ind } A + \text{ind } B$.

Theorem 6.2 will be related to theorem 2.8, in its proof. We need a few preparations, however.

<u>Corollary 6.4.</u> An unbounded closed operator A is Fredholm if and only if there exist two bounded operators $A', A'' \in L(\mathfrak{X}, \mathcal{D}(A))$ such that

(6.21) $\quad A'A \subset 1 + E' \quad , \quad E' \in E(\mathfrak{X})$,

and

(6.22) $\quad AA'' = 1 + E'' \quad , \quad E'' \in E(\mathfrak{X})$.

If $\mathfrak{M} \subset \mathfrak{X}$ and $\mathfrak{M}^* \subset \mathfrak{X}^*$ are a dense sub-set of $\mathfrak{X}$, and a total sub-set of $\mathfrak{X}^*$, respectively, then it is possible to select A' and A" in such a way that E' of (6.21) and E" of (6.22) satisfy

(6.23) $\quad \text{im } E'' \subset \mathfrak{M} \quad , \quad (\ker E')^{\perp} \subset \mathfrak{M}^*$,

and that

(6.24) $\text{im } E' \subset \vartheta(A)$, $(\ker E'') \subset \vartheta^*(A)$.

In the above, a total subset of $\mathcal{X}^*$ by definition is a set $\mathcal{M}^* \subset \mathcal{X}^*$ with the property that

(6.25) $\langle x^*, x \rangle = 0$, $x^* \in \mathcal{M}^*$ $\Rightarrow$ $x = 0$.

By $\vartheta^*(A)$ we mean the set denoted by $\vartheta(A^*)$ in (6.7). We have mentioned that $\vartheta^*(A)$ is the domain of the adjoint A^* of A, if it is dense in $\mathcal{X}^*$, and that this always is the case if $\mathcal{X}$ is a reflexive Banach space. Also we mentioned the existence of operators in non-reflexive spaces, which are closed but do not have $\vartheta^*(A)$ dense in $\mathcal{X}^*$. On the other hand, please compare the lemma below.

Lemma 6.5. For any pre-closed operator A: $\vartheta(A) \to \mathcal{X}$ the space $\vartheta^*(A)$ always is total.

The proof results from a simple change of the proceedure after formula (6.7). The element v^*, defined uniquely by (6.7), for any $u^* \in \vartheta^*(A)$ has the property that the collection

(6.26) $\mathcal{Z}^\circ = \{ [-v^*, u^*] : u^* \in \vartheta^*(A) \}$

is the orthogonal complement of the graph $\mathcal{G}(B)$, with the closure B of A. Since $[0,v] \in \mathcal{G}(B)$ holds if and only if $v = 0$, we conclude that $\langle u^*, v \rangle = 0$ for all $u^* \in \vartheta^*(A)$ implies v=0, or we have (6.25) for $\mathcal{M}^* = \vartheta^*(A)$. Thus indeed $\vartheta^*(A)$ is total , q.e.d..

Again the operators A' and A" of (6.21) and (6.22), respectively, will be called left and right Fredholm inverses of A. An operator B=A'=A" satisfying both relations will be called a Fredholm inverse of A. Note that a Fredholm inverse B of the unbounded closed operator A necessarily must satisfy the relations of Theorem 6.1. In particular, since necessarily im B = im A" $\subset \vartheta(A)$, by (6.22), we get B: $\mathcal{X} \to \vartheta(A)$, and $\|Bx\|_A \leq \|Bx\| + \|BAx\| \leq (1 + \|B\| + \|E'\|) \|x\|$, $x \in \vartheta(A)$, so that B: $\mathcal{X} \to \vartheta(A)$ is bounded. We have E'x = BAx - x $\in \vartheta(A)$ for x $\in \vartheta(A)$, and for any y $\in$ im E' a vector x $\in \vartheta(A)$ can be found, which is in $\vartheta(A)$ and satisfies y = E'x, because E' has finite rank but $\vartheta(A)$ has infinite dimension. Accordingly, im E' $\subset \vartheta(A)$, and $E' \in E(\mathcal{X}, \vartheta(A))$, so that indeed every Fredholm inverse satisfies the conditions of Theorem 6.1.

We even get

ABAx = Ax + FAx = Ax + AF'x, x $\in \vartheta(A)$,

with F, F' as in (6.11), so that we have

(6.27) $FA \subset AF' \in E(\mathcal{X})$,

which implies that we also have $(\ker F)^\perp \subset \vartheta^*(A)$, or, that also the second condition (6.24) (with E" replaced by F', in notation only) is necessarily true for any Fredholm inverse.

This, in particular means that an operator A is Fredholm if and only if a Fredholm inverse exists.

A left (right) Fredholm operator is one which has a left (right) Fredholm inverse, and a semi-Fredholm operator is a closed operator, which is either left or right Fredholm.

Proof of corollary 6.4. If A' and A" are a left and right Fredholm inverse, respectively, then we get

$$(6.28) \qquad A'AA'' = A'' + E'A'' = A' + A'E''$$

with E', E" as in (6.21) and (6.22), so that

$A'' = A' + A'E'' - E'A''$, which yields $A''A = 1 + A'E''A + E'(1-A''A)$. The expression $H' = A'E''A+E'(1-A''A) = A''A-1$ takes $\mathfrak{D}(A)$ to itself, and has finite rank, hence the continuous linear map $A'': \mathfrak{X} \to \mathfrak{D}(A)$ is a Fredholm inverse of the continuous linear map $A: \mathfrak{D}(A) \to \mathfrak{X}$, in the sense of section 3. Thus A is Fredholm, also as an unbounded closed operator of $\mathfrak{X}$. Vice versa, if A is Fredholm, then a Fredholm inverse exists, by theorem 6.1, which also represents a left and a right Fredholm inverse.

Starting from a Fredholm inverse B of A (satisfying (6.11)), let us finally construct the special left and right Fredholm inverses A', A", satisfying (6.23) and (6.24) by setting

$$(6.29) \qquad A' = B - \sum f_j >< x_j^* \quad , \quad A'' = B - \sum y_j >< g_j^*$$

with f_j, g_j^* determined by

$$(6.30) \qquad F = \sum f_j >< f_j^* \quad , \quad G = F' = \sum g_j >< g_j^*$$

as in (6.11). We get

$$(6.31) \qquad \begin{aligned} E' &= A'A - 1 = \sum f_j > (< f_j^* - < x_j^* A) \\ E'' &= A\,A'' - 1 = \sum (g_j - Ay_j) >< g_j^* \end{aligned} .$$

Since B is a Fredholm inverse, we may choose $f_j \varepsilon \mathfrak{D}(A)$, $g_j^* \varepsilon \mathfrak{D}^*(A)$, as was seen above. Then (6.23) is an immediate consequence of the lemma below, and theorem 6.4 is established.

Lemma 6.6. Given (a) a closed space $\mathfrak{Y} \subset \mathfrak{X}$, with finite co-dimension, and a dense space $\mathfrak{M}$. Or, given (b) a space $\mathfrak{Z}$ of finite dimension, $\mathfrak{Z} \subset \mathfrak{X}$, and a space $\mathfrak{M}^*$ total in $\mathfrak{X}^*$.

Then (a) implies $\mathfrak{X} = \mathfrak{Y} + \mathfrak{M}$, and (b) implies $\mathfrak{X}^* = \mathfrak{Z}^\perp + \mathfrak{M}^*$.

We know that im $A = \mathfrak{Y}$ is a space of finite co-dimension, and that im $A^* = (\ker A)^\perp$, by Lemma 3.3, (easily generalized to unbounded operators).

Therefore indeed we may decompose $g_j = Ay_j + h_j$, $f_j^* = x_j^* + h_j^*$, with $h_j \in \mathfrak{M}$, $h_j^* \in \mathfrak{M}^*$, so that (6.23) follows.

Let us prove lemma 6.6 in the stronger form, below.

Lemma 6.7. Under the assumption (a) the space $\mathfrak{Y}$ has a complement $\mathfrak{W} \subset \mathfrak{M}$; under the assumption (b) the space $\mathfrak{Y}^\perp$ has a complement $\mathfrak{W}^* \subset \mathfrak{M}^*$.

Proof. (a) Select a basis x_j $j=1,\ldots,N$ of $\mathfrak{X}$ mod $\mathfrak{Y}$, and construct $y_j \in \mathfrak{M}$ with small $\|x_j - y_j\|$. Show that y_j, $j=1,\ldots,N$, still form a basis of $\mathfrak{X}$ mod $\mathfrak{M}$, and define $\mathfrak{W}$ as the span of that basis.

(b) For a basis x_j, $j=1,\ldots,N$ of $\mathfrak{Y}$ construct x_j^*, $j=1,\ldots,N$, with $\langle x_j^*, x_l\rangle = \delta_{jl}$, by (6.25). For a general $x^* \in \mathfrak{X}^*$ confirm that $s^* = x^* - \sum_{l=1}^{N} x_l^* \langle x^*, x_l\rangle \in \mathfrak{Y}^\perp$, so that the space $\mathfrak{W}^*$ spanned by $\{x_j^*\}$ is the desired complement of $\mathfrak{Y}^\perp$, q.e.d..

Corollary 6.8. Under the assumptions (a) above we have $\mathfrak{M} \cap \mathfrak{Y}$ dense in $\mathfrak{Y}$.

Proof. For $x \in \mathfrak{Y}$ there exists $x_m \in \mathfrak{M}$ with $x_m \to x$. Decompose $x_m = y_m + z_m$, with $y_m \in \mathfrak{M} \cap \mathfrak{Y}$, $z_m \in \mathfrak{W}$, $z_m \to 0$. Conclude that $y_m \to x$, q.e.d..

Proof of theorem 6.3. Let A, B be unbounded Fredholm operators, and use corollary 6.4 to construct left and right Fredholm inverses A', B' and A", B" of A and B, respectively, satisfying (6.23), with the spaces

$\mathfrak{M} = \mathfrak{D}(B)$, $\mathfrak{M}^* = \mathfrak{D}^*(B)$ for the operator A, and with $\mathfrak{M} = \mathfrak{D}(A)$, $\mathfrak{M}^* = \mathfrak{D}^*(A)$ for the operator B. Accordingly,

$$A'A = 1 + E',\ AA'' = 1 + E'',\ B'B = 1 + F',\ BB'' = 1 + F'',$$

(6.32)

$$AE'B,\ AF''B,\ BE''A,\ BF'A \in E(\mathfrak{X}).$$

Now, using corollary 6.8 for $\mathfrak{M} = \mathfrak{D}(A)$, $\mathfrak{Y} = \text{im } B$, conclude that $C = AB$ defined by (6.2) has dense domain $\mathfrak{D}(C)$. This is a closed operator, since $u_m \to u$, $Cu_m \to v$ implies $Bu_m = A'Cu_m - E'Bu_m \to A'v - (E'B)^{cl}u$ using (6.32), so that $u \in \mathfrak{D}(B)$, $Bu \in \mathfrak{D}(A)$, $ABu = v$. In the above "cl" denotes the closure of the operator.

By (6.32) it is evident that the two bounded operators

$$(6.33) \qquad C' = B'A' \quad , \quad C'' = B''A''$$

are a left and a right Fredholm inverse of C, so that C is Fredholm, by corollary 6.4.

To prove (6.20), let the two linear maps B_0: $\mathfrak{D}(C) \to \mathfrak{D}(A)$ and B_0'': $\mathfrak{D}(A) \to \mathfrak{D}(C)$ be defined as restrictions of B and B", respectively. The linear map C: $\mathcal{D}(C) \to \mathfrak{X}$ coincides with the composition AB_0 (of linear maps). Confirm that ker B = ker B_0, and that $\mathfrak{X}$/im B is canonically isomorphic to $\mathcal{D}(A)$/im B_0, so that B_0 is a Fredholm operator in the algebraic sense (without topology), and has the same index as B_0. Then (6.20) follows from the composition theorem for algebraic Fredholm maps (Theorem 2.8). This completes the proof of Theorem 6.3.

Finally we wish to mention that all above discussions have trivial generalizations to the case of unbounded operators A from $\mathcal{D}(A) \subset \mathfrak{X}$ to another space $\mathfrak{Y}$. We will occasionally use such things, but leave their proof to the reader.

7. Fredholm pairs and unbounded operators.

For unbounded closed operators of a Banach space $\mathfrak{X}$ a variety of topologies have been introduced (Rellich [19], Newburgh [16], Cordes-Labrousse [3], Neubauer [14], [15]), the most important of which is defined by the aperture of graphs (c.f. Gohberg-Krein [8], Berkson [1], Kato [12], for Hilbert spaces also [3]).

In this section we take a somewhat more pragmatic approach, asking only for a perturbation of closed Fredholm operators more general than that of section 6, so that, for example, the case of continuously changing boundary conditions with an elliptic differential operator on a compact manifold represents a continuous family of closed operators.

Example. The unbounded pre-closed operator

$$\mathfrak{D}(L_\varepsilon) = \{u: u\in C^\infty([0,1]),\ u(0) = 0,\ u'(1)+\varepsilon u(1) = 0\} \quad ,$$

(7.1)

$$L\,u = -u'',\ 0 \le x \le 1 \quad .$$

has a closure with graph changing continuously in the aperture topology of $\mathfrak{H}\times\mathfrak{H}$, with $\mathfrak{H} = L^2([0,1])$, but since the domain changes with ε this is not a case of L_0-bounded perturbation, as in theorem 6.2.

Let us observe that the Fredholm properties of a closed operator A may be entirely expressed in terms of its graph $\mathfrak{G}(A)$. For any closed operator A from $\mathfrak{X}$ to $\mathfrak{Y}$, with domain $\mathfrak{D}(A)$, we get

$$\text{(7.2)} \qquad \ker A = \mathfrak{X} \cap \mathfrak{G}(A),\ \mathfrak{X} + \operatorname{im} A = \mathfrak{X} + \mathfrak{G}(A),$$

where $\mathfrak{X}$ and ker A are identified with subspaces of the graph space $\mathfrak{Z}^0 = \mathfrak{X}\times\mathfrak{Y}$, by $x \leftrightarrow [x,0]$ for $x\in\mathfrak{X}$. Similar $y\leftrightarrow [0,y]$ for $y\in\mathfrak{Y}$.

Note that a closed operator A from $\mathfrak{X}$ to $\mathfrak{Y}$ is Fredholm if and only if the pair $[\mathfrak{X},\mathfrak{G}(A)]$ of closed subspaces of the graph space $\mathfrak{Z}$ satisfies the properties

$$\text{(7.3)} \qquad \alpha = \dim\,(\mathfrak{X}\cap\mathfrak{G}(A)) < \infty,\ \beta = \dim\,\mathfrak{Z}/(\mathfrak{X}+\mathfrak{G}(A)) < \infty,\quad \operatorname{ind} A = \alpha - \beta \quad .$$

Following Kato [12], we shall call a Fredholm pair any pair $[\mathfrak{S},\mathfrak{T}]$ of closed subspaces of a Banach space $\mathfrak{X}$ with the property that $\mathfrak{S}+\mathfrak{T}$ is closed, and that

$$\text{(7.4)} \qquad \alpha = \dim\,(S\cap T) < \infty, \qquad \beta = \operatorname{codim}\,(\mathfrak{S}+\mathfrak{T}) < \infty \quad .$$

The Fredholm index of the pair $[\mathfrak{S},\mathfrak{T}]$ is defined by

$$\text{(7.5)} \qquad \operatorname{ind}\,[S,T] = \alpha - \beta \quad .$$

By the above remarks it is evident that a closed operator A with graph $\mathfrak{G}(A)$ is Fredholm if and only if $[\mathfrak{X},\mathfrak{G}(A)]$ is a Fredholm pair of the graph space $\mathfrak{Z} = \mathfrak{X}\times\mathfrak{Y}$, and that

$$\alpha(A) = \alpha([\mathfrak{X},\mathfrak{G}(A)]) \ ,\quad \beta(A) = \beta([\mathfrak{X},\mathfrak{G}(A)])$$

(7.6)

$$\operatorname{ind} A = \operatorname{ind}\,[\mathfrak{X},\mathfrak{G}(A)] \quad .$$

If $\mathfrak{S} + \mathfrak{T}$ is closed, but only one of the dimensions (7.4) is finite, we shall call $[\mathfrak{S}, \mathfrak{T}]$ (a) semi-Fredholm (pair). Especially "lower (upper) semi-Fredholm" means that the first (second) dimension (7.4) is finite, respectively.

In the discussion, below, we shall always assume that $\mathfrak{S}, \mathfrak{T}$ are closed subspaces of a Banach space $\mathfrak{X}$, and that P,Q are (bounded) projections onto $\mathfrak{S}$, $\mathfrak{T}$, respectively. Especially this means that $\mathfrak{S}$, $\mathfrak{T}$ must have a complement each, whenever P and Q exist.

Theorem 7.1. The following conditions are equivalent.

(i) $[\mathfrak{S}, \mathfrak{T}]$ is Fredholm ;

(ii) there exists P,Q which commute, and such that PQ and (1-P)(1-Q) are in $E(\mathfrak{X})$;

(iii) there exists P,Q with $1-P-Q \in K(\mathfrak{X})$;

(iv) there exist P,Q with $1-P-Q \in \mathfrak{J}$ with any Riesz ideal $\mathfrak{J}$ of $L(\mathfrak{X})$;

(v) there exist P,Q such that P + Q and (1-P)+(1-Q) are Fredholm operators of $\mathfrak{X}$.

Corollary 7.2. For P,Q satisfying (ii) above we have

(7.7) $$\mathfrak{S} \cap \mathfrak{T} = \text{im } PQ, \quad (\mathfrak{S}+\mathfrak{T}) \oplus \text{im } (1-P)(1-Q) = \mathfrak{X} .$$

Proof. Let $\mathfrak{S}$, $\mathfrak{T}$ be a Fredholm pair, and let $\mathfrak{S}'$, $\mathfrak{T}'$ be complements of $\mathfrak{S} \cap \mathfrak{T} = \mathfrak{M}$ in $\mathfrak{S}$, $\mathfrak{T}$, respectively. Such complements exist, since $\mathfrak{M}$ is finite dimensional. For similar reason there exists a complement $\mathfrak{N}$ of $\mathfrak{S} + \mathfrak{T}$ in $\mathfrak{X}$. Then we get

(7.8) $$\mathfrak{X} = \mathfrak{S}' \oplus \mathfrak{T}' \oplus \mathfrak{M} \oplus \mathfrak{N} ,$$

a direct decomposition, leading to the partition of identity

(7.9) $$1 = P' + Q' + M + N ,$$

with commuting projections P', Q', M, N onto the spaces at the right side of (7.8), respectively. P' projects onto $\mathfrak{S}'$ and annihilates $\mathfrak{T}' \oplus \mathfrak{M} \oplus \mathfrak{N}$, for example, etc., and P'Q' = Q'P' = 0, etc.. Confirm that

(7.10) $$P = P' + M, \quad Q = Q' + M$$

satisfy the conditions of (i), specifically, that

(7.11) $$PQ = M, \quad (1-P)(1-Q) = N .$$

If P,Q with the property of (ii) are given, write

(7.12) $$P' = P(1-Q), \quad Q' = (1-P)Q , \quad M = PQ , \quad N = (1-P)(1-Q) ,$$

and confirm that all these operators are projections, and that (7.10) as well as the decomposition of identity (7.9) hold. Note that

$$Mx = PQx = QPx = x$$

holds if and only if $x \in \mathfrak{S} \cap \mathfrak{T}$, so that im $\mathfrak{M} = \mathfrak{S} \cap \mathfrak{T}$. Also, it is trivial that im N is a complement of $\mathfrak{S}+\mathfrak{T}$, by (7.9). So, the equivalence of (i) and (ii) as well as the corollary are established.

Note that (ii) $\Rightarrow$ (iii) and (ii) $\Rightarrow$ (iv) since $1-P-Q = (1-P)(1-Q) - PQ$, and that (iv) $\Rightarrow$ (v) is trivial. Accordingly the remaining conclusions are a consequence of the lemma, below.

Lemma 7.3. If $P + Q \in F(\mathfrak{X})$, then $\mathfrak{S}+\mathfrak{T}$ is closed, and has finite codimension. If $(1-P)+(1-Q) \in F(\mathfrak{X})$, then $\dim (\mathfrak{S}\cap\mathfrak{T}) < \infty$.

Proof. In the first case we get $\mathfrak{S}+\mathfrak{T}\subset \operatorname{im}(P+Q)$, so that $\mathfrak{S}+\mathfrak{T}$ is closed, and has finite co-dimension. In the second case, we get

$$\mathfrak{S}\cap\mathfrak{T} = \{x: Px = Qx = x\} \subset \{x: (2-P-Q)x = 0\} = \ker ((1-P)+(1-Q)),$$

of finite dimension, q.e.d..

Corollary 7.4. If $P+Q$ is right Fredholm, then $[\mathfrak{S},\mathfrak{T}]$ is upper semi-Fredholm. If $2-P-Q$ is left Fredholm then $[\mathfrak{S},\mathfrak{T}]$ is lower semi-Fredholm. If either $P+Q$ is right Fredholm, or $2-P-Q$ is left Fredholm, then $[\mathfrak{S},\mathfrak{T}]$ is semi-Fredholm.

Proof. The only additional point to be studied, after the proof of lemma 7.3 is the closedness of $\mathfrak{S}+\mathfrak{T}$. Let $2-P-Q$ be left Fredholm, and let $\mathfrak{S}'$ be a complement of $\mathfrak{M}=\mathfrak{S}\cap\mathfrak{T}$ in $\mathfrak{S}$ again. Since again $\mathfrak{M}$ has finite dimension, such complement will exist. Replace P by P', the projection onto $\mathfrak{S}'$, annihilating $\ker P \oplus \mathfrak{M}$. Note that $2-P'-Q$ differs from $(2-P-Q)$ by an operator of finite rank, hence again is left Fredholm. This means that we may assume $\mathfrak{M} = 0$, without loss of generality, because $\mathfrak{S}' \oplus \mathfrak{T} = \mathfrak{S}+\mathfrak{T}$, evidently.

Assuming $\mathfrak{M} = 0$, for the closedness of $\mathfrak{S}+\mathfrak{T}$ we only must show that the projection $R: s+t \to s$, from $\mathfrak{S}+\mathfrak{T}$ onto $\mathfrak{S}$, annihilating $\mathfrak{T}$, is bounded, so that $u^n = s^n + t^n \to 0$, $s^n \in \mathfrak{S}$, $t^n \in \mathfrak{T}$, implies $s^n \to 0$. However, $(1-Q)u^n = (1-Q)s^n = ((1-P)+(1-Q))s^n \to 0$. Since we have $(2-P-Q)$ left Fredholm we conclude that either $\|s^n\| \to \infty$, or that there exists a convergent sub-sequence $s^{n_k} \to s \in \mathfrak{S}$. For reason of symmetry we have $s \in \mathfrak{S}\cap\mathfrak{T}$, thus $s = 0$. If $\|s^n\| \to \infty$ then we may repeat the procedure with $r^n = s^n/\|s^n\|$, which then must have a subsequence, converging to 0, a contradiction, since $\|r^n\| = 1$.

Corollary 7.5. $[\mathfrak{S},\mathfrak{T}]$, a pair of closed sub-spaces of $\mathfrak{X}$, is a Fredholm pair, if and only if $[\mathfrak{S}^\perp,\mathfrak{T}^\perp]$ is a Fredholm pair, and then we have

$$\text{(7.13)} \qquad \operatorname{ind}[\mathfrak{S}^\perp,\mathfrak{T}^\perp] + \operatorname{ind}[\mathfrak{S},\mathfrak{T}] = 0 ,$$

as well as

$$\text{(7.14)} \qquad \alpha([\mathfrak{S}^\perp,\mathfrak{T}^\perp]) = \beta([\mathfrak{S},\mathfrak{T}]) , \quad \beta([\mathfrak{S}^\perp,\mathfrak{T}^\perp]) = \alpha([\mathfrak{S},\mathfrak{T}]) .$$

Proof. Apply theorem 2.1, for a given Fredholm pair $[\mathfrak{S},\mathfrak{T}]$, to select a pair of projections, onto $\mathfrak{S}$ and $\mathfrak{T}$, respectively, with $PQ = QP$, and PQ, $(1-P)(1-Q) \in E(\mathfrak{X})$.

Note that $P^{\perp} = 1-P^*$ and $Q^{\perp} = 1-Q^*$ project onto $\mathfrak{X}^{\perp}$ and $\mathfrak{Y}^{\perp}$, respectively, and that $P^{\perp}Q^{\perp} = Q^{\perp}P^{\perp}$ as well as $P^{\perp}Q^{\perp} = ((1-P)(1-Q))^*$, $(1-P^{\perp})(1-Q^{\perp}) = (PQ)^*$. It follows that condition (ii) of theorem 7.1 for P,Q, and for $P^{\perp}$, $Q^{\perp}$ mean the same. Accordingly the corollary follows, especially also due to

$$\alpha([\mathfrak{X}^{\perp}, \mathfrak{Y}^{\perp}]) = \dim \operatorname{im} (P^{\perp}Q^{\perp}) = \dim \operatorname{im} ((1-P)(1-Q))^* = \beta([\mathfrak{X}, \mathfrak{Y}])$$

Corollary 7.6. For a Fredholm pair $[\mathfrak{X},\mathfrak{Y}]$ we have

$$(7.15) \qquad (\mathfrak{X}+\mathfrak{Y})^{\perp} = \mathfrak{X}^{\perp}\cap\mathfrak{Y}^{\perp}, \quad (\mathfrak{X}\cap\mathfrak{Y})^{\perp} = \mathfrak{X}^{\perp}+\mathfrak{Y}^{\perp} \quad .$$

Proof. The first relation is evident. As to the second, note that, in terms of corollary 7.2 we get

$$(\mathfrak{X}\cap\mathfrak{Y})^{\perp} = (\operatorname{im} PQ)^{\perp} = \operatorname{im} (1-(PQ)^*) =$$
$$= \operatorname{im} (P^{\perp} + Q^{\perp} - P^{\perp}Q^{\perp}) \qquad = \mathfrak{X}^{\perp} + \mathfrak{Y}^{\perp}, \quad \text{q.e.d.}\ .$$

Corollary 7.7. If a closed operator A is given which has an adjoint then A is Fredholm if and only if its adjoint is Fredholm, and we have

$$(7.16) \qquad \operatorname{ind} A + \operatorname{ind} A^* = 0, \qquad \alpha(A) = \beta(A^*), \qquad \beta(A) = \alpha(A^*)$$

Proof. For any closed operator A from $\mathfrak{X}$ to $\mathfrak{Y}$, and its adjoint (from Y^* to X^*) the graphs $\mathfrak{G}(A)$ and $\mathfrak{G}(A^*)$ as well as the spaces $\mathfrak{X}$ and $\mathfrak{Y}^*$ are orthogonal complements, respectively, therefore the above is a consequence of corollary 7.5.

Let us denote by A and A_c the collections of all closed subspaces of $\mathfrak{X}$, and of all closed complementable subspaces, respectively. On A_c we introduce a topology, referred to as the projection topology defined as the topology generated by the sets

$$(7.17) \qquad S = S_r(Q) = \{\mathfrak{X} : \mathfrak{X} \in A_c, \ \|P_{\mathfrak{X}} - Q\| < r\}$$

Here Q and r denote a given fixed bounded projection, and a positive number; $P_{\mathfrak{X}}$ denotes any bounded projection with image $\mathfrak{X}$.

The result below is a simple application of theorem 7.1, and the discussion of sections 2, 3 and 4.

Theorem 7.8. The Fredholm pairs form an open sub-set of $A_c \times A_c$, under the projection topology, and the Fredholm index is a constant on connected components of A_c.

Proof. By Theorem 7.1 (v) $[\mathfrak{X}, \mathfrak{Y}]$ is Fredholm if the two bounded operators P+Q and 2-P-Q are Fredholm operators. Thus sections 2, 3, 4 may be applied, q.e.d..

The closed operators A from $\mathfrak{X}$ to $\mathfrak{Y}$ are in 1-1-correspondence with their graphs $\mathfrak{G}(A)$, which are closed subspaces of the graph space $\mathfrak{Z}$. By theorem 7.1 the graphs of Fredholm operators are always complementable. We therefore may introduce a topology on closed operators with complementable graph, also called the projection topology, simply by taking the relative topology for graphs. Then we get the result below.

Corollary 7.9. The unbounded Fredholm operators are an open subset of the closed operators with complementable graph,under the projection topology, and the Fredholm index is constant on the connected components of this set.
Proof evident.

Remark. We note here that the projection topology coincides with the relative topologies induced by the aperture topology ([3], [8], [15]) and the invertible mapping topology [1] on the subset A_c .Proofs are omitted.

8. References

[1] E. Berkson, Metrics on the subspaces of a B-space; Pac. J. Math. 13 (1973), 7-22

[2] J.W. Calkin, Two sided iedeals in the ring of bounded operators of a Hilbert space; Ann. of Math. (2) 42 (1941), 839-873.

[3] H.O. Cordes and J.P. Labrousse, Invariance of the index of closed operators; J. Math. Mech. 12 (1963), 693-720.

[4] N. Dunford and J. Schwartz, Linear operators, vol. 1, Interscience 1958.

[5] J. Feldman and T. Gamelin, c.f. bibliography of Gamelin, below.

[6] L. Fredholm, Class d'équations fonctionelles; Acta Math. 27 (1903), 365-390.

[7] T. Gamelin, Decomposition theorems for Fredholm operators; Pac. J. Math. 15 (1965), 97-106.

[8] I. Gohberg and M. Krein, Defect numbers, root numbers and indices of linear operators. Amer. Math. Soc. Transl. Ser. 2, 13 (1960), 185-264.

[9] I. Gohberg and A. Markus, Öffnung von Unterräumen eines Banachraumes. Uspehi Mat. Nauk. 14 (1959) 5 (89), 135-140.

[10] M. Kaeshoek, Closed linear operators on B-spaces; Thesis Univ. of Leiden 1964.

[11] T. Kato, Perturbation theory for nullity, deficiency. J. d'analyse Math. 6 (1958), 261-322.

[12] --"-- , Perturbation theory for linear operators; 2nd ed. Berlin, Heidelberg, New York, Springer 1976.

[13] G. Köthe, Topologische lineare Räume; Berlin, Göttingen, Heidelberg 1960.

[14] G. Neubauer, Index abgeschlossener Operatoren in Banachräumen I; Math. Ann. 160 (1965), 93-130.

[15] ---"--- , Part II, Math. Ann. 162 (1965), 92-119.

[16] J. Newburgh, Topology for closed operators; Ann. of Math. 53 (1951), 250-255.

[17] F. Nöther, Eine Klasse singulärer Integralgleichungen, Math Ann. 82 (1920), 42-63.

[18] F. Rellich, Störungstheorie der Spektralzerlegung; Part II, Math. Ann. 113 (1936), 677-685.

[19] --"-- , Part IV; Math. Ann. 118 (1942), 462-484.

[20] C.E. Rickart, General theory of B-algebras, Princton, v. Nostrond 1960.

Appendix AII: Preparations on Commutative Banach Algebras.

In appendix AII we have collected the essential facts on the Gel'fand-Naïmark representation of a commutative Banach algebra with unit. This material will play a central part in all of chapter IV. It is contained in standard texts on Banach algebras, like [3] or [6], but we have isolated the points we need here for the convenience of the reader.

In section 1 we discuss elementary definitions and properties Banach algebras, commutative or not. Section 2 discusses matters of the spectrum and the spectral radius for an element of a Banach algebra, as preparation for the Gel'fand theorem in section 4. Likewise section 3 discusses maximal ideals of a commutative Banach algebra with unit, and the space of all maximal ideals together with its topology, which makes it a compact Hausdorff space, called the maximal ideal space or structure space (or spectrum) of the algebra.

Section 4 then establishes a continuous homomorphism between an arbitrary commutative B-algebra $\mathfrak{A}$ and the space $C(\mathbf{M})$ of continuous complex-valued functions over the maximal ideal space $\mathbf{M}$ of $\mathfrak{A}$. In case of a C^*-algebra, that is, in essence of an adjoint invariant sub-Banach algebra of $L(\mathfrak{H})$, for a Hilbert space $\mathfrak{H}$, this Gel'fand homomorphism proves to be an isometric isomorphism, as is proven in section 7.

As a preparation for section 7 and also as a general aid we require the Stone-Weierstrass theorem proven in section 6.

Finally, in section 5, we discuss the associate dual map of a homomorphism between commutative Banach algebras with unit, which is a continuous map of the corresponding Gel'fand spaces. The associate dual map also will find essential use in chapter IV.

In section 8 we discuss a proof of the fact that the quotient algebra of a C^*-algebra,modulo one of its closed two-sided ideals always is a C^*-algebra again. This will become important in chapters IV and V :The quotient of the Laplace comparison algebra ,modulo the ideal of compact operators , is a C^*-algebra again, hence must be a function algebra,by the Gel'fand- Naimark theorem.

1. Fundamentals of B-algebras

A Banach algebra (abbreviated B-algebra) is a Banach space $\mathfrak{A}$ with elements x, y, z, ... and norm $\| x \|$ in which also a (not necessarily commutative) product $x \cdot y = xy$ is defined which again represents an element of $\mathfrak{A}$. Scalar multiplication cx, sum x + y and product xy meet all the usual algebra axioms so that $\mathfrak{A}$ also is an algebra over the complex field $\mathbb{C}$ (sometimes only over the reals $\mathbb{R}$).

We have

$$\| xy \| \leq \| x \| \ \| y \| \quad , \tag{1.1}$$

and if there exists a unit element e of the algebra we insist on

$$\| e \| = 1. \tag{1.2}$$

The words homomorphism, isomorphism, endomorphism, automorphism shall be used for mapping between B-algebras preserving the algebra operation, exactly as conventional.

In general we do not require that such map preserves the norm nor even that it is continuous.

We speak of an isometric isomorphism or contractive homomorphism etc., if the map also leaves the norm invariant or if it does not increase the norm, resp.

If a B-algebra $\mathfrak{A}$ does not have a unit element, we always may adjoin one, that is imbedd $\mathfrak{A}$ isometrically in another Banach algebra $\mathfrak{A}_1$ which has a unit.

In fact, let $\mathfrak{A}_1$ denote the cartesian product $\mathbb{C} \times \mathfrak{A}$ of the two Banach spaces $\mathbb{C}$ and $\mathfrak{A}$, that is the collection of ordered pairs (c,x), with norm

$$\| (c,x) \| = |c| + \| x \| \quad , \tag{1.3}$$

and introduce the product

$$(c,x) \cdot (d,y) = (cd, \ cy + dx + xy).$$

It is easily verified that the algebra axioms hold and that (1,0) is a unit element of $\mathfrak{A}_1$. Clearly (1.1) and (1.2) hold as well, and the mapping $\mathfrak{A} \to \mathfrak{A}_1$ defined by $x \to (0,x)$ imbeds $\mathfrak{A}$ isometrically and isomorphically into $\mathfrak{A}_1$.

Note that the same procedure works also if $\mathfrak{A}$ already has a unit e. Then $\mathfrak{A}_1$ and $\mathfrak{A}$ form an example for the fact that an algebra $\mathfrak{A}_1$ and its proper sub-algebra $\mathfrak{A}$ need not to have the same unit. For (0,e) is not unit in $\mathfrak{A}_1$, only in $\mathfrak{A}$. We get $(1,0) \cdot (0,e) = (0,e) \neq (1,0)$.

For a B-algebra $\mathfrak{A}$ with unit e we call x and y <u>inverses</u> of each other if xy = e and yx = e. If only the first relation holds then x is called a <u>left inverse</u> of y or y is called a <u>right inverse</u> of x. If there exist a (left) (right) inverse for an element then it will be called (left) (right) <u>invertible</u> or (left) (right) <u>regular</u>. If an inverse y of x exists it is unique and will be denoted by $y = x^{-1}$. The same statement for left or right inverses does not hold.

We will frequently tend to the habit of denoting the unit e of a B-algebra simply by 1 and its multiples λe, $\lambda \in \mathbb{C}$, by λ.

Even if $\mathfrak{A}$ has no unit we might refer to $x + \lambda$, $x \in \mathfrak{A}$, $\lambda \in \mathbb{C}$, and then always mean the element $(\lambda,x) \in \mathfrak{A}_1$ above.

Examples: (i) The algebra $\mathfrak{M}_n$ of all n×n-matrices $A = ((a_{ik}))$ with complex coefficients, under the ordinary matrix sum and product, and with Schmidt-norm

$$(1.4) \qquad |||A||| = \{\sum_{i,k=1}^{n} |a_{ik}|^2\}^{1/2}$$

satisfies all conditions for a B-algebra, except (1.2). In fact, we get $|||\, I\, ||| = |||\, ((\delta_{ik}))\, ||| = \sqrt{n} > 1$ unless $n = 1$, and therefore $\mathfrak{M}_n$ under $|||\, A\, |||$ is not a B-algebra.

This cannot be repaired by introducing a multiplicative constant into the norm, because then (1.1) will get violated.

(ii) On the other hand the norm $|||\cdot|||$ is equivalent to the "operator norm"

$$(1.5) \qquad ||A|| = \sup_{x \neq 0} \{\sum_i |\sum_k a_{ik}x_k|^2 / \sum_i |x_i|^2\}^{1/2}$$

under which $\mathfrak{M}_n$ is a B-algebra. The equivalence follows from the fact that

$$(1.6) \qquad |||A||| = (\text{trace } A^*A)^{1/2} = (\sum_i \lambda_i)^{1/2}$$

with the eigenvalue λ_i of A^*A while

$$(1.7) \qquad ||A|| = (\operatorname{Max}_{i=1}^{n} \lambda_i)^{1/2} \quad ,$$

so that

$$(1.8) \qquad ||A|| \leq |||A||| \leq \sqrt{n}\, ||A|| \quad .$$

(iii) More generally, the algebra $L(\mathfrak{X})$ of all bounded linear operators A of a given Banach space $\mathfrak{X}$ always is a B-algebra under the operator norm

$$(1.9) \qquad ||A|| = \sup_{x \neq 0} \{\, ||Ax|| \,/\, ||x||\}.$$

We only discuss completeness of this algebra: If $\{A_n\}_{n=1,2,...}$ is a Cauchy sequence so that $||A_n - A_m|| < \varepsilon$, $n, m > N(\varepsilon)$, then for any $x \in \mathfrak{X}$ the sequence $\{A_n x\}$ is Cauchy in $\mathfrak{X}$, so will converge to some element y. The assignment $x \to y$ defines a linear operator A which is the strong limit of the sequence A_n. Then

$$\begin{aligned} ||(A_n - A)x|| &\leq ||(A_n - A_l)x|| + ||A_l x - Ax|| \\ &\leq ||A_n - A_l||\ \ ||x|| + \varepsilon_l(x), \end{aligned}$$

where $\varepsilon_l(x) \to 0$, for fixed x, as $l \to \infty$. If $n > N(\varepsilon)$ we get

$$||(A_n - A)x|| \leq \varepsilon\, ||x|| + \varepsilon_l(x)$$

for all $l > N(\varepsilon)$ and $l \to \infty$ gives

$$||(A_n - A)x|| \leq \varepsilon\, ||x||\, , \; n > N(\varepsilon)$$

for all x. It follows that

$$\|A_n - A\| \leq \varepsilon,\ n > N(\varepsilon)$$

or that $A_n \to A$ in $L(\mathfrak{X})$. Hence $L(\mathfrak{X})$ is complete.

(iv) The collection of all compact (completely continuous) operators

$$C \in L(\mathfrak{X})$$

forms a B-algebra as well, again under the operator norm. See Ch. I, sec. 4.

(v) If $\mathbb{M}$ is a compact Hausdorff space then the collection $C(\mathbb{M})$ of all continuous complex-valued functions $f: \mathbb{M} \to \mathbb{C}$ forms a B-algebra under the norm

$$(1.10) \qquad \|f\|_0 = \sup_{x \in \mathbb{M}} |f(x)|$$

and with sum and product defined pointwise, for every x, as sum and product of complex numbers.

For any Cauchy sequence f_n under the norm (1.10) we have pointwise convergence to a function f and

$$|f_n(x) - f(x)| = \lim_{l \to \infty} |f_n(x) - f_l(x)| < \varepsilon$$

whenever $n > N(\varepsilon)$.

So

$$\|f_n - f\|_0 < \varepsilon,\ n > N(\varepsilon),$$

or the sequence $\{f_n\}$ of continuous functions converges uniformly to f, which implies $f \in C(\mathbb{M})$. Hence $C(\mathbb{M})$ is complete.

Note that the algebra $C(\mathbb{M})$ is commutative - its product $\phi(x) \cdot \psi(x)$ is commutative.

(vi) If $\mathbb{M}$ is a compact Haüsdorff space and $\mathfrak{A}$ is a B-algebra then the collection $C(\mathbb{M}, \mathfrak{A})$ of all continuous functions

$$F: \mathbb{M} \to \mathfrak{A}$$

again is a Banach algebra under

$$(1.11) \qquad \|F\|_0 = \sup_{m \in \mathbb{M}} \|F(m)\| .$$

This is true for similar reason as under (v).

Of specific importance for later discussions of structure theory will be the algebra

$$C(\mathbb{M}, K)$$

with the compact ideal K of some Hilbert or Banach space.

We remind of the concept of (left) (right) (two-sided) ideal $\mathfrak{J}$ of an algebra $\mathfrak{A}$ which is defined as any linear subspace invariant under (left) (right) (both sided) multiplication by elements of $\mathfrak{A}$. We will mean a two-sided ideal, unless we explicitely refer to left or right.

A closed ideal of a Banch algebra is a closed such subspace. Any B-algebra is its own ideal. An ideal of $\mathfrak{A}$ different from $\mathfrak{A}$ is called a proper ideal. In particular $\{0\}$ always is a proper ideal, unless $\mathfrak{A} = \{0\}$.

If $\mathfrak{J}$ is an ideal of the algebra $\mathfrak{A}$ we may write $x \equiv y (\mathrm{mod}\ \mathfrak{J})$ if $x - y \in \mathfrak{J}$ and then have an equivalence relation which leads to a subdivision of $\mathfrak{A}$ into equivalence classes mod $\mathfrak{J}$. For $x \in \mathfrak{A}$ let $x^\vee = x + \mathfrak{J}$ denote its equivalence class.

The collection $\mathfrak{A}/\mathfrak{J}$ of all such classes forms an algebra again, with operations meaningfully defined by

$$cx^\vee = (cx)^\vee ,\ x^\vee + y^\vee = (x+y)^\vee ,\ x^\vee y^\vee = (xy)^\vee .$$

If $\mathfrak{A}$ is a Banach algebra and $\mathfrak{J}$ is a proper closed ideal then $\mathfrak{A}/\mathfrak{J}$ contains an element $\neq 0$ and can be provided with the norm

$$(1.12) \qquad \|x^\vee\| = \inf \{ \|x+j\| : j \in \mathfrak{J} \}$$

under which it becomes a B-algebra again.

Indeed, the triangle inequalities for sum and product are easily verified. Also, if e is an identity in $\mathfrak{A}$ then $e^\vee = e + \mathfrak{J}$ is the identity in $\mathfrak{A}/\mathfrak{J}$ which must have norm ≥ 1, due to $\|e^\vee\| = \|(e^\vee)^2\| \leq \|e^\vee\|^2$, but also must have norm ≤ 1 since

$$\|e^\vee\| = \inf \|e+j\| \leq \|e\| = 1.$$

Hence $\|e^\vee\| = 1$.

To establish completeness let

$$(1.13) \qquad \|x_n^\vee - x_m^\vee\| = \inf_j \|x_n - x_m + j\| < \varepsilon ,$$

$n, m > N(\varepsilon)$, for a sequence $\{x_n\} \in \mathfrak{A}$. By induction we may construct a subsequence x_{n_l}, $l = 1,2,\ldots$ and a sequence $j_l \in \mathfrak{J}$ such that $y_l = x_{n_l} + j_l$ satisfies

$$\|y_l - y_{l+1}\| \leq 2^{-l}.$$

For this simply select n_l as an increasing sequence of integers such that $\|x_{n_l}^\vee - x_m^\vee\| < 2^{-l}$, for all $m > n_l$, then conclude existence of $k_l \in \mathfrak{J}$ with $\|x_{n_l} - x_{n_{l+1}} + k_l\| < 2^{-l}$ and define $j_1 = 0$

$$j_l = -(k_1+k_2+\ldots+k_{l-1}),\ l \geq 2.$$

Next, conclude convergence of the sequence $\{y_l\}$, because

$$\|y_{l+m} - y_l\| \leq \sum_{j=1}^{m} \|y_{l+j} - y_{l+j-1}\|$$

$$\leq \sum_{j=1}^{m} 2^{-j-l+1} \leq 2^{-l+1}$$

and since $\mathfrak{A}$ is complete. Then define $y = \lim y_l$. Note that

$$\|y - x_n + j\| = \lim_{l\to\infty} \|x_{n_l} + j_l - x_n + j\|$$

$$\leq \limsup_{l\to\infty} \|x_{n_l} - x_n\|$$

for all $j \in \mathfrak{J}$. Hence

$$\|y^\vee - x_n^\vee\| \leq \limsup_{l\to\infty} \|\check{x}_{n_l} - x_n^\vee\| \to 0,\ n \to \infty.$$

Hence $y^\vee = \lim x_n^\vee$ and $\mathfrak{A}/\mathfrak{J}$ is complete.

We now can state with greater precision the first objective of the present chapter.

Definition:

An element x of B-algebra $\mathfrak{A}$ such that

$$\lim \sqrt[n]{\|x^n\|} = 0$$

will be called quasi-nilpotent.

Definition:

Let A denote a collection of continuous complex-valued functions over a space $\mathbb{M}$. We will say that A separates $\mathbb{M}$ if for any pair m_1, $m_2 \in \mathbb{M}$ there exists a function $\phi \in A$ such that $\phi(m_1) \neq \phi(m_2)$. If in addition for every $m \in \mathbb{M}$ there exists $\phi \in A$ with $\phi(m) \neq 0$ we will say that A separates $\mathbb{M}$ strongly.

Theorem 1.1. (Gel'fand)

Let $\mathfrak{A}$ be a commutative B-algebra with unit and let the only quasi-nilpotent element of $\mathfrak{A}$ be the zero-element.

Assertion:

There exists a compact Hausdorff space $\mathbb{M}$ satisfying the following conditions.

(i) $\mathfrak{A}$ is isomorphic to a sub-algebra A of $C(\mathbb{M})$, the algebra of continuous complex-valued functions over $\mathbb{M}$; the isomorphism carries the unit e into the funtion equal to 1 at every point and the norm of the associated function ϕ_x (as defined by (1.10)) is not larger than the norm of the element $x \in \mathfrak{A}$.

(ii) A separates the points of $\mathbb{M}$.

(iii) Any compact Hausdorff space $\mathbb{N}$ such that conditions (i) and (ii) are met by an isomorphism $\psi: \mathfrak{A} \to B \subset C(\mathbb{N})$ may be homeomorphically imbedded into $\mathbb{M}$ in such a way that ψ_x is the restriction of the function ϕ_x to the compact subset $\mathbb{N}$ of $\mathbb{M}$, for every $x \in \mathfrak{A}$.

The space $\mathbb{M}$ is uniquely determined by the above conditions (i), (ii) and (iii). Theorem 1.1 will be proven in section 4, in a more general setting.

2. Spectrum and spectral radius

Definition 2.1.

For any element x of a B-algebra $\mathfrak{A}$ the spectral radius $\rho = \rho(x)$ is defined by

(2.1) $$\rho(x) = \lim_{n\to\infty} \|x^n\|^{1/n} = \inf_n \|x^n\|^{1/n} .$$

The existence of the limit and its equality with the infimum is a direct consequence of the proposition below.

Proposition 2.2.

If a sequence $\{a_n\}_{n=1,2,\dots}$ of reals satisfies

$$a_{n+m} \le a_n + a_m$$

then a_n/n is either convergent or diverges to $-\infty$, and $\lim a_n/n = \inf a_n/n$, as $n \to \infty$.

For the proof we refer to Polya Szegö [5], p. 178.

If proposition 2.2 is applied to $a_n = \log \|x^n\|$ the existence of the limit in (2.1) and equality with the infimum is immediate.

As already pointed out in section 1, an element x with $\rho(x) = 0$ is called quasi - nilpotent. Nilpotent elements (that is those $x \in \mathfrak{A}$ with $x^n = 0$ for some n) are trivially quasi - nilpotents.

Simplest example of a nilpotent element is the matrix

$$\begin{pmatrix} 0 & 1 \\ 0 & 0 \end{pmatrix}$$

of the algebra $\mathfrak{M}_2$. The simplest example of a nontrivial quasi-nilpotent element is, perhaps, the Volterra type integral operator

$$K: C([0,1]) \to C([0,1])$$

defined by

$$Ku = v, \quad v(s) = \int_0^s k(s,t)u(t)dt ,$$

with a given kernel $k \in C(\{0 \le t \le s \le 1\})$. One checks easily that

$$\|K^n\|^{1/n} \le \|k\|_0 (n!)^{-1/n} \to 0, \quad n \to \infty ,$$

with

$$\|k\|_0 = \sup \{|k(s,t)| : 0 \le t \le s \le 1\}.$$

As another such example we mention the operator of $L(l^2)$ defined by the infinite matrix

$$A = ((k^{-1}\,\delta_{i,k-1})).$$

Proposition 2.4. (Porperties of the spectral radius).

(i) $\rho(cx) = |c|\rho(x)$;

(ii) $\rho(xy) \leq \rho(x)\rho(y)$ whenever $xy = yx$;

(iii) $\rho(x+y) \leq \rho(x) + \rho(y)$ whenever $xy = yx$;

(iv) $\rho(e) = 1$ if $\mathfrak{A}$ has a unit e;

(v) $0 \leq \rho(x) \leq \|x\|$.

Proof.

Points (i), (ii), (iv) and (v) are trivial. To prove (iii) let α, β be reals with $\alpha > \rho(x)$, $\beta > \rho(y)$ amd introduce $a = x/\alpha$, $b = y/\beta$. Using commutativity of x and y we then get

$$\|(x+y)^n\|^{1/n} = \|\sum_{k=0}^{n}\binom{n}{k}x^k y^{n-k}\|^{1/n} \tag{2.2}$$

$$\leq \{\sum_{k=0}^{n}\binom{n}{k}\alpha^k\beta^{n-k}\|a^k\| \ \|b^{n-k}\|\}^{1/n}.$$

For every $n = 1,2,\ldots$ select $n' = n'(n)$ with $0 \leq n' \leq n$ such that

$$\|a^{n'}\| \ \|b^{n-n'}\| = \text{Max}\{\|a^k\| \ \|b^{n-k}\| : 0 \leq k \leq n\}. \tag{2.3}$$

Then the right hand side of (2.2) is not larger than

$$(\alpha+\beta)\|a^{n'}\|^{1/n}\|b^{n-n'}\|^{1/n}.$$

Select a subsequence n_k of integers such that

$$\delta = \lim_{k\to\infty} n'(n_k)/n_k \tag{2.4}$$

exists. This is possible, because of $0 \leq n'(n)/n \leq 1$. It follows that $0 \leq \delta \leq 1$.

Suppose first that $\delta > 0$. Then $n'(n_k) \to \infty$ and

$$\lim_{k\to\infty}\|a^{n'(n_k)}\|^{1/n_k} = \lim_{k\to\infty}\{\|a^{n'}\|^{1/n'}\}^{n'/n_k} = \rho(a)^{\delta} \leq 1.$$

Next, if $\delta = 0$, it follows that

$$\limsup_{k\to\infty}\|a^{n'(n_k)}\|^{1/n_k} \leq \limsup_{k\to\infty}\|a\|^{n'/n_k} \leq \|a\|^0 = 1.$$

Accordingly

$$\limsup_{k\to\infty}\|a^{n'(n_k)}\|^{1/n_k} \leq 1 \tag{2.5}$$

holds in each of the two cases. For reason of symmetry we also get

$$\limsup_{k\to\infty}\|b^{n_k-n'(n_k)}\|^{1/n_k} \leq 1. \tag{2.6}$$

Let $n = n_k$ in (2.2) and use (2.3), (2.5) and (2.6) to derive

$$\rho(x+y) = \lim_{k\to\infty}\|(x+y)^{n_k}\|^{1/n_k} \leq (\alpha+\beta).$$

Since $\alpha > \rho(x)$, $\beta > \rho(y)$ were arbitrary, it follows that (iii) is correct, and proposition 2.4 is established.

We observe that proposition 2.4 implies that the spectral radius defines a semi-norm on $\mathfrak{A}$ whenever $\mathfrak{A}$ is commutative. That semi-norm becomes a norm if and only if there are no quasi-nilpotents except zero.

From now on through sections 2, 3, 4 and 5 we always assume that $\mathfrak{A}$ has an identity.

Definition 2.5.

The spectrum $Sp(x) = Sp_{\mathfrak{A}}(x)$ of an element x of the B-algebra $\mathfrak{A}$ is defined as the collection of all complex numbers λ such that $x - \lambda e$ is not invertible.

Note that generally the spectrum of $x \in \mathfrak{A}$ may change if one restricts the algebra to a sub-algebra. In this respect, a good example is the algebra $A_h \subset C(S_1)$ as discussed in section 4, example 2.

Proposition 2.6.

The invertible elements of $\mathfrak{A}$ form a topological group $G(\mathfrak{A})$ (with algebra multiplication as group operation).

$G(\mathfrak{A})$ is an open subset of $\mathfrak{A}$.

Proof.

$G(\mathfrak{A})$ trivially is a group and the product is continuous, by (1.1). To show that G is open in $\mathfrak{A}$ note that for $y \in \mathfrak{A}$ close to $x \in G$ we may write

$$y = x(1 - x^{-1}(x - y)). \tag{2.7}$$

It thus will be sufficient to show that elements of the form $1 - z$ with small z can be inverted. Also we must show continuity of the inverse operation and to that extent will write

$$||x^{-1} - y^{-1}|| \leq ||x^{-1}||\ ||y - x||\ ||y^{-1}|| \tag{2.8}$$

and derive a bound for $||y^{-1}||$ whenever $||y - x||$ is small. These remarks make the proof of proposition 2.6 entirely a matter of the next following two propositions.

Proposition 2.7.

If $\rho(z) < 1$ then $1 - z$ is invertible and

$$(1 - z)^{-1} = 1 + z + z^2 + \ldots \quad . \tag{2.9}$$

Proof.

Consider the power series

$$1 + \lambda ||z|| + \lambda^2 ||z^2|| + \ldots + \lambda^n ||z^n|| + \ldots \tag{2.10}$$

and notice it has radius of convergence equal to

$$(\limsup \sqrt[n]{||z^n||})^{-1} = (\rho(z))^{-1},$$

by the well known formula. Hence the series at right in (2.9) converges absolutely because $\lambda = 1$ is within the circle of convergence of (2.10). This implies its convergence, by

$$\|\sum_N^{\infty} z^n\| \le \sum_N^{\infty} \|z^n\| .$$

It is verified at once that

$$(1 - z)(\sum_0^{\infty} z^n) = (\sum_0^{\infty} z^n)(1 - z) = 1.$$

Thus proposition 2.7 is established.

Proposition 2.8.

Let $x \in G(\mathfrak{A})$ and $0 < \eta < \|x^{-1}\|^{-1}$, then the closed ball

(2.11) $$\mathcal{D}_{x,\eta} = \{y: \|y - x\| \le \eta\}$$

is contained in $G(\mathfrak{A})$, and we have

(2.12) $$\|y^{-1}\| \le \frac{\|x^{-1}\|}{1-\eta\|x^{-1}\|}, \quad y \in \mathcal{D}_{x,\eta} .$$

Proof.

Let $z = x^{-1}(x-y)$ and note that $\rho(z) \le \|z\| \le \|x^{-1}\|\eta < 1$, so that

(2.13) $$y^{-1} = (1 - z)^{-1}x^{-1}$$

exists, by proposition 2.7, and (2.7). Moreover, (2.9) gives

$$\|(1 - z)^{-1}\| \le \sum_{n=0}^{\infty} \eta^n \|x^{-1}\|^n = (1 - \eta\|x^{-1}\|)^{-1} ,$$

hence (2.12) results.

Corollary 2.9.

We have

(2.14) $$\|y^{-1}-x^{-1}\| \le \|x^{-1}\|^2(1-\eta\|x^{-1}\|)^{-1}\|y-x\|$$

for $y \in \mathcal{D}_{x,\eta}$.

The proof is a consequence of (2.8) and (2.12).

Proposition 2.10.

The spectrum Sp(x) of an element is a closed subset of $\mathbb{C}$.

The proof is evident, after proposition 2.6.

Proposition 2.11.

The spectrum Sp(x) of an element $x \in \mathfrak{A}$ is never empty, and we have

(2.15) $$\rho(x) = \sup\{|\lambda|: \lambda \in Sp(x)\}.$$

The proof of proposition 2.11, presented in the following, draws heavily on theory of functions of a complex variable and therefore makes use of the fact that the scalars of a B-algebra are the (field of) complex numbers. We will first discuss a few properties of the function $F: (\mathbb{C} - Sp(x)) \to \mathfrak{A}$ defined by

(2.16) $$F(\lambda) = (x - \lambda)^{-1}.$$

Much of it is strictly analogous to conventional results on (complex-valued) holomorphic functions. We shall come back to the proof of proposition 2.11 after a few other remarks.

Proposition 2.12.

We have existence of the limit in (2.17) and

$$\lim_{\mu\to\lambda,\mu\neq\lambda,\mu\in\mathbb{C}} \frac{F(\mu)-F(\lambda)}{\mu-\lambda} = (x-\lambda)^{-2}, \tag{2.17}$$

for every $\lambda \notin Sp(x)$.

Proof.

Write

$$(x-\mu)^{-1} - (x-\lambda)^{-1} = (x-\mu)^{-1}\left[(x-\lambda) - (x-\mu)\right](x-\lambda)^{-1}$$
$$= (\mu-\lambda)(x-\mu)^{-1}(x-\lambda)^{-1}$$

and use proposition 2.6 to establish (2.17).

Proposition 2.13.

If $|\lambda| \geq \rho(x)$ then $\lambda \notin Sp(x)$ and

$$F(\lambda) = (x-\lambda)^{-1} = -\frac{1}{\lambda}(1 + \frac{x}{\lambda} + \frac{x^2}{\lambda^2} + \ldots). \tag{2.18}$$

If a fortiori $|\lambda| \geq \|x\|$, then

$$\|F(\lambda)\| = \|(x-\lambda)^{-1}\| \leq (|\lambda| - \|x\|)^{-1}. \tag{2.19}$$

Proof.

Write

$$(x-\lambda)^{-1} = -\lambda^{-1}(1-x/\lambda)^{-1}$$

and apply proposition 2.7 and 2.8 using that $\rho(x/\lambda) \leq \frac{\|x\|}{\|\lambda\|} < 1$.

Proposition 2.14.

Sp(x) is not empty. Indeed, for any continuous linear functional x^* over $\mathfrak{A}$ define

$$\phi(\lambda) = \langle x^*, F(\lambda)\rangle, \quad \lambda \notin Sp(x). \tag{2.20}$$

If Sp(x) is empty, then $\phi(\lambda)$ is an entire analytic function vanishing at $\lambda = \infty$, by (2.19). Hence we obtain $\phi(\lambda) = 0$, by Liouville's theorem. This is a contradiction, because $F(\lambda)$ is never zero, because it is an inverse at every point, hence a continuous linear functional may be constructed which does not vanish at $z = F(\lambda)$. Thus $Sp(x) \neq \phi$.

Proposition 2.15.

Let

$$\rho' = \{\inf |\lambda| : \lambda \in Sp(x)\}. \tag{2.21}$$

Then the function G defined by

$$G(\lambda) = F(1/\lambda), \ 0 < |\lambda| < 1/\rho' ,$$

$G(0) = 0$, admits Cauchy's integral formula in the form

$$(2.22) \qquad G(\lambda) = (2\pi i)^{-1} \int_{|\xi|=\eta} G(\xi)/(\xi-\lambda)\,d\xi \quad ,$$

$$|\xi| < \eta < 1/\rho' \quad ,$$

where the right hand side's integral exists as Riemann-integral in norm convergence taken in the positive orientation of $|\xi| = \eta$.

Proof.

Recall that the Riemann integral is defined as $\lim_{\delta_{\mathfrak{z}}\to 0} S_{\mathfrak{z}}$ where $\mathfrak{z}$ is a partition of its integration curve (here the circle $|\xi| = \eta$) into finitely many segments $\Delta_1,\ldots,\Delta_N$ while $l(\Delta_j)$ = length of Δ_j, $\xi_j \in \Delta_j$,

$$S_{\mathfrak{z}} = \sum_{j=1}^{N} \frac{G(\xi_j^*)}{\xi_j^*-\lambda}\, l(\Delta_j)$$

$$\delta_{\mathfrak{z}} = \mathrm{Max}\,\{l(\Delta_j) : j = 1,\ldots,N\}.$$

Existence of this limit, interpreted as norm convergence in $\mathfrak{A}$, is a consequence of the fact that the integrand $G(\xi)/(\xi-\lambda)$ is uniformly continuous over the compact arc $|\xi| = \eta$, following a standard argument.

Accordingly let $\eta < 1/\rho'$ be fixed and let $W(\lambda)$ denote the difference between the left and right hand side of (2.22), which now is a well defined element of $\mathfrak{A}$ for $|\lambda| < \eta$. Let $x^* \in \mathfrak{A}^*$ again and let $\psi(\lambda) = \langle x^*, G(\lambda)\rangle$. This function is holomorphic in $0 < |\lambda| < 1/\rho'$ and has a removable singularity at $\lambda = 0$, hence is also holomorphic at $\lambda = 0$. Confirm that

$$\langle x^*, W(\lambda)\rangle = \psi(\lambda) - \langle x^*, \lim S_{\mathfrak{z}}\rangle$$

$$= \psi(\lambda) - (2\pi i)^{-1} \lim \langle x^*, S_{\mathfrak{z}}\rangle$$

$$= \psi(\lambda) - (2\pi i)^{-1} \int_{|\xi|=\eta} \frac{\psi(\xi)}{\xi-\lambda}\,d\xi = 0$$

using the conventional Cauchy-integral formula. Hence $W(\lambda) \equiv 0$ and (2.22) is established.

Proposition 2.16.

We have the right hand side in (2.23) convergent and

$$(2.23) \qquad -G(\lambda) = \lambda + \lambda^2 x + \lambda^3 x^2 + \ldots + \lambda^{n+1} x^n + \ldots$$

for all $|\lambda| < 1/\rho'$, with ρ' as in (2.21).

Proof.

The power series of $G(\lambda)$ around $\lambda = 0$ is unique, because that of $\psi(\lambda)$ above is unique, for any $x^* \in \mathfrak{A}^*$. Accordingly we must show only that $G(\lambda)$ has a convergent power series in any circle $|\lambda| < \eta' < 1/\rho'$, and then use (2.18), to verify (2.23).

Now in any such circle use (2.22) with some η, $\eta' < \eta < 1/\rho'$ and expand

$$1/(\xi-\lambda) = 1/\xi(1-\lambda/\xi)^{-1} = \sum_{0}^{\infty} \lambda^n/\xi^{n+1} .$$

Substituting this into (2.22) and using uniform convergence of $\sum_{0}^{\infty}\lambda^n G(\xi)/\xi^{n+1}$ on $|\xi| = \eta$, whenever $|\lambda| \leq \eta'$, we get

$$G(\lambda) = \sum_{n=0}^{\infty} a_n\lambda^n, \quad a_n = (2\pi i)^{-1}\int_{|\xi|=\eta} G(\xi)/\xi^{n+1} d\xi$$

convergent in $|\lambda| \leq \eta'$. This proves the proposition.

Proof of proposition 2.11.

After propositions 2.13 and 2.14 we are left with the proof of the inequality $\rho(x) \leq \rho'$ only. For any $|\lambda| < 1/\rho'$ we have the series in (2.23) convergent. If S_n denotes the partial sum then

$$\|\lambda^{n+1}x^n\| = \|S_n - S_{n-1}\| < \varepsilon, \ n > N(\varepsilon).$$

Hence $|\lambda|^n \|x^n\| \leq c$ for all n with a suitable c. Accordingly the series

$$\sum_{n=0}^{\infty} \mu^n \|x^n\| \tag{2.24}$$

converges for all $|\mu| < |\lambda|$, since it is dominated by the geometric series $\sum c\left|\frac{\mu}{\lambda}\right|^n$. Since this holds for all $|\lambda| < 1/\rho'$ the power series (2.24) of complex forms has radius of convergence $r \geq 1/\rho'$. Or,

$$(\limsup_{n\to\infty} \|x^n\|^{1/n}) \geq 1/\rho'$$

by the well known formula. Hence we get

$$\rho(x) = \lim_{n\to\infty} \|x^n\|^{1/n} \leq \rho'$$

and proposition 2.11 is established.

3. The maximal ideal space

Proposition 3.1.

A B-algebra $\mathfrak{A}$ with unit, for which all elements $x \neq 0$ are regular consists precisely of the complex multiples of its unit element.

Proof.

For every $x \in \mathfrak{A}$ there is a point $\lambda \in Sp(x)$, due to proposition 2.11. Then $x - \lambda$ is singular and hence $= 0$ by assumption. It follows that every element of $\mathfrak{A}$ is of the form $\lambda = \lambda e$ and since $e \in \mathfrak{A}$ we conclude that $\mathfrak{A} = \mathbb{C}e$.

Corollary 3.2.

A Banach algebra is a field if and only if it is (isometrically isomorphic to) the field $\mathbb{C}$ of complex numbers.

Proof.

This is only another way of expressing proposition 3.1.

For the remainder of this section as well as for section 4 and 5 we automatically assume that $\mathfrak{A}$ is a commutative B-algebra with unit. This implies that all ideals of $\mathfrak{A}$ are necessarily two-sided.

Definition 3.3.

A maximal ideal of $\mathfrak{A}$ is defined to be a proper ideal $\mathfrak{m}$ of $\mathfrak{A}$ which is not properly contained in any other proper ideal of $\mathfrak{A}$.

Proposition 3.4.

A proper ideal cannot contain invertible elements.

Indeed if $\mathfrak{J}$ contains an invertible element x, then $\mathfrak{J}$ contains every $a \in \mathfrak{A}$ because we can write $a = (ax^{-1})x \in \mathfrak{J}$. So $\mathfrak{J} = \mathfrak{A}$ is not proper, q.e.d.

Proposition 3.5.

The closure of a proper ideal is proper.

Indeed, if $\mathfrak{J}$ is an ideal whose closure equals $\mathfrak{A}$, then the identity is limit of a sequence $\{x_n\}$ of elements in $\mathfrak{J}$. Hence $\|1-x_n\| < 1$, for sufficiently large n and proposition 2.7 then implies invertibility of $x_n = 1 - (1-x_n)$, so $\mathfrak{J}$ is not proper.

Proposition 3.6.

A maximal ideal is closed.

Indeed, the closure of $\mathfrak{m}$ is a proper ideal, by 3.5, and hence cannot be a proper extension of $\mathfrak{m}$, or $\mathfrak{m}$ is closed, q.e.d.

Proposition 3.7.

Every singular element $x \in \mathfrak{A}$ (and every proper ideal $\mathfrak{J}$ of $\mathfrak{A}$) is contained in a maximal ideal $\mathfrak{m}$.

Proof.

If $x \in \mathfrak{A}$ is singular, then

$$(3.1) \qquad \mathfrak{J}_x = \{xy: y \in \mathfrak{A}\}$$

is an ideal (using that $\mathfrak{A}$ is commutative) and does not contain 1, because there is no inverse of x.

If $\mathfrak{J}_0$ is any proper ideal, then let $\mathcal{K}$ denote the collection of all proper ideals containing $\mathfrak{J}_0$. $\mathcal{K}$ is partially ordered by inclusion.

Suppose $\mathcal{K}_0$ is any linearly ordered subcollection so that for any pair $\mathfrak{J}, \mathfrak{I}$ $\in \mathcal{K}_0$ we either have $\mathfrak{J} \subset \mathfrak{I}$ or $\mathfrak{I} \subset \mathfrak{J}$. Then it follows that $\bigcup\{\mathfrak{J}: \mathfrak{J} \in \mathcal{K}_0\}$ is a linear subspace of $\mathfrak{A}$, and therefore an ideal. This ideal is proper, because 1 is not in any of the $\mathfrak{J} \in \mathcal{K}_0$, so it cannot be in the union.

This shows that every chain in $\mathcal{K}_0$ has an upper bound. Zorn's Lemma will therefore imply existence of a maximum of $\mathcal{K}$. That is, $\mathcal{K}$ contains a maximal ideal, q.e.d.

Corollary 3.8.

There always exist maximal ideals.

Indeed, every commutative B-algebra with unit has the trivial proper ideal $\{0\}$.

Proposition 3.9.

A proper ideal $\mathfrak{J}$ of $\mathfrak{A}$ is maximal if and only if

$$(3.2) \qquad \mathfrak{A}/\mathfrak{J} = \{\mathbb{C}e^{\vee}\}$$

with $e^{\vee} = \{e + \mathfrak{J}\}$ denoting the unit of the quotient.

Proof.

Let $\mathfrak{J} \subset \mathfrak{I}$ be proper ideals of $\mathfrak{A}$, then $\mathfrak{J}$ also is an ideal of $\mathfrak{I}$ (trivially) and (considered as algebra for itself)

$$(3.3) \qquad \mathfrak{I}/\mathfrak{J} \subset \mathfrak{A}/\mathfrak{J}$$

by natural identification. Also it follows that $\mathfrak{I}/\mathfrak{J}$ is a proper ideal of $\mathfrak{A}/\mathfrak{J}$.

The algebra $\mathbb{C}$ has only one proper ideal - the trivial ideal $\{0\}$, because by proposition 3.4 no proper ideal may contain an invertible element and all elements $\neq 0$ of $\mathbb{C}$ are invertible. Therefore (3.2) implies that $\mathfrak{I}/\mathfrak{J} = \{0\}$, or that $\mathfrak{I} = \mathfrak{J}$. Or, $\mathfrak{J} = \mathfrak{m}$ is maximal. Vice versa, if $\mathfrak{J} = \mathfrak{m}$ is a maximal ideal then no element $x^{\vee}$ of $\mathfrak{A}/\mathfrak{m}$ can be singular, except $x^{\vee} = 0$. Indeed, such singular element $x^{\vee} \neq 0$ would lead to a proper extension

$$\mathfrak{I} = \{xy + m: y \in \mathfrak{A}, m \in \mathfrak{m}\}$$

of $\mathfrak{J}$ where $x \in x^{\vee}$ is a representative. Accordingly, the assumptions of proposition 3.1 are satisfied and (3.2) follows q.e.d.

For any pair $(x, \mathfrak{m})$ of an element $x \in \mathfrak{A}$ and a maximal ideal $\mathfrak{m}$ there is a unique complex number, denoted by $\phi_x(\mathfrak{m})$, defined by the relation

(3.4) $$x \equiv \phi_x(\mathfrak{m})\, e \pmod{\mathfrak{m}} ,$$

or, in details,

(3.5) $$x - \phi_x(\mathfrak{m}) \in \mathfrak{m}.$$

This indeed is a consequence of proposition 3.9.

We denote the collection of all maximal ideals by $\mathbb{M}$ and notice that for every fixed $x \in \mathfrak{A}$ a function $\phi_x : \mathbb{M} \to \mathbb{C}$ is defined by the assignment $\mathfrak{m} \to \phi_x(\mathfrak{m})$. Let A denote the collection of all such complex-valued functions over $\mathbb{M}$. Introduce on $\mathbb{M}$ the "weak topology with respect to A". That is, introduce as basis of a topology the collection $\{N\}$ with

$$N = N_{\mathfrak{m}_0; x_1,\dots,x_n; \varepsilon} = \{\mathfrak{m} : |\phi_{x_j}(\mathfrak{m}) - \phi_{x_j}(\mathfrak{m}_0)| < \varepsilon,\ j = 1,\dots,n\}.$$

The open sets in $\mathbb{M}$ are the unions of such sets N. It is known (and obvious) that all functions in A are continous with respect to this topology. We have the Hausdorff separation axiom satisfied. For if $\mathfrak{m}_1, \mathfrak{m}_2 \in \mathbb{M}$ are different, then there exists an $x \in \mathfrak{m}_1$ wich is not in $\mathfrak{m}_2$, because otherwise $\mathfrak{m}_1 \subset \mathfrak{m}_2$, hence $\mathfrak{m}_1 = \mathfrak{m}_2$, because $\mathfrak{m}_1$ is maximal and $\mathfrak{m}_2$ is proper. Hence we get $\phi_x(\mathfrak{m}_1) = 0$, but $\phi_x(\mathfrak{m}_2) \neq 0$. Therefore, if $\eta < \frac{1}{2}|\phi_x(\mathfrak{m}_2)|$ the two sets $N_{\mathfrak{m}_j; x; \eta}$ are disjoint open neighbourhoods of $\mathfrak{m}_j$, $j = 1,2$, as required.

<u>Definition 3.10.</u>

A multiplicative linear functional (abbreviated mlf) over $\mathfrak{A}$ is defined to be a bounded linear functional x^* over the B-space $\mathfrak{A}$ satisfying the conditions

(i) $\langle x^*, xy\rangle = \langle x^*, x\rangle\langle x^*, y\rangle$, $x, y \in \mathfrak{A}$

(ii) $\langle x^*, e\rangle = 1$.

Note that (ii) may be replaced by (iii) $x^* \neq 0$, because the existence of an x with $\langle x^*, x\rangle \neq 0$ and (i) imply $\langle x^*, x\rangle = \langle x^*, ex\rangle = \langle x^*, e\rangle\langle x^*, x\rangle$ or (ii).

Clearly "mlf" is just a synonymous expression for "nonvanishing continuous algebra homomorphism from $\mathfrak{A}$ to $\mathbb{C}$, the field of complex numbers".

<u>Proposition 3.11.</u>

For any mlf x^* over $\mathfrak{A}$ the kernel

(3.6) $$\mathfrak{m} = \{x : x \in \mathfrak{A},\ \langle x^*, x\rangle = 0\}$$

is a maximal ideal of $\mathfrak{A}$.

Indeed, since x^* is a nontrivial homomorphism, its kernel is a proper ideal. If $x \in \mathfrak{A}$ is arbitrary then also we calculate that $x_0 = x - \langle x^*, x\rangle e$ is in $\mathfrak{m}$, which means that

(3.7) $$x \equiv \langle x^*, x\rangle \bmod \mathfrak{m}$$

for every $x \in \mathfrak{A}$.

Accordingly $\mathfrak{A}/\mathfrak{m} = \{\mathbb{C}e^{\vee}\}$ or $\mathfrak{m}$ is maximal, by proposition 3.9.

Proposition 3.12.

For any maximal ideal $\mathfrak{m}$ of $\mathfrak{A}$ the linear functional x^* defined by

$$\text{(3.8)} \qquad \langle x^*, x\rangle = \phi_x(\mathfrak{m})$$

is an mlf.

Proof.

Recall that $\phi_x(\mathfrak{m})$ was defined by (3.4) or (3.5).

Hence (3.8) defines a homomorphism $\mathfrak{A} \to \mathbb{C}$ which is contractive, because

$$\begin{aligned} |\langle x^*,x\rangle| &= |\phi_x(\mathfrak{m})| = \|\phi_x(\mathfrak{m})e^{\vee}\| \\ &= \inf\{\|x+m\| : m \in \mathfrak{m}\} \le \|x\| . \end{aligned}$$

Hence x^* is a continuous linear functional, so an mlf.

Corollary 3.13.

The propositions 3.11 and 3.12 establish a 1-1-correspondence between the space $\mathbb{M}$ of maximal ideals and the collection $\mathbb{H}$ of all mlf.

Proof.

We only must confirm that the maps of the diagram below invert each other.

$$\begin{array}{ccc} & \text{ker} & \\ & \longrightarrow & \\ \mathbb{H} & & \mathbb{M} \\ & \longleftarrow & \\ & \phi & \end{array}$$

But $x \in \mathfrak{m} \iff \phi_x(\mathfrak{m}) = 0$ is evident, hence $\ker(\phi_{\cdot}(\mathfrak{m})) = \mathfrak{m}$. Also comparison of (3.7) and (3.4) shows that

$$\phi_x(\ker x^*) = \langle x^*,x\rangle, \text{ q.e.d.}$$

We now will use the 1-1-correspondence of corollary 3.13 to carry over the topology from $\mathbb{M}$ to $\mathbb{H}$. This gives for $\mathbb{H}$ the basis $\{\mathcal{U}\}$ with

$$\mathcal{U} = \mathcal{U}_{x_0^*;x_1,\dots,x_n;\varepsilon} = \{x^* : x^* \in \mathbb{H}, |\langle x^* - x_0^*, x_j\rangle| < \varepsilon .$$

We can say now that $\mathbb{M}$ and $\mathbb{H}$ are homeomorphic.

Proposition 3.14.

The space $\mathbb{H}$ is naturally identified with a closed subset of the unit sphere $\mathcal{T}^*$ in the adjoint space $\mathfrak{A}^*$ of the underlying B-space of $\mathfrak{A}$, under weak-*-topology.

Recall that $\mathcal{T}^*$ consists of all bounded linear functionals x^* over the B-space $\mathfrak{A}$ with

$$\text{(3.9)} \qquad \|x^*\| = \sup\{\langle x^*,x\rangle : \|x\| = 1\}.$$

The weak-$*$-topology in $\mathcal{X}^*$ is introduced by the base $\{V\}$ with

$$V = V_{x_0^*;x_1,\ldots,x_n;\varepsilon} = \{ x^* \in \mathcal{X}^*: |\langle x^*-x_0^*,x_j\rangle| < \varepsilon,\ j = 1,\ldots,n\}$$

We already showed that all mlf's have $\|x^*\| \le 1$ and now notice that (ii) of Definition 3.10 implies that $\|x^*\| = 1$.

Hence indeed $\mathbb{H}$ is a subcollection of $\mathcal{X}^*$ and we get

$$U_{x_0^*,\ldots} = V_{x_0^*,\ldots} \cap \mathbb{H}$$

whenever $x_0^* \in \mathbb{H}$. If for $x_0^* \notin \mathbb{H}$ the neighbourhood $V_{x_0^*,\ldots}$ contains a point $x_1^* \in \mathbb{H}$, then it still contains a suitable neighborhood $V_{x_1^*,\ldots}$ of x_1^*. Accordingly, the topology introduced by $\{U\}$ in $\mathbb{H}$ is the relative topology of $\mathbb{H}$ as a subset of $\mathcal{X}^*$.

To show that $\mathbb{H}$ is a closed subset, let x_0^* be in the closure. Then for $\varepsilon > 0$ and $x, y \in \mathfrak{A}$ there exists $x^* \in \mathbb{H}$ in $V_{x_0^*;x,y,xy,e;\varepsilon}$.

Using that $x^* \in \mathbb{H}$ is multiplicative we then confirm that

$$|\langle x_0^*,xy\rangle - \langle x_0^*,x\rangle\langle x_0^*,y\rangle| < \varepsilon(1 + \|x\| + \|y\|)$$

and that

$$|\langle x_0^*,e\rangle - 1| < \varepsilon.$$

Accordingly x_0 is multiplicative and $\mathbb{H}$ is closed.

Proposition 3.15.

$\mathbb{M}$ is a compact Hausdorff space.

Proof.

After the above this is an immediate consequence of a result by Alaoglu [1].

Theorem.

$\mathcal{X}^*$ is compact.

We will not offer the proof of this theorem. As a side remark notice that convergence of a sequence $\{x_n^*\}$ in weak-$*$-topology is equivalent to weak convergence of the sequence of linear functionals.

Recall the old result of D. Hilbert that any bounded sequence in separable Hilbert space has a weakly convergent subsequence. In other words, for a separable Hilbert space $\mathfrak{H}$ the unit sphere $\mathcal{X}^*$ is sequentially compact.

Alaoglu's result now appears as a generalization of the old result of Hilbert.

The compact Hausdorff space $\mathbb{M}$ as defined above is called the spectrum of $\mathfrak{A}$ or the structure space of $\mathfrak{A}$. We also will call it the maximal ideal space on the Gel'fand space of $\mathfrak{A}$.

4. The Gel'fand Theorem

We may summarize the achievements of section 3 and specify some further details in the following form.

Theorem 4.1.

The map $x \to \phi_x$ of (3.4) establishes a homomorphism $\phi: \mathfrak{A} \to C(\mathbb{M})$, with $\phi_e(\mathfrak{m}) = 1$.

We have the following properties of ϕ.

(i) $\phi_x(\mathfrak{m}) = 0$ if and only if $x \in \mathfrak{m}$;

(ii) $Sp(x) = Sp_{\mathfrak{A}}(x) = \phi_x(\mathbb{M}) = \text{im}\,\phi_x$;

(iii) $\rho(x) = \lim\limits_{n\to\infty} \|x^n\|^{1/n} =$

$= \|\phi_x\|_0 = \sup\{|\phi_x(\mathfrak{m})| : \mathfrak{m} \in \mathbb{M}\} \leq \|x\|$;

(iv) $\phi_x(\mathfrak{m}) = 0$ for all $\mathfrak{m} \in \mathbb{M}$ if and only if x is quasi nilpotent (that is $\rho(x) = 0$);

(v) $\mathbb{M}$ is separated strongly by $\mathcal{A} = \text{im}\,\phi$;

(vi) x is regular if and only if $\phi_x(m) \neq 0$ over all of $\mathbb{M}$.

Proof.

For every fixed maximal ideal $\mathfrak{m}$ the assignment $x \to \phi_x(\mathfrak{m})$ is a homomorphism into $\mathbb{C}$, hence $x \to \phi_x$ also is a homomorphism into $C(\mathbb{M})$. By definition $\phi_x(\mathfrak{m})e^{\vee}$ denotes the congruence class of x mod $\mathfrak{m}$, hence $\phi_x(\mathfrak{m}) = 0$ if and only if $x \in \mathfrak{m}$, proving (i).

Proof of (vi).

A singular element x of $\mathfrak{A}$ is contained in some maximal ideal $\mathfrak{m}_0$ (proposition 3.7) so for a singular element x we must have $\phi_x(\mathfrak{m}) = 0$ for at least one point $\mathfrak{m}_0$. A maximal ideal cannot contain regular elements, on the other hand (proposition 3.4). So, a regular element x is in <u>no</u> maximal ideal, hence $\phi_x(\mathfrak{m}) \neq 0$ for all $\mathfrak{m} \in \mathbb{M}$, by (i). This proves (vi). Now (ii) becomes evident: $x - \lambda e$ is singular if and only if $\phi_x(\mathfrak{m}) - \lambda = 0$ for some $\mathfrak{m}$ that is, if $\lambda = \phi_x(\mathfrak{m}) \in \text{im}\,\phi_x$. It follows that

$$\sup\{|\lambda| : \lambda \in Sp(x)\} = \sup\{|\phi_x(\mathfrak{m})| : \mathfrak{m} \in \mathbb{M}\}.$$

At left we have the spectral radius (proposition 2.11), at right we have $\|\phi_x\|_0$ the norm of the function ϕ_x in $C(\mathbb{M})$. This proves (iii). Also (iv) now becomes evident. To prove (v) we note that the function $\phi_e(\mathfrak{m}) \equiv 1$ vanishes nowhere on $\mathbb{M}$. Also for any two distinct maximal ideals $\mathfrak{m}_1$ and $\mathfrak{m}_2$ there must be an $x \in \mathfrak{m}_1$ not contained in $\mathfrak{m}_2$ (otherwise $\mathfrak{m}_1 \subset \mathfrak{m}_2$ and thus $\mathfrak{m}_1 = \mathfrak{m}_2$). Hence we get $\phi_x(\mathfrak{m}_1) = 0$, $\phi_x(\mathfrak{m}_2) \neq 0$ which proves (v). The theorem is established.

The kernel of the homomorphism ϕ is a closed (2-sided) ideal $\mathfrak{R}$ since ϕ is continuous (in fact, a contraction), by theorem 4.1, (iii).

By theorem 4.1, (iv) it follows that $\mathfrak{R}$ consists precisely of all quasi-nilpotent elements.

Definition 4.2.

The ideal $\mathfrak{R}$ is called the radical of $\mathfrak{A}$. If $\mathfrak{R}$ is trivial, that is if 0 is the only quasi-nilpotent element of the algebra then $\mathfrak{A}$ is called semi-simple.

If the algebra $\mathfrak{A}$ is semi-simple (that is, has no nonvanishing quasi-nilpotents) then the Gel'fand homomorphism ϕ is an isomorphism.

In the general case it may be emphasized that ϕ only represents $\mathfrak{A}/\mathfrak{R}$ not $\mathfrak{A}$ (isomorphically) as a function algebra. Nevertheless we notice that even the Gel'fand homomorphism does not annihilate information concerning invertibility of elements in A, as manifested by theorem 4.1, (vi):

Proposition 4.3.

If $x \in \mathfrak{A}$ is invertible and y is quasi-nilpotent, then x + cy is invertible for any complex number c.

This may be concluded from theorem 4.1 (vi) but also more directly from proposition 2.7, since

$$x + cy = x(1 + cx^{-1}y)$$

and

$$\rho(cx^{-1}y) \leq |c|\rho(x^{-1})\rho(y) = 0.$$

Proposition 4.4.

If the B-algebra with unit A is a sub-algebra of some $C(\mathfrak{X})$, where $\mathfrak{X}$ is a compact Hausdorff space and if A separates $\mathfrak{X}$ strongly then for any $p_0 \in \mathfrak{X}$ the collection

$$(4.1) \qquad \mathfrak{m}_{p_0} = \{x\colon x \in A,\ x(p_0) = 0\}$$

forms a maximal ideal of A. We have

$$(4.2) \qquad \phi_x(\mathfrak{m}_{p_0}) = x(p_0) \quad \text{for all } x \in A.$$

Indeed $\mathfrak{m}_{p_0}$ is a proper ideal since A separates $\mathfrak{X}$ strongly and hence contains x with $x(p_0) \neq 0$. The unit e of A must be the function $\equiv 1$ because $e(p)x(p) = x(p)$ and $x(p) \neq 0$ exists. Every element $x \in A$ satifies

$$(4.3) \qquad x \equiv x(p_0)e \pmod{\mathfrak{m}_{p_0}}$$

which shows that $A/\mathfrak{m}_{p_0} = \{\mathbb{C}e^{\vee}\}$ and thus that $\mathfrak{m}_{p_0}$ is a maximal ideal (proposition 3.9). Also we have proved that (4.2) holds.

Proposition 4.5.

The map defined by $p_0 \to \mathfrak{m}_{p_0}$ defines a homeomorphism between $\mathfrak{X}$ and a compact subset of $\mathbb{M}$.

If it is used to identify $\mathfrak{X}$ with its image in $\mathbb{M}$, then the function $x \in A$ is equal to the restriction of ϕ_x to $\mathfrak{X}$.

Proof.

The map $p \to \mathfrak{m}_p$ is 1 - 1 since $\mathfrak{m}_{p_1} \neq \mathfrak{m}_{p_2}$ whenever p_1, p_2 are distinct, as a consequence of the fact that A separates $\mathfrak{X}$. Then (4.2) shows that indeed the function x appears as the restriction of ϕ_x to $\mathfrak{X}$. The relative topology of $\mathfrak{X}$ as a subset of $\mathbb{M}$ is the weak topology with respect to A. Since all functions of A are continuous in the topology of $\mathfrak{X}$, that topology must be stronger than the relative topology. Therefore the map $p \to \mathfrak{m}_p$ is continuous and hence a homeomorphism, q.e.d.

For $x \in A$ as above we define the norm

$$(4.4) \qquad \|x\|_0 = \operatorname{Max} \{|x(p)| : p \in \mathfrak{X}\}$$

which needs not to coincide with the norm $\|x\|$ of x in A. However, we have

$$(4.5) \qquad \|x\|_0 \leq \|\phi_x\|_0 = \rho(x) \leq \|x\|$$

since we know that x is a restriction of ϕ_x. Of special interest is the case where the norm in A is the sup norm, i.e. $\|x\|_0 = \|x\|$. Then it follows from (4.5) that

$$(4.6) \qquad \|x\|_0 = \rho(x) = \|\phi_x\|_0 = \|x\| .$$

Accordingly, the Gel'fand homomorphism is an isometry. In the examples below we will see that such cases can actually occur where $\mathfrak{X}$ is a proper subset of $\mathbb{M}$ and $\|x\|_0 = \|x\|$.

Examples:

1) Given the algebra $A = C(\mathfrak{X})$ of example (v), section 1, with a compact Hausdorff space $\mathfrak{X}$. It is trivial that $C(\mathfrak{X})$ separates $\mathfrak{X}$ strongly, thus we have the natural imbedding (4.1) of $\mathfrak{X}$ into $\mathbb{M}$.

Proposition 4.6.

We have $\mathbb{M} = \mathfrak{X}$ in the above example.

Proof.

Suppose $\mathfrak{m} \in \mathbb{M}$ is arbitrary but $\neq \mathfrak{m}_p$ for any $p \in \mathfrak{X}$. This means that for every $p \in \mathfrak{X}$ there exists some function $x = x_p$ such that $x_p(p) \neq 0$, $x_p \in \mathfrak{m}$. But if $\bar{x}$ denotes the complex conjugate function (which is also in $C(\mathfrak{X})$) then $\bar{x}_p x_p$ is ≥ 0 over X and in $\mathfrak{m}$. Also $\bar{x}_p x_p > 0$ at p and hence in a neighborhood N_p of p. The compact space $\mathfrak{X}$ can be covered by finitely many such neighborhoods N_{p_j}, $j = 1,\dots,n$. Then $y = \sum_j \bar{x}_{p_j} x_{p_j} \in \mathfrak{m}$ is positive on all of $\mathfrak{X}$ and hence invertible. This is a contradiction.

Remark.

The very same conclusion $\mathbb{M} = \mathfrak{X}$ is true whenever $A \subset C(\mathfrak{X})$ is any sub-algebra such that

(i) A strongly separates points

(ii) A contains the complex conjugate of each of its elements.

This is at once verified by the same conclusion as in proposition 4.6.

Example 2).

Let $A_h = A_h(\mathcal{D}_1)$ denote the collection of all continuous complex-valued functions over

$$\mathcal{D}_1 = \{z: z \in \mathbb{C}, |z| \leq 1\}$$

which are holomorphic in the interior of $\mathcal{D}_1$. Then A_h is a commutative B-algebra with unit ($\equiv 1$) under the norm of $C(\mathcal{D}_1)$. A_h is naturally imbedded in $C(\mathcal{D}_1)$ and strongly separates the points of $\mathcal{D}_1$.

All this is an immediate consequence of the fact that the uniform limit of a sequence of holomorphic functions is holomorphic again.

Again the natural imbedding (4.1) identifies $\mathcal{D}_1$ with a compact subset of $\mathbb{M} = \mathbb{M}(A_h)$.

Proposition 4.7.

We have $M(A_h) = \mathcal{D}_1$.

Proof.

Notice that A_h is the closure of its subalgebra consisting of all polynomials in z. For if $f \in A_h$ then f_n, defined by $f_n(z) = f(z(1 - \frac{1}{n}))$, $z \in \mathcal{D}_1$, $n = 1,2,\dots$, defines a sequence of functions all holomorphic in all of $\mathcal{D}_1$ converging uniformly to f, since f is uniformly continuous on $\mathcal{D}_1$. Every $f_n(z)$ is uniform limit of the partial sums of its power series, which are polynomials. So there exists a sequence $p_n(z)$ of polynomials in z with $\|f - p_n\|_0 \to 0$, $n \to \infty$.

Suppose that $\mathfrak{m}$ is any maximal ideal of A_h. Then let $\xi = \phi_z(\mathfrak{m})$, where "z" stands for the identity function $z \to z$. The spectrum of this function is equal to $\mathcal{D}_1$, because $z - \lambda$ is not invertible if and only if $|\lambda| \leq 1$ (when it vanishes somewhere in $\mathcal{D}_1$). Using theorem 4.1, (ii) we conclude that $\xi \in \mathcal{D}_1$. We want to prove that $\mathfrak{m} = \mathfrak{m}_\xi$. But if $f \in \mathfrak{m}$ and p_n is an approximating sequence of polynomials we get

$$\phi_{p_n}(\mathfrak{m}) = p_n(\phi_z(\mathfrak{m})) = p_n(\xi). \tag{4.7}$$

As $n \to \infty$ we get $p_n \to f$, hence $\phi_{p_n} \to \phi_f$. Since $f \in \mathfrak{m}$ the left hand side of (4.7) tends to zero, the right hand side to $f(\xi)$. Hence $f(\xi) = 0$ for all $f \in \mathfrak{m}$. That is, $\mathfrak{m} \subset \mathfrak{m}_\xi$, or $\mathfrak{m} = \mathfrak{m}_\xi$, since both ideals are maximal.

Notice that the restriction map

$$\kappa : A_h \to C(S_1),$$

defined by $x \to x|S_1$, where $S_1 = \{z \in \mathbb{C} : |z| = 1\}$ is the unit circle, is an isometric isomorphism from A_h onto a sub-algebra of $C(S_1)$, the isometry being a consequence of the maximum modules principle, which states that any function f in A_h must assume its maximum at the boundary S_1 of $\mathcal{D}_1$. The image $B \subset C(S_1)$ of A_h under κ

supplies an example for the fact that a sub-algebra $\mathcal{B}$ of an algebra $C(\mathbb{X})$, strongly separating points, may have a structure space $\mathbb{M}$ which contains $\mathbb{X}$ as a proper subset.

Theorem 4.8.

The structure space $\mathbb{M}$ of a semi-simple commutative B-algebra $\mathfrak{A}$ with unit is uniquely characterized by the following properties:

(i) $\mathbb{M}$ is compact Hausdorff ;

(ii) $\mathfrak{A}$ can be isomorphically mapped into $C(\mathbb{M})$ by an isomorphism such that it strongly separates the points of $\mathbb{M}$;

(iii) For any other isomorphism

$$\psi: \quad \rightarrow C(\mathbb{X})$$

satisfying (i) and (ii) above there exists an injection $\kappa: \mathbb{X} \xrightarrow{\text{into}} \mathbb{M}$ such that

$$\psi_x = \phi_x \circ \kappa, \; x \in \mathfrak{A}.$$

Proof.

We have seen in theorem 4.1, proposition 4.4 and proposition 4.5, that the structure space $\mathbb{M}$ of $\mathfrak{A}$ indeed has the properties (i), (ii), (iii) with $\phi: \mathfrak{A} \rightarrow C(\mathbb{M})$ denoting the Gel'fand homomorphism, with κ defined by (4.1).

Also, if $\mathbb{N}$ and the isomorphism $\psi: \mathfrak{A} \rightarrow C(\mathbb{N})$ have all the properties (i), (ii), (iii) then we have two homeomorphisms

$$K: \mathbb{N} \rightarrow \mathbb{M}, \; \lambda: \mathbb{M} \rightarrow \mathbb{N}$$

such that

$$\psi_x = \phi_x \circ \kappa \quad \phi_x = \psi_x \circ \lambda.$$

If $\lambda \circ \kappa$ is not the identity then $\lambda \circ \kappa(m) \neq m$ for some m. But $\psi(\mathfrak{A})$ separates points, hence $\psi_x(m) \neq \psi_x(\lambda \circ \kappa(m))$ for some x .This is a contradiction, hence $\lambda \circ \kappa$ is the identity map. Similar for $\kappa \circ \lambda$. Hence κ and λ are inverse, and the spaces $\mathbb{M}$ and $\mathbb{N}$ are homeomorphic. q.e.d.

Remark.

The proof of theorem 4.1 now is a simple consequence of theorem 4.1 and theorem 4.8.

Let us finally remark without further discussion that for any algebra $A \subset C(\mathbb{X})$ with norm being that of $C(\mathbb{X})$ there also exists a smallest compact set $\mathbb{S} \subset \mathbb{M}$ such that A allows an isomorphic representation on $C(\mathbb{S})$. We have seen in the discussion after (4.5) that such isomorphism is also an isometry. Hence we have a maximum principle for $\mathbb{S}$: Every function in A must take on the maximum of its modulus at $\mathbb{S}$.

The uniquely determined $\mathbb{S}$ is called the Šilov boundary of $\mathbb{X}$ (with respect to A).

5. The associate dual map

Suppose we have a homomorphism

(5.1) $$h: \mathfrak{A}_1 \longrightarrow \mathfrak{A}_2, \quad h(e_1) = e_2 \quad ,$$

between two commutative B-algebras $\mathfrak{A}_j$ with unit e_j, $j = 1,2$. h is not assumed to be continuous. Let $\mathfrak{A}_j$ have the structure space $\mathbb{M}_j$ and Gel'fand homomorphism ϕ^j, respectively.

Now, if $\mathfrak{m} \in \mathbb{M}_2$ is arbitrary, define $\mathfrak{m}' \in \mathfrak{A}_1$ by

(5.2) $$\mathfrak{m}' = \{ x \in \mathfrak{A}_1 : h(x) \in \mathfrak{m} \} .$$

Proposition 5.1.

$\mathfrak{m}'$ is a maximal ideal of $\mathfrak{A}_1$, and

(5.3) $$\phi^2_{h(x)}(\mathfrak{m}) = \phi^1_x(\mathfrak{m}') \quad .$$

It is trivially verified that $\mathfrak{m}'$ is an ideal and $\mathfrak{m}'$ is proper, since $h(e_1) = e_2 \notin \mathfrak{m}$ so $e_1 \notin \mathfrak{m}'$. For arbitrary $x \in \mathfrak{A}_1$ we get

$$h(x) - \phi^2_{h(x)}(\mathfrak{m}) e_2 \in \mathfrak{m} \quad ,$$

and this implies that

$$z = x - \phi^2_{h(x)}(\mathfrak{m}) e_1$$

is in $\mathfrak{m}'$, since $h(z) \in \mathfrak{m}$.

Conclude that

(5.4) $$x \equiv \phi^2_{h(x)}(\mathfrak{m}) e_1 \pmod{\mathfrak{m}'}$$

for all $x \in \mathfrak{A}_1$.

Hence $\mathfrak{m}'$ is a maximal ideal, by proposition 3.9 and also (5.3) follows, q.e.d.

Definition 5.2.

We denote the mapping $h' : \mathbb{M}_2 \to \mathbb{M}_1$ defined by $\mathfrak{m} \to \mathfrak{m}'$ with $\mathfrak{m}'$ as in (5.2) as the associated dual map of the homomorphism h. From (5.3) it is evident that for fixed x we get

(5.5) $$\phi^1_x \circ h' = \phi^2_{h(x)} \quad .$$

Every continuous mapping $\tau' : \mathbb{M}_2 \to \mathbb{M}_1$ induces a homomorphism $\tau'' : C(\mathbb{M}_1) \to C(\mathbb{M}_2)$, defined by

(5.6) $$f_1 \to f_2 = f_1 \circ \tau' \quad .$$

For $\tau' = h'$ in (5.6) we obtain a map h'' which also takes $C(\mathbb{M}_1)$ into $C(\mathbb{M}_2)$, although it is not yet known whether h' is continuous. This is a consequence of the fact that the diagramm

(5.7) $$\begin{array}{ccc} \mathfrak{A}_1 & \xrightarrow{\ h\ } & \mathfrak{A}_2 \\ \downarrow \phi^1 & & \downarrow \phi^2 \\ C(\mathbb{M}_1) & \xrightarrow{\ h''\ } & C(\mathbb{M}_2) \end{array}$$

is commutative, as manifested by (5.5).

Proposition 5.3.

h' is continuous (regardless whether h is continuous or not).

The topologies in $\mathbb{M}_1$ and $\mathbb{M}_2$ are the weak topologies of the images A_j under ϕ^j and τ'' takes A_1 into A_2. Continuity of h' follows from the fact that

$$(5.8) \qquad h'^{-1} N_{h'(\mathfrak{m}_0^2);x^1,\dots,x^n;\varepsilon} \cap (\text{im } h') = N_{\mathfrak{m}_0^2;h(x_1)\dots h(x_n);\varepsilon}$$

as easily verfified using (5.5).

Proposition 5.4.

For any surjective isomorphism $h: \mathfrak{A}_1 \to \mathfrak{A}_2$ the associate dual map h' is a homeomorphism from $\mathbb{M}_2$ onto $\mathbb{M}_1$.

Proof.

We have h' and $h^{-1'}$ defined by

$$\mathfrak{m}' = h'(\mathfrak{m}) = \{x: h(x) \in \mathfrak{m}\}, \ \mathfrak{m} \in \mathbb{M}_2$$

$$\mathfrak{m} = h^{-1'}(\mathfrak{m}') = \{y: h^{-1}(y) \in \mathfrak{m}'\}, \mathfrak{m}' \in \mathbb{M}_1 \ .$$

It follows that $\mathfrak{m}' = h^{-1'}\mathfrak{m}$ and $\mathfrak{m} = h'\mathfrak{m}$, hence h' and $h^{-1'}$ invert each other. Accordingly h is a homeomorphism form $\mathbb{M}_2$ onto $\mathbb{M}_1$, q.e.d.

Proposition 5.5.

If a homomorphism $h: \mathfrak{A}_1 \to \mathfrak{A}_2$ is surjective, it follows that h' is injective.

Proof.

Suppose

$$\mathfrak{m}' = h'(\mathfrak{m}) = h'(\tilde{\mathfrak{m}}) \quad \text{for } \mathfrak{m}, \tilde{\mathfrak{m}} \in \mathbb{M}_2 \ .$$

This means that for $x \in \mathfrak{A}_1$ the two conditions $h(x) \in \mathfrak{m}$ or $h(x) \in \tilde{\mathfrak{m}}$ mean the same.

Let $y \in \mathfrak{m}$ be given. There exists $x \in \mathfrak{A}_1$ with $y = h(x)$, because h is surjective. But then $y = h(x) \in \mathfrak{m}$ and so also $y = h(x) \in \tilde{\mathfrak{m}}$. This implies $\mathfrak{m} \subset \tilde{\mathfrak{m}}$ or that both coincide, since they are maximal, q.e.d.

Before our next proposition we again consider the example 2) at the end of section 4.

The isomorphic imbedding $A_h \to C(S_1)$ has ist associate dual map defined by

$$\mathfrak{m}_{z_0}^2 \to \mathfrak{m}' = \{f: f \in A_h, \ f|S_1 \in \mathfrak{m}_{z_0}^2\}$$
$$= \{f: f \in A_h, \ f(z_0) = 0\} = \mathfrak{m}_{z_0}^1$$

for $|z_0| = 1$. Clearly this is the injection of S_1 into $\mathcal{D}_1$, containing it as a subset.

This example shows that the associate dual map of an isomorphism needs not to be surjective.

Definition 5.6.

An algebra A of continuous functions over a compact Hausdorff space $\mathfrak{X}$ will be called completely regular, if for any closed set $\mathfrak{Q} \varepsilon \mathfrak{X}$ and any $x \notin \mathfrak{Q}$ there exists a function in A vanishing at $\mathfrak{Q}$ but not at x.

It is clear that A_h above is not completely regular, since a holomorphic function vanishing at any open set in $\mathcal{D}_1$, must vanish identically.

Definition 5.7.

A commutative B-algebra $\mathfrak{A}$ with unit is called completely regular, if the image $\phi(\mathfrak{A}) = A$ of $\mathfrak{A}$ under its Gel'fand homomorphism is completely regular.

Proposition 5.8.

Let $\tau: \mathfrak{A}_1 \to \mathfrak{A}_2$ be an isomorphism and let $\mathfrak{A}_1$ be completely regular and $\mathfrak{A}_2$ be semi-simple. Then the associate dual map τ' is surjective.

Proof.

Consider the diagramm

(5.10)
$$\begin{array}{ccc} A_1 & \xrightarrow{\quad\tau\quad} & A_2 \\ \phi^1 \downarrow & & \downarrow \phi^2 \\ C(\mathbb{M}_1) & \xrightarrow{\quad\tau''\quad} & C(\mathbb{M}_2) \\ & \tau' & \\ \mathbb{M}_1 & \longleftarrow & \mathbb{M}_2 \end{array} .$$

The map τ' is continuous, hence its image in $\mathbb{M}_1$ is a compact set. If it is not all of $\mathbb{M}_1$ there exists $\mathfrak{m}_1 \notin$ im τ' and $x_0 \varepsilon \mathfrak{A}_1$ such that $\phi^1_{x_0}(\mathfrak{m}_1) \neq 0$ but $\phi^1_{x_0} = 0$ on im τ'. It follows that $x_0 \neq 0$, because $\phi^1_{x_0}$ does not vanish identically, hence $h(x_0) \neq 0, \phi^2_{h(x_0)} \neq 0$ since h and ϕ^2 are isomorphisms. On the other hand

$$\phi^2_{h(x_0)} = \phi^1_{x_0} \circ \tau'$$

and the right hand side vanishes identically since $\phi^1_{x_0} = 0$ in im τ'. Thus we have a contradiction.

Corollary 5.9.

If in proposition 5.8 the condition that $\mathfrak{A}_2$ be semi-simple is replaced by the condition

(5.11) $$\|h(x)\| \geq \eta \|x\| \; , \; x \,\varepsilon\, \mathfrak{A}_1$$

with some fixed $\eta > 0$ then the assertion holds as well.

Proof.

With the construction of proposition 5.8 note that also $\rho(x_0) = \|\phi^1_{x_0}\|_0 \neq 0$ because $\phi^1_{x_0} \neq 0$. Use (5.11) to conclude that

$$\rho(h(x)) = \lim \|(h(x))^n\|^{1/n} = \lim \|h(x^n)\|^{1/n}$$
$$\geq \lim c^{1/n} \|x^n\|^{1/n} = \rho(x),\ x \in \mathfrak{A}_1 .$$

Accordingly we get $\rho(h(x_0)) > 0$ and so $\phi^2_{h(x_0)} \neq 0$ regardless of the condition that $\mathfrak{A}_2$ be semi-simple. Q.e.d.

Proposition 5.10.

If $\mathfrak{A}$ is completely regular then for any isomorphism ψ: $\mathfrak{A} \to C(\mathfrak{X})$ with a compact Hausdorff space $\mathfrak{X}$ we have $\mathfrak{X} = \mathbb{M}$ and $\psi = \phi$.

Proof.

We have the canonical imbedding $\mathfrak{X} \to \mathbb{M}$ defined by (4.1) and all assumptions of proposition 5.8 hold for ψ: $\mathfrak{A} \to C(\mathfrak{X})$. Hence the associate dual map ψ' is surjective. Also notice that for p ε $\mathfrak{X}$ we get

$$\psi'(p_0) = \{x \in \mathfrak{A} : \psi_x(p_0) = 0\} = \mathfrak{m}_{p_0} .$$

Accordingly, ψ': $\mathfrak{X} \to \mathbb{M}$ coincides with the canonical imbedding. So the canonical imbedding is a homeomorphism between $\mathfrak{X}$ and $\mathbb{M}$. Q.e.d.

6. The Stone-Weierstrass theorem.

Definition 6.1.

A unction algebra $A \subset C(X)$ will be called self-adjoint if $a \in A$ implies that the complex conjugate function $\bar{a}$ also is in A.

For a self-adjoint function algebra A it is true that $a \in A$ implies that Re a and Im a both are in A. The collection

$$\text{(6.1)} \qquad \operatorname{Re} A = \{\operatorname{Re} a : a \in A\}$$

forms an algebra over the field of real numbers. If A separates (strongly) the points of X, so does Re A separate (strongly) the points of X. Note that $C(X)$ itself is clearly self-adjoint.

Theorem 6.2. (The Stone-Weierstrass theorem).

Let X be a compact Hausdorff space. Then any self-adjoint subalgebra A of $C(X)$, which separates strongly the points of X is dense in $C(X)$.

Proof.

It is sufficient to show that Re A is dense in Re $C(X)$, or even that the closure B of Re A is dense in $C(X)$. This will be accomplished by the series of propositions below.

Proposition 6.3.

If $f \in B$ then $|f| \in B$.

Proof.

By the classical Weierstrass approximation theorem, there exists, for any $\varepsilon > 0$, a polynomial $p_\varepsilon = p_\varepsilon(t)$ of one real variable t such that $|p_\varepsilon(t) - |t|| < \varepsilon$ in $|t| \le \|f\|$. Define g_ε by $g_\varepsilon(x) = p_\varepsilon(f(x))$ and notice that $\||f| - g_\varepsilon\|_0 < \varepsilon$. Hence $|f| \in B$.

Proposition 6.4.

If $f_1,\ldots,f_n \in B$ then h and k defined by $h(x) = \operatorname{Max}_{j=1}^{n} f_j(x), k(x) = \operatorname{Min}_{j=1}^{n} f_j(x)$ are in B.

It suffices to prove this for two functions $f_1 = f$, $f_2 = g$ and for the Maximum only. But then

$$\text{(6.2)} \qquad \max\{f(x),g(x)\} = \{g + \tfrac{1}{2}\{(f-g) + |f-g|\}\}(x)$$

is in B, q.e.d.

Propostion 6.5.

Given $x,y \in X$ $x \neq y$. There exists $h = h_{xy} \in B$ with $h(x) = 1$, $h(y) = 0$.

Indeed, first construct $f, g \in B$ with $f(x) \neq f(y)$ and $g(x) \neq 0$. Define $k = f$ if $f(x) \neq 0$, $k = g$ if $g(x) \neq g(y)$ but $f(x) = 0$ and $k = f + g$ if $f(x) = 0$ and $g(x) = g(y)$. Then in every case $k(x) \neq 0$, $k(x) \neq k(y)$. Next define $h = k/k(x)$ if

$k(y) = 0$. If $k(y) \neq 0$ define $l = k/k(y) - (k/k(y))^2$ and $h = 1/l(x)$. Q.e.d.

Proposition 6.6.

Given $f \in \mathrm{Re}\, C(\mathcal{X})$ and $x,y \in \mathcal{X}$. There exists a function $f_{xy} \in \mathcal{B}$ such that $f_{xy} = f$ at x and y.

Proof.

If $x = y$ set $f_{xy} = f(x)g/g(x)$ with any function $g \in \mathcal{B}$ with $g(x) \neq 0$. It $x \neq y$ then set

$$f_{xy} = f(x)h_{xy} + f(y)h_{yx}$$

with the function h constructed in proposition 6.5. Q.e.d.

Proof of Theorem 6.2.

Let $f \in \mathrm{ReC}(X)$ and $\varepsilon > 0$ be given. Let also $y \in \mathcal{X}$ be fixed. For $x \in \mathcal{X}$ define

$$\mathcal{N}_{xy} = \{\omega \in \mathcal{X}: f(\omega) - \varepsilon < f_{xy}(\omega)\}.$$

This is an open set containing x. From the cover $\{\mathcal{N}_x : x \in \mathcal{X}\}$ of the compact space $\mathcal{X}$ we may extract a finite subcover $\{\mathcal{N}_{x_1y} \ldots \mathcal{N}_{x_ny}\}$ and then define

$$f_y = \mathrm{Max}\,\{f_{x_jy}: j = 1,\ldots,n\}.$$

The function f_y is in $\mathcal{B}$ (proposition 6.4) and since every $\omega \in \mathcal{X}$ is in at least one $\mathcal{N}_{x_jy}$ it follwos that $f(\omega) - \varepsilon < f_{x_{j_0}y}(\omega) \leq f_y$. Next define

$$\mathcal{N}_y = \{\omega: \omega \in \mathcal{X},\ f_y(\omega) < f(\omega) + \varepsilon\}.$$

Again this constitutes an open cover of the compact space $\mathcal{X}$ and a finite subcover $\mathcal{N}_{y_1}, \ldots \mathcal{N}_{y_m}$ can be found. Define $g = \mathrm{Min}\,\{f_{y_1}, \ldots f_{y_m}\}$. Clearly then $\omega \in \mathcal{X}$ lies in some $\mathcal{N}_{y_{1_0}}$ and we get

$$f(\omega) - \varepsilon < \mathop{\mathrm{Min}}_{1}^{m} f_{y_j}(\omega) = g(\omega) < f_{y_{1_0}}(\omega) < f(\omega) + \varepsilon.$$

We have proved that

$$|f(\omega) - g(\omega)| < \varepsilon.$$

It follows that $\mathcal{B} = \mathrm{Re}\, C(\mathcal{X})$ and theorem 6.2 is established.

7. C^*-algebras.

Definition 7.1.

An involution on a B-algebra $\mathfrak{A}$ is a map $*: \mathfrak{A} \to \mathfrak{A}$ with the following properties.

(i) $x^{**} = x$; (ii) $(cx)^* = \bar{c}x^*$

(iii) $(x+y)^* = x^* + y^*$; (iv) $(xy)^* = y^*x^*$

for all $x, y \in \mathfrak{A}$, $c \in \mathbb{C}$.

Definition 7.2.

An involution $*$ is said to have the C^*-property if

$$(7.1) \qquad \|x^*x\| = \|x\|^2$$

for all $x \in \mathfrak{A}$.

Definition 7.3.

A C^*-algebra is a B-algebra (with unit or without) which has an involution satisfying the C^*-property. A $*$-homomorphism, $*$-isomorphism, etc. is a homomorphism, isomorphism, etc., between C^*-algebras leaving the involution invariant.

Ramark.

If $\mathfrak{A}$ has a unit e then any involution satisfies $e^* = e$, since $e^*x = (x^*e)^* = x^{**} = x$, for $x \in \mathfrak{A}$ and similar for xe^*.

Remark.

Note that it is possible to consider involutions for arbitrary algebras over $\mathbb{C}$, not necessarily B-algebras. Only the C^*-property involves the norm of the algebra.

Example 7.4.

For any compact space $\mathfrak{X}$ the algebra $C(\mathfrak{X})$ with sup norm and involution $f \to \bar{f}$ defined by $\bar{f}(x) = \overline{f(x)}$ is a C^*-algebra with unit.

Example 7.5.

For a locally compact space $\mathfrak{X}$ the algebra $CO(\mathfrak{X})$ of all continuous complex-valued functions over $\mathfrak{X}$ vanishing at ∞ (that is becoming arbitrarily small outside very large compact sets) is a C^*-algebra, with norm and involution defined as in Example 7.4. $CO(\mathfrak{X})$ has a unit if and only if $\mathfrak{X}$ is compact.

Example 7.6.

Let $\mathfrak{H}$ be a Hilbert space. Then any sub-algebra $\mathfrak{A}$ of $L(\mathfrak{H})$ which is closed under the operator norm (1.5) and contains all adjoints of its elements, is a C^*-algebra under operator norm and operator adjoint. To see that the C^*-property holds, note that $\|A\| = \|A^*\|$ for all bounded linear operators between B-spaces, so we get $\|A^*A\| \le \|A^*\| \, \|A\| = \|A\|^2$. Also, for $x \in \mathfrak{H}$ and the inner product $(\cdot,\cdot)$ of $\mathfrak{H}$, $\|Ax\|^2 = (x, A^*Ax) \le \|x\|^2 \|A^*A\|$ which implies $\|A\|^2 \le \|A^*A\|$. Hence we have

(7.1) satisfied.

The above three examples exhaust all the possibilities for a C^*-algebra, as may be seen from the results below.

Theorem 7.7.

For a commutative C^*-algebra $\mathfrak{A}$ with unit the Gel'fand homomorphism is a surjective isometric $*$-isomorphism. That is, in particular,

(i) $\|x\| = \underset{m}{\mathrm{Max}}\{|\phi_x(m)|\} = \|\phi_x\|_0$;

(ii) $\mathfrak{A}$ is semi-simple ;

(iii) $A = \mathrm{im}\ \phi = C(\mathbb{M})$.

Theorem 7.8.

A commutative C^*-algebra $\mathfrak{A}$ without unit is isometrically $*$-isomorphic to $CO(\mathbb{M})$ with a locally compact, non-compact space $\mathbb{M}$ and $CO(\mathbb{M})$ as in Example 7.5.

Theorem 7.9.

Every C^*-algebra, whether commutative or not, is isometrically $*$-isomorphic to some norm-closed, adjoint invariant sub-algebra $\mathfrak{A}$ of $L(\mathfrak{H})$ with a suitable Hilbert space $\mathfrak{H}$.

We shall not discuss the proof of theorem 7.9, since the result will never be used here. In a C^*-algebra $\mathfrak{A}$ we distinguish hermitian, unitary and normal elements, as for operators in Hilbert space, defined as elements $x \in \mathfrak{A}$ satisfying $x^* = x$, $x^* = x^{-1}$ and $x^*x = xx^*$, respectively. Hermitian and unitary elements are always normal.

Proposition 7.10.

For any normal element x of a C^*-algebra we have $\rho(x) = \|x\|$.

Proof.

First let $x = h$ be hermitian and observe that

$$\|h^{2^n}\| = \|h^{2^{n-1}*}h^{2^{n-1}}\| = \|h^{2^{n-1}}\|^2 = \ldots = \|h\|^{2^n}$$

using the C^*-property. Accordingly,

$$\rho(h) = \lim \|h^{2^n}\|^{1/2^n} = \|h\|.$$

If x is normal then $h = x^*x$ is hermitian, and x, x^* commute.
Hence

$$\begin{aligned}\rho(x)^2 &= \lim (\|x^n\|^2)^{1/n} \\ &= \lim (\|x^{n*}x^n\|)^{1/n} = \lim (\|h^n\|^{1/n}) \\ &= \rho(h) = \|h\| = \|x^*x\| = \|x\|^2 ,\end{aligned}$$

q.e.d.

Corollary 7.11.

In a commutative C^*-algebra the spectral radius coincides with the norm.

Corollary 7.12.

For a commutative C^*-algebra the Gel'fand homomorphism is an isometry.

Both corollaries are evident consequence of proposition 7.10.

From now on let $\mathfrak{A}$ be always a commutative C^*-algebra with unit.

Proposition 7.13.

Any maximal ideal of $\mathfrak{A}$ is invariant under the involution.

Proof.

Clearly $\mathfrak{m}^* = \{x^*: x \in \mathfrak{m}\}$ is also a maximal ideal. If not $\mathfrak{m}^* = \mathfrak{m}$ then the algebraic sum $\mathfrak{m}+\mathfrak{m}^* = \{x + y^*: x,y \in \mathfrak{m}\}$ must contain e, because it is a proper extension ideal of $\mathfrak{m}$, and thus equal to $\mathfrak{A}$. Hence $e = x_1 + x_2^*$, $x_j \in \mathfrak{m}$. Taking invclution and the average of both equations get

(7.2) $$e = y + y^*,\ y = (x_1 + x_2)/2 \in \mathfrak{m}.$$

We claim that

$$z = \int_0^\infty e^{-\alpha y}\, d\alpha$$

is an inverse of y, which is a contradiction, since $y \in \mathfrak{m}$ cannot be invertible. First, the integral exists, because $e^{-\alpha y}$, defined by its power series, is uniformly continuous in every $[0,\eta]$, and since $\|e^{-\alpha y}\|^2 = \|e^{-\alpha(y+y^*)}\| = e^{-\alpha}$ using (7.1) and (7.2) and commutativity of y with y^*. Also we get

$$yz = zy = \int_0^\infty y\, e^{-\alpha y} d\alpha = \int_0^\infty \frac{d}{d\alpha}(e^{-\alpha y})\, d\alpha = e\ ,$$

q.e.d.

Proof of Theorem 7.7.

It is evident from corollary 7.12 that $\phi: \mathfrak{A} \to C(\mathbb{M})$ is an isometry, and from proposition 7.13 that $\phi_{x^*} = \overline{\phi}_x$. So ϕ is a $*$-homomorphism and its image $A \in C(\mathbb{M})$ is self-adjoint. From theorem 4.1 (v) we know that A strongly separates points of $\mathbb{M}$. Therefore the Stone-Weierstrass theorem (theorem 6.2) implies that A is dense in $C(\mathbb{M})$ under the norm of $C(\mathbb{M})$ which is the norm of $\mathfrak{A}$. As a B-algebra, A is closed, hence $A = C(\mathbb{M})$, q.e.d.

Proof of Theorem 7.8.

For a C^*-algebra $\mathfrak{A}$ without 1 let $\mathfrak{A}_1 = \{(c,x): c \in \mathbb{C},\ x \in \mathfrak{A}\}$ be as in section 1, with involution $(c,x)^* = (\bar{c},x^*)$, and norm $\|(c,x)\| = \sup\{\|cy-xy\| : \|y\| \le 1\}$ (in contrast to (1.3)). One concludes that

$$\|x\| \ge \sup\{\|xy\| : \|y\| \le 1\} = \|(0,x)\| \ge \|xx^*\| / \|x^*\| = \|x\|$$

and

$$\|cy+xy\|^2 = \|(cy+xy)^*(cy+xy)\| = \|y^*(c\ x)^*(c,x)y\| \leq \|y\|^2 \|(c,x)^*(c,x)\| .$$

The first relation implies $\|(0,x)\| = \|x\|$. The second yields $\|(c,x)\|^2 \leq \|(c,x)^*(c,x)\| \leq \|(c,x)\|^2$ so that $\mathfrak{A}_1$ is a C^*-algebra with unit, and with $\mathfrak{A}$ isometrically imbedded, of codimension 1.

Clearly, if $\mathfrak{A}$ is commutative so is $\mathfrak{A}_1$, and then we have $\mathfrak{A} = \mathfrak{m}_0$, a maximal ideal of $\mathfrak{A}_1$. We get $A_1 \cong C(\mathbb{M}_1)$ with a compact $\mathbb{M}_1$, by Theorem 7.7. Let $\mathbb{M} = \mathbb{M}_1 - \{\mathfrak{m}_0\}$. Then $\mathbb{M}$ is locally compact, and $A \cong CO(\mathbb{M})$, q.e.d.

Corollary 7.14.

A commutative C^*-algebra with unit is completely regular.

Since $\operatorname{im}\phi$ contains all continuous functions this is a consequence of Uryson's lemma asserting that even for any two disjoint closed sets K_1, K_2 in a compact space $\mathbb{M}$ a continuous function $\mathbb{M} \to \mathbb{C}$ can be constructed, vanishing in K_1 and equal to 1 in K_2.

Lemma 7.15.

Let $\mathfrak{X}_1$, $\mathfrak{X}_2$ be C^*-algebras with 1 where $\mathfrak{X}_1 \subset \mathfrak{X}_2$. Assume $\mathfrak{X}_1$ and $\mathfrak{X}_2$ have the same involution. Then $x \in \mathfrak{X}_1$ is invertible in $\mathfrak{X}_2$ if and only if it is invertible in $\mathfrak{X}_1$.

Proof.

It is no loss to assume $\mathfrak{X}_1$ commutative. For $x \in \mathfrak{X}_1$, invertible in $\mathfrak{X}_2$, we get $h = (x^*x)^{1/2} \in \mathfrak{X}_1$ invertible in $\mathfrak{X}_2$, by Corollary 8.5, below, and $x = uh$, with $u = xh^{-1}$ unitary. Here the hermitian element h is contained in a commutative C^*-algebra (with 1) $\mathfrak{X}_3 \subset \mathfrak{X}_1 \subset \mathfrak{X}_2$. Hence $h^{-1} \in \mathfrak{X}_3 \subset \mathfrak{X}_1$. Then $u \in \mathfrak{X}_1$ is contained in a commutative C^*-algebra (with 1) $\mathfrak{X}_4 \subset \mathfrak{X}_1 \subset \mathfrak{X}_2$. Hence $u^{-1} \in \mathfrak{X}_4 \subset \mathfrak{X}_1$, and $x^{-1} = h^{-1}u^{-1} \in \mathfrak{X}_1$ follows, if we can prove the assertion for commutative $\mathfrak{X}_1$.

Let $z \in \mathfrak{X}_2$ be an inverse of $x \in \mathfrak{X}_1$. Clearly z commutes with all of $\mathfrak{X}_1$ and so does z^*. Accordingly we may adjoin z and z^* to $\mathfrak{X}_1$ to obtain a commutative C^*-algebra $\mathfrak{X}_3$ which contains $\mathfrak{X}_1$. Let $\mathbb{M}_j$, $j = 1,3$ and ϕ^j, $j = 1,3$, be the maximal ideal spaces and Gel'fand homomorphisms of $\mathfrak{X}_j$. Let $\rho: \mathbb{M}_3 \to \mathbb{M}_1$ be the associate dual map of the injection $\mathfrak{X}_1 \to \mathfrak{X}_3$. It follows that ρ is surjective (prop. 5.8). Also we get

(7.3) $$\phi^1_x(\rho\mathfrak{m}) = \phi^3_x(\mathfrak{m}) \quad , \mathfrak{m} \in \mathbb{M}_3 .$$

The right hand side is different from zero since x is invertibel in $\mathfrak{X}_3$. Since ρ is surjective we also get $\phi^1_x(\mathfrak{m}) \neq 0$, $\mathfrak{m} \in \mathbb{M}_1$. It follows that x is invertible in $\mathfrak{X}_1$ and then z must be in $\mathfrak{X}_1$, since the inverse in $\mathfrak{X}_3$ is unique. Q.e.d.

Corollary 7.16.

A C^*-subalgebra of $L(\mathfrak{H})$, for a Hilbert space $\mathfrak{H}$, is Fredholm closed, in the sense of AI, 5, if ever it contains the ideal $E(\mathfrak{H})$ of operators of finite rank.

Proof.

Such a C^*-algebra $\mathfrak{A}$ must contain $K(\mathfrak{H})$ as well which is the closure of $E(\mathfrak{H})$ in $L(\mathfrak{H})$.

Hence we get the inclusion $\mathfrak{A}/K(\mathfrak{H}) \subset L(\mathfrak{H})/K(\mathfrak{H})$, where both quotients are C^*-algebras, by theorem 8.1. For any Fredholm inverse $B \in L(\mathfrak{H})$ of $A \in \mathfrak{A}$ the cosets $\check{A} = A+K(\mathfrak{H})$ and $\check{B} = B+K(\mathfrak{H})$ invert each other. Thus by lemma 7.15 it follows that $\check{B} \in \mathfrak{A}/K(\mathfrak{H})$. Since $K(\mathfrak{H}) \subset \mathfrak{A}$ we conclude that $B \in \mathfrak{A}$. Thus $\mathfrak{A}$ contains all Fredholm inverses of its elements, q.e.d.

Theorem 7.17.

Every isomorphism $\tau : \mathfrak{X} \to \mathfrak{Y}$, with C^*-algebras $\mathfrak{X}$ and $\mathfrak{Y}$, which preserves involutions is an isometry.

Proof.

For $x \in \mathfrak{X}$ let $y = \tau(x)$. We get $\|y\|^2 = \|y^*y\| = \|\tau(h)\|$, with $h = x^*x$, and $\|h\| = \|x\|^2$, using the C^*-condition in $\mathfrak{X}$ and $\mathfrak{Y}$. Thus it suffices to prove that $\|\tau(h)\| = \|h\|$ for hermitian elements, or that $\rho(k) = \rho(h)$ with $k = \tau(h)$, by Proposition 7.10.

Let $\mathfrak{X}_1$ and $\mathfrak{Y}_1$ be the commutative C^*-algebras generated by norm closing the polynomials $\{P(h)\}$ and $\{P(k)\}$, respectively. Then $\mathfrak{X}_1$ and $\mathfrak{Y}_1$ are isomorphic under τ, and the associate dual map $\tilde{\tau}'$ of $\tilde{\tau} = \tau|\mathfrak{X}_1$ provides a homeomorphism $\mathbb{M} \leftrightarrow \mathbb{N}$ between the Gel'fand spaces of $\mathfrak{X}_1$ and $\mathfrak{Y}_1$ by Proposition 5.4. Also we have $\phi_k(\mathfrak{m}) = \phi_h(\tilde{\tau}'\mathfrak{m})$, $\mathfrak{m} \in \mathbb{N}$, which gives $\rho(h) = \sup |\phi_h(\mathfrak{m})| = \sup |\phi_k(\mathfrak{m})| = \rho(k)$. This proves that $\|h\| = \|k\|$, q.e.d.

8. Closed ideals of a C^*-algebra.

In this section we shall prove the result below. It will find applications in chapter IV and V.

Theorem 8.1.

Every closed 2-sided ideal $\mathfrak{J}$ of a C^*-algebra $\mathfrak{A}$ is a $*$-ideal: we have $\mathfrak{J}^* = \mathfrak{J}$. Then the quotient algebra $\mathfrak{A}/\mathfrak{J}$, under its natural norm and involution, is a C^*-algebra again.

The proof will require some preparations, partly already used in section 7. From the proof of Theorem 7.8 we summarize:

Proposition 8.2.

For a C^*-algebra $\mathfrak{A}$ without unit there exists an isometric imbedding $\mathfrak{A} \to \mathfrak{A}_1$ into a C^*-algebra $\mathfrak{A}_1$ with 1, having codimension 1. If $\mathfrak{A}$ is commutative so is $\mathfrak{A}_1$. Any ideal of $\mathfrak{A}$ is also an ideal of $\mathfrak{A}_1$.

As a consequence we may assume, for Theorem 8.1, that $\mathfrak{A}$ has a unit.

A hermitian element $h \in \mathfrak{A}$ will be called positive (written as $h \geq 0$) if $Sp(h) \in [0, \infty)$ (written as $Sp(h) \geq 0$). Similarly $h \leq k$, for hermitian h, k shall mean that $k - h \geq 0$.

Lemma 8.3.

For every $x \in \mathfrak{A}$ we have $x^*x \geq 0$.

Proof.

$h = x^*x$ is hermitian, and contained in a commutative C^*-algebra (with 1) $\mathfrak{A}_0 \subset \mathfrak{A}$. We have $\mathfrak{A}_0 \cong C(M_0)$, by Theorem 7.7 and $h \cong \phi_h \in C(M_0)$. Define the hermitian elements h_+, h_- by $\phi_{h_\pm} = \mathrm{Max}\,\{\pm\phi_h, 0\}$, and conclude that $Sp(h_\pm) \geq 0$, so that $h_\pm \geq 0$, while $h = h_+ - h_-$, $h_+h_- = h_-h_+ = 0$. Let $x_- = xh_-$ and get $-x_-^*x_- = -h_-hh_- = h_-^3 \geq 0$.

Now write $x_- = s + it$, with $s = (x_- + x_-^*)/2$, $t = (x_- - x_-^*)/2i$ hermitian. Get $x_-^* = s - it$, and $x_-^*x_- = -x_-x_-^* + 2s^2 + 2t^2 = a_1 + a_2 + a_3$, with a_j, $j = 1,2,3$ all positive. This is trivial for a_2 and a_3 and follows from proposition 8.4, below, for $-x_-x_-^* = a_1$. We claim that $a_1 + a_2 + a_3 \geq 0$. Indeed assume without loss that at least one $a_j \neq 0$ and let $\lambda = \mathrm{Max}\ \|a_j\|$ and $a_j' = a_j/\lambda$. Observe that $0 \leq Sp(a_j') \leq 1$, hence $0 \leq Sp(1-a_j') \leq 1$, so that $\rho(1-a_j') = \|1 - a_j'\| \leq 1$. This yields $\rho(1-\sum_1^3 a_j'/3) = \|1-\sum_1^3 a_j'/3\| \leq \sum_1^3 \|1-a_j'\|/3 \leq 1$. Or, $-1 \leq Sp(-1+\sum a_j/3\lambda) \leq 1$. Therefore $Sp(x_-^*x_-) = 3\lambda\ Sp(\sum_1^3 a_j/3\lambda) \geq 0$.

But also we have $Sp(x_-^*x_-) \leq 0$, hence $Sp(x_-^*x_-) = \{0\} \Rightarrow \rho(x_-^*x_-) = \|x_-^*x_-\| = 0$. Accordingly $h_- = 0$ and $h = h_+ \geq 0$, q.e.d.

We are left with the proposition, below.

Proposition 8.4.

For $x \in \mathfrak{A}$ we have

(8.1) $\quad Sp(x^*x) - \{0\} = Sp(xx^*) - \{0\}.$

Proof.

We have $0 \neq \lambda \in Sp(x^*x)$ if and only if $p = (1 - x^*x/\lambda)^{-1}$ exists. Then also $q = (1 - xx^*/\lambda)^{-1}$ exists, hence $0 \neq \lambda \in Sp(xx^*)$, because p and q are related by the formulas

(8.2) $$p = 1 + x^*qx/\lambda \quad , \qquad q = 1 + xpx^*/\lambda \quad ,$$

by a calculation. Q.e.d.

Corollary 8.5.

For $x \in \mathfrak{A}$ the hermitian element $h = x^*x$ admits a positive hermitian square root $l \in \mathfrak{A}$, such that $l^2 = h$, $l \geq 0$. If h is invertible so is l

Proof.

In the above algebra $\mathfrak{A}_0 \cong C(M_0)$ define l by $\phi_l = |\sqrt{\phi_h}|$. Then l satisfies all requirements, q.e.d.

Proof of Theorem 8.1.

For $x \in \mathfrak{A}$ let $h = x^*x$ and $k = \alpha h/(1+\alpha h)$ (which exists for $\alpha > 0$, by Lemma 8.3). For $x \in \mathfrak{J}^*$ get $h \in \mathfrak{J}$ and $k = \alpha h - \alpha kh \in \mathfrak{J}$, hence $xk \in \mathfrak{J}$. Also

(8.3) $$\|x - xk\|^2 = \|(1-k)x^*x(1-k)\| = \|h/(1-\alpha h)^2\|$$
$$= \mathrm{Max}\{|\phi_h(m)/(1+\alpha\phi_h(m))^2|\} \leq 1/4\alpha \to 0, \ \alpha \to \infty,$$

using that $\phi_h(m) \geq 0$, and that

$$\gamma/(1+\alpha\gamma)^2 \leq 1/4\alpha, \qquad \text{as } \gamma \geq 0, \ \alpha > 0.$$

Clearly (8.3) implies that $x = \lim_{\alpha\to\infty} xk \in \mathfrak{J}$, since $\mathfrak{J}$ is closed. This proves that $\mathfrak{J}^* \subset \mathfrak{J}$. For reason of symmetry we also get $\mathfrak{J} \subset \mathfrak{J}^*$, hence $\mathfrak{J}^* = \mathfrak{J}$.

Now, since $\mathfrak{J} = \mathfrak{J}^*$, the algebra $\mathfrak{A}^\vee = \mathfrak{A}/\mathfrak{J}$ has a natural involution $x^\vee = x + \mathfrak{J} \to x^* + \mathfrak{J} = x^{\vee *}$. It suffices to show that it has the C^*-property. Let $x \in \mathfrak{A}$, and let h and k be as above. We get

(8.4) $$\|x^\vee\| \leq \|x(1-k)\| + \|x^\vee\| \, \|k^\vee\| \leq 1/2\sqrt{\alpha} + \|x^\vee\| \, \|k^\vee\| ,$$

using (8.3).

Assume without loss that $h \notin \mathfrak{J}$. In order to estimate $\|k^\vee\|$ let us focus on the commutative algebra $\mathfrak{A}_0/\mathfrak{J}_0$, with the closed ideal $\mathfrak{J}_0 = \mathfrak{J} \cap \mathfrak{A}_0$ of $\mathfrak{A}_0$. We get $\mathfrak{A}_0 \cong C(M_0)$ and $\mathfrak{J}_0 \cong \{\phi_a \in C(M_0) : \phi_a = 0 \text{ on } \mathcal{Q}\}$, where $\mathcal{Q}$ is the closed set of all maximal ideals containing $\mathfrak{J}_0$. This follows by application of proposition 3.7. Then $\mathfrak{A}_0/\mathfrak{J}_0 \cong CO(M_0 - \mathcal{Q})$, with sup norm. Now the map $\mathfrak{A}_0^\wedge = \mathfrak{A}_0/\mathfrak{J}_0 \to \mathfrak{A}^\vee$ given by $x^\wedge = x + \mathfrak{J}_0 \to x^\vee = x + \mathfrak{J}$ is a contractive isomorphism, since the infimum at right is taken over a larger set. Also $x^\vee = 0$, for $x^\wedge \in \mathfrak{A}_0^\wedge$ yields $x \in \mathfrak{A}_0 \cap \mathfrak{J} = \mathfrak{J}_0$, hence $x^\wedge = 0$. Accordingly a norm on $\mathfrak{A}_0^\wedge$ is defined by $|x^\wedge| = \|x^\vee\|$. We also get

$$(8.5) \qquad \|x^{\vee}\| = |x^{\wedge}| \geq \rho(x^{\wedge}) = \|x^{\wedge}\| \geq \|x^{\vee}\| ,$$

since $x^{\wedge}$ is normal, as an element of a commutative C^*-algebra, and since every norm is not smaller than the spectral radius.

We conclude that $\|x^{\vee}\| = \|x^{\wedge}\|$ for all $x \in \mathfrak{A}_0$. Therefore for $x = k$ we get

$$(8.6) \qquad \|k^{\vee}\| = \|k^{\wedge}\| = \sup \{|\phi_k(\mathfrak{m}) : \mathfrak{m} \in \mathbf{M}_0 - \mathcal{Q}\}$$
$$= \sup \alpha\phi_h/(1+\alpha\phi_h) = \alpha\|h^{\wedge}\| /(1+\alpha \|h^{\wedge}\|).$$

Substitute this into (8.4) and set $\alpha = \|h^{\wedge}\|^{-1}$ for

$\| x^{\vee}\| \leq 1/2 \|h^{\wedge}\|^{1/2} + 1/2 \|x^{\vee}\| \Rightarrow \|x^{\vee}\|^2 \leq \|x^{\vee *}x^{\vee}\| \leq \|x^{\vee}\|^2$, which amounts to the C^*-property for $\mathfrak{A}/\mathfrak{J}$, q.e.d.

References for Appendix AII

[1] L. Alaoglu, Weak topologies of normed linear spaces, Ann. of Math. 41 (1940), p. 252-267.

[2] J. Dixmier, Les C*-algèbres et leurs représentations, (1964), Gauthier-Villars, Paris.

[3] E. Hille - R. Phillips, Functional Analysis and Semi-groups, (1957) AMS, Providence, R.I.

[4] L. Loomis, Abstract harmonic Analysis, (1953) van Nostrand, New York, N.Y.

[5] G.Polya-G.Szegö, Aufgaben und Lehrsätze aus der Analysis, Vol. I, (1925) Springer, Berlin.

[6] C. Rickart, General theory of Banach Algebras, (1960) van Nostrand, Princeton, N.J.

Index

Vol. 580: C. Castaing and M. Valadier, Convex Analysis and Measurable Multifunctions. VIII, 278 pages. 1977.

Vol. 581: Séminaire de Probabilités XI, Université de Strasbourg. Proceedings 1975/1976. Edité par C. Dellacherie, P. A. Meyer et M. Weil. VI, 574 pages. 1977.

Vol. 582: J. M. G. Fell, Induced Representations and Banach *-Algebraic Bundles. IV, 349 pages. 1977.

Vol. 583: W. Hirsch, C. C. Pugh and M. Shub, Invariant Manifolds. IV, 149 pages. 1977.

Vol. 584: C. Brezinski, Accélération de la Convergence en Analyse Numérique. IV, 313 pages. 1977.

Vol. 585: T. A. Springer, Invariant Theory. VI, 112 pages. 1977.

Vol. 586: Séminaire d'Algèbre Paul Dubreil, Paris 1975–1976 (29ème Année). Edited by M. P. Malliavin. VI, 188 pages. 1977.

Vol. 587: Non-Commutative Harmonic Analysis. Proceedings 1976. Edited by J. Carmona and M. Vergne. IV, 240 pages. 1977.

Vol. 588: P. Molino, Théorie des G-Structures: Le Problème d'Equivalence. VI, 163 pages. 1977.

Vol. 589: Cohomologie l-adique et Fonctions L. Séminaire de Géométrie Algébrique du Bois-Marie 1965–66, SGA 5. Edité par L. Illusie. XII, 484 pages. 1977.

Vol. 590: H. Matsumoto, Analyse Harmonique dans les Systèmes de Tits Bornologiques de Type Affine. IV, 219 pages. 1977.

Vol. 591: G. A. Anderson, Surgery with Coefficients. VIII, 157 pages. 1977.

Vol. 592: D. Voigt, Induzierte Darstellungen in der Theorie der endlichen, algebraischen Gruppen. V, 413 Seiten. 1977.

Vol. 593: K. Barbey and H. König, Abstract Analytic Function Theory and Hardy Algebras. VIII, 260 pages. 1977.

Vol. 594: Singular Perturbations and Boundary Layer Theory, Lyon 1976. Edited by C. M. Brauner, B. Gay, and J. Mathieu. VIII, 539 pages. 1977.

Vol. 595: W. Hazod, Stetige Faltungshalbgruppen von Wahrscheinlichkeitsmaßen und erzeugende Distributionen. XIII, 157 Seiten. 1977.

Vol. 596: K. Deimling, Ordinary Differential Equations in Banach Spaces. VI, 137 pages. 1977.

Vol. 597: Geometry and Topology, Rio de Janeiro, July 1976. Proceedings. Edited by J. Palis and M. do Carmo. VI, 866 pages. 1977.

Vol. 598: J. Hoffmann-Jørgensen, T. M. Liggett et J. Neveu, Ecole d'Eté de Probabilités de Saint-Flour VI – 1976. Edité par P.-L. Hennequin. XII, 447 pages. 1977.

Vol. 599: Complex Analysis, Kentucky 1976. Proceedings. Edited by J. D. Buckholtz and T. J. Suffridge. X, 159 pages. 1977.

Vol. 600: W. Stoll, Value Distribution on Parabolic Spaces. VIII, 216 pages. 1977.

Vol. 601: Modular Functions of one Variable V, Bonn 1976. Proceedings. Edited by J.-P. Serre and D. B. Zagier. VI, 294 pages. 1977.

Vol. 602: J. P. Brezin, Harmonic Analysis on Compact Solvmanifolds. VIII, 179 pages. 1977.

Vol. 603: B. Moishezon, Complex Surfaces and Connected Sums of Complex Projective Planes. IV, 234 pages. 1977.

Vol. 604: Banach Spaces of Analytic Functions, Kent, Ohio 1976. Proceedings. Edited by J. Baker, C. Cleaver and Joseph Diestel. VI, 141 pages. 1977.

Vol. 605: Sario et al., Classification Theory of Riemannian Manifolds. XX, 498 pages. 1977.

Vol. 606: Mathematical Aspects of Finite Element Methods. Proceedings 1975. Edited by I. Galligani and E. Magenes. VI, 362 pages. 1977.

Vol. 607: M. Métivier, Reelle und Vektorwertige Quasimartingale und die Theorie der Stochastischen Integration. X, 310 Seiten. 1977.

Vol. 608: Bigard et al., Groupes et Anneaux Réticulés. XIV, 334 pages. 1977.

Vol. 609: General Topology and Its Relations to Modern Analysis and Algebra IV. Proceedings 1976. Edited by J. Novák. XVIII, 225 pages. 1977.

Vol. 610: G. Jensen, Higher Order Contact of Submanifolds of Homogeneous Spaces. XII, 154 pages. 1977.

Vol. 611: M. Makkai and G. E. Reyes, First Order Categorical Logic. VIII, 301 pages. 1977.

Vol. 612: E. M. Kleinberg, Infinitary Combinatorics and the Axiom of Determinateness. VIII, 150 pages. 1977.

Vol. 613: E. Behrends et al., L^p-Structure in Real Banach Spaces. X, 108 pages. 1977.

Vol. 614: H. Yanagihara, Theory of Hopf Algebras Attached to Group Schemes. VIII, 308 pages. 1977.

Vol. 615: Turbulence Seminar, Proceedings 1976/77. Edited by P. Bernard and T. Ratiu. VI, 155 pages. 1977.

Vol. 616: Abelian Group Theory, 2nd New Mexico State University Conference, 1976. Proceedings. Edited by D. Arnold, R. Hunter and E. Walker. X, 423 pages. 1977.

Vol. 617: K. J. Devlin, The Axiom of Constructibility: A Guide for the Mathematician. VIII, 96 pages. 1977.

Vol. 618: I. I. Hirschman, Jr. and D. E. Hughes, Extreme Eigen Values of Toeplitz Operators. VI, 145 pages. 1977.

Vol. 619: Set Theory and Hierarchy Theory V, Bierutowice 1976. Edited by A. Lachlan, M. Srebrny, and A. Zarach. VIII, 358 pages. 1977.

Vol. 620: H. Popp, Moduli Theory and Classification Theory of Algebraic Varieties. VIII, 189 pages. 1977.

Vol. 621: Kauffman et al., The Deficiency Index Problem. VI, 112 pages. 1977.

Vol. 622: Combinatorial Mathematics V, Melbourne 1976. Proceedings. Edited by C. Little. VIII, 213 pages. 1977.

Vol. 623: I. Erdelyi and R. Lange, Spectral Decompositions on Banach Spaces. VIII, 122 pages. 1977.

Vol. 624: Y. Guivarc'h et al., Marches Aléatoires sur les Groupes de Lie. VIII, 292 pages. 1977.

Vol. 625: J. P. Alexander et al., Odd Order Group Actions and Witt Classification of Innerproducts. IV, 202 pages. 1977.

Vol. 626: Number Theory Day, New York 1976. Proceedings. Edited by M. B. Nathanson. VI, 241 pages. 1977.

Vol. 627: Modular Functions of One Variable VI, Bonn 1976. Proceedings. Edited by J.-P. Serre and D. B. Zagier. VI, 339 pages. 1977.

Vol. 628: H. J. Baues, Obstruction Theory on the Homotopy Classification of Maps. XII, 387 pages. 1977.

Vol. 629: W. A. Coppel, Dichotomies in Stability Theory. VI, 98 pages. 1978.

Vol. 630: Numerical Analysis, Proceedings, Biennial Conference, Dundee 1977. Edited by G. A. Watson. XII, 199 pages. 1978.

Vol. 631: Numerical Treatment of Differential Equations. Proceedings 1976. Edited by R. Bulirsch, R. D. Grigorieff, and J. Schröder. X, 219 pages. 1978.

Vol. 632: J.-F. Boutot, Schéma de Picard Local. X, 165 pages. 1978.

Vol. 633: N. R. Coleff and M. E. Herrera, Les Courants Résiduels Associés à une Forme Méromorphe. X, 211 pages. 1978.

Vol. 634: H. Kurke et al., Die Approximationseigenschaft lokaler Ringe. IV, 204 Seiten. 1978.

Vol. 635: T. Y. Lam, Serre's Conjecture. XVI, 227 pages. 1978.

Vol. 636: Journées de Statistique des Processus Stochastiques, Grenoble 1977, Proceedings. Edité par Didier Dacunha-Castelle et Bernard Van Cutsem. VII, 202 pages. 1978.

Vol. 637: W. B. Jurkat, Meromorphe Differentialgleichungen. VII, 194 Seiten. 1978.

Vol. 638: P. Shanahan, The Atiyah-Singer Index Theorem, An Introduction. V, 224 pages. 1978.

Vol. 639: N. Adasch et al., Topological Vector Spaces. V, 125 pages. 1978.

Vol. 640: J. L. Dupont, Curvature and Characteristic Classes. X, 175 pages. 1978.

Vol. 641: Séminaire d'Algèbre Paul Dubreil, Proceedings Paris 1976–1977. Edité par M. P. Malliavin. IV, 367 pages. 1978.

Vol. 642: Theory and Applications of Graphs, Proceedings, Michigan 1976. Edited by Y. Alavi and D. R. Lick. XIV, 635 pages. 1978.

Vol. 643: M. Davis, Multiaxial Actions on Manifolds. VI, 141 pages. 1978.

Vol. 644: Vector Space Measures and Applications I, Proceedings 1977. Edited by R. M. Aron and S. Dineen. VIII, 451 pages. 1978.

Vol. 645: Vector Space Measures and Applications II, Proceedings 1977. Edited by R. M. Aron and S. Dineen. VIII, 218 pages. 1978.

Vol. 646: O. Tammi, Extremum Problems for Bounded Univalent Functions. VIII, 313 pages. 1978.

Vol. 647: L. J. Ratliff, Jr., Chain Conjectures in Ring Theory. VIII, 133 pages. 1978.

Vol. 648: Nonlinear Partial Differential Equations and Applications, Proceedings, Indiana 1976–1977. Edited by J. M. Chadam. VI, 206 pages. 1978.

Vol. 649: Séminaire de Probabilités XII, Proceedings, Strasbourg, 1976–1977. Edité par C. Dellacherie, P. A. Meyer et M. Weil. VIII, 805 pages. 1978.

Vol. 650: C*-Algebras and Applications to Physics. Proceedings 1977. Edited by H. Araki and R. V. Kadison. V, 192 pages. 1978.

Vol. 651: P. W. Michor, Functors and Categories of Banach Spaces. VI, 99 pages. 1978.

Vol. 652: Differential Topology, Foliations and Gelfand-Fuks-Cohomology, Proceedings 1976. Edited by P. A. Schweitzer. XIV, 252 pages. 1978.

Vol. 653: Locally Interacting Systems and Their Application in Biology. Proceedings, 1976. Edited by R. L. Dobrushin, V. I. Kryukov and A. L. Toom. XI, 202 pages. 1978.

Vol. 654: J. P. Buhler, Icosahedral Golois Representations. III, 143 pages. 1978.

Vol. 655: R. Baeza, Quadratic Forms Over Semilocal Rings. VI, 199 pages. 1978.

Vol. 656: Probability Theory on Vector Spaces. Proceedings, 1977. Edited by A. Weron. VIII, 274 pages. 1978.

Vol. 657: Geometric Applications of Homotopy Theory I, Proceedings 1977. Edited by M. G. Barratt and M. E. Mahowald. VIII, 459 pages. 1978.

Vol. 658: Geometric Applications of Homotopy Theory II, Proceedings 1977. Edited by M. G. Barratt and M. E. Mahowald. VIII, 487 pages. 1978.

Vol. 659: Bruckner, Differentiation of Real Functions. X, 247 pages. 1978.

Vol. 660: Equations aux Dérivée Partielles. Proceedings, 1977. Edité par Pham The Lai. VI, 216 pages. 1978.

Vol. 661: P. T. Johnstone, R. Paré, R. D. Rosebrugh, D. Schumacher, R. J. Wood, and G. C. Wraith, Indexed Categories and Their Applications. VII, 260 pages. 1978.

Vol. 662: Akin, The Metric Theory of Banach Manifolds. XIX, 306 pages. 1978.

Vol. 663: J. F. Berglund, H. D. Junghenn, P. Milnes, Compact Right Topological Semigroups and Generalizations of Almost Periodicity. X, 243 pages. 1978.

Vol. 664: Algebraic and Geometric Topology, Proceedings, 1977. Edited by K. C. Millett. XI, 240 pages. 1978.

Vol. 665: Journées d'Analyse Non Linéaire. Proceedings, 1977. Edité par P. Bénilan et J. Robert. VIII, 256 pages. 1978.

Vol. 666: B. Beauzamy, Espaces d'Interpolation Réels: Topologie et Géometrie. X, 104 pages. 1978.

Vol. 667: J. Gilewicz, Approximants de Padé. XIV, 511 pages. 1978.

Vol. 668: The Structure of Attractors in Dynamical Systems. Proceedings, 1977. Edited by J. C. Martin, N. G. Markley and W. Perrizo. VI, 264 pages. 1978.

Vol. 669: Higher Set Theory. Proceedings, 1977. Edited by G. H. Müller and D. S. Scott. XII, 476 pages. 1978.

Vol. 670: Fonctions de Plusieurs Variables Complexes III, Proceedings, 1977. Edité par F. Norguet. XII, 394 pages. 1978.

Vol. 671: R. T. Smythe and J. C. Wierman, First-Passage Perculation on the Square Lattice. VIII, 196 pages. 1978.

Vol. 672: R. L. Taylor, Stochastic Convergence of Weighted Sums of Random Elements in Linear Spaces. VII, 216 pages. 1978.

Vol. 673: Algebraic Topology, Proceedings 1977. Edited by P. Hoffman, R. Piccinini and D. Sjerve. VI, 278 pages. 1978.

Vol. 674: Z. Fiedorowicz and S. Priddy, Homology of Classical Groups Over Finite Fields and Their Associated Infinite Loop Spaces. VI, 434 pages. 1978.

Vol. 675: J. Galambos and S. Kotz, Characterizations of Probability Distributions. VIII, 169 pages. 1978.

Vol. 676: Differential Geometrical Methods in Mathematical Physics II, Proceedings, 1977. Edited by K. Bleuler, H. R. Petry and A. Reetz. VI, 626 pages. 1978.

Vol. 677: Séminaire Bourbaki, vol. 1976/77, Exposés 489–506. IV, 264 pages. 1978.

Vol. 678: D. Dacunha-Castelle, H. Heyer et B. Roynette. Ecole d'Eté de Probabilités de Saint-Flour. VII-1977. Edité par P. L. Hennequin. IX, 379 pages. 1978.

Vol. 679: Numerical Treatment of Differential Equations in Applications, Proceedings, 1977. Edited by R. Ansorge and W. Törnig. IX, 163 pages. 1978.

Vol. 680: Mathematical Control Theory, Proceedings, 1977. Edited by W. A. Coppel. IX, 257 pages. 1978.

Vol. 681: Séminaire de Théorie du Potentiel Paris, No. 3, Directeurs: M. Brelot, G. Choquet et J. Deny. Rédacteurs: F. Hirsch et G. Mokobodzki. VII, 294 pages. 1978.

Vol. 682: G. D. James, The Representation Theory of the Symmetric Groups. V, 156 pages. 1978.

Vol. 683: Variétés Analytiques Compactes, Proceedings, 1977. Edité par Y. Hervier et A. Hirschowitz. V, 248 pages. 1978.

Vol. 684: E. E. Rosinger, Distributions and Nonlinear Partial Differential Equations. XI, 146 pages. 1978.

Vol. 685: Knot Theory, Proceedings, 1977. Edited by J. C. Hausmann. VII, 311 pages. 1978.

Vol. 686: Combinatorial Mathematics, Proceedings, 1977. Edited by D. A. Holton and J. Seberry. IX, 353 pages. 1978.

Vol. 687: Algebraic Geometry, Proceedings, 1977. Edited by L. D. Olson. V, 244 pages. 1978.

Vol. 688: J. Dydak and J. Segal, Shape Theory. VI, 150 pages. 1978.

Vol. 689: Cabal Seminar 76–77, Proceedings, 1976–77. Edited by A.S. Kechris and Y. N. Moschovakis. V, 282 pages. 1978.

Vol. 690: W. J. J. Rey, Robust Statistical Methods. VI, 128 pages. 1978.

Vol. 691: G. Viennot, Algèbres de Lie Libres et Monoïdes Libres. III, 124 pages. 1978.

Vol. 692: T. Husain and S. M. Khaleelulla, Barrelledness in Topological and Ordered Vector Spaces. IX, 258 pages. 1978.

Vol. 693: Hilbert Space Operators, Proceedings, 1977. Edited by J. M. Bachar Jr. and D. W. Hadwin. VIII, 184 pages. 1978.

Vol. 694: Séminaire Pierre Lelong – Henri Skoda (Analyse) Année 1976/77. VII, 334 pages. 1978.

Vol. 695: Measure Theory Applications to Stochastic Analysis, Proceedings, 1977. Edited by G. Kallianpur and D. Kölzow. XII, 261 pages. 1978.

Vol. 696: P. J. Feinsilver, Special Functions, Probability Semigroups, and Hamiltonian Flows. VI, 112 pages. 1978.

Vol. 697: Topics in Algebra, Proceedings, 1978. Edited by M. F. Newman. XI, 229 pages. 1978.

Vol. 698: E. Grosswald, Bessel Polynomials. XIV, 182 pages. 1978.

Vol. 699: R. E. Greene and H.-H. Wu, Function Theory on Manifolds Which Possess a Pole. III, 215 pages. 1979.

Vol. 700: Module Theory, Proceedings, 1977. Edited by C. Faith and S. Wiegand. X, 239 pages. 1979.

Vol. 701: Functional Analysis Methods in Numerical Analysis, Proceedings, 1977. Edited by M. Zuhair Nashed. VII, 333 pages. 1979.

Vol. 702: Yuri N. Bibikov, Local Theory of Nonlinear Analytic Ordinary Differential Equations. IX, 147 pages. 1979.

Vol. 703: Equadiff IV, Proceedings, 1977. Edited by J. Fábera. XIX, 441 pages. 1979.

Vol. 704: Computing Methods in Applied Sciences and Engineering, 1977, I. Proceedings, 1977. Edited by R. Glowinski and J. L. Lions. VI, 391 pages. 1979.

Vol. 705: O. Forster und K. Knorr, Konstruktion verseller Familien kompakter komplexer Räume. VII, 141 Seiten. 1979.

Vol. 706: Probability Measures on Groups, Proceedings, 1978. Edited by H. Heyer. XIII, 348 pages. 1979.

Vol. 707: R. Zielke, Discontinuous Čebyšev Systems. VI, 111 pages. 1979.

Vol. 708: J. P. Jouanolou, Equations de Pfaff algébriques. V, 255 pages. 1979.

Vol. 709: Probability in Banach Spaces II. Proceedings, 1978. Edited by A. Beck. V, 205 pages. 1979.

Vol. 710: Séminaire Bourbaki vol. 1977/78, Exposés 507–524. IV, 328 pages. 1979.

Vol. 711: Asymptotic Analysis. Edited by F. Verhulst. V, 240 pages. 1979.

Vol. 712: Equations Différentielles et Systèmes de Pfaff dans le Champ Complexe. Edité par R. Gérard et J.-P. Ramis. V, 364 pages. 1979.

Vol. 713: Séminaire de Théorie du Potentiel, Paris No. 4. Edité par F. Hirsch et G. Mokobodzki. VII, 281 pages. 1979.

Vol. 714: J. Jacod, Calcul Stochastique et Problèmes de Martingales. X, 539 pages. 1979.

Vol. 715: Inder Bir S. Passi, Group Rings and Their Augmentation Ideals. VI, 137 pages. 1979.

Vol. 716: M. A. Scheunert, The Theory of Lie Superalgebras. X, 271 pages. 1979.

Vol. 717: Grosser, Bidualräume und Vervollständigungen von Banachmoduln. III, 209 pages. 1979.

Vol. 718: J. Ferrante and C. W. Rackoff, The Computational Complexity of Logical Theories. X, 243 pages. 1979.

Vol. 719: Categorial Topology, Proceedings, 1978. Edited by H. Herrlich and G. Preuß. XII, 420 pages. 1979.

Vol. 720: E. Dubinsky, The Structure of Nuclear Fréchet Spaces. V, 187 pages. 1979.

Vol. 721: Séminaire de Probabilités XIII. Proceedings, Strasbourg, 1977/78. Edité par C. Dellacherie, P. A. Meyer et M. Weil. VII, 647 pages. 1979.

Vol. 722: Topology of Low-Dimensional Manifolds. Proceedings, 1977. Edited by R. Fenn. VI, 154 pages. 1979.

Vol. 723: W. Brandal, Commutative Rings whose Finitely Generated Modules Decompose. II, 116 pages. 1979.

Vol. 724: D. Griffeath, Additive and Cancellative Interacting Particle Systems. V, 108 pages. 1979.

Vol. 725: Algèbres d'Opérateurs. Proceedings, 1978. Edité par P. de la Harpe. VII, 309 pages. 1979.

Vol. 726: Y.-C. Wong, Schwartz Spaces, Nuclear Spaces and Tensor Products. VI, 418 pages. 1979.

Vol. 727: Y. Saito, Spectral Representations for Schrödinger Operators With Long-Range Potentials. V, 149 pages. 1979.

Vol. 728: Non-Commutative Harmonic Analysis. Proceedings, 1978. Edited by J. Carmona and M. Vergne. V, 244 pages. 1979.

Vol. 729: Ergodic Theory. Proceedings, 1978. Edited by M. Denker and K. Jacobs. XII, 209 pages. 1979.

Vol. 730: Functional Differential Equations and Approximation of Fixed Points. Proceedings, 1978. Edited by H.-O. Peitgen and H.-O. Walther. XV, 503 pages. 1979.

Vol. 731: Y. Nakagami and M. Takesaki, Duality for Crossed Products of von Neumann Algebras. IX, 139 pages. 1979.

Vol. 732: Algebraic Geometry. Proceedings, 1978. Edited by K. Lønsted. IV, 658 pages. 1979.

Vol. 733: F. Bloom, Modern Differential Geometric Techniques in the Theory of Continuous Distributions of Dislocations. XII, 206 pages. 1979.

Vol. 734: Ring Theory, Waterloo, 1978. Proceedings, 1978. Edited by D. Handelman and J. Lawrence. XI, 352 pages. 1979.

Vol. 735: B. Aupetit, Propriétés Spectrales des Algèbres de Banach. XII, 192 pages. 1979.

Vol. 736: E. Behrends, M-Structure and the Banach-Stone Theorem. X, 217 pages. 1979.

Vol. 737: Volterra Equations. Proceedings 1978. Edited by S.-O. Londen and O. J. Staffans. VIII, 314 pages. 1979.

Vol. 738: P. E. Conner, Differentiable Periodic Maps. 2nd edition, IV, 181 pages. 1979.

Vol. 739: Analyse Harmonique sur les Groupes de Lie II. Proceedings, 1976-78. Edited by P. Eymard et al. VI, 646 pages. 1979.

Vol. 740: Séminaire d'Algèbre Paul Dubreil. Proceedings, 1977–78. Edited by M.-P. Malliavin. V, 456 pages. 1979.

Vol. 741: Algebraic Topology, Waterloo 1978. Proceedings. Edited by P. Hoffman and V. Snaith. XI, 655 pages. 1979.

Vol. 742: K. Clancey, Seminormal Operators. VII, 125 pages. 1979.

Vol. 743: Romanian-Finnish Seminar on Complex Analysis. Proceedings, 1976. Edited by C. Andreian Cazacu et al. XVI, 713 pages. 1979.

Vol. 744: I. Reiner and K. W. Roggenkamp, Integral Representations. VIII, 275 pages. 1979.

Vol. 745: D. K. Haley, Equational Compactness in Rings. III, 167 pages. 1979.

Vol. 746: P. Hoffman, τ-Rings and Wreath Product Representations. V, 148 pages. 1979.

Vol. 747: Complex Analysis, Joensuu 1978. Proceedings, 1978. Edited by I. Laine, O. Lehto and T. Sorvali. XV, 450 pages. 1979.

Vol. 748: Combinatorial Mathematics VI. Proceedings, 1978. Edited by A. F. Horadam and W. D. Wallis. IX, 206 pages. 1979.

Vol. 749: V. Girault and P.-A. Raviart, Finite Element Approximation of the Navier-Stokes Equations. VII, 200 pages. 1979.

Vol. 750: J. C. Jantzen, Moduln mit einem höchsten Gewicht. III, 195 Seiten. 1979.

Vol. 751: Number Theory, Carbondale 1979. Proceedings. Edited by M. B. Nathanson. V, 342 pages. 1979.

Vol. 752: M. Barr, *-Autonomous Categories. VI, 140 pages. 1979.

Vol. 753: Applications of Sheaves. Proceedings, 1977. Edited by M. Fourman, C. Mulvey and D. Scott. XIV, 779 pages. 1979.

Vol. 754: O. A. Laudal, Formal Moduli of Algebraic Structures. III, 161 pages. 1979.

Vol. 755: Global Analysis. Proceedings, 1978. Edited by M. Grmela and J. E. Marsden. VII, 377 pages. 1979.

Vol. 756: H. O. Cordes, Elliptic Pseudo-Differential Operators – An Abstract Theory. IX, 331 pages. 1979.